AF522275

ENCYCLOPAEDIC DICTIONARY OF SCIENCE

ENCYCLOPAEDIC DICTIONARY OF SCIENCE

Vol. 3
[L–Q]

Edited by
SATISH ANAND

ANMOL PUBLICATIONS PVT. LTD.
NEW DELHI - 110 002 (INDIA)

ANMOL PUBLICATIONS PVT. LTD.
4374/4B, Ansari Road, Daryaganj
New Delhi - 110 002
Ph.: 23261597, 23278000
Visit us at: www.anmolpublications.com

Encyclopaedic Dictionary of Science

First Published, 1995

Reprint 1997

Reprint, 2004

ISBN 81-7488-003-8 (Set)

PRINTED IN INDIA

Published by J.L. Kumar for Anmol Publications Pvt. Ltd., New Delhi - 110 002 and Printed at Mehra Offset Press, Delhi.

Preface

The Encyclopaedic Dictionary of Science is an effort to keep pace with continuing rapid developments and expanding vocabularies in various branches of science. This dictionary has been encyclopaedic in its contents and style of representation. Throughout the process of compilation and editing the volumes, the main effort was to make the dictionary serve as a ready reference for undergraduate and postgraduate students of science and for those appearing for competitive examinations, research scholars, teachers and authors. The entries have been written in a clear and explanatory style to provide both straight forward definitions and invaluable background information. This approach, combined with an extensive cross-reference system, enables the reader to place the each entry into a broader scientific context. At certain appropriate places, the diagrams have been included whenever the meaning of a word can be best understood by means of a diagram.

The presentation throughout has been aimed at sustaining the interest of readers while enriching their vocabulary and comprehension of technical terms and expressions of various branches of science.

This encyclopaedic dictionary will be of immense value to the students of science and to scientists and research scholars working on science projects.

In compiling a set of this kind it becomes necessary to draw upon the work of many authorities and seek the advice of colleagues to all of whom the editor is deeply indebted. All the comments from users on omissions or shortcomings will be most welcome.

The Editor

Vol. 3

L — Q

Label. A radioactive atom in a molecule, used to determine the mechanism of a chemical reaction.

Labile. Denoting a chemical compound that is unstable and likely to break down or change.

Lace. Patterned open work fabric made by plaiting, knotting, looping, or twisting threads. Needlepoint lace, done with a needle in variations of the buttonhole stitch, arose in Italy during the Renaissance. The art later passed to France, Belgium, England, and Ireland. Bobbin or pillow lace (so named because the threads are wound onto bobbins resting on a pillow and are then twisted around pins) reached Flanders from Italy in the late 15th cent. Laces are often named for their places of origin, e.g., Alencon, Honiton, and Maltese. Machine-made lace appeared c. 1760. Linen (and, less often, silk, gold, or silver) thread was used in lace-making; in the 1830s cotton somewhat replaced linen. Knitting, Macrame, tatting, and Crochet can also be used to produce lace; the finest crocheted lace was made in Ireland. Today relatively little lace is handmade.

Lacquer. Solution of film-forming materials, natural or synthetic, usually applied as an ornamental or protective coating. Quick-drying synthetic lacquers are used to coat such products as automobiles, furniture, and textiles. Lacquer-work was one of the earliest industrial arts of the Orient. It was highly developed in India; the Chinese inlaid lacquer-ware with ivory, jade, coral, or abalone. The art spread to Korea, then to Japan, where it took new forms. The ware, which is often given more than 40 coats of lacquer, may be decorated in color, gold, or silver and enhanced by relief, engraving, or carving. In the 17th cent. the technique known as japanning was used to make Western European imitations of lacquer-ware. Commercial production of lacquer-ware in the 19th cent. resulted in a decline in quality.

Lactam. Any of a class of cyclic aliphatic compounds derived from amino acids by elimination of water between the amino group of the molecule and the carboxyl group. Lactams contain the group-NH.CO-.

Lactate. Any salt or ester of lactic acid.

Lactic Acid (2-hydroxypropanoic acid). A colourless hygroscopic syrupy liquid, $CH_3CHOHCOOH$, prepared by fermentation of starch or molasses. Lactic acid is used as an acidulant of foods, preservative, and in making plasticizers and adhesives. M.pt. 18°C ; b.pt 122°C (15 mmHg) ; r.d. 1.2.

Lactone. Any of a class of cyclic aliphatic compounds derived from hydroxy carboxylic acids by elimination of water between the hydroxyl group of the molecule and the carboxyl group. Lactones thus contain the group-O.CO- and are internal esters.

Lactose or **Milk Sugar.** (Empirical formula : $C_{12}H_{22}O_{11}$), white crystalline Sugar formed in the mammary glands of all lactating animals and present in their milk. A disaccharide, lactose can be broken down by Hydrolysis into Glucose and galactose. When milk sours, the lactose in it is converted by bacteria to lactic acid. Lactose is less sweet-tasting than Sucrose and, unlike sucrose, is not found in plants and is not fermented by ordinary yeast.

Lactose (milk sugar). A white crystalline disaccharide sugar, $C_{12}H_{22}O_{11}$, with a sweet taste. It occurs in the milk of mammals and may be prepared from whey. Lactose is used in infant foods, margarine manufacture, and medicine. Decomposes at 203°C ; r.d. 1.5.

Ladenburg Benzene. Benzene. [After Albert Ladenburg (1842-1911), German chemist.]

Laetrile. Name given the chemical amygdalin, found in the kernels of many fruits, notably Apricots, bitter Almonds, and Peaches. The subject of controversy for many years, laetrile has been purported by some to be a cure for Cancer. In 1981 the U.S. National Cancer Institute reported laetrile to be ineffective against cancer.

Laevorotatory (laevorotary). Denoting a chemical compound that is capable of imparting an anticlockwise rotation to the plane of polarization of polarized light, as observed by someone facing the light source.

Lagrange, Joseph Louis, Comte. (LagraNzh'), 1736-1813, French mathematician and astronomer; b. Italy. He was director of mathematics (1766-87) at the Berlin Academy of Sciences. After moving to Paris, he became (1793) chairman of the French commission on weights and measures and was influential in the adoption of the decimal system as the basis for the metric system. Under Napoleon, Lagrange was made senator and count. His contributions to mathematics included work on the Calculus, the calculus of variations, number theory, the solutions of equations, and the application of calculus to probability theory. In astronomy, he made theoretical calculations on the liberation of the moon and on the motions of the planets and satellites of Jupiter. He also did research on the nature and propagation of sound and on the vibration of strings. His chief work was the *Mecanique analytique* (1788).

Lagrangian Function. Symbol : *L*. The kinetic energy of a system minus its potential energy. [After Joseph Louis Lagrange (1736-1813), Italian-French mathematician.]

LAH. Lithium aluminium hydride.

Lake. Body of standing water occupying a depression in the earth.

Most lakes are freshwater bodies; a few (*e.g.*, the Great Salt Lake, the Dead Sea), however, are more salty than the oceans. The Great Lakes of the U.S. and Canada are the world's largest system of freshwater lakes. The Caspian Sea is the world's largest lake. Most lake basins were formed by the erosive action of Glaciers on bedrock. Other sources include volcanic calderas and natural and human-made dams in streams and Rivers. Lakes are transient geologic features, eventually disappearing because of several factors, *e.g.*, climatic changes, erosion of an outlet, and Eutrophication.

Lake. An insoluble pigment formed by the combination of an organic dyestuff with a metallic compound (salt, oxide, or hydroxide). The organic substance is either chemically combined with the inorganic compound or absorbed by the compound. Lakes are used in paints and printing inks.

Lamarck, Jean Baptiste Pierre Antoine de Monet, Chevalier de. 1744-1829, French naturalist. Regarded as the founder of invertebrate paleontology, he is noted for his study and classification of invertebrates and for his evolutionary theories; the latter were first made public in his *Systeme des animaux sans vertebres* (1801). *Lamarck's theory of evolution,* or *Lamarckism,* asserted that all life forms have arisen by a continual process of gradual modification throughout geological history. It was based on the theory of Acquired Characteristics, which held that new traits in an organism develop because of a need created by the environment and that they are transmitted to its offspring. Although the latter hypothesis was rejected as the principles of heredity were established, Lamarck's theory was an important forerunner of Charles Darwin's theory of Evolution.

Lambda Particle. Symbol : **A**. An elementary particle with zero charge and a mass 2183 times that of the electron. It is classified as a hyperon.

Lambda Point. 1. The temperature (2.186 K) at which helium I is in equilibrium with helium II.

2. The temperature at which a second-order phase change occurs, such as the change of an alloy from an ordered to a disordered crystal structure or the change from a ferromagnetic to a paramagnetic state. A graph of specific heat capacity against temperature shows a sharp peak at the lambda point.

Lambert. A unit of luminance equal to the luminance of a surface emitting one lumen per square centimetre. [After Johann H. Lambert (1782-77), German mathematician and physicist.]

Lambert's Law (cosine law). L. The principle that, for a perfectly diffusing surface, the luminous intensity in any direction is proportional to the cosine of the angle between the direction and the normal.

2. (of illumination). The principle that the illumination of a surface by a point source is universely proportional to the square of the distance from the source to the surface, the illumination occurring normal to the surface. If the normal to the surface is at an angle θ to the light rays, then illumination is proportional to cos θ.

Lamb Shift. A small energy difference observed between two levels in the hydrogen-atom spectrum ($^2S_{1/2}$ and $^2P_{1/2}$). On normal theories of fine structure, the two levels should have the same energy. The shift is explained by interaction between the electron and the radiation field. '[After Willis Eugene Lamb (b. 1913), American physicist.]

Laminar Flow. Steady flow of a fluid in parallel layers, with little or no mixing between adjacent layers. *Compare* turbulent flow.

Lamination. The use of thin strips of metal to form the core of a choke, transformer, etc., thus increasing the electrical resistance and reducing eddy currents. The strips are often oxidized or varnished.

Lamp Black. The soot (amorphous carbon) produced by burning certain organic compounds, such as oil or turpentine. It is used as a pigment.

Lamprey. Primitive marine and freshwater Fish (family Petromyzontidae) lacking a sympathetic nervous system, spleen, and scales. Although unrelated, the lamprey externally resembles the Eel. Most lampreys are parasitic bloodsuckers, attaching themselves to other fish by means of horny teeth set in a circular, jawless mouth; an anticoagulant in the saliva keeps the blood fluid. Some freshwater lampreys eat flesh as well. They are abundant in the Great Lakes, where they are considered pests by the fishing industry.

Lamp Shell. Marine invertebrate animal (phylum Brachiopoda) with a bivalve shell. In appearance it resembles the clam, but is not related to it. Found in shallow seas, lamp shells attach themselves to objects by means of a short stalk (pedicel) and feed by means of a characteristic tentacled organ (lophophore) surrounding the mouth. The lophophore creates currents that draw water, with food and oxygen, into the shell. Lamp shells are usually from 1 to 2 in. (2.5-5 cm) across.

Lancelet. Small, fishlike lower chordate related to the Vertebrates. Usually about 1 in (2.5 cm) long, with a transparent body tapering at both ends, the lancelet has no distinct head and no paired fins. A filter feeder that lives in shallow marine waters, it is usually found buried in the sand with only the mouth end projecting. The lancelet has a nerve cord but no heart, no brain, and no eyes. Lamprey larvae resemble the lancelet.

Land, Edwin Herbert, 1909—. American inventor and industrialist; b. Bridgeport, Conn. As a student at Harvard, he became interested in Polarized Light and invented (1932) a material to eliminate glare, now known by the trademark Polaroid. In 1937, Land established the Polaroid Corp. to manufacture such products as sunglasses, camera filters, and headlights. He also developed a process for three-dimensional pictures (1941); invented the Polaroid Camera (1947), which takes and prints photos in one step; and introduced "instant" color photographs (1963) and motion pictures 1977).

Landau, Lev Davidovich. 1908-68, Soviet physicist. He was head of the theoretical department of the USSR Academy of Sciences. Landau was a leader in Soviet space technology and helped make the first Soviet atomic bomb. For his contributions to Low-Temperature Physics, especially his development of a theory of Superfluidity, and his pioneering studies on gases, he received the 1962 Nobel Prize in physics.

Langmuir. Irving. 1881-1957, American chemist. He introduced atomic-hydrogen welding, invented a gas-filled tungsten lamp, and contributed to the development of the radio vacuum tube. For his work in surface chemistry, in which he developed an important technique for studying layers of molecules one molecule thick, he won the 1932 Nobel Prize in chemistry. He discovered that particles of dry ice and iodide added to certain clouds could produce rain or snow.

Lanthanide (lanthanon). Any of a series of fifteen metallic elements with atomic numbers from 57 (lanthanum) to 71 (lutetium) inclusive, in which the elements have an incomplete *f* sub-shell. Barium, which immediately precedes lanthanum in the periodic table, has the outer electron configuration $4s^2 4p^6 4d^{10} 5s^2 5p^6 6s^2$. The subsequent elements could, in theory, form a fourth transition series by filling up the 5*d* sub-level. In fact, they form the lanthanide series by filling up the 4*f* level, which can hold a maximum of 14 electrons. Sometimes lanthanum itself is not considered a member of the series.

All are strongly electropositive mainly trivalent metals which form M^{3+} ions. They all have similar chemical properties and mixtures of their compounds occur naturally, especially in monazite. Lanthanide compounds can be separated by ion-exchange chromatography. The series of trivalent metal ions show a decrease in their ionic radius as the atomic number increases: the lanthanum ion has a radius of 0.106 nm and lutetium has 0.085 nm. This is known as the *lanthanide contraction.* It is caused by the extra nuclear charges not being shielded by extra electron charges: a consequence of the shapes of *f* orbitals. The lathanides are often called the *rare earths,* or more correctly the *rare-earth elements*— the former term strictly applies to the oxides. Scandium and yttrium, which occupy the same group as lanthanum in the periodic table, are also often classified with the lanthanides.

Lanthanide Series. Rare-earth Metals with atomic numbers 58 through 71 in group IIIb. of the Periodic Table. They are, in order of increasing atomic number, Cerium, Praseodymium, Neodymium, Promethium, Samarium, Europium, Gadolinium, Terbium, Dysprosium, Holmium, Erbium, Thulium, Ytterbium, and Lutetium. Although they closely resemble Lanthanum and each other in their chemical and physical properties, lanthanum (at. no. 57) is not always considered a member of the series.

Lanthanon. Lanthanide.

Lanthanum. Symbol: La. A soft silvery-white ductile metallic element belonging to the lanthanide series. A.N. 57; A.W. 138.9055; m.pt. 920°C; b.pt. 3454°C; r.d. 6.2; valency 3.

Lanthanum (La). Metallic element, discovered in 1839 by C.G. Mosander. It is a silver-white, soft, malleable, ductile, and chemically active Rare-Earth Metal. Lanthanum is used in making ductile cast Iron and as an alloy in cigarette-lighter flints. It occurs in Monasite.

Lapis Lazuli. Gem composed of lazurite and other minerals in shades of blue, usually flecked with Pyrite. Most often found in massive form in metamorphosed limestones, it has been used since ancient times for beads and small ornaments. Lapis lazuli was the original pigment for ultramarine and was the "sapphire" of the ancients.

Laplace, Pierre Simon, Marquis de. 1749-1827. French astronomer and mathematician. On the basis of Isaac Newton's gravitational theory, he made mathematical studies of the motions of comets, the moon, Saturn, Jupiter, and Jupiter's satellites, as well as of the theory of tides. His research results, which together with those of Joseph Lagrange and earlier mathematicians established Newton's theory beyond a doubt, were published in his famous *Mecanique celeste* (5 vol., 1799-1825). His more popular *Exposition du systeme du monde* (1796) gave scientific form to the nebular hypothesis of the origin of the solar system.

Lark. Perching Bird of the mainly Old World Alaudidae family. Skylarks, the best known of the larks, are c.$7^1/_4$ in. (18.2 cm) long and similar in coloration (grays and browns above and light underneath) and nesting habits (meadows, plains, and other open areas). The horned lark belongs to the only species native to North America, *Eremophila alpestris*; the prairie lark is a subspecies. On the ground, these larks run rather than hop; in flight, they have a melodious song. The Meadowlark belongs to another family

Larmor Precession. The precession of a charged particle when it moves under the influence of an applied magnetic field and a central force. For instance, the orbit of an electron in an atom will precess about an external field, such that the normal to the plane of the orbit sweeps out the surface of a cone with the field's direction as axis. The frequency of precession (the *Larmor frequency*) is $eH/4\pi mc$, where H is the field strength, m the mass, e the electron's charge, and c the electron's velocity. [After Sir Joseph Larmor (1857-1942), British physicist.]

Larynx. Organ of voice in mammals. An extension of the trachea, the human larynx is a small, boxlike chamber with walls of cartilage bound by muscles and membranes. The vocal cords, a pair of elastic folds in the mucous-membrane lining, lie within the larynx. During speech the cords are stretched across the larynx; outgoing breath, forced between them, vibrates and produces voice. The sound varies with the tension of the cords and the space between them.

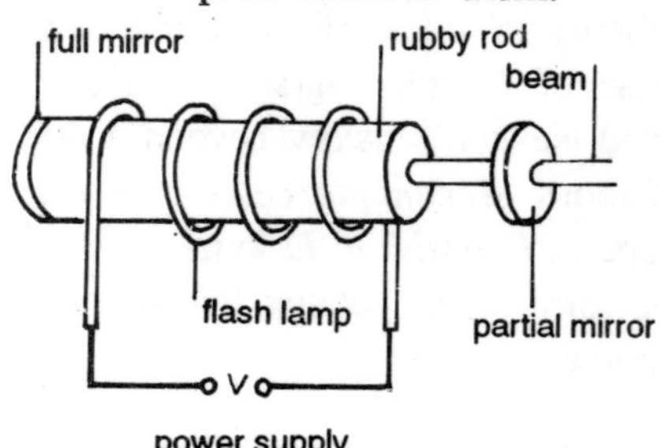

Fig. L-1. Ruby laser.

Laser. A device for producing intense light or infrared or ultraviolet radiation by stimulated emission. The electro-magnetic radiation from a laser is coherent and monochromatic and can be obtained in a parallel beam with very little spread. The principle of laser action can be demonstrated by the earliest type of model, the *ruby laser*, which consists of a rod of ruby with a silvered mirror at one end and a partially silvered mirror at the other. A flash lamp is used to excite chromium ions in the ruby above their ground state. After a short time each ion spontaneously reverts to the ground state with the emission of a photon. If power is continuously supplied by the external light source, a high proportion of the ions are excited at any time—a situation known as *population inversion*. The process of producing large numbers of excited states in this way is called *optical pumping*. The emitted photons can collide with other excited ions and produce stimulated emission of other photons. The light is reflected backwards and forwards up and down the tube and the beam, which in the ruby laser has a wavelength of 694.3 nm, emerges through the partially reflecting end. It is monochromatic because all the photons correspond to the same energy-level change and it is coherent because in stimulated emission the emitted light has the same phase as the incident light from previous stimulated emissions.

Moreover, in stimulated emission the emitted photons move in the same direction as the incident photons—giving the beam a highly directional quality. Other types of solid laser often use lanthanide elements, such as neodymium and yttrium. The name is an acronym for *light amplification by stimulated emission of radiation.* Laser action can also be shown by gases and liquids. In the carbon dioxide laser a mixture of carbon dioxide and nitrogen is used in a gas-discharge tube and population inversion is achieved for vibrationally excited CO_2 molecules. The device gives infrared radiation with a wavelength of 10.6 micrometres. Other types of laser produce population inversion by chemical action.

Lasers can be made with a range of power outputs up to about 100 kilowatts. They are used for cutting metals, surgery, surveying, hologrpahy, and scientific research.

Laser. [Acronym for light amplification by stimulated emission of radiation], device for the creation and amplification of a narrow, intense beam of monochromatic and coherent Light. In a laser, the atoms or molecules are excited so that more of them are at higher energy levels than are at lower energy levels. If a Photon whose frequency corresponds to the energy difference between the excited and ground states strikes an excited atom, the atom is stimulated, as it falls back to a lower energy state, to emit a second photon of the same frequency, in phase with and in the same direction as the bombarding photon. This process is called stimulated emission. The bombarding photon and the emitted photon may then each strike other excited atoms, stimulating further emission of photons, all of the same frequency and phase. This process produces a sudden burst of coherent radiation as all the atoms discharge in a rapid chain reaction. First built in 1960, lasers are widely used in industry, medicine, communications, scientific research, and Holography.

Latent Heat. Symbol : *L.* The total heat absorbed or produced during a change of phase (fusion, vaporization, etc.) at a constant temperature. The heat change per unit mass is the *specific latent heat.*

Lateral Inversion. Inversion from left to right, as seen in an image in a plane mirror.

Latex. Rubber.

Latitude. 1. The distance of a point on the earth from the equator measured as the angle between a normal to the earth at that point and the plane of the equator.

2. (celestial latitude). the angular distance of a body on the celestial sphere from the ecliptic; considered as positive if the body is north of the ecliptic.

Lattice. 1. A regular array of points in two or three dimensions. A crystal lattice is the array of points on which the atoms, molecules, or ions are centred in a crystal.

2. The regular arrangement of fissile material and moderator in a nuclear reactor.

Lattice Energy. The energy that would be produced if the ions forming a given crystal were brought together from infinite separation to their positions in the lattice.

Laughing Gas. Nitrogen oxide.

Lauric Acid (dodecanoic acid). A white crystalline carboxylic acid, $CH_3(CH_2)_{10}COOH$, obtained from coconut oil. It is used in making alkyd resins, detergents, and insecticides. M.pt. 44°C; b.pt. 225°C; r.d. 0.8.

Lava. Molten Rock erupted on the earth's surface by a Volcano or through a fissure in the earth. It solidifies into igneous rock that is also called lava. Before reaching the surface, lava is known as magma.

Laveran, Charles Louis Alphonse. 1845-1922, French physician. While an army surgeon in Algiers, he discovered (1880) the parasite that causes Malaria. For his work on protozoa in the causation of disease he received the 1907 Nobel Prize in physiology or medicine.

Lavoisier, Antoine Laurent. (Lavwazya'), 1743-94, French chemist and physicist. A founder of modern chemistry, he was one of the first to use effective quantitative methods in the study of reactions. His classification of substances is the basis of the modern distinction between chemical elements and compounds and of the system of chemical nomenclature. He proposed the oxygen theory of combustion, thereby discrediting the Phlogiston theory, and described oxygen's role in respiration. Concerned with improving social and economic conditions in France, he held various government posts; he was guillotined during the Reign of Terror.

Lawes, Sir John Bennet. 1814-1900, English agriculturist. He founded the experimental farm at Rothamsted, where, with chemist Sir J.H. Gilbert, he experimented with plants and animals. His development of the superphosphate marked the beginning of the chemical fertilizer industry.

Lawrence Berkeley Laboratory, (LBL; est. early 1930s) and **Lawrence Livermore Laboratory,** (LLL; est. 1952). Nuclear-science research centers, founded by Ernest Lawrence and located, respectively, in Berkeley and Livermore, Calif. The Univ. of California operates them with funds provided by the U.S. Dept. of Energy. Most of LBL's work centers on the study of atomic nuclei and makes use of four major particle accelerators; in addition, LBL formally began research on environmental problems in 1971. LLL carries out applied research on nuclear weaponry, peaceful uses of nuclear explosives, effects of radiation on living organisms, and controlled thermonuclear reactions.

Lawrencium (Lr). Radioactive element, first prepared in 1961 by A. Ghiorso and co-workers by boron nuclei bombardment of claifornium. It is an Actinide-Series element. In 1965 a Soviet group prepared a different isotope by the reaction of oxygen-18 with americium-243.

Lawrencium. Symbol: Lr. A transuranic actinide element obtained in trace amounts by bombarding californium with boron nuclei. The isotope produced, ^{257}Lr, has a half-life of 8 seconds. A.N. 103. [After E. O. Lawrence (1901-58), U.S. physicist who invented the cyclotron.]

Layer lattice. A type of crystal lattice in which the atoms are strongly bound in layers with relatively weak bonding between the layers. Graphite is an example of a substance with a layer lattice.

L.C.M. Least common multiple.

L-dopa. Drug used to alleviate symptoms of Parkinson's Disease, particularly rigidity, slow movements, and trembling, resulting from a deficiency of dopamine in the brain. Introduced into the bloodstream, l-dopa is probably converted to dopamine by neurons in the brain.

Leaching. The process of extracting soluble constituents from a mixture by washing them out with a percolating solvent.

Lead. Symbol: Pb. A very soft bluish-white metallic element existing chiefly as galena (PbS), from which it is obtained by roasting. It is used in some alloys, such as Solder and Babbitt metal, and in pipes, cable coverings, and accumulators. It is an efficient shielding material for radiation and X-rays. The element is also used in the manufacture of tetraethyl lead, which is used as a petrol additive. Lead is very resistant to corrosion and will not dissolve in dilute sulphuric or concentrated hydrochloric acid. It is attacked by alkalis to form plumbites. Lead forms covalently bonded lead IV (*plumbic*) compounds and more ionic lead II (*plumbous*) compounds. A.N. 82; A.W. 207.19; m.pt. 327.5°C; b.pt. 1744°C; r.d. 11.35.

Lead (Pb). Metallic element, one of the earliest known metals, used by the ancient Egyptians and Babylonians. A poor conductor of heat and electricity, lead is silver-blue, dense, relatively soft, and malleable, with low tensile strength. It is used in lead-acid storage batteries, Solder, and plumbing and as protective shielding against X-rays and radiation from nuclear reactors. The principal lead ores are Galena, cerussite, and anglesite. Lead compounds (all poisonous) include tetraethyl lead (a gasoline antiknock additive) and oxides used in mordants and pigments. Continued exposure to lead—through inhalation of fumes or sprays and ingestion of food containing lead—can result in a cumulative chronic disease called *lead poisoning*. It was once a serious occupational hazard, put protective equipment and other precautionary measures have reduced the incidence in industry. A frequent cause of lead poisoning in children, especially in poor areas, is ingestion of paint chips from peeling walls or pipes.

Lead Acetate (sugar of lead). A white crystalline poisonous solid, $Pb(C_2H_3O_2)_2.3H_2O$, made by the action of acetic acid on lead II oxide. It is a soluble lead salt, used in medicine, textiles, and as an analytical reagent. B.pt.280°C (anhydrous); r.d.2.5.

Lead Carbonate. A white poisonous powder, $PbCO_3$, made by adding sodium hydrogen carbonate to lead nitrate solution. It is used as a pigment in paints. r.d. 6.6.

Lead-Chamber Process. A process for the manufacture of sulphuric acid by oxidizing sulphur dioxide with nitrogen dioxide in large lead chambers. The first step is the production of sulphur trioxide: $SO_2 + NO_2 = SO_3 + NO$. the sulphur trioxide is reacted with water to give sulphuric acid, $SO_3 + H_2O = H_2SO_4$, and the nitrogen dioxide is regenerated by the reaction of nitric oxide with oxygen in the air: $2NO + O_2 = 2NO_2$. The process has been replaced by the contact process.

Lead Chromate. A bright yellow insoluble salt, PbCrO made from a solution of a lead salt and potassium dichromate. It is the basis of various yellow pigments (*chrome yellow*).

Lead Dioxide. Lead oxide.

Lead Monoxide. Lead oxide.

Lead Nitrate. A white crystalline poisonous water-soluble solid, $PbNo_3$, made by dissolving lead in nitric acid and used in preparing lead salts and as an oxidizer in the tanning industry. Decomposes at 470°C; r.d. 4.5.

Lead oxide. Any of three oxides of lead. *Lead II oxide* (plumbous oxide, massicot, litharge, lead monoxide) is a yellow crystalline powder, PbO, made by controlled heating of lead. It is an amphoteric substance, used in storage batteries, ceramics, pigments, and oil refining. M.pt. 888°C; r.d. 9.53. *Trilead textroxide* (red lead, minimum) is a red powder, Pb_3O_4, made by heating lead II oxide in air. It is a mixed oxide, used in storage batteries, paints, and ceramics. Decomposes above 500°C; r.d.8.3-9.2. *Lead IV oxide* (lead dioxide, lead peroxide, plumbic oxide) is a brown powder, PbO_2, prepared by adding bleaching powder to an alkaline solution of lead hydroxide. Its principal use is as an oxidizing agent. Decomposes at 290°C; r.d. 9.4.

Lead Peroxide. Lead oxide.

Lead-Plate Accumulator. A type of accumulator consisting of two lead plates dipping into dilute sulphuric acid. When the cell is fully discharged both plates are coated with lead sulphate. In charging the cell, a current is passed from the cell's positive electrode to the negative electrode, through the electrolyte. At the positive plate the lead sulphate is converted to a layer of lead oxide, PbO_2, and at the negative plate the lead sulphate is reduced to a layer of spongy lead. During discharge the reverse processes occur: lead oxide changes to lead sulphate with loss of negative charge from the electrode and lead changes to lead sulphate with gain of negative charge. The overall reaction is $PbO_2 + Pb + 2H_2SO_4 = 2PbSO_4 + 2H_2O$.

Lead-plate accumulators are used in car batteries. The relative density of the acid falls as the cell is discharged and should normally be in the range 1.2-1.28. If the battery is allowed to go "flat" an insoluble sulphate may form.

Lead Sulphate. A white insoluble crystalline salt, $PbSO_4$. M.pt. 1170°C; r.d. 6.2.

Lead Sulphide. A black insoluble compound, PbS, precipitated from solutions of lead salts by hydrogen sulphide. It occurs naturally as galena. M.pt. 1114°C; r.d. 7.5.

Lead Tetraethyl. Tetraethyl lead.

Leaf. The chief food-manufacturing organ of higher plants, a lateral outgrowth of the Stem. The typical leaf consists of a stalk, or petiole, and a thin, flat, expanded portion (needle-like in most Conifers), or blade. The blade, veined with sap-conducting tubes (xylem and phloem), consists of upper and lower layers of epidermal cells, including cells that control the size of tiny pores (stomata) that are used in gas exchange and Transpiration. Between the two layers are cells, rich in Chilorophyll, that conduct Photosynthesis.

Leaf Insect. Tropical herbivorous Insect, about 4 in. (10 cm) long, whose flattened, green irregularly shaped body gives it a leaflike appearance. The related Walking Stick resembles a twig.

Leakage. Flow of a small electric current through imperfect insulation.

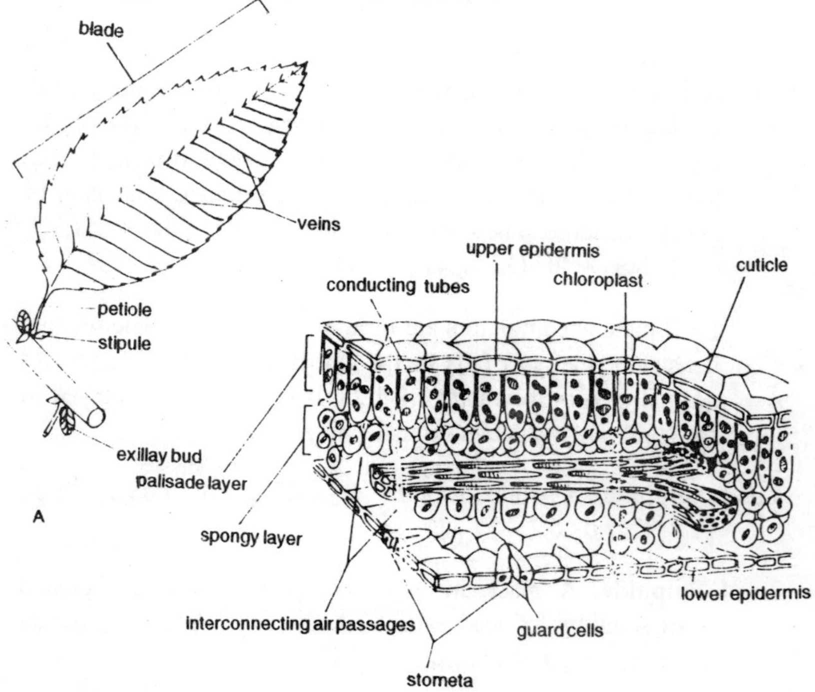

Fig. L-2. A. General structure of a leaf. B. Microscopic cross section of the leaf blade.

Least Common Multiple (LCM). The smallest number that every number of a given set of numbers can divide into exactly. For example, the LCM of 2, 7, and 5 is 70.

Least Squares. A method of obtaining an equation to fit experimental data. A typical case would be one in which a number of measured values of a dependent variable x are known for values of the independent variable y. In general, the values will show a scatter resulting from experimental error and it is necessary to find the best line, say $y = mx + c$, to fit the data. Values of the constants, m and c, are chosen and the deviations of the experimental values of x from calculated values are obtained. The best fit is given for values of m and c for which the sum of the squares of these deviations has a minimum.

Leather. Skin or hide of animals, cured by Tanning to prevent decay and to impart flexibility and toughness. Early peoples used pelts preserved with grease or smoke for garments, tents, and containers. Since the 18th cent. machines have been used to split the tanned leather into the desired thicknesses of flesh layers and grain (hair-side) layers. Pelts are prepared by dehairing, cleaning, tanning, and treating with fats to ensure pliability. Finishes include glazing, staining or dye coloring, enameling or lacquering (as for patent leather), and sueding (buffing to raise a nap). Artificial leather, made since c.1850, is now mainly manufactured from vinyl Plastic.

Leblanc Process. A former process for the manufacture of sodium carbonate by heating sodium chloride with sulphuric acid to give sodium sulphate, which is then heated with coke and limestone to yield sodium carbonate and calcium sulphide. [After Nicolas Leblanc (1742-1806), French chemist.]

Le Chatalier's Principle. The principle that if a disturbance is applied to displace a system from chemical equilibrium, the system tends to change in such a way as to compensate for the disturbance. The principle can be illustrated by the reaction: $N_2 + 3H_2 = 2NH_3$. The production of ammonia molecules is exothermic and their decomposition is endothermic. If the temperature is increased, the equilibrium is displaced to the left: ammonia molecules decompose as if to restore the original temperature. Thus, high yields of ammonia are favoured at low temperatures although in practice the rate of attainment of equilibrium must also be considered. If the total pressure is increased, the reaction shifts to the right: more ammonia molecules are produced as if to decrease the total number of molecules and thus restore the total pressure. [After Henri Louis Le Chatalier (1850-1936), French chemist.]

Leclanche Cell. A type of primary cell consisting of a carbon anode surrounded by a fabric bag containing carbon powder mixed with manganese dioxide. The electrolyte is ammonium chloride solution and the cathode is a zinc rod. At the cathode zinc atoms change to zinc ions, leaving a negative charge.

Ammonium ions from the elctrolyte gain electrons at the anode, leaving a net positive charge, and decompose to nitrogen and hydrogen. The manganese dioxide helps to prevent polarization of the anode by oxidizing the hydrogen. Common dry batteries are types of Leclanche cell: the e.m.f. is about 1.5 volts. [After Georges Leclanche (1839-82), French chemist.]

Leech. Segmented or Annelid Worm with a cylindrical or slightly flattened body having suckers at both ends; it usually feeds on blood, which it stores in pouched large enough to hold several months' supply. Most are aquatic. Leeches were once used to bleed patients suffering form almost any ailment; they are still used in some regions to treat bruises.

Leeuwenhoek, Antony Van. (La'Venhook'), 1632-1723, Dutch student of natural history. He made over 247 Microscopes, some of which magnified objects 270 times. He examined microorganisms and tissue samples and gave the first complete descriptions of bacteria, protozoa (which he called animalcules), spermatozoa, and striped muscle. He also studied capillary circulation and observed red blood cells.

Legendre, Adrien Marie. (LezhaN'dre), 1752-1833, French mathematician, noted especially for his work on number theory and elliptic integrals. He invented, independently of Carl Gauss, and was the first to state in print (1806) the method of least squares.

Legionnaire's disease. Infectious, sometimes fatal, disease characterized by high fever, dry cough, lung congestion, and subsequent Pneumonia. The disease struck over 180 people attending an American Legion convention in Philadelphia in July, 1976—hence the name. The causative bacterium, later identified as *Legionella pneumophilla,* is thought to spread through air conditioning and ventilation systems and thus infect many people simultaneously. The disease is treated with the Antibiotic erythromycin.

Leibniz or **Leibnitz, Gottfried Wilhelm, Beron Von.** 1646-1716, German philosopher and mathematician. His career as a scholar embraced the physical sciences, law, history, diplomacy, and logic, and he held diplomatic posts (from 1666) under various German princes. Leibniz also invented the Calculus, concurrently with but independently of Newton. His philosophical writings, including *Theodicy* (1710) and *Monadology* (1714), popularized by the philosopher Christian von Wolf, were orthodox and optimistic, claiming that a divine plan made this the best of all possible worlds (a view satirized by Voltaire in *Candide*). According to Leibniz, the basic constituents of the universe are simple substances he called monads, infinite in number, nonmaterial, and hierarchically arranged. His major work, *New Essays on Human Understanding*, a treatise on John Locke's *Essay concerning Human Understanding*, was written in 1704 but because of Locke's death published only in 1765. A critique of Locke's theory that the mind is a blank at birth, it exerted great influence on Kant and the German Enlightenment. Modern studies have tended to focus on Leibniz's contributions to mathematics and logic; manuscripts published in the 20th cent. show him to be the founder of symbolic logic.

Lemma. A result proved as a preliminary to the proof of a theorem.

Lemming. Mouselike Rodent of arctic or northern regions, inhabiting tundra or open meadows. All are about 5 in. (13 cm) long, with stout bodies, thick fur, and short tails. Two or three times per decade, Norway lemmings (*Lemmus lemmus*) undergo a population explosion that forces them to set out in search of food. Crossing bodies of water by swimming, some reach the ocean and drown—giving rise to folklore about lemmings committing mass suicide.

Lemon. Yellow-skinned Citrus Fruit of a small tree (*Citrus limon*) of the Rue family. High in Vitamin C, lemons prevent scurvy. Products include Citric Acid, juice, oil, polish, pectin, and flavorings. Lemons grow best in a mild climate, *e.g.*, the Mediterranean, California, and Florida.

Lemur. Prosimian, or lower Primate, of the related families Lemuridae and Indriidae, found only on Madagascar and adjacent islands. Lemurs have monkeylike bodies, long, bushy tails; pointed muzzles, large eyes, and flat nails, except the second toe, which has a stout claw. Most are arboreal. Best known is the ring-tailed lemur (*Lemur catta*), which is a typically terrestrial.

Lens. A device for forming an image of an object by the Refraction, or bending, of light. In its simplest form it is a disk of transparent substance, commonly glass, with its two surfaces curved or with one surface plane and the other curved. Generally each curved surface—called *convex* if curved outward and *concave* if curved inward—of a lens is made as a portion of a spherical surface; the center of the sphere is called the center of curvature (C) of the surface. All rays of light passing through a lens are refracted except those that pass directly through a point called the optical center. A divergent lens (thicker at the edges than at the center) bends parallel light rays passing through it away from each other. The image formed by a diverging lens is always erect (upright), smaller than the object, and virtual (located on the same side of the lens as the object). A convergent lens (thicker at the center than at the edges) bends parallel light rays toward one another; if they are parallel to the principal axis of the lens, they converge to a common point, or focus (F), behind the lens. The image formed by a converging lens depends on the position of the object relative to the lens's focal length (distance between the focus and the optical center) and its center of curvature.

Lens. 1. A device for converging or deverging a beam of light by refraction, consisting of a flat piece of transparent material with one or two curved faces. Usually the faces are spherical although other shapes are used for special purposes. Lenses

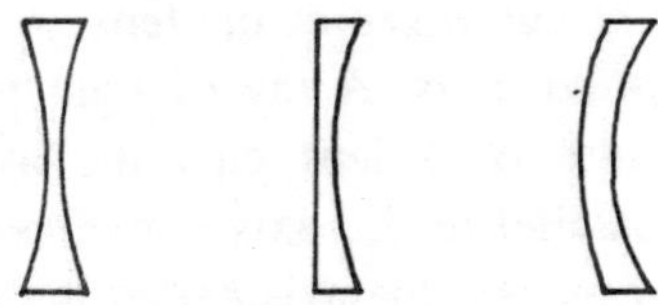

Fig. L-3. Ray diagrams.

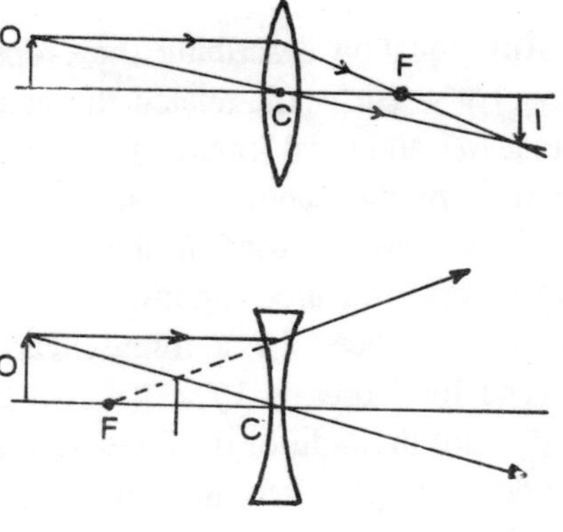

L-4. Types of lens.

are classified into *converging lenses* (which converge the beam) and *diverging lenses*. Converging lenses are *convex*: *i.e.* they are thicker at their centre than at their edge. Convex lenses are further classified into *biconvex* lenses, which have both faces curved, *plano-convex* lenses, with one plane face, and *concavo-convex* (or positive *meniscus)* lenses, which have both faces curving in the same direction. Diverging lenses are *concave*: they are thinner at their centre than their edge. Concave lenses are similarly classified into *biconcave, plano-concave,* and *convexo-concave* or (negative *meniscus*) lenses. A *thin lens* is one whose thickness is small compared with its focal length: its properties are described by a simple lens formula. The formula cannot be applied to a *thick lens.*

The centres of the spheres of which the lens surfaces are part are the *centres of curvature* of the lens. A line through these points is the *optical axis.* A ray of light passing undeviated through the centre of a lens cuts the axis at the *optical centre.* A ray parallel to the axis converges to or appears to diverge from a point on the axis called the *principal focus* or *focal point.* The distance from the optical centre of a thin lens to the principal focus is the *focal length.*

2. Any device for diverging or converging a beam.

Lens Formula. Any equation describing the properties of a lens. The equation $1/v + 1/u = 1/f$, related the image distance (v), object distance (u), and focal length (f) of a thin lens. Distances are measured from the centre of the lens and the "real is positive" sign convention is used. In the other sign convention, in which distances measured against the light are positive, the plus sign is replaced by a minus. The other equation commonly used for lenses is $1/f = [(n_2 - n_1)/n_1][1/r_1 - 1/r_2]$, where r_1 and r_2 are the radii of the faces and n_2 and n_1 are the refractive indexes of the lens material and of the medium surrounding the lens.

Lenticular. Relating to a lens.

Lentil. Old World annual *(Lens culinaris)* of the Pulse family. Its pods contain two dark seeds—lentils—unusually high in Protein content; the seeds are ground into meal or used in soups. One of the first food plants cultivated in Europe, lentils are increasingly grown for food in the U.S.

Lenz's Law. The principle that if the magnetic flux changes with respect to a conductor the current induced in the conductor tends to flow in such a direction as to oppose the change (by producing an opposing field). [After Heinrich Lenz (1804-65), German physicist.]

Leopard. Large carnivore *(Panthera pardus)* of the Cat family, found in Africa and Asia. Its yellowish fur is patterned with black spots and rings. Black leopards, a color variant, are called Panthers. The largest male leopards are about 7 ft (2.3 m) long, including the tail. They live mainly in forests and are solitary and nocturnal, preying on small animals and livestock.

Leprosy or **Hansen's Disease.** Chronic infectious disease, caused by *Mycobacterium leprae,* affecting the skin and superficial nerves. It is found mainly, but not exclusively, in tropical regions. The disease produces numerous skin and nerve lesions, which, if left unteated, enlarge and may result in severe disfigurement. Leprosy is treated—successfully in most cases—with dapsone and other drugs.

Lepton. Any of a class of elementary particles that have half-integral spin and take part in weak interactions. The leptons are fermions: they include the electron, the muon, and the neutrino, and their antiparticles. It is possible to assign a quantum number (the *lepton number, 1*) to each particle, defined to be 1 for the electron, negative muon, and neutrino and -1 for their antiparticles *(i.e.* for the positron, negative muon, and antineutrino). Particles that are not leptons have lepton number 0. In any reaction the total lepton number is unchanged. For example, in beta decay: $n \longrightarrow p + e + v$, the neutron has 1 = 0 and the total lepton number of the products, 1 + 0 + (-1), is also 0. *Compare* baryon.

Leslie's cube. A large cubic metal container with four of its sides having different finishes or colours, filled with boiling water and used to demonstrate the effect of the surface on the emission of radiant heat. A thermopile or similar detector is used to measure the radiation from each face. [After Sir John Leslie (1766-1832), British physicist.]

Lettuce. Garden annual *(Lactuca sativa* and varieties) of the Composit family. Long cultivated as a salad plant and unknown in the wild state, lectuce is possibly derived from the weed called wild lettuce *(L. scariola)*. Three types of lettuce are grown: head, leaf, and Cos, or romaine (the most heat-tolerant).

Leukemia. Term for any of a variety of cancerous disorders of blood-forming tissues (bone marrow, lymphatics, spleen), characterized by the abnormal proliferation of blood cells (commonly white blood cells). The disease may be chronic or acute. The cause of leukemia is unknown, but genetics, certain viruses, and exposure to radiation may play a role. Symptoms include weakness, fever, bleeding and susceptibility to infection. Chemotherapy is effective against some forms of leukemia, especially those occurring in children.

Level. The ratio of a value of a quantity to a reference value of that quantity. Often a logarithmic scale is used.

Lever. A simple machine consisting of a rigid bar turning about a pivot (the *fulcrum*). There are three types, depending on the relative positons of load, effort, and fulcrum. The mechanical advantage is the ratio of the distance from the load to the fulcrum to the distance from the effort to the fulcrum.

Lewis Acid. A compound that can accept a pair of electrons in the formation of a coordinate bond. According to this extension of the theory of acids and bases the electron acceptor is the acid and the compound donating the lone pair is a *Lewis base*. In the reaction $BC1_3 + :NH_3 = C1_3B:NH_3$, boron trichloride is the acid and ammonia is the base. [After Gilbert N. Lewis (1875-1946), American chemist.]

Lewis base. Lewis acid.

Lewisite. A colourless volatile liquid, $ClCH:CHAsCl_2$, used as a vesicant war gas M.pt. 18.2°C; b.pt. 190°C (decomposes); r.d. 1.89. [After Winford Lee Lewis (1878-1943), American chemist.]

Leyden Jar. An early form of capacitor, consisting of a glass jar coated with tinfoil on part of its outer and inner surfaces. [After Leyden, the town in Holland where it was invented.]

LF. *See* Low frequency.

Libration. *See* Moon.

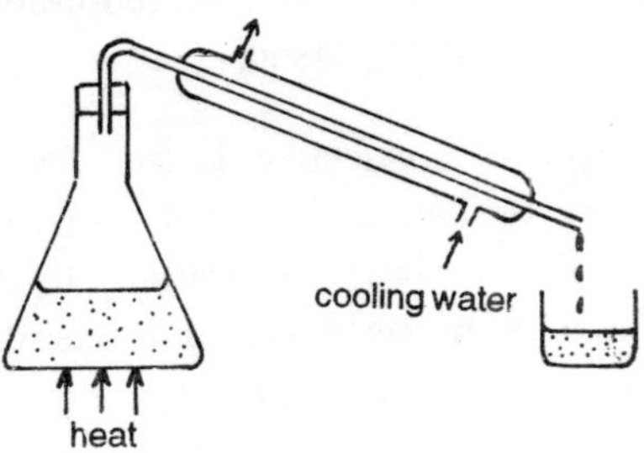

Fig. L-5. Liebig condenser.

Lichen. Simple plant consisting of blue-green or green Algae living symbiotically with Fungi, usually sac fungi. Lichens commonly grow on rocks or trees. The body, or thallus, is composed of fungal filament, or hyphae. The fungi obtains food from the algae and, in turn, absorbs and retains the water that is used by the algae for Photosynthesis. Reproduction of the algae and fungi is usually simultaneous. Lichens can withstand great extremes in temperature and can be found in deserts and the polar regions.

Licorice. European, blue-flowered perennial *(Glycyrrhiza glabra)* of the Pulse family; also, the sweet substance obtained from its roots, used medicinally and as a flavoring. It is cultivated chiefly in the Near East.

Liebig condenser. A simple condenser used in Laboratory distillations, consisting of a straight glass tube surrounded by a glass cooling jacket through which water is passed.[After Baron von Liebig (1803-73), German chemist.]

Lie detector. Instrument designed to determine whether a subject is lying or telling the truth, generally by detecting evidence of the slight increase in body tension believed to occur when a person knowingly lies. Various devices developed in the 20th cent. and used mainly in police work measure blood pressure, respiration, pulse, electrical changes on the skin, and vaice frequencies, Lie detectors are not considered infallible, and test results are not usually accepted as evidence in U.S. courts. The use of lie detectors to screen employees and job applicants is highly controversial.

Ligament. Strong band of connective Tissue that joins Bones to other bones or to cartilage in the Joint areas. Ligaments tend to be pliable but not elastic, permitting limited movement while holding attached bones in place. Fibrous sheets supporting internal organs are also ligaments.

Ligand. A molecule, ion, or group that coordinates to a metal atom or ion in a complex.

Light. Electromagnetic radiation that produces a visual sensation when it strikes the human eye. Light has the wavelength range 400-740 nm. Its velocity, in a vacuum, is $2.997\ 925 \times 10^{8}$ m s^{-1}. Several theories of light have been used. According to the *corpuscular theory* light consist of small elastic particles emitted by a luminous body. The theory can explain geometrical optics but not interference and polarization. The *wave theory* of light is that light is a wave motion propagated in the ether. Maxwell showed that light is a form of electromagnetic wave motion, consisting of transverse varying electric and magnetic fields. Some phenomena, such as the photoelectric effect, are best explained by considering the light to be streams of photons.

Light. That part of Electromagnetic Radiation to which the human eye is sensitive. The wavelengths of visible light range from c. 400 to c. 750 nanometers. If white light, which contains all wavelengths, is separated into a Spectrum, each wavelength is seen to correspond to a different Color. The scientific study of the behaviour of light is called Optics; it covers Reflection of light by a Mirror or other object, Refraction of light by a Lens or Prism, and Diffractions of light as it passes by an opaque object. Christiaan Huygens proposed (1690) a theory that explained light as a Wave phenomenon. Isaac Newton, however, held (1704) that light is composed of tiny particles, or corpuscules, emitted by luminous bodies. By combining his corpuscular theory with his laws of mechanics, he was able to explain many optical phenomena. Newton's corpuscular theory of light was favored over the wave theory until important experiments, which could be interpreted only in terms of the wave theory, were done on the diffraction and Interference of light by Thomas Young (1801) and A.J. Fresnel (1814-15). In the 19th cent. the wave theory became the dominant theory of the nature of light. The electromagnetic theory of James Clerk Maxwell (1864) supported the view that visible light is a form of Electromagnetic Radiation. With the acceptance of the electromagnetic theory of light, only two general problems remained. It was assumed that a massless medium, the Ether, was the carrier of light waves, just as air or water carries sound waves. The famous experiments (1881-87) by A.A. Michelson and E.W. Morley, in which they tried unsuccessfully to measure the velocity of the earth with respect to this medium, failed to support the ether hypothesis. With his special theory of Relativity, Albert Einstein showed (1905) that the ether was unnecessary to the electromagnetic theory. Also in 1905, Einstein, in order to explain the Photoelectric Effect, suggested that light, as well as other forms of electromagnetic radiation, travel as tiny bundles of energy, called light quanta, or Photons, that behave as particles. Light thus behaves as a wave, as in diffraction and interference phenomena, or as a stream of particles, as in the photoelectric effect. The theory of relativity predicts that the speed of light in a vacuum (186,282 mi/sec = 299,792.458 km/sec) is the limiting velocity for material particles.

Lightness. The property of a colour of a pigment, dye, etc., determined by the amount of light it reflects. Colours of the same hue that differ in lightness are *shades.*

Lightning. Electrical discharges in the atmosphere, resulting from the formation of regions of positive and negative charge within a cloud. The spark produced can occur to the ground (cloud to ground), between clouds (cloud to cloud), or within a cloud. The potential difference initiating the flash is about 10^8 volts, this being sufficient to cause electrical breakdown of the air. In cloud to ground lightning there is an initial faintly luminous leader stroke passing towards the ground. An upward discharge occurs to meet this leader about 50 metres above the ground, thus inducing the bright return stroke. A typical lightning flash consists of four or five strokes about 40 ms apart, each propagated at about 5×10^7 m s^{-1}. Enormous amounts of energy are dissipated: the return stroke can be 5 km in length dissipating 10^5 joules per metre and currents of 10000 A and temperatures of 30 000 K are produces. The heating and cooling of the air produced the pressure waves causing thunder. Flashes of this type constitute "forked lightning." "Sheet lightning" is observed when a flash occurs within a cloud, thus lighting up the whole cloud without the channel being visible. *Ball lightning*—the production of a small slowly moving luminous ball of plasma—is the rarest and least understood form of lightning.

Lightning. Electrical discharge accompanied by Thunder, commonly occurring during a Thunderstorm. The discharge may take place between two parts of the same cloud, between two clouds, or between a cloud and the earth. Lightning may appear as a jagged streak (forked lightning), as a vast flash in the sky (sheet lightning), or, rarely as a brilliant ball (ball lightning). The electrical nature of lightning was proved by Benjamin Franklin in his famous kite experiment of 1752.

Lightning conductor. A sharply pointed metal rod attached to the top of a building and connected to earth, used to protect the building from lightning. The electric field induced by an electrically charged cloud has a high gradient in the region of the sharp point. This causes ionization of the air and prevents the build up of the high potential differences necessary for lightning to occur.

Light pen. A device connected to the visual-display unit of a computer. It is capable of sensing the information on the screen and can be used to input information by "drawing" lines on the screen.

Light-year. A unit of length equal to the distance travelled by electromagnetic radiation in one year. It is used for expressing astronomical distances and is equal to 9.4607×10^{15} metres ($5.878\ 48 \times 10^{12}$ miles).

Lignite or **Brown Coal.** Carbon—containing fuel intermediate between Coal and Peat, brown or yellowish in color and woody in texture. Lignite contains more moisture than coal and tends to dry and crumble when exposed to air. It burns with a long, smoky flame but little heat.

Lignum Vitae. Tropical American evergreen tree (genus *Guaiacum).* Its dense, durable wood, chiefly from *G. sanctum* and *G. officinale,* is used where strength and hardness are required, *e.g.,* in ship construction and for butcher blocks.

Ligroine. *See* petroleum ether.

Lilac. Old World shrub or small tree (genus *Syringa)* of the Olive family, noted for its fragrant, cone-shaped masses of lavender or white flowers. Many variations in form, *e.g.,* double flowers, and color, *e.g.,* rosy pink, have been hybridized from the familiar common lilac.

Lily. Common name for the family Liliaceae, perennial plants having showy flowers and erect clusters of narrow, grasslike leaves. The lily family is distributed worldwide but is particularly abundant in warm temperate and tropical regions. Most species grow from Bulbs or other forms of enlarged underground Stems. Common wildflowers in the family are Asphodel, Dogtooth Violet, Lily of the Valley, and Trillium. Ornamentals of commercial importance include lilies, Hyacinths, Meadow Saffron, Squill, and Tulips; food plants of commercial importance are Asparagus and plants of the Onion genus. Yucca and Aloe species are popular Succulents. True lilies include the Madonna lily *(Lilium candidum)* of Europe; the white trumpet lily *(L. longiflorum)* of Japan, which includes the Easter, or Bermuda, lily (var *eximium);* the tiger lily *(L. tigrinum)* of China; and the Turk's-cap lily *(L. superbum)* and leopard lily *(L. pardalinum),* both of North America.

Fig. L-6. *Wood lily,* Lalium philadelphicum.

Lily of the Valley. Fragrant, spring-blooming perennial (genus *Convallaria)* of the Lily family. It has dainty, bell-shaped white flowers on a stalk between two shiny leaves and grows in the shade. There are two species: the widely cultivated *C.*

majalis, native to Europe, and *C. montana,* which grows in the Appalachian Mts.

Lime. Small shrublike tree *(Citrus aurantifolia)* of the Rue family. Its bright-green fruit, smaller and more acid than the Lemon, has long been used to prevent scurvy. The plant grows well in rocky or sandy soils. It is the most frost-sensitive of the Citrus Fruits, and chief production areas are in tropical regions.

Limestone. Sedimentary rock composed of calcium carbonate. It is ordinarily white but may be colored brown, yellow, or red by iron oxide and blue, black, or gray by carbon impurities. Most limestones are formed from the skeletons of marine invertebrates; a few are chemically precipitated from solution. Organic acids acting on underground deposits lead to formations such as those in Carlbad Caverns and Mammoth Cave National Parks. Limestone is used in iron extraction, in cements and building stones, and as a source of lime. Limestone varieties include Chalk, Dolomite, Marble, Oolite, and travertine.

Limestone. A rock composed of calcite, $CaCo_3$, used in making carbon dioxide, calcium oxide, cement, and other calcium compounds. It is also used as a flux in smelting metals and as a building stone.

Lime Water. A clear aqueous solution of calcium hydroxide. It is used as a test for carbon dioxide, which produces a milky precipitate of calcium carbonate.

Limit. The value approached by a mathematical function as its independent variable approaches some specified value.

Limonite or **Brown Hematite.** Yellowish to dark brown mineral [FeO(OH). nH_2O] occurring worldwide in deposits formed by the alteration of other minerals containing iron. It is used as a pigment (in ocher) and as an ore of Iron. Both iron rust and bog iron ore are limonite.

Limonite. A dark brown or black mineral consisting of hydrated iron II oxide, FeO(OH). nH_2O, used as an ore of iron and a yellow pigment.

Limpet. Gastropod mollusk with a flattened conical shell and a muscular foot with which it clings tightly to rocks. Although occasional specimens reach 4 in. (10 cm) in length, most are smaller. Limpets are found mainly in cooler waters of the Atlantic and Pacific oceans.

Linac. *See* linear accelerator.

Lindane. *See* hexachlorocyclohexane.

Linden. A woody shrub or tree of the family Tiliaceae, including the tropical genus *Corchorus,* from which Jute is obtained. The name most often refers to deciduous trees (genus *Tilia)* known as linden, lime tree, or basswood, valued for ornament and shade. Their light, strong wood is useful for woodenware and excelsior and in Bee culture; their flowers yield an excellent honey. Fiber made from the tough inner Bark, or bast (hence the name basswood), is used in caning and wickerwork.

Linde Process. A process for liquefying air by expansion through a nozzle. The air is compressed by a pump and expanded through a valve into an expansion chamber. The expanding air is cooled as a result of work done against intermolecular forces. The cooled air is used to cool the incoming compressed air: eventually the temperature falls to the point at which the air is liquefied. [After Carl von Linde (1842-1934), German engineer.]

Linear. 1. In a straight line; characterized by one dimension only.

2. Pertaining to a relationship of direct proportionality: *i.e.* one that would be represented by a straight line on a graph. A *linear equation* is one in which the terms containing the variables are all of the first degree, as in $y = 7x + 5$. A *linear scale* is one in which the intervals are equally spaced. A device, component, or circuit is said to be linear if its output is directly proportional to its input, as in a *linear amplifier*.

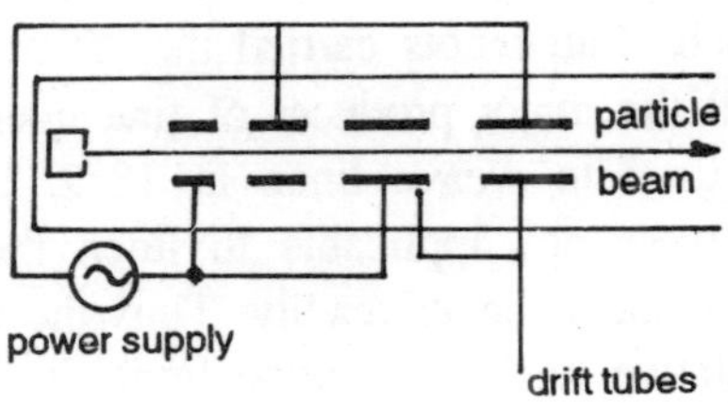

Fig. L-7. Linear accelerator.

Linear Accelerator (linac). A type of particle accelerator in which the particles travel in straight lines along a long evacuated tube. In low energy machines the particles are accelerated by a series of cylindrical electrodes separated by gaps. Radiofrequency alternating potentials are applied to the electrodes in such a way that there is an accelerating field present when the particles are travelling through the gaps. The lengths and spacing of the electrodes are arranged so that as the particles gain energy in moving along the tube they always reach the inter-electrode gaps at the correct point in time for acceleration to occur. They move at constant velocity inside the cylindrical electrodes (*drift tubes*).

Higher energy linear accelerators do not contain electrodes. Microwave fields are applied and the particles move down the tube accelerated by a travelling wave in a waveguide. The energy obtained depends on the length, typical energy gains being 7 MeV per metre for electrons and 1.5 MeV per metre for protons. The linear accelerator at Stanford, California, is two miles long and produces particles in the 10-20 GeV range.

Line defect. *See* defect.

Linen. Fabric or yarn made from Flax, probably the first vegetable Fiber known. Linen fabric dating from 5000 B.C. has been found in Egyptian tombs. Egyptian, greek, and Jewish priests wore linen to symbolize purity. Brought to N Europe by the Romans, it became the chief European textile of the Middle Ages. French Hunguenots carried the art of working flax to Ireland, still the major producer of fine linen. Power Looms were first used to weave linen in 1812, but many textile inventions were not applicable to linen thread because its inelasticity made it break readily. Thus the expense of linen weaving relative to that of cotton limits its use. It is woven into fabrics ranging from heavy canvas to sheer handkerchief linen.

Line of Force. A line in a magnetic or electric field, whose direction at any point is the direction of the field at that point.

Line Printer. A device that prints the output of a computer a whole line at a time, rather than printing individual characters. Line printers can work at speeds as high as 3000 lines per minute.

Line spectrum. A spectrum consisting of discrete lines. Line spectra are produced by atoms—the lines correspond to emission or absorption of photons as a result of electron transitions between energy levels.

Linkage. The amount of magnetic flux passing through a coil of wire or other electric circuit.

Linnaeus, Carolus. 1707-78, Swedish botanist and taxonomist, considered the founder of the binomial system of nomenclature and the originator of modern scientific Classification of plants and animals. In *Systema naturae* (1735) and *Genera plantarum* (1937) he presented and explained his classification system, which remains the basis of modern taxonomy. His more than 180 works also include *Species plantarum* (1753), books on the flora of Lapland and Sweden, and the *Genera morborum* (1763), a classification of diseases.

Linseed Oil. Amber-coloured oil extracted from linseed, the seed of the Flax plant. The oil obtained from hydraulically pressed seeds is pale in colour and practically odorless and tasteless. Oil that has been boiled or extracted by application of heat and pressure is darker, with a bitter taste and unpleasant odor. Linseed oil is used as a drying oil in paints and Varnishes and in making linoleum, oilcloth, and certain inks.

Lion. Large carnivore *(Panthera leo)* of the Cat family, found in open country in Africa, with a few surviving in India. The tawny-coated male lion usually has a long, thick mane and may reach 9 ft (2.7 m) in length and 400 lb (180 kg) in weight. Lions live in prides of up to 30 individuals. Females do most of the hunting, preying on zebra, antelope, and domestic livestock.

Lipids. Natural products in living systems that are insoluble in water but soluble in organic solvents. Major classes of lipids include fatty acids, glycerol-derived lipids (including fats and cils), sphingosine-derived lipids associated with the nervous system, steroids, terpenes, certain aromatic compounds, and long-chain alcohols and waxes. The fat-soluble Vitamins can be classified as lipids.

Lip Reading. Method by which the deaf are able to read the speech of other from movements of the lips and mouth. A medium of education in many schools for deaf children, it came into wide use after World War I in the rehabilitation of shell-shocked and otherwise deafened soldiers.

Liquation. The process of separating solid mixtures by heating them to a temperature at which one component melts.

Liquefaction. The change of a substance into the liquid state. The term is often applied to the liquefaction of gases. If the substance has a critical termperature above room temperature, it can be liquefied by pressure alone. Otherwise some cooling process must be used.

Liqueur. Strong alcoholic beverage made of nearly neutral spirits flavored with herbs, fruits, or other materials, and usually sweetened. The alcoholic content ranges from c.27% to 80%. Cordials are prepared by steeping fruit pulps or juices in sweetened alcohol. Well-known liqueurs include anisette, benedictine, chartreuse, creme de menthe, kirsch, and kummel.

Liquid. A state of matter intermediate between a gas and a solid, characterized by ease of flow combined with incompressibility. Thus a liquid, like a gas, will take the shape of its container, offering little resistance to shear stress. Unlike a gas, it will not change its volume to fill the container. The intermolecular forces involved are larger than those in gases and smaller than those in solids, being large enough to prevent spontaneous expansion or significant compression but too small to maintain the order and relative rigidity found in solids. In liquids, molecular order is short range and transient. Theories of liquids are less well developed than theories of solids or gases.

Liquid Air. A pale blue liquid made by the liquefaction of air and used as a refrigerant. It is mainly liquid oxygen (b.pt. 182.9°C) and liquid nitrogen (b.pt. 195.7°C).

Liquid Crystal. A state of certain molecules that flow like liquids but have an ordered arrangement of molecules. Substances that form liquid crystals have long molecules and the order results from intermolecular forces. Three types of liquid crystal exist. In *nematic* crystals the molecules are randomly arranged but all their axes are parallel. *cholesteric* crystals have molecules arranged in layers with their axes all parallel and lying in the plane of the layer. *Smetic crystals* also have molecules arranged in layers; in this case the axes of the molecules are perpendicular to the plane.

Liquid Crystal. Liquid whose component particles, atoms or molecules, tend to arrange themselves with a degree of order far exceeding that of ordinary liquids and approaching that of solid crystals. As a result, liquid crystals have many of the optical properties of solid crystals. Moreover, because its atomic or molecular order is not as firmly fixed as that of a solid crystal, a liquid can be easily modified by electromagnetic radiation, mechanical stress, or temperature, with corresponding changes in its optical properties. This characteristic has made possible liquid crystal displays ((LCD) such as those used on some digital clocks, electronic calculators, and personal computers.

Liquid-drop Model. A model of the atomic nucleus in which it is visualized as a drop of liquid, with its nucleons behaving like the molecules in a drop of water. The impact of a neutron causes the drop to oscillate and split into two: *i.e.* nuclear fission occurs. The model can be used to predict fission energies for nuclei.

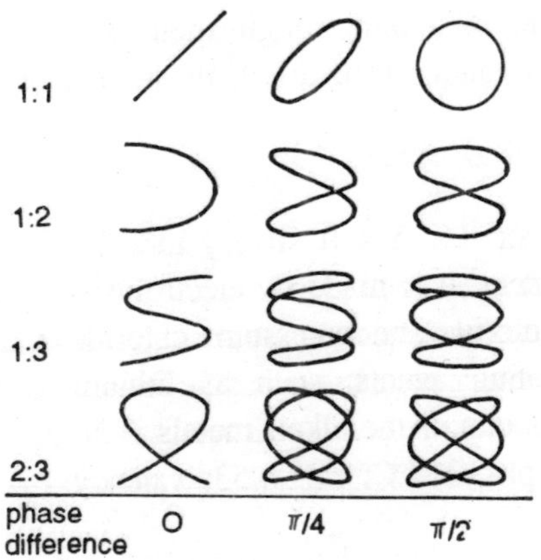

Fig. L-8. Lissajous figures.

Lissajous' figures. Figures produced by the path of a point that moves in two dimensions with components that are simple harmonic motions. The figures can be produced on an oscilloscope by deflecting the spot horizontally with one sinusoidal signal and vertically with another. Two components with the same frequency and phase give a straight line. If they are out of phase, ellipses are formed—a 90° phase difference gives a circle. If the two signals have different phases and frequencies more complicated figures result. [After Jules-Antoine Lissajous (1822-80), French mathematician.]

Lister, Joseph Lister, 1st **Baron,** 1827-1912. English surgeon. He introduced to surgery the principle of antisepsis, an outgrowth of Louis Pasteur's theory that bacteria cause infection, and he founded (1865) modern antiseptic surgery. Using carbolic acid as an antiseptic agent in conjunction with heat sterilization of instruments, he dramatically decreased post-operative fatalities. He developed absorbable ligatures and the drainage tube, both now in general use for wounds and incisions.

Litchi. Chinese tree *(Litchi chinensis)* of the soapberry family, also cultivated in other warm countries. It has a small, aromatic pulpy fruit in a thin, rough shell. The best-known Chinese fruit, it is eaten fresh, dried, preserved, or canned.

Litharge. *See* lead oxide.

Lithium. Symbol: Li. A soft silvery metallic element; the lightest solid element. It is made by electrolysis of a fused mixture of lithium chloride and potassium chloride and used in making hydrogenating agents, such as lithium aluminium hydride. Lithium is one of the alkali metals A.N. 3; A.W. 6.939; m.pt. 179°C; b.pt. 1317°C; r.d. 0.53; valency 1.

Lithium. (Li), metallic element, discovered in 1817 by J.A. Arfvedson. A soft silver-white corrosive Alkali Metal, lithium is the least dense metal. Lithium compounds are used in lubricating greases, special glasses, and ceramic glazes; as brazing and welding fluxes; and in the preparation of plastics and synthetic rubber. Lithium is also a medical antidepressant.

Lithium Aluminium Hydride (LAH). A white or grey powder, $LiAlH_4$, prepared by reacting aluminium chloride with lithium hydride. It is extensively used in preparative organic chemistry as a reducing agent for many functional groups. Decomposes above 130°C; r.d. 0.92.

Lithium Chloride. A white deliquescent crystalline solid, LiCl, used in soldering flux and mineral waters. M.pt. 614°C; b.pt. 1360°C; r.d. 2.07.

Lithium Hydride. A white crystalline solid, LiH, made by direct combination of the elements at high temperature. It is used as a reducing agent in organic synthesis. M.pt. 680°C; r.d. 0.82.

Lithium Nitride. A brownish-red powder, Li_3N, used as a nitriding agent in metal-lurgy and as a reducing agent in organic chemistry. M.pt. 845°; r.d. 1.3.

Litmus. A soluble compound obtained from certain lichens. It has a red colour in acid solutions and a blue colour in alkaline solutions: hence its use as an indicator.

Lithography. Type of planographic or surface printing used as an art process and in commercial Printing, where the term is synonymous with offset printing. Lithography was invented c. 1796 by Aloys Senefelder, and the Bavarian limestone he used is still considered the best material for art printing. Lithography is based on the antipathy of oil and water. A drawing is made in reverse on the ground (flat) surface of the stone with a crayon or ink the contains soap or grease. The image produced on the stone will accept printing ink and reject water. Once the grease in the ink has penetrated the stone, the drawing is washed off and the stone kept moist. It is then inked with a roller and printed on a lithographic press. As a process, lithography is probably the most unrestricted, allowing a wide range of tones and effects. Several hundred fine prints can be taken from a stone. The medium was employed by many 19th-cent. artists, including Goya, Delacroix, Daumier, Degas, Whistler, and Toulouse-Lautrec. Among American artists noted for their lithographs are A.B. Davies, George Bellows, and Currier and Ives. The medium remains popular with contemporary artists. *Photolithography* is frequently used in the commercial reproduction of art works. With the process, a photographic negative is exposed to light over a gelatin-covered paper, and those portions of the gelatin that are exposed become insoluble. The soluble portions are washed away, and the pattern to be printed is transferred to a stone or metal plate. In colour lithography or color photolithography, a stone or plate is required for each colour used.

Litmus. Organic dye usually used as an indicator of acidity or alkalinity. Naturally pink in colour, it turns blue in alkaline solutions and red in acids. Litmus paper is paper treated with the dye.

Litre. Symbol: 1. A metric unit of volume equal to one cubic decimetre. The name is not used in precision measurements. The original definition of the litre, in 1901, was the volume of one kilogram of pure water at the temperature of its maximum density (4°C) and at 760 mmHg pressure, making it equal to 1000.028 cubic centimetres.

Liver. Largest glandular organ of the body. It lies on the right side of the abdominal cavity, beneath the diaphragm, and is made up of four unequal lobes. Liver tissue consists of thousands of tiny lobules, in turn made up of hepatic cells, the basic metabolic cells. The liver is thought to perform over 500 functions involving the Digestive System, Excretion, blood chemistry and detoxification, and the storage of vitamins and minerals. Of the liver's many digestive system functions, the production of Bile (for fat digestion) and storage of glucose (see Glycogen) are particularly important.

Liver of Sulphur. A brown mixture of potassium polysulphides and potassium thiosulphate made by heating potassium carbonate with sulphur.

Liverwort. Small, flowerless, primitive, green land plant (division Bryophyta), characterized by horizontal growth and related to the Mosses. Usually growing in moist places, liverworts are considered intermediate between the aquatic Algae and the terrestrial mosses and Ferns. The ancients believed that liverworts could cure diseases of the liver, hence their name.

Lixiviation. The process of separating solid mixtures by washing soluble components out with water.

Lizard. Reptile of the order Squamata, which also includes the Snake, distributed worldwide (except for the Arctic) but most common in warm climates. Lizards typically have four legs with five toes on each foot, although a few are limbless, retaining internal vestiges of legs. They also differ from snakes in having ear openings, movable eyelids, and less flexible jaws. Several, most notably Chameleons, undergo colour changes under the influence of environmental and emotional stimuli. Lizards range in size from species under 3 in. (7.6 cm) long to the 10-ft (3-m) Komodo dragon.

Load. *See* machine.

Loaded Concrete. Concrete with added amounts of compounds of heavy metals, such as barium or lead. These are efficient absorbers of radiation and the material is used for shielding nuclear reactors.

Lobefin. Name for several lunged, fleshy-finned, bony Fishes, predecessors of the Amphibians. Lobefins were considered extinct until 1938, when a live coelocanth *(Latimeria chalumne),* a marine lobefin, was caught off S. Africa. Coelocanths are brown to blue and 5 ft (150 cm) long, with circular, overlapping scales, a laterally flattened three-lobed tail, a spiny dorsal fin, and a vestigial lung.

Lobster. Large marine Crustacean with five pairs of jointed legs, the first pair bearing large pincerlike claws of unequal size adapted to crushing the shells of its prey. The darkgreen common American lobster *(Homarus americanus)* is found from Labrador to North Carolina, but especially along the New England coast. When lobster is cooked, the shell turns bright red; the meat is considered a delicacy.

Locomotive. Vehicle used to a pull a train of unpowered Rail-Road cars. From their invention in the early 19th cent. until the early years of the 20th, all locomotive were powered by Steam Engines fueled by wood or coal, with a rod-and-piston arrangement to move the drive wheels. The front wheel assembly (the swivel truck) and the wheels mounted under the cab were unpowered. *Electric locomotives,* introduced c. 1895, obtained their power from an electric trolley, or pantograph, running on an overhead wire, or from a third rail. Electric locomotives are used chiefly on steep grades and on runs of high traffic density. Although very efficient, they are not move widely used because of the cost of electrifying larger railroad systems. In parts of Europe and in Japan, however, high-speed electric locomotives are widely used. American railroads today are largely powered by *diesel-electric locomotives (introduced* c. 1924). These use a Diesel Engine to drive an electric generator, which feeds electric motors that turn the driving wheels. *Gas turbine-electric locomotives* are similar to the diesel-electric but use a gas Turbine to drive the generator.

Locus. The curve traced by a point moving so as to satisfy a specified condition. A circle, for example, is the locus of a point that moves so as to be equidistant from a fixed point.

Locust. In botany, deciduous tree or shrub (genus *Robinia)* of the Pulse family, native to the U.S. and Mexico. The black locust *(R. pseudoacacia),* a popular ornamental, has fragrant flowers; its durable wood is used for treenails in shipbuilding and for fenceposts, turning, and fuel. The carob, thought to be the biblical locust tree, and the Honey Locust belong to other genera in the same family.

Locust. In zoology, migratory Insect of the short-horned Grasshopper family. Locust migration is an occasional event. Under certain environmental conditions, which also lead to population increases, young locusts develop into a short-winged migratory form, gather in huge swarms, and, at maturity, take to the air. The swarms can include more than 100 billion insects. When they finally settle, the resulting agricultural devastation is enormous.

Lodestone. *See* magnetite.

Loewi, Otto. 1873-1961, American physiologist and pharmacologist; b. Germany. For his discovery of the chemical transmission of nerve impulses he shared with Sir Henry Dale the 1936 Nobel Prize in physiology or medicine. Loewi investigated the physiology and pharmacology of metabolism, the kidneys, the heart, and the nervous system. He was professor of pharmacology at the Univ. of Graz, Austria (1909-38), and New York Univ. (1940-61).

Logarithm. The power to which a number, called the base, must be raised in order to obtain a given positive number. For example, the logarithm of 100 to the base 10 is 2, because $10^2 = 100$. Common logarithms use 10 as the base; natural, or Napierian, logarithms (for John Napier) use the number *e*.

Logarithm. The power to which a fixed number (the *base*) must be raised to produce a given number. Thus, if a is the logarithm of c to base b, written $a = \log_b c$, then $c = b^a$. It follows from the laws of exponents that $\log m + \log n = \log mn$, $\log m - \log n = \log m/n$, and $\log m^n - n \log m$. These rules are used in performing calculations with tabulated values of logarithms of numbers to the base 10 *(common logarithms)*. Usually tables only contain the logarithms of numbers between 0 and 10. The logarithm of any number is then found by expressing it as a number between 1 and 10 multiplied by a power of 10. For example, 2570.1 is 2.5101×10^3. $\log_{10} (2.5701 \times 10^3)$ is log 2.5701 + 3, or 3,4099. The integer, in this case 3, is the *characteristic* and the decimal, in this case 0.4099, is the *mantissa.* This system is also used for numbers between 0 and 1. For example, $\log_{10}0.001670$ is $\log_{10}$ (1.670×10^{-3}) = 0.2227 - 3. Usually this is written as 3.2227, where 3 is the mantissa and 0.2227 is the characteristic. *Natural logarithms* use the base *e*(2.7182...) Log_e *a* is often written as In *a.* They are also called *hyperbolic* or *Naperian* logarithms.

Logarithmic. Pertaining to a relationship in which one variable is proportional to the logarithm of another. The scale marked on a slide rule, for example, is logarithmic: the distance of each number from the zero is proportional to the logarithm of the number. Logarithmic scales of this type are found in certain measuring instruments. Some physical measurements are defined by logarithmic relationships. The pH of a solution, for example, is a measure of the reciprocal of the hydrogen-ion concentration, a difference of 1 in pH corresponding to a factor of 10 in concentration.

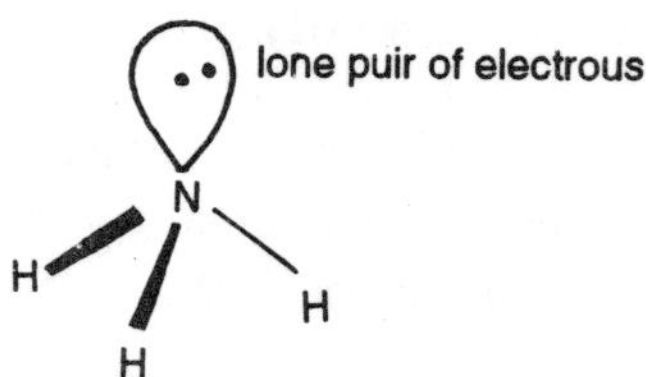

Fig. L-9. Ammonia.

Logic. Systematic study of valid inference. Classical, or Aristotelian, logic is concerned with the formal properties of an argument, not its factual accuracy. Aristotle, in his *Organon,* held that any logical argument could be reduced to a sequence of 3 propositions (2 premises and a conclusion), known as a Syllogism, and posited 3 laws as basic to all logical thought: the law of identity *(A is A);* the law of contradiction *A cannot be both A and not A);* and the law of the excluded middle *(A must be either A or not A).* Aristotle assumed a correspondence linking the structures of reality, the mind, and language, a position known in the Middle Ages as Realism.

The opposing school of thought, Nominalism, represented by William of Occam, maintains that language and logic correspond to the structure of the mind only, not to that of reality. John Stuart Mill in the 19th cent. helped to formulate the scientific method of Induction, *i.e.*, movement from specific perceptions to generalizations. Aristotelian logic basically held sway in the Western world for 2,000 years, but since the 19th cent. it has been largely supplanted as a field of study by symbolic logic, which replaces ordinary language with mathematical symbols. Symbolic logic draws on the concepts and techniques of mathematics, notably Set theory, and in turn has contributed to the development of the foundations of mathematics. In the early 20th cent. Bertrand Russell and Alfred North White head attempted to develop logical theory as the basis for mathematics.

Logical Positivism. Also known as scientific Empiricism, modern school of philosophy that in the 1920s attempted to introduce the methodology and precision of mathematics to the study of philosophy, much as had been done in symbolic logic. Led by the Vienna Circle, a group including the philosophers Rudolf Carnap and Moritz Schlick and the mathematician Kurt Godel, the logical positivists held that metaphysical speculation is nonsensical; that logical and mathematical propositions are tautological; and that moral and value statements are merely emotive. The function of philosophy, they maintained, is to clarify concepts in both everyday and scientific language. The movement received its inspiration from the work of Frege, Bertrand Russell, Wittgenstein, and G.E. Moore. The Vienna Circle disintegrated in the late 1930s after the nazis took Austria, but its influence spread throughout Europe and America, and its concept, particularly its emphasis on the analysis of language as the function of philosophy, has been carried on throughout the West.

Logic Circuit. Electric Circuit whose output depends upon the input in a way that can be expressed as a function in symbolic

Logic; it has one or more binary inputs (capable of assuming either of two states, *e.g.*, "on" or "off") and a single binary output. Logic circuits that perform particular functions are called gates. Basic logic circuits include the AND gate, the OR gate, and the NOT gate, which perform the logical functions *AND*, *OR* and *NOT*. Logic circuits, which are mainly used in digital Computers, can be built from any binary electric or electronic devices, including Switches, Relays, Electron Tubes, solid-state Diodes, and Transistors.

Lone Pair. A pair of electrons with opposite spins occupying the same orbital in an atom. A lone pair of electrons has a position in space in the same way that a chemical bond is directional. This is used in explaining the shapes of simple molecules. Ammonia (NH_3), for example, is not planar but has a pyramidal structure because of the position of the lone pair.

Longitude. Angular distance on the earth's surface measured along the Equator east or west of the Prime Meridian, which is at 0°. All other points have longitudes from 0° to 180° east or west. Meridians of longitude (imaginary lines drawn from pole to pole) and parallels of Latitude form a grid by which any position on the earth's surface can be specified.

Longitude. 1. The angular distance between one of the earth's meridians and the standard meridian through Greenwich.

2. (celestial longitude). The angular distance of a body on the celestial sphere measured anticlockwise from the first point of Aries (the vernal equinox).

3. *See* polar coordinates.

Longitudinal wave. *See* wave.

Long sight. *See* hypermetropia.

Loom. Frame or machine used for Weaving; used since 4400 B.C.

Looms on which the warp threads are stretched horizontally have several fundamental parts: a warp beam, on which the warp threads are wound; heddles, each with an eye through which a warp thread is drawn; harnesses, frames that contain the heddles and that, when raised or lowered, form a shed between the warp threads for insertion of the weft; a comblike reed, which separates the warp threads; a beater, which pushes the weft against the cloth after each row of weaving; a breast beam, over which the cloth is wound before being rolled onto the cloth beam; and a brake, which maintains the tension of the warp threads. The shuttle is a tool that carries the weft through the shed. The foot loom operates the harnesses by treadles. Looms on which the warp threads are stretched vertically, *e.g.,* Tapestry and Navaho Indian looms, are more simply constructed. Edmund Cartwright patented (1785) the first practical power loom, and Joseph Marie Jacquard perfected (1804) a device using punched cards to weave complicated designs on a power loom.

Loon. Migratory aquatic Bird, found in fresh and salt water in the colder parts of the Northern Hemisphere. Its strange, laughing call carries for great distances. Expert swimmers and divers, loons walk on the land with difficulty. Their long, sharp beaks are well adapted for catching fish.

Loran. [Long-range navigation], long-range, accurate radio navigational system used by a ship or aircraft to determine its geographical position. The measured time-of-arrival difference between signals transmitted from two geographically separated ground stations determines the hyperbolic curve on which the receiver is situated. By taking a similar time-difference reading from a second pair of stations whose curve intersects that of the first pair, a definite geographical fix may be obtained.

Lorentz-Fitzgerald Contraction. A contraction in length by a factor $(1 - v^2/c^2)$, postulated to occur to a body moving through the ether with a velocity v (c) is the velocity of light). The contraction takes place in the direction of the body's motion. The phenomenon was suggested as an explanation of the negative result obtained in the Michelson-Morley experiment: the dimensions of the interferometer change in the direction of the earth's rotation in such a way that the two light beams take the same time to cover their paths. An explanation of the contraction is given by the theory of relativity. [After Hendrik Antoon Lorentz (1853-1928), Dutch physicist, and George Francis Fitzgerald (1851-1901), Irish physicist.]

Lorentz, Hendrik Antoon. 1853-1928, Dutch physicist. For his explanation of the Zeeman effect (a change in spectral lines in a magnetic field), which was based on his postulating the existence of Electrons, he shared with Pieter Zeeman the 1902 Nobel Prize in physics. He extended the hypothesis of George Fitzgerald, an Irish physicist, that a body's length contracts as its speed increases (the Fitzgerald-Lorentz contraction) and formulated the Lorentz transformation, by which space and time coordinates of one moving system can be correlated with the known space and time coordinates of any other system. This work influenced, and was confirmed by, Albert Einstein's special theory of Relativity.

Lorentz Transformation. A transformation defined by a set of equations that relate the coordinates of space and time in two frames of reference when one frame moves at a constant velocity relative to the other. They are used in the theory of special relativity.

Lorenz, Konrad. 1930-, Austrian zoologist and ethologist. For his work in Ethology, particularly his studies of the organization of individual and group behaviour patterns, he shared the 1973 Nobel Prize in physiology of medicine with Karl von Frisch and Nikolaas Tinbergen. With Oscar Heinroth, Lorenz discovered imprinting, a rapid and nearly irreversible learning process occurring early in life. His controversial book *On Aggression* (1966) maintains that aggressive impulses are to a degree innate, and draws analogies between human and animal behaviour.

Loschmidt's Number. Symbol *L*. The number of molecules in one cubic centimetre of an ideal gas at standard temperature and pressure. It has the value 2.687 19 X 10^{19}. [After Joseph Loschmidt (1821-95), Austrian physicist.]

Loudness. The sensation that enables a listener to judge the intensity of a sound wave. The loudness is approximately proportional to the logarithm of the sound intensity.

Loudspeaker. A device for converting changing electric currents into sounds, usually consisting of a small coil that is attached to a cardboard cone and is free to move in the field of a strong permanent magnet. The current produces vibrating motion of the coil, thus vibrating the cone and producing sound. The effect is the opposite of that in the microphone, although loudspeakers work at higher power levels.

Loudspeaker or **Speaker.** Device used to convert electrical energy into sound. It consists essentially of a thin flexible sheet called a diaphragm that is made to vibrate by an electric signal from an Amplifier. The vibrations create sound waves in the air around the speaker. In a common dynamic speaker, the diaphragm has a cone shape and is attached to a wire coil suspended in magnetic field. A signal current in the suspended coil creates another magnetic field that interacts with a already existing field, causing the coil and the diaphragm attached to it to vibrate. Quality sound systems employ three different sized speakers. The largest one, the woofer, reproduces low frequencies; the medium-sized one, called a mid-range speaker, reproduces middle frequencies; the smallest one, called a tweeter, reproduces high frequencies.

Low Frequency (LF). A frequency in the range 30 kilohertz to 300 kilohertz.

Low-temperature Physics or Cryogenics. Science concerned with the production and maintenance of very low temperatures, and with the effects that occur under such conditions. Although it is impossible to reach absolute zero, a temperature as low as about one millionth of a degree on the Kelvin scale above absolute zero can be attained. Low temperatures are achieved by removing energy from a substance. By using a succession of liquefied gases. a substance may be cooled to as low as 4.2° K, the boiling point of liquid helium. Still lower temperatures may be reached by successive magnetization and demagnetization. Some unusual conditions, notably Superconductivity and Superfluidity, prevail at cryogenic temperatures.

Low Tension. Low voltage.

LSD or **Lysergic Acid Diethylamide.** Hallucinogenic Drug, an extremely potent drug causing physiological and behavioral changes. Reactions to LSD, such as heightened sense perceptions, anxiety, and hallucinations, are influenced by the amount of the drug taken and the user's personality and expectations. Prolonged psychic disturbances have been reported with LSD use, and there is some evidence linking it with chromosome damage.

Luge. Type of small sled in which one or two persons race down snowy hillsides or steeply banked, curving chutes. Steering is accomplished by shifting weight, pulling straps attached to the runners, and use of the feet. Lugeing is an Olympic event for both men and women.

Lumen. Symbol: lm. An SI unit of luminous flux equal to the amount of light emitted per second in a cone of one steradian solid angle by a point source of one candela.

Luminance. Symbol: *L.* The brightness in a particular direction of a surface that is emitting or reflecting light. It is given, at a particular point, by the luminous intensity *(I)* per unit of area projected onto an area at right angles to the direction. Thus for a uniformly reflecting surface $L = I/A \cos \theta$, where θ is the angle the direction makes with the surface.

Luminescence. The emission of light by sources other than a hot, incandescent body. It is caused by the movement of electrons within a substance from more energetic states to less energetic states. Among several types are chemilumi-nescence, electroluminescence, and triboluminescence, which are produced, respectively, by chemical reactions, electric discharges, and the rubbing or crushing of crystals.

Luminescence. The emission of light by any mechanism that does not depend on the body having a high temperature: *i.e.*, it is distinguished from incandescence. Luminescence is the result of transitions of electrons in atoms and ions whereas incandescence is the effect of vibrations of the atoms in a solid. In luminescence, atoms or molecules have to be produced in an excited state. They then change from that state to the ground state and emit light. Many different mechanisms are possible: the excited state may be produced by other electromagnetic radiation (photoluminescence), by electron bombardment (electroluminescence), or by chemical or biological reactions (chemiluminescence and bioluminescence). Luminescence is also induced by such processes as friction (triboluminescence), and radioactive decay (radioluminescence). The term is applied both to the process and to the light emitted and is also used for analogous effects yielding ultraviolet radiation and X-rays.

Luminosity. The rate at which energy of all types is radiated by a Star in all directions. A star's luminosity varies approximately as the square of its radius and the fourth power of its absolute surface temperature.

Luminosity. 1. The brightness of a source of light,

2. The absolute brightness of a star, depending on the amount of light it emits and independent of the distance from the earth.

Luminous Flux. The rate of flow of light energy from a source as measured by its visual sensation. The luminous flux is related to the total power emitted corrected according to the sensitivity of the eye to light of different wave-lengths.

Luminous Intensity. Symbol : *I*. The amount of light emitted per second per unit of solid angle from a point source of light in a given direction. It is measured in candelas. Formerly, the quantity was measured in candles and called *candlepower*.

Luminous Paint. A type of paint containing a phosphor mixed with a small amount of radioactive material such as radium. Light is emitted by radiolum-inescence.

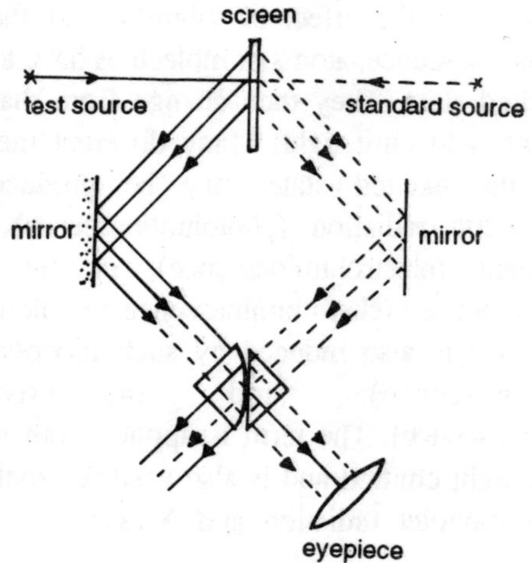

Fig. L-10. Lummer-Brodhun photometer.

Lummer-Brodhum Photometer. A type of photometer in which the light source investigated is compared with a standard source using a combination of prisms which allow the illumination from the test source to be seen as a spot in the centre of the illumination from the standard. The positions of the lamps are varied until the field of view is evenly illuminated.

Lunar caustic. *See* silver nitrate.

Lunar Crater. Any of numerous typically saucer-shaped depressions on the surface of the moon. Their evolution is somewhat mysterious though it is now generally thought that they arose from a combination of meteroid impacts and lunar volcanic disturbances.

Lunar Time. Time measured with respect to the moon. The *lunar month* which is also called the *lunation* or *synodic month)* is the time between successive new moons. The *lunar year* is defined as 12 lunar months. 1 lunar year = 354.3671 mean solar days.

Lunation. *See* lunar time.

Lungfish. Lung-bearing Fish, often resembling an Eel, found in rivers in South America, Africa, and Australia. Like the Lobefin, it is ancestrally related to the four-footed land animals. The most primitive living lungfish is a stout-bodied Australian species, 5 ft .(150 cm) long with paired fins set on short stumps. African species, which hibernate in hard clay during the dry season, breathe through gills in water. Other species will drown if held under water.

Lungs. Pair of elastic organs used for breathing in vertebrate animals. In humans, they are located on either side of the heart, filling much of the chest cavity. Air enters each lung through a large tube, or bronchus, which divides and subdevides into a network of bronchioles. These tiny tubules lead to cup-shaped air sacs known as alveoli, each of which is surrounded by a net of capillaries. As blood flows through the capillary net, carbon dioxide passes into the alveoli and oxygen diffuses into the bloodstream. Covered by a thin membrane, the pleura, which allows them to move freely during breathing, the lungs are expanded (inhalation) and contracted (exhalation) by the combined movement of the diaphragm and the rib cage. Diseases of the lungs include bronchitis, pleurisy, and pneumonia.

Lutetium (Lu). Metallic element, discovered independently by Georges Urbain in 1907 and by Carl Auer von Welsbach in 1908. A silver-white Rare-Earth Metal and Lanthanideseries element with few commercial uses, it is found in Monazite.

Lutetium (cassiopeium). Symbol : Lu. A soft silvery-white metallic element belonging to the lanthanide series: A.N. 71; A.W. 174.79; m.pt. 1656°C; b.pt. 3315°C; r.d. 9.8; valency 3.

Lux (metre-candle). Symbol : lx. An SI unit of intensity of illumination equal to the illumination produced by a flux of one lumen falling uniformly on a surface one square metre in area.

Lwoff. Andre. 1902—, French microbiologist. In the 1920s his study of the morphogenesis of protozoa led to the discovery of extranuclear inheritance in these organisms. After World War II he conducted experiments that enabled him to explain the phenomena of lysogenic bacteria. For his contributions to molecular biology, Lwoff shared with Francois Jacob and Jacques Monod the 1965 Nobel Prize in physiology or medicine.

Lyman Series. A spectral series in the ultraviolet emission of hydrogen with wavelengths given by $1/\lambda = R(1/1^2 — 1/m^2)$, where R is the Rydberg constant and m is 2, 3, 4, etc. [After Theodore Lyman (1874-1954), U.S. physicist.]

Lyophilic. Having an affinity for a solvent.

Lymphatic System. Network of vessels carrying lymph, or tissue fluid, from the Tissues into the veins of the Circulatory System. Lymph, a colourless fluid similar to blood but containing no red blood cells or platelets and considerably less protein, is continuously passing through the walls of the capillaries. It transports nutrients and oxygen to the cells and collects waste products. It also transports large molecules, such as proteins, and bacteria, which cannot enter the small pores of the circulatory system. Most of the lymph passes back into the venous capillaries; however, a small amount is returned to the blood via the lymphatic system. The lymphatic system is composed of fine capillaries that lie adjacent to the blood capillaries. These merge first into larger tributaries called trunks, and then into two still larger vessels called ducts. Ducts feed into the circulatory system in the region of the collarbone, returning the lymph to the bloodstream there. Along the lymphatic network, in the neck, armpit, groin, abdomen, and chest, are small reservoirs, the lymph nodes, which collect bacteria and other deleterious agents and act as a barrier against the entrance of these substances into the blood. In addition, the lymphatic system, like the circulatory system, absorbs nutrients from the small intestine.

Lynx. Any of several small, ferocious Cats found in N North America and N Eurasia. Lynxes have small heads with tufted ears and heavy bodies with long legs and short tails; they have yellow-brown to grayish fur. Nocturnal hunters, they prey mainly on small animals. The Canadian lynx may be over 3 ft (90 cm) in length and weigh up to 40 lb (18 kg). The smaller North American spotted lynx *(Felis rufa)* is also known as a bobcat.

Mach, Ernst (makh), 1838-1916. Austrian physicist and philosopher. He did his major work in the philosophy of science, striving to rid science of metaphysical assumptions, and influenced the development of Logical Positivism. He also did work in ballistics; the Mach Number is named for him.

Machine. Any of various simple devices in which a small force (the *effort*) is used to overcome a larger force (the *load*). The *velocity ratio* of the machine is the distance moved by the effort divided by the distance moved by the load. The *mechanical advantage* is the load divided by the effort. The *efficiency* is the work done on the load divided by the work done by the effort, often expressed as a percentage. In a perfect machine, with no friction, the efficiency is 1 and the mechanical advantage is equal to the velocity ratio.

Machine. Any arrangement of stationary and moving mechanical parts used to perform some useful work or a specialized task. By means of a machine, a small force can be applied to move a much greater resistance or load; the force, however, must be applied through a much greater distance than it would if it could move the load directly. The mechanical advantage of a machine is the factor by which it multiplies any applied force. The simplest machines are (1) the *lever,* consisting of a bar supported at some stationary point (the fulcrum) along its length, and used to overcome resistance at a second point by application of force at a third point; (2) the *pulley,* consisting of a wheel over which a rope, belt, chain, or cable runs; (3) the *inclined plane,* consisting of a sloping surface, whose purpose is to reduce the force that must be applied to raise a load; (4) *screw,* consisting essentially of a solid cylinder around which an inclined plane winds spirally, whose purpose is to fasten one object to another, to lift a heavy object, or to move an object by a precise amount; and (5) the *wheel* and *axle,* consisting of a wheel mounted rigidly upon an axle or drum of smaller diameter, the wheel and axle having the same axis. The more complicated machines are merely combinations of these simple machines. Machines used to transform other forms of energy (as heat) into mechanical energy are known as engines, *e.g.,* the Steam Engine or the Internal Combustion Engine. The electric Motor transforms electrical energy into mechanical energy; its operation is the reverse of that of the electric Generator. In the past, the first machines, *e.g.,* the catapult, were built to improve war-making capacity. The first manufacturing machines, powered by steam engines, appeared during the 18th cent., causing the onset of the Industrial Revolution.

Machine Tool. Power-operated tool used for shaping or finishing metal parts by removing chips, shavings, large pieces, or extremely small particles. Machine tools vary in size from hand-held devices used for drilling and grinding to large stationary machines that perform a number of different operations. The lathe, for example, can turn, face, thread, and drill. The working surfaces of a machine tool are made of such substances as high-speed steels, sintered carbides, and diamonds—substances that can withstand the great heat generated by the action of the working surface against the workpiece.

Mach Number. Ratio between the speed of an object, usually an airplane, and the speed of Sound in the medium in which the object is traveling. A plane traveling at Mach 3.0 is traveling at three times the speed of sound.

Mach Number. The ratio of the speed of a fluid or of a body in a fluid to the speed of sound in the fluid. *Mach 1* is the speed of sound in the fluid. *Mach 5* is five times the speed of sound, etc. [After Ernst Mach (1838-1916), German physicist.]

Mackerel. Open-sea Fish of the family Scombridae, including the albacore, bonito, and Tuna. Mackerel have deeply forked tails, narrowed where they join the body, finlets behind the dorsal and anal fins, and streamlined bodies. They are superb, swift swimmers of generally large size and are important commercially as food. They travel in schools feeding on fish, especially Herring, and Squid, and migrate between deep and shallow waters. The largest mackerel is the tuna (up to 3/4 ton/680 kg); among the smallest ($1^1/_2$ lb/ 0.675 kg) is the common mackerel of the Atlantic.

Macrame. Decorative knotting technique. Named for an Arabic word for knotted fringe, it arose in the 13th cent. and reached Europe during the next hundred years. A traditional sailors' pastime, after decades of obscurity it was revived in the 1960s for wall hangings, jewelry, and other objects.

Macromolecule. A very large molecule, as found in polymers or in compounds such as haemoglobin.

Macroscopic. 1. Large enough to be seen without the use of a microscope.

2. Having properties determined by the statistical behaviour of large numbers of atoms or molecules.

Madder. The dyestuff alizarin, obtained from madder roots.

Madder. Common name for the family Rubiaceae, chiefly tropical and subtropical trees, shrubs, and herbs, especially abundant in N. South America. The family is important economically for several tropical crops, *e.g.,* Coffee and Quinine (from Cinchona), and for many ornamentals, *e.g.,* the madder, Gardenia and bedstraw. True madder (*Rubia tinctorum*), also called turkey red, is a dye plant native to S. Europe. The herb's long, fleshy root was the principal source of various brilliant red dye pigments until artificial production of alizarin, the pigment chemical in madder. The bed-straws (genus (*Galium*), formerly used for mattress filling because of their pleasing odor, have clusters of tiny white or yellow flowers.

Magellanic Clouds (clouds of Magellan, Nubeculae). Two galaxies, generally considered to be satellites of the Milky Way system, visible to the naked eye as conspicuous hazy patches of light in the southern sky. The Larger Cloud (*Nubecula Major*) has a diameter of about 30,000 light-years while the Smaller Cloud (*Nebucula Minor*) is 25,000 light-years in diameter. Both galaxies contain innumerable observable stars (many variable), clusters, and nebulae. The two galaxies are classified as irregular, though the Larger Cloud may be a spiral system like the Milky Way system. Both Clouds lie at a distance of some 120000 light-years from the sun. [Named after Ferdinand Magellan (1480-1521), Portuguese navigator.]

Magellanic Clouds. Two irregular Galaxies that are the nearest extragalactic objects (nearly 2,00,000 light-years distant). They are visible to the naked eye in the southern skies. The Large Magellanic cloud, about 7° in angular diameter, is located mostly in the constellation Dorado; the Small Magellanic Cloud, about 4° in diameter, is almost completely in the constellation Tucana.

Magic. Practice of manipulating the course of nature by controlling supernatural forces through ritual and spell. The spell, or incantation, unlocks the full power of the ritual. Black magic is intended to harm or destroy; white magic is to benefit the community (as in fertility rites) or an individual, especially one suffering the effects of black magic. Sympathetic magic treats an image (sticking pins in a dolly, representing one's enemy), while contiguous magic deals with things that have touched a person (clothing, hair, or even a footprint); the two can be used together.

Magic Numbers. The numbers 2, 8, 20, 28, 50, 82, and 126.

Atomic nuclei that contain these numbers of protons or numbers of neutrons are more stable than other nuclei: elements such as tin and calcium with magic atomic numbers have relatively large numbers of isotopes. The numbers correspond to filled shells in the nucleus.

Magnesia. Common name for the chemical compound magnesium oxide (MgO). The fine powder is used in soaps, cosmetics, pharmaceuticals, and as a filler for rubber goods. Because of its refractory properties (it melts at c.2800°C), it is used in crucibles and ceramics. Crude magnesia is prepared by roasting Dolomite or Magnesite. Magnesia is also extracted from seawater.

Magnesite. White, yellow, or gray magnesium carbonate mineral ($MgCO_3$). It is formed through the alteration of olivine or Serpentine by waters carrying carbon dioxide; through the replacement of calcium by magnesium in Dolomite or Limestone; and through precipitation from magnesium-rich water that has reacted with sodium chloride. It is used for floorings, as a stucco, and to make firebrick, Epsom salts, face powder, boiler wrappings, and disinfectants.

Magnesite. The naturally occurring mineral magnesium carbonate. $MgCO_3$, used in producing magnesium oxide.

Magnesium. Symbol: Mg. A silvery malleable metallic element found in several minerals, including carnallite, kainite, magnesite, and dolomite. It is produced by electrolysis of fused magnesium chloride. The metal burns with an intense white flame and is used in pyrotechnics and flash tubes. Other uses include the manufacture of lightweight alloys, the production of some magnesium compounds, such as Grignard reagents, and the reduction of certain metal oxides to yield the metal. It is an alkaline-earth metal. A.N. 12; A.W. 24.321; m.pt. 651°C; b.pt. 1107°C; r.d. 1.74; valency 2.

Magnesium (mg). Metallic element, discovered as an oxide by

Sir Humphry Davy in 1808. A ductile, silver-white, chemically active Alkaline-Earth Metal, it is the eighth most abundant element in the earth's crust. Its commercial uses include lightweight alloys in aircraft fuselages, jet-engine parts, rockets and missiles, cameras, and optical instruments. The metal is used in pyrotechnics. Magnesium is found in plant chlorophyll and is necessary in the diet of animals and humans.

Magnesium Carbonate. A light white powder, $MgCO_3$, used in heat insulation and as a pigment dentifrice. Decomposes at 350°C; r.d. 3.0.

Magnesium Chloride. A white deliquescent crystalline solid, $MgCl_2.6H_2O$, extracted from carnallite, sea water, and brines. It is used in the manufacture of metallic magnesium and for fireproofing wood. M.pt. 708°C (anhydrous); b.pt. 1412°C; r.d. 2.3.

Magnesium Hydroxide. A white powder, $Mg(OH)_2$, manufactured by precipitation from sea water with lime. It is alkaline in water and is used in the refining of sugar and as an antacid. Decomposes at 350°C; r.d. 2.4.

Magnesium Oxide (magnesia). A white powder, MgO, prepared by heating magnesium carbonate. It is a basic oxide, used in high-temperature refractory materials, electrical insulation, pharmaceuticals, and cosmetics. M.pt. 2800°C; b.pt. 3600°C; r.d. 3.6.

Magnesium Sulphate (Epsom Salt). A colourless crystalline solid, $MgSO_4.\ 7H_2O$, with a saline bitter taste. It occurs naturally or may be prepared by the action of sulphuric acid on the hydroxide or carbonate. Magnesium sulphate is used in dressing textiles, fireproofing, and as a purgative. Decomposes at 1124°C; r.d. 1.7.

Magnet. A device for producing a magnetic field. A permanent magnet is simply a piece of magnetized ferromagnetic material.

In an electromagnet the field is produced by a flow of electric current.

Magnetic Bottle. An arrangement of magnetic fields used to contain the plasma in a controlled thermonuclear reaction.

Magnetic Circuit. A closed circuit of lines of magnetic flux.

Magnetic Constant. The constant μ_0 appearing in the equation derived from Ampere's law for the magnetic flux density produced by an electric current. It arises from the choice of units: in SI units has the value $4\pi \times 10^{-7}$ henry per metre. It is sometimes called the *permeability of free space.*

Magnetic Declination. *See* Declination.

Magnetic Dipole. A small magnet, either a permanent magnet or a loop carrying an electric current, that can produce a magnetic field and is influenced by the direction of an external magnetic field. In a bar magnet, for example, the field produced comes from two regions near each end—the *poles.* By analogy with an electric dipole, this can be regarded as a magnetic dipole consisting of two opposite monopoles separated by a distance. The system will have a *magnetic moment* given by the product of pole strength and distance. In fact, the analogy between magnetic monopoles and isolated electric charges is a weak one because magnetic poles are always found in pairs. It is now more usual to regard the magnetic dipole as the fundamental entity. Its magnetic moment—or *magnetic dipole moment* (symbol: m)—is the torque experienced by a magnetic dipole in a field of unit magnetic flux. Its units are weber-metres (Wb m). A small loop of wire carrying a current *(I)* has a magnetic moment given by *IA*, where *A* is the coil's area. This is sometimes called the *electromagnetic moment* and given in units of ampere-metres squared (A m^2), which are equivalent to weber-metres.

Magnetic Dipole Moment. *See* Magnetic Dipole.

Magnetic Equator (aclinic line). A line around the earth close to and crossing the true equator, on which all points have zero dip.

Magnetic Field. A field characterized by the fact that a small loop of wire carrying a current will experience a force in the field. A magnetic field can be produced by a current flowing in a conductor or by a permanent magnet. A field is characterized by its magnetic flux density, B, or by a magnetic field strength, H.

Magnetic Field Strength. Symbol: H. A measure of magnetic field, equivalent to the part of the total field that is not caused by magnetization of the material. It is defined as $H = B/\mu_o - M$, where B is the magnetic flux density, μ_o the magnetic constant, and M the magnetization. Magnetic field strength was formerly called *magnetic intensity*. It is measured in amperes per metre.

Magnetic Flux. Symbol: Ø The total magnetic field produced by a current in a wire or coil or by a permanent magnet. It is simply the product of the magnetic flux density and the area. The unit is the weber.

Magnetic Flux Density. Symbol: B. A measure of the magnetic field at a point by the effect it has on a current-carrying conductor. A small loop of wire carrying a current has a magnetic moment m directed along its axis. The value of B is given by $T = m \times B$, where T is the torque produced on the coil when its axis is perpendicular to the field direction. The magnetic flux density is sometimes thought of as the number of lines of force per unit area. It was formerly called *magnetic induction*. The unit is the tesla.

Magnetic Induction. 1. The magnetization of a ferromagnetic material in a magnetic field. An unmagnetized bar of iron, for example, placed close to a bar magnet will itself act like a bar magnet with a north pole close to the south pole of the original magnet (or vice versa) and be attracted to the original magnet. The effect is caused by alignment of magnetic dipoles within the material by the external field.

2. Magnetic flux density.

Magnetic Intensity. Magnetic field strength.

Magnetic Meridian. One of a set of lines on the earth's surface joining points that have the same horizontal component of field.

Magnetic Mirror. A strong magnetic field with a particular configuration suitable for reversing the direction of charged particles. Such fields are used in magnetic bottles.

Magnetic Moment. Magnetic dipole.

Magnetic Monopole. A hypothetical particle that would be an isolated north or south magnetic pole. The particle has been postulated by analogy with charged particles. Despite many attempts to detect magnetic monopoles no evidence for their existence has yet been found.

Magnetic Permeability. Relative permeability.

Magnetic Pole. One of the regions near the ends of a permanent magnet into which the lines of force of the field converge or from which they diverge. If a magnet is freely suspended it will come to rest along the earth's lines of force with its *north pole* (or north-seeking pole) pointing north and its *south pole* (or south-seeking pole) pointing south. The earth's *magnetic North Pole* and *magnetic South Pole* are the points at which the earth's magnetic flux is strongest. Unlike magnetic poles attract one another and like poles repel in the same way that opposite electric charges attract and similar charges repel. By analogy, it is possible to make use of the idea of *magnetic pole strength* (symbol: P). In electromagnetic units, electrical measurements are based on the concept of a unit magnetic pole. In SI units the pole strength is defined by means of a magnetic dipole. If a magnetic pole of strength P is situated a distance l away from an unlike pole of equal magnitude, the magnetic dipole moment is given by $m = P\,l$. A law of forces between poles can be obtained: $F = \mu_o P_1 P_2 / 4\pi r^2$, where r is the distance between poles of strength P_1 and P_2 and μ_o is the magnetic constant.

Magnetic Pole. Either of two points on the earth, one in the Northern Hemisphere and one in the Southern; each point attracts one end of a compass needle and repels the opposite end. Studies of magnetism in rocks indicate that in the geological past the earth's magnetic field has reversed its polarity often and that rock movement (due to Continental Drift and Plate Tectonics) relative to the magnetic poles has occurred.

Magnetic Quantum Number. Atom.

Magnetic Resonance. In physics and chemistry, phenomenon produced by simultaneously applying a steady magnetic field and Electromagnetic Radiation (usually radio waves) to a sample of atoms and then adjusting the frequency of the radiation and the strength of the magnetic field to produce absorption of the radiation. The resonance refers to the enhancement of the absorption that occurs when the correct combination of field and frequency is reached. Most magnetic resonance phenomena depend on the fact that both the proton and the electron behave like microscopic magnets—a property that can be ascribed to an intrinsic rotation, or spin. Types of magnetic resonance include electron paramagnetic resonance (EPR), involving the magnetic effect of electrons, and nuclear magnetic resonance (NMR), involving the magnetic effects of protons and neutrons in the nuclei of atoms. The NMR resonant frequency provides information about the molecular material in which the nuclei reside. For the use of NMR in medicine.

Magnetic Storm. Disturbances of the earth's magnetic field caused by electrical disturbances resulting from sunspot activity.

Magnetic Susceptibility. Susceptibility.

Magnetic Tape. Plastic tape coated with a layer of iron oxide, which can be magnetized to record information for computers or to record sound in tape recorders.

Magnetic Variation. Declination.

Magnetism. The phenomenon produced by magnetic fields, involving interaction at a distance caused by the motion or spin of charged particles. Several different types of magnetic behaviour are distinguished, depending on the susceptibility of the material and its temperature dependence.

Magnetism. Force of attraction or repulsion between various substances, especially those containing iron and certain other metals, such as nickel and cobalt; ultimately it is due to the motion of electric charges. Any object that exhibits magnetic properties is called a magnet. An ordinary magnet has two poles where the magnetic forces are the strongest; these poles are designated as a north (north-seeking) pole and a south (south-seeking) pole, because a magnet freely rotating in the earth's magnetic field tends to orient itself along a north-south line. The like poles of different magnets repel each other, and the unlike poles attract each other. Whenever a magnet is broken, a north pole appears at one of the broken faces and a south pole at the other, such that each piece has its own north and south poles. In the 18th cent. Charles Coulomb found that the magnetic forces of attraction and repulsion are directly proportional to the product of the strengths of the poles and inversely proportional to the square of the distances between them. As with electric charges, the effect of this magnetic force acting at a distance is expressed in terms of a field of force. A picture of the magnetic field lines can be obtained by placing a piece of paper over a magnet and sprinkling iron filing on it. The individual pieces of iron become magnetized by entering a magnetic field, *i.e.,* they act like tiny magnets, lining themselves up along the magnetic field lines. The connection between magnetism and Electricity was discovered in the early 19th century. Hans Oersted found (1820) that a wire carrying an electrical current deflects the needle of a magnetic compass because a magnetic field is created by the moving electric charges constituting the current. Andre Ampere showed (1825) that magnets exert forces on current-carrying conductors. In 1831 Michael Faraday and Joseph Henry independently discovered electromagnetic Induction—the production of a current in a conductor by a change in the magnetic field around it. The magnetic propoerties of matter are also explained by the motion of charges. Because the electron has both an electric charge and a spin, it can be considered a charge in motion, giving rise to a tiny magnetic field. In many atoms, all the electrons are paired within energy levels, so that the electrons in each pair have opposite (antiparallel) spins, and their magnetic fields cancel. In some atoms there are more electrons with spins in one direction than the other, resulting in a net magnetic field for the atom as a whole. Placed in an external field, the individual atoms will tend to align their fields with the external one. Because of thermal vibrations the alignment is not complete, and materials, called paramagnetic substances, that contain such atoms react only weakly to a magnetic field. Materials such as iron, nickel, or cobalt that respond strongly to a magnetic field are called ferromagnetic. In a ferromagnetic substance there are also more electrons with spins in one direction than in the other. The individual magnetic fields of the atoms in a given region, called a domain, tend to line up in one direction, so that they reinforce each other. Materials such as bismuth and antimony that are repelled by a magnetic field are called diamagnetic. In a diamagnetic substance, an external magnetic field accelerates the electrons moving in one direction and retards those moving in the opposite direction; this situation produces an induced magnetization opposite in direction to the external field.

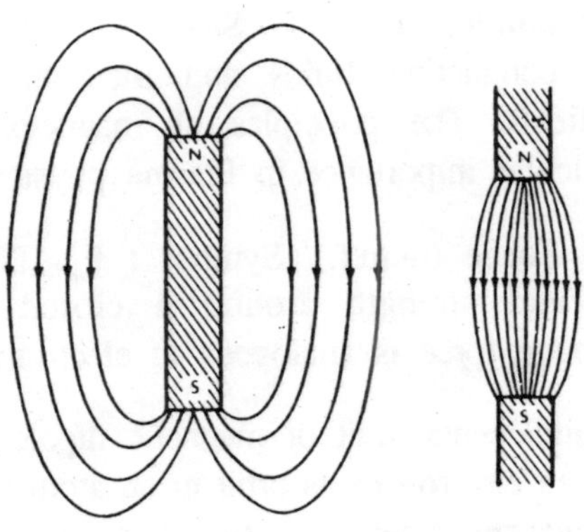

Fig. M-1. Magnets: Lines of induction around a single bar magnet (left) and between opposite poles of different magnets (right).

Magnetite. Lustrous, black, magnetic iron mineral (Fe_3O_4), occurring as crystals, masses, and sand in Sweden, South Africa, Italy, and parts of the U.S. It is an important ore of Iron. Lodestone is a naturally magnetic variety that exhibits polarity.

Magnetite (lodestone). A black magnetic mineral consisting of ferroferric oxide, Fe_3O_4. It is used as an ore of iron.

Magnetization. Symbol: *M*. The effect of an external magnetic field on a sample of material, in which the sample acts as a magnet itself and changes the value of the external field. If the field in free space has a magnetic flux density $\mathbf{B}_0$ and a sample is introduced, then the flux density is changed to B, where $B = \mathbf{B}_0 + \mathbf{B}_M$. $\mathbf{B}_M$ is an extra contribution arising in the material and resulting from the alignment of small magnetic dipoles. The magnetization can be defined as the magnetic dipole moment per unit volume. It is related to the magnetic field strength (H) by the equation $M = B/\mu_o - H$, where μ_o is the magnetic constant.

Magnetohydrodynamics (MHD). The study of the motion of electrically conducting fluids and their behaviour in magnetic fields.

Magnetohydrodynamics (MHD). Study of the motions of electrically conducting fluids and their interactions with magnetic fields. The principles of magnetohydrodynamics are of particular importance in Plasma physics.

Magnetomotive Force (m.m.f.). Symbol : F_m. The integral of magnetic field strength around a closed path *ØHdx*. Magnetomotive force is analogous to electromotive force.

Magneton. A fundamental unit of magnetic dipole moment. The moment of an electron in its orbit in an atom is quantized in values of $eh/4\pi m$, where e is the electron charge, m is its mass, and h is the Planck constant. This is called the *Bohr magneton* (symbol: m_B): its value is $9.274\ 096 \times 10^{-24}$ A m^2. The Bohr magneton is also used for electron spin. The *nuclear magneton,* m_N is given by $m_e m_B/m_p$, where m_e is the electron mass and m_p is the proton mass. Its value is 5.0503×10^{-27} A m^2.

Magnetron. An electron tube for producing microwaves. It contains a hot cathode surrounded by a coaxial cylindrical anode. The device is placed in a magnetic field which causes the electrons emitted by the cathode to spiral around and induce radiofrequency fields in cavities in the anode.

Magnification. A measure of the performance of a microscope, telescope, or other optical instrument equal to the ratio of the size of the final image to the size of the object as seen with the unaided eye.

Magnitude. The brightness of a celestial object as measured on a scale of numerical values. In ancient times orders of brightness were established for naked-eye objects and Ptolemy fixed the brightest stars as being of the first magnitude and the faintest as the sixth magnitude. In modern times the scale has been fixed so that a difference of five magnitudes gives a brightness ratio of 100:1. Thus, a star of magnitude *m* is the fifth root of one hundred (2.512) times brighter than one of magnitude *m* + 1. The largest telescopes have been used to photograph objects as faint as magnitude 22.

The *apparent magnitude* is the observed brightness established by eye (*visual magnitude*) or by photography (*photographic magnitude*). The *absolute magnitude* is the apparent magnitude that a star would have if it were placed at a standard distance of 10 parsecs (32.6 light-years) away from the earth. The absolute magnitude measures the intrinsic brightness or luminosity of a star.

Magnitude. Measure of the brightness of a celestial object. Apparent magnitude is that determined on the basis of an object's relative brightness as seen from the earth. Objects differing by one magnitude differ in brightness by a factor of 2.512 (the 5th root of 100). The brightest stars have a magnitude of about + 1; the sun's magnitude is -26.8. Absolute magnitude, a measure of the intrinsic luminosity, or true brightness, of an object, is the apparent magnitude an object would have if located at a standard distance of 10 Parsecs.

Magpie. Name for certain Birds of the Crow and Jay family. The black-billed magpie (*Pica pica*), of W North America, has iridescent black plumage and white wing patches and abdomen. Other species are found in Europe, Asia, and Africa. Magpies are scavengers and often collect small, bright objects. In captivity they can learn to imitate some words.

Mahogany. Common name for the Meliaceae, a family of chiefly tropical shrubs and trees, from which the valuable hardwood called mahogany is obtained. Principal sources of the hardwood, often used for furniture, are trees of the American genus *Swietenia* and the W. African genus *Khaya.* Varying in color from golden to deep red-brown, the woods are usually scented, close-grained, and resistant to termites.

Main-sequence Star. A star that lies along the narrow diagonal distribution band (the main sequence) on the Hertzsprung-Russell diagram. The vast majority of observed stars, including the sun, are main-sequence stars.

Majority Carrier. The type of carrier responsible for transporting more than one half of the current in a semiconductor.

Malachite. A green basic carbonate of copper, $CuCO_3.Cu(OH)_2$, occurring naturally and used as an ore and a pigment.

Malachite. Green copper carbonate mineral $[Cu_2CO_3(OH)_2]$, found in crystal and, more commonly; massive form. It is used as a Gem, a Copper ore, and, when ground, a pigment. It occurs associated with other copper ores in the U.S., Chile, the USSR, Zimbabwe, Zaire, and Australia.

Malaria. Infectious parasitic disease characterized by high fever, severe chills, enlargement of the spleen, and sometimes Anemia and Jaundice. It can be acute or chronic and is frequently recurrent. Widespread throughout tropical and subtropical areas of the world, malaria is transmitted by the *Anopheles* mosquito, which picks up the causative *Plasmodium* parasite from the blood of an infected person and transfers it to that of a healthy person. Quinine, the traditional treatment, has largely been replaced by modern anti-malarials, including chloroquine and primaquine.

Malate. Any salt or ester of malic acid.

Maleate. Any salt or ester of maleic acid.

Maleic Acid (cis-butenedioic acid). A colourless crystalline toxic carboxylic acid, HCOOCH:CHCOOH, with an astringent taste, used in dyeing and in the manufacture of synthetic resins. Maleic and fumaric acids are geometrical isomers. M.pt. 130°C; r.d. 1.59.

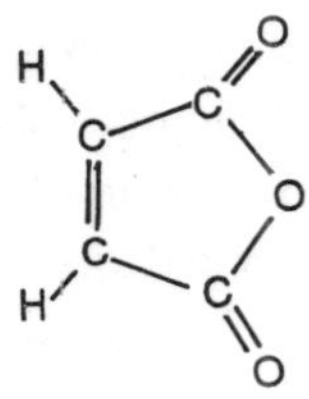

Fig. M-2. Maleic anhydride.

Maleic Anhydride. A colourless crystalline solid, $C_4H_2O_3$, made by the catalytic oxidation of benzene. It is used in the manufacture of polyester resins and dyestuffs. M.pt. 53°C; b.pt. 202°C; r.d. 1.5.

Malic Acid. A colourless crystalline optically active solid, $HOOCCH_2CH(OH)COOH$, occurring naturally in apples and many other fruits. It is used as a food additive and chelating agent. M.pt. 99°C (*l*-); b.pt. 140°C (decomposes); r.d. 1.6.

Mallow. Common name for the Malvaceae, a family of widely distributed shrubs and herbs, most abundant in the American tropics and typified by mucilaginous sap and showy flowers with a prominent column of fused stamens. The family includes the true mallows (genus *Malva*) of the Old World, and the false mallows (genus *Malvastrum*), and rose, or swamp, mallows (genus *Hibiscus*) of North America. Introduced *Hibiscus* species include the rose of Sharon (*H. syriacus*), a popular ornamental, and okra, or gumbo (*H. esculentus*), whose mucilaginous pods are used as a vegetable. The most popular ornamental of the family, the hollyhock (*Althea rosea*), originally a Chinese perennial, is widely cultivated in many varieties and colors. The European marsh mallow (*A. officinalis*) is used medicinally and was formerly used in the confection marshmallow, now usually made from syrup, gelatin, and other ingredients. Economically the most important plant of the family is Cotton.

Malnutrition. Insufficiency of one or more nutrients necessary for health. Primary malnutrition is caused by a lack of essential foodstuffs (usually Vitamins, Minerals, or Proteins) in the diet. Such a lack can be caused by regional conditions such as drought and overpopulation or by poor eating habits. Secondary malnutrition is caused by failure to absorb or utilize nutrients, as in disease of the gastrointestinal tract, Kidney, or Liver; by failure to satisfy increased nutritional requirements, as during Pregnancy; or by excessive excretion, as in diarrhea. Malnutrition can cause such conditions as Anemia and Kwashiorkor.

Malonic Acid (propanedioic acid). A colourless crystalline dicarboxylic acid, $CH_2(COOH)_2$, found in sugar beet. M.pt. 135.6°C; decomposes at 140°C; r.d. 1.62.

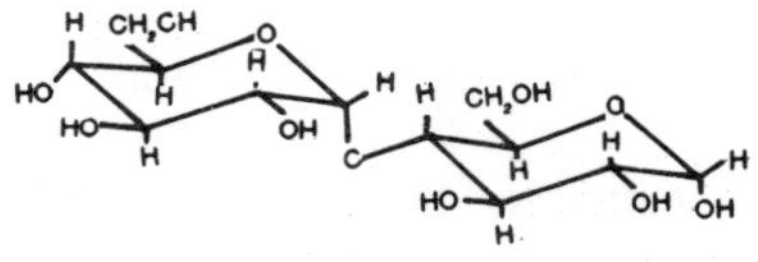

Fig. M-3. Maltose.

Malt. A grain (usually Barley) steeped in water, partially germinated, then dried and cured. It is used in brewing to convert cereal starches to sugars by means of the Enzymes (chiefly diastase) produced during germination.

Maltese. Very small Toy Dog; shoulder height, c.5 in. (12.7 cm); weight, 2-7 lb (.9-1.4 kg). Its silky white coat hangs down almost to the ground. Probably an ancient breed, closely resembling lap dogs of ancient Greece and Rome, it was very popular in Europe by the 19th cent.

Maltose or **Malt Sugar.** (Empirical formula: $C_{12}H_{22}O_{11}$), crystalline Sugar involved in brewing beer. Maltose can be produced from Starch by Hydrolysis in the presence of diastase, an enzyme found in Malt. A disaccharide, maltose is hydrolized to Glucose by maltase, an enzyme present in Yeast. The glucose thus formed can be fermented by another enzyme in yeast to produce Ethanol.

Maltose (malt sugar). A colourless crystalline sugar, $C_{12}H_{22}O_{11}$, made by the action of enzymes on starch. It is used as a nutrient and sweetner. M.pt. 1.2°C.

Mammal. Warm-blooded animal of the class Mammalia, the highest class of Vertebrates, found in terrestrial and aquatic habitats. The female has mammary glands, which secrete milk for the nourishment of the young after birth. In most mammals the body is partially or wholly covered with hair, the heart has four chambers, and a muscular diaphragm separates the chest from the abdominal cavity. Except for the egg-laying Monotremes *(e.g.,* the Platypus), mammals give birth to live young. In some Marsupials and higher mammals the young receive prenatal nourishment through a placenta. Terrestrial mammals include carnivores *(e.g.,* Cat, Dog, Bear); rodents *(e.g.* Beaver, Squirrel); hoofed animals *(e.g.,* Horse, Rhinoceros, Deer Swine, Cattle); primates *(e.g.,* human being, Monkey, Lemur); and others, such as the Bat and Elephant. Aquatic mammals include the carnivorous Seal and Walrus and the omnivorous Whale and Dolphin.

Mammary Gland or Breast. Organ of the female mammal that produces milk for nourishment of the young. Mammary glands develop during puberty and distend during pregnancy in preparation for nursing. They are sometimes considered part of the Reproductive System.

Mammoth. Name for several prehistoric Elephants of the extinct genus *Mammuthus,* found in Eurasia and North America in the Pleistocene epoch. The imperial mammoth of North America was about $13^1/_2$ ft (4.1 m) high at the shoulder. As depicted in Cro-Magnon cave paintings in S. France, the mammoth had a shaggy coat, complex molar teeth, slender tusks, and a long trunk.

Mandrake. Herbaceous perennial plant (genus *Mandragora)* of the Nightshade family, native to the Mediterranean and Himalayan regions. True mandrakes contain several Alkaloids of medicinal value, and they have been used as pain-killers. Magical powers have often been attributed to the root, which crudely resembles the human form.

Mandrill. Large Monkey *(Mandrillus sphinx)* found in the forests of central W. Africa, related to the Baboon. The fur of the mandril is mostly dark brown, but the bare face and buttocks are patterned in coloures particularly spectacular in the adult male -- bright red, blue, black, purple, and pale yellow.

Manganate. A salt containing the ion MnO_4^{2}. Manganates are derived from the hypothetical *manganic acid,* H_2MnO_4. The manganate ion is dark green. It is found in the crystalline sodium and potassium salts and exists in strongly basic solutions. In acid or neutral solution it changes to the permanganate ion.

Maganese. Symbol: Mn. A grey-white hard brittle transition element found in many ores, especially pyrolusite (MnO_2) and rhodochrosite ($MnCO_3$). It is extracted by reduction of the oxide with magnesium or aluminium. Manganese is a ferromagnetic metal and is used in producing ferromagnetic alloys as well as in many important steels. The element is fairly electropositive, reacting with many nonmetals at higher temperatures and dissolving in nonoxidizing acids. It forms manganese II *(manganous)* and the less stable manganese III *(manganic)* compounds, including ionic salts and many complexes. It also exists in higher oxidation states, as in the manganates and permanganates. A.N. 25; A.W. 54.938; m.pt. 1247°C; b.pt. 2097°C; r.d. 7.21; valency 1-7.

Manganese (Mn). Metallic element, first isolated by J.G. Gahn in 1774. It is pinkish-gray and chemically active, and resembles iron. It is an unique deoxidizing and desulfurizing agent in the manufacture of steel and is widely used in making alloys.

Manganese dioxide. *See* manganese oxide.

Manganese Oxide. Any of five oxides of manganese. *Manganese II oxide* (manganous oxide) is a green powder, MnO, formed by reducing higher oxides with hydrogen. It is a basic substance. M.pt. 1650°C; r.d. 5.4. *Manganese IV oxide* (manganese dioxide) is a black powder occurring naturally as pyrolusite. It is used in making manganese steel and as an oxidizing agent and a depolarizer in the Leclanche cell. R.d. 5.03. Above 535°C it decomposes to *manganese III oxide* (manganic oxide), Mn_2O_3. The other oxides of manganese are Mn_3O_4 (trimanganese tetroxide) and Mn_2O_7 (manganese heptoxide), which is the anhydride of permanganic acid.

Manganic. *See* manganese.

Manganic Acid. *See* manganate.

Manganin. An alloy of copper (70-85%), manganese (15-25%), and nickel. Its high resistivity and low dependence of resistance on temperature make it suitable for use in resistors.

Manganous. *See* Manganese.

Mange. Contagious skin disease of animals, caused by parasitic Mites that burrow into the skin, Hair follicles, or sweat glands. This leads to itching, inflammation, hair loss, and secondary bacterial infection. The disease is also called scabies, scab, and barn itch.

Mango. Evergreen tree of the Sumac family, native to tropical E Asia but now grown in both hemispheres. The trees grow rapidly and can attain heights of up to 90 ft (27 m); they are densely covered with glossy leaves and bear small, fragrant yellow or red flowers. The aromatic, slightly acid fruit, a fleshy drupe with a thick, greenish to yellow-red skin, is an important food in the tropics.

Mangrove. Large, tropical evergreen tree (genus *Rhizophora)* found on muddy tidal flats and along shorelines, most abundant in tropical Asia, Africa, and the SW Pacific. Aerial roots produced from the trunk, become embedded in the mud and form a tangled network that serves as a prop for the tree and a means of aerating the roots. The fruit is a conical, reddish-brown berry with a single seed; it germinates inside the fruit while it is still on the tree, forming a primary root that quickly anchors the seedling in the mud.

Manometer. Any device used for measuring pressure. A simple type consists of a U-shaped glass tube containing mercury. One arm is open to the atmosphere and the other is connected to the system to be measured: the pressure difference is obtained from the difference in heights of the mercury columns in each arm of the tube.

Man, Prehistoric, or **Early Man.** About 14-12 million years ago, man and apes began to develop along separate lines. *Ramapithecus* was probably the earliest genus of the modern homind family, followed in Africa by the short, bipedal, small-brained *Australopithecus* (five to one million years ago). Some archaeologist classify *Homo habilis* as the first true human being, while others assign those remains to *Australopithecus* and claim that *Homo erectus* introduced the human species c. 1.5 million years ago. The early Cultures of the genus *Homo* were generally distinguished by regular use of stone tools and other artifacts, and by a hunting-and-gathering economy in volving division of labour; *Homo erectus* used fire and the hand axe. *Homo sapiens* evolved about 300,000 years ago, with a larger brain and a smaller jaw. There is disagreement about whether such archaic forms as Neanderthal Man (c. 75,000 years ago) evolved directly into biologically modern man *(Homo sapiens sapiens)* or were displaced as the more modern species expanded its range. The first human types indistinguishable from modern man, including Cro-magnon Man, appeared around 40,000 years ago.

Mantissa. *See* logarithm.

Maple. Common name for the genus *Acer* of the Aceraceae, a family of deciduous trees and shrubs of the Northern Hemisphere, characterized by winged seeds. Maples are popular as shade trees and noted for their brilliant fall colours. Several species provide close-grained, hardwood timber, *e.g.*, the sugar maple *(A. saccharum)* and the black maple *(A. nigrum)*; these trees are also the main source of maple syrup. In the spring, their sap is drawn off, strained, and concentrated by boiling to produce syrup and sugar. Other well-known species include the box elder *(A. negundo)*, a shade tree, and the swamp, or red, maple *(A. rubrum)*.

Map Projection. Transfer of the features of the earth's surface or those of another spherical body onto a flat sheet of paper. Only a globe can represent surface features correctiy with reference to area, shape, scale, and direction. Projection from a globe to a flat map always causes some distortion. A grid or net of two intersecting systems of lines corresponding to parallels and meridians must be drawn on a plane suface. Some projections (equidistant) aim to keep correct distances in all directions from the center of the map. Others show areas (equal-area) or shapes (conformal) equal to those on the globe of the same scale. Projections are cylindrical, conical, or azimuthal in geometric origin.

Marble. A Rock formed by the Metamorphism of Limestone. The term is loosely applied to any limestone or Dolomite that takes a good polish and is otherwise suitable as a building or ornamental stone. Its colour varies depending on the types of impurities present. It has been used since ancient times for statuary, monuments, and facing stones. Like all limestones, it is corroded by water and acid fumes and is therefore ultimately uneconomical for use in exposed places.

Marconi, Guglielmo, Marchese. 1874-1937, Italian physicist. For his development of wireless telegraphy, he shared with

Karl Braun the 1909 Nobel Prize in physics. Marconi sent (1895) long-wave signals over a distance of more than a mile and received (1901) the first tromsatlantic wireless signals.

Marigold. Plant (genus *Tagetes)* of the Composite family, mostly Central and South American herbs cultivated as garden flowers. Two common annuals are the large-folowered, strong-scented, yellow or orange African marigold *(T. erecta)* and the smaller, yellow or orange and red French marigold *(T. patula)*. Both are native to Mexico and Guatemala.

Marijuana or **Marihuana.** Relatively mild, nonaddictive drug with hallucinogenic properties, obtained from the flowering tops, stems, and leaves of the Hemp plant. Resins found on the surface of the female plant are used to prepare the most potent form of Marijuana, hashish. The primary active substance is tetrahydrocannabinol. Marijuana produces a dreamy, euphoric state of altered consciousness, with feelings of detachment and gaiety. The appetite is usually enhanced, while the sex drive may increase or decrease. Adverse reactions are relatively rare, and most can be attributed to adulterants frequently found in marijuana preparations. Marijuana has been used experimentally to reduce nausea from cancer Chemotherapy and in the treatment of Glaucoma. In the U.S. there have been efforts to reduce criminal penalties for possession and use of marijuana.

Marine Biology. Study of ocean plants and animals and their ecological relationships. Marine organisms are classified according to their mode of life as nektonic (free-swimming), planktonic (floating), or benthic (bottom-dwelling). Their distribution depends on the chemical and physical properties of seawater *(e.g.*, temperature, salinity, and dissolved nutrients), ocean current and penetration of light.

Marjoram or **Sweet Marjoram.** Old World perennial aromatic herb *(Marjorana hortensis)* of the Mint family, cultivated for flavoring and for origanum oil, used in perfumed soaps.

The closely related European wild marjoram *(Origanum vulgare)* is the spice usually solid as Oregano.

Marlin. Open-sea Fish related to the Sailfish and Swordfish (family Istiophoridae), prized by sportsmen. The blue marlin (genus *Makaira)* of the Gulf Stream may reach 1,000 lb (454 kg) in weight. The marlin's upper jaw extends into a long spike, which it uses to club the fish on which it feeds.

Mars. In astronomy, 4th Planet from the sun, at a mean distance of 141.6 million mi (227.9 million km). It has a diameter of 4,223 mi (6,796 km) and a thin atmosphere composed largely of carbon dioxide. The surface of Mars is highly diverse: the younger, lower terrain of the northern hemisphere is sparsely cratered, whereas the older, higher terrain of the southern hemisphere is often densely cratered and contains numerous channels tens of miles wide and hundreds of miles long. It also has numerous volcanoes—including Olympus Mons (c.370 mi/600 km in diameter and 16 mi/26 km tall), the largest in the solar system—and lava plains. Dust storms, often local but sometimes global in extent, have been observed moving across the Martian surface. Mars has two known satellites, *Phobos* and *Deimos,* both discovered by the American astronomer Asaph Hall in 1877; both are very small, irregular ellipsoids. Space Probes that have encountered Mars include *Mariners 4, 6,* and 7 and 9 (1965, 1969, 1971); *Vikings* 1 and 2 (1976); and several Soviet Mars spacecraft. Experiments on the Viking landing craft detected no definite evidence of life.

Mars. The fourth planet in the solar system in outward succession from the sun. A characteristically red-coloured planet with polar ice caps that recede and advance with the Martian seasons, it lies at a mean distance of 227,800,000 kilometres from the sun, orbiting it at an average velocity of 24.1 km s^{-1} once every 687 earth days. Its period of axial rotation is 24 hours 39 minutes. Mars is 6786 km in diameter and has a mean relative density of 3.96. Its gravity is about one third and its mass about one tenth those of the earth. Mars' orangered surface has dark markings on it and is cratered like the surface of the moon. Its tenuous atmosphere is composed mainly of carbon dioxide and trace amounts of water vapour and carbon monoxide. The pressure at the surface is about 0.01 atmosphere. The polar caps are probably frozen carbon dioxide, though they might have cores of frozen water. Daytime temperatures on Mars reach —40°C at the equator, dropping to —70°C at night. At its closest, Mars is less than 56 million km from earth. The eccentricity of Mars orbit is 0.093 and its inclination is 1.85° to the plane of the ecliptic. It has two satellites, *Phobos* and *Deimos*.

Marsh gas. *See* methane.

Marsh's Test. An analytical test for arsenic. The sample is mixed with hydrochloric acid and zinc and any arsenic present is reduced to arsine. The gas evolved is passed through a hot tube where the arsine is decomposed to arsenic, which condenses as a brown deposit on the cool part of the tube. Antimony gives a similar result but can be distinguished by the fact that it is insoluble in sodium hypochlorite.

Marsupial. Member of the order Marsupialia, or pouched Mammals. All but the New World Opossums and an obscure South American family are found only in Australia, Tasmania, and New Guinea, and on a few nearby islands. Unlike that of higher mammals, the marsupial embryo is generally not connected to its mother by a placenta. The young are born in an undeveloped state and crawl to the mother's nipples which are in a pouch, or marsupium, formed by a fold of abdominal skin. The order includes the Kangaroo, Koala, Tasmanian Devil, and Wombat.

Martensite. A solid solution of carbon or iron carbide in alpha iron. It is the main constituent of many steels, formed from austenite by rapid quenching from high temperatures.

Marx, Karl. 1818-83, German social philosopher and revolutionary; with Friedrich Engels, a founder of modern Socialism and Communism. The son of a lawyer, he studied law and philosophy; he rejected the idealism of G.W.F. Hegel but was influenced by Ludwig Feuerbach and Moses Hess. His editorship (1842-43) of the *Rheinische Zeitung* ended when the paper was suppressed. In 1844 he met Engels in Pairs, beginning a lifelong collarboation. With Engels he wrote the *Communist Manifesto* (1848) and other works that broke with the tradition of appealing to natural rights to justify social reform, invoking instead the laws of history leading inevitably to the triumph of the working class. Exiled from Europe after the Revolutions of 1848, Marx lived in London, earning some money as a correspondent for the New York *Tribune* but dependent on Engels's financial help while working on his monumental work *Das Kapital* (3 vol., 1867-94), in which he used Dialectical Materialism to analyze economic and social history; Engels edited vol. 2 and 3 after Marx's death. With Engels, Marx helped found (1864) the International Workingmen's Association, but his disputes with the anarchist Mikhail Bakunin eventually led to its breakup. Marxism has greatly influenced the development of socialist thought; further, many scholars have considered Marx a great economic theoretician and the founder of economic history and sociology.

Mascon. One of a number of localized regions of high gravity found on the moon.

Maser.[Acronymn for microwave amplification by stimulated emission of radiation], device, first operated in 1954, for the creation and amplification of high-frequency radio waves. The waves produced by the maser are coherent, *i.e.,* all of the same frequency, direction, and phase relationship. Used as an oscillator, the maser provides a very sharp, constant signal and thus serves as a time standard for atomic clocks. The maser can also serve as a relatively noise-free amplifier. The optical maser is now called a Laser.

Maser. A type of amplifier for producing intense monochromatic coherent highly directional beams of microwaves by stimulated emission. Masers work on the same principle as the laser; the name is an acronym for *microwave amplification by stimulated emission of radiation.*

Mask. Face or head covering used for disguise or protection. Masks have been worn by primitive peoples since ancient times in Magic and religious ceremonies. Notable are those of W and central Africa, of the Iroquois Indians and tribes of the Pacific Northwest, and of the Aztecs. Masks have been integral to drama in East and West, particularly Oriental Drama and the Commedia Dell'arte. Protective masks are worn by warriors, athletes, and surgeons.

Mass. In physics, the quantity of matter in a body regardless of its volume or of any forces acting on it. There are two ways of referring to mass, depending on the laws of physics defining it. The *gravitational* mass of a body may be determined by comparing the body on a beam balance with a set of standard masses; in this way the gravitational factor is eliminated. The *inertial* mass of a body is a measure of the body's resistance to acceleration by some external force. All evidence seems to indicate that the gravitational and inertial masses are equal. According to the special theory of Relativity, mass increases with speed according to the formula $m = m_0\sqrt{1-v^2/c^2}$, where m_0 is the rest mass (mass at zero velocity) of the body, v its speed, and c the speed of light in vacuum. The theory also leads to the Einstein mass-energy relation $E = mc^2$. where E is the energy and m the relativistic mass.

Mass. Symbol : *m*. The property of a body that determines the acceleration that would be produced by application of a force. According to Newton's law the acceleration is proportional to the force applied: the mass of the body is the constant of proportionality. Thus the mass is a measure of the body's tendency to resist a change in mition : *i.e.* its inertia. The quantity is sometimes called the *inertial mass*.

Mass is also the property that determines the mutual attraction of two bodies by gravitational interaction *(see* gravitation). Thus, it is the property determining the weight of the object: the *gravitational mass,* which is equivalent to the inertial mass. The unit of mass is the kilogram.

The *law of conservation of mass* states that the total mass of a system is constant no matter what changes occur in the system. This principle has been modified by the theory of relativity in which matter is regarded as a form of energy.

Mass Action, Law of. The principle that the rate of a chemical reaction is proportional to the active masses (concentrations) of the reacting substances.

Mass Defect. The difference between the sum of the masses of the individual nucleons in an atomic nucleus and the mass of the nucleus. This difference in mass is equivalent to the energy holding the nucleons together.

Mass-Energy. Mass and energy considered as interconvertible, connected by Einstein's equation $E = mc^2$, where c is the velocity of light. Matter can be created or destroyed with corresponding decrease or increase of the energy. Thus the laws of conservation of mass and of energy are not strictly true. However, the total energy, calculated by considering matter as a form of energy, is conserved. This is known as the *law of conservation of mass-energy.*

Massicot. *See* lead oxide.

Mass Number (nucleon number). Symbol: *A*. The number of nucleons in the nucleus of an atom.

Mass Number. Represented by the symbol *A,* the total number of nucleons (Neutrons and Protons) in the nucleus of an Atom. All atoms of a chemical Element have the same Atomic Number but may have different mass numbers (from having different numbers of neutrons in the nucleus). Atoms of an element with the same mass number make up an Isotope of the element. Isotopes of different elements may have the same mass number but different numbers of protons.

Mass Spectrometer. An instrument for producing and identifying ions by deflecting them with magnetic or electric fields. Several types exist. In the simplest a gaseous sample at low pressure is ionized by a beam of electrons and the ions produced are accelerated into the evacuated analyser by an electric field. Here the moving ions are deflected into circular paths by a magnetic field at right angles to their direction of motion. The magnetic or electric field can be continuously varied and successive ions are focused onto a detector: the deflection of an ion depends on its charge-to-mass ratio. A *mass spectrum* is produced consisting of a series of peaks, each corresponding to a different fragment ion. The mass spectrum of a compound can be used to find its formula and chemical structure. The technique is also an accurate method of measuring atomic weights.

Mastectomy. Surgical removal of all or part of the breast. For small tumors, a partial mastectomy may be performed, removing only the involved breast tissue. When Cancer is present, a modified radical mastectomy may be performed, excising the breast and lymph nodes of the adjacent armpit.

Mastodon. Name for several prehistoric Mammals of the extinct genus *Mammut,* from which Elephants are believed to have evolved. Long-jawed mastodons about $4^1/_2$ ft (137 cm) high, with four tusks, live in Africa during the Oligocene epoch. Later forms were larger, with long flexible trunks and two tusks. Forest dwellers, mastodons fed by browsing.

Match. Small stick whose chemically coated tip bursts into flame when struck on a rough surface. The first friction match was devised in 1827 by the English apothecary John Walker, and the first phosphorus match was invented in 1831 by the French student Charles Sauria. The safety match (invented in

Sweden in 1855) allowed ignition only when struck on a special surface. Modern, mass-produced matches use nontoxic phosphorus sesquisulfide as the igniting agent.

Materialism. In philosophy, a widely held system of thought that explains the nature of the world as entirely dependent on matter, the final reality. Early Greek teaching, *e.g.*, that of Democritus, Epicurus, and the proponents of Stoicism, conceived of reality as material in nature. The theory was renewed and developed beginning in the 17th cent., especially by Hobbes, and in the 18th cent. Locke's investigations were adapted to the materialist position. The system was developed further from the middle of the 19th cent., particularly in form of Dialectical Materialism and in the formulations of Logical Positivism.

Mathematics. Deductive study of numbers, geometry, and various abstract constructs.

Branches. Mathematics is very broadly divided into foundations, algebra, analysis, geometry, and applied mathematics. The term *foundations* is used to refer to the formulation and analysis of the language, Axioms, and logical methods on which all of mathematics rest; Set theory, originated by Georg Cantor, now constitutes a universal mathematical language. Algebra, historically, is the study of solutions of one or several algebraic equations, involving Polynomial functions of one or several variables; Arithmetic and Number Theory are areas of algebra concerned with special properties of the integers. Analysis applies the concepts and methods of the Calculus to various mathematical entitles. Geometry is concerned with the spatial side of mathematics, *i.e.,* the properties of and relationships between points, lines, planes, figures, solids, and surfaces. Topology studies the structures of geometric objects in a very general way. The term *applied mathematics* loosely designates a wide range of studies with significant current use in the empirical sciences. It includes

Computer science, mathematical physics, Probability theory, and mathematical Statistics.

History. The earliest records indicate that mathematics arose in response to the practical needs of agriculture, business, and industry in the 3d and 2d millennia B.C. in Egypt and Mesopotamia and, possibly, India and China. Between the 6th and 3d cent. B.C., the Greeks Thales, Pythagoras, Plato, Aristotle, Euclid, Archimedes, and Zeno of Elea profoundly changed the nature of mathematics, introducing abstract notions such as Infinity and irrational Numbers and a deductive system of proof. Their work was carried on in the 2d and 3d cent. A.D. by Hero of Alexandria, Ptolemy, and Diophantus. With the decline of learning in the West, the development of mathematics was continued in the East by the Chinese, the Indians (who invented the Numerals now used throughout the civilized world), and the Arabs. Their writings began to reach the West in the 12th cent., and by the end of the 16th cent. Europeans had made advances in algebra, Trigonometry, and such areas of applied mathematics as mapmaking. In the 17th cent. decimal fractions and Logarithms were invented, and the studies of projective geometry and probability were begun. Blaise Pascal, Pierre de Fermat, Galileo, and Johannes Kepler made fundamental contributions. The greatest advances of the century, however, were the invention of Analytic Geometry by Rene Descartes and of the calculus by Sir Isaac Newton and, independently, G.W. Leibniz. The history of mathematics in the 18th cent. Is dominated by the development of the methods of the calculus and their application to physical problems, both terrestrial and celestial, with leading roles being played by the Bernoulli family, Leonhard Euler, Joseph Lagrange, and Pierre de Laplace. The modern period of mathematics dates from the beginning of the 19th cent., and its dominant figure is Carl Gauss, who made fundamental contributions to algebra, arithmetic, geometry, number theory, and analysis. In that century Non-Euclidean Geometry was invented independently by Nikolai Lobachevsky, Janos Bolyai, and, in another form, G.F.B. Riemann, whose work was of great importance in the development of the general theory of relativity. Number theory and abstract algebra received significant contributions from Sir William Rowan Hamilton, M.S. Lie, Georg Cantor, Julius Dedekind, and Karl Weierstrass. Weierstrass and Augustin Cauchy brought new rigor to the foundations of the calculus and of analysis. In the 20th cent. there have been two main trends. One is toward increasing generalization and abstraction, exemplified by investigations into the foundations of mathematics by David Hilbert, Bertrand Russell and Alfred North Whitehead, and Kurt Godel. The other trend is toward concrete applications to such areas as linguistics and the social sciences, as well as computer science, made possible by the work of John Von Neumann, Norbert Wiener, and others.

Matrix. 1. The continuous phase in a heterogeneous solid in which other phases are contained.

2. A mathematical entity consisting of an array of numbers in rows and columns. Matrices obey certain defined mathematical rules. They can be used in solving sets of simultaneous equations.

Matte. A mixture of metal sulphides obtained at an intermediate stage in smelting sulphide ores. Copper matte contains a mixture of iron and copper sulphides.

Matter. Anything that has mass. Because of its mass, all matter has Weight, if it is in a gravitational field, and Inertia. The three common States of Matter are solid, liquid, and gas; scientists also recognize a fourth, Plasma. Ordinary matter consists of Atoms and Molecules.

Maximum and Minimum Thermometer. A type of thermometer designed to record the maximum and minimum temperatures reached during a period of time. Two small steel spring-loaded indicators are pushed along the tube by the mercury column and remain in position when the mercury recedes.

The instrument is reset by moving the indicators with a magnet.

Maxwell. Symbol: Mx. A CGS-electromagnetic unit of magnetic flux equal to a flux of one gauss per square centimetre. One maxwell is equivalent to 10^{-8} weber. [After James Clerk Maxwell (1831-79), Scottish physicist.]

Maxwell's Demon. A small imaginary creature who, according to Maxwell, could separate a gas at uniform temperature into a hot region and a colder region by opening and closing a shutter to select the molecules with higher kinetic energy. Such a process would violate the second law of thermodynamics.

Maxwell's Equations. Four general differential equations for the varying electric and magnetic fields at a point. The equations are curl $H = \delta D/\delta t + j$; div $B = O$; curl $E = - \delta B/\delta t$; and div D P.H is the magnetic field strength, D the displacement, j the current density, E the electric field strength, and P the volume charge density. Maxwell showed that electromagnetic waves propagated through free space at a velocity equal to that of light and deduced that light was an electromagnetic wave.

Maxwell, James Clerk. 1831-79, Scottish physicist. In 1871 he became the first professor of experimental physics at Cambridge, where he organized the Cavendish Laboratory. Maxwell's notable work in Electricity and Magnetism was summarized in his *A Treatise on Electricity and Magnetism* (1873). He developed the theory of the electromagnetic field on a mathematical basis and concluded that electric and magnetic energy travel in transverse waves that propagate at a speed equal to that of light; light is thus only one type of electromagnetic Radiation. Maxwell's theoretical study (1859) of Saturn's rings foreshadowed his later investigations of heat and the kinetic theory of gases.

Mayer, Johann Tobias. 1723-62, German mathematician and

astronomer. His highly accurate lunar tables (1752) were used to compute the lunar ephemerides in the early editions of the *Nautical Almanac,* anabling navigators to determine precisely longitude at sea. He also made improvements in mapmaking and invented the repeating circle, later used in measuring the arc of the meridian.

Meadowlark. North American meadow Bird of the family Icteridae, which includes the Blackbird, Oriole, and Cowbird. Unlike other members of the family, the meadowlark (also called meadow starling) does not travel in flocks, and eats harmful insects rather than grain. The eastern species, *Sturmella magna,* is known for its clear, whistling song. It is c. 10 in. (25 cm) long and has black-streaked brown colouring above and yellow below.

Meadow Saffron or **Autumn Crocus.** Perennial garden ornamental *(Colchicum autumnale)* of the Lily family, native to Europe and N Africa and now naturalized in the U.S. Its poisonous corms and seeds yield the drug colchicine. Thę purplish fall flowers resemble those of the unrelated true Crocus and true Saffron (of the Iris family).

Mean. The average value of a set of numbers.

Mean Free Path. The average distance moved by a molecule in a gas between collisions. It is equal to $1/(2^{1/2}\pi nd^2)$, where n is the number of molecules in unit volume and d is the molecular diameter.

Mean Life. Symbol : T. The average lifetime of a nucleus of a radioactive isotope.

Measles or **Rubeola.** Highly contagious viral disease of young children spread by droplet spray from the mouth, nose, and throat during the infectious stage (beginning two to four days before the rash appears and lasting two to five days thereafter). Early symptoms (fever, redness of eyes) are followed by characteristic whitespots in the mouth and a facial rash that spreads to the rest of the body. Although one attack confers lifelong immunity, immunization is advisable because of the possibility of serious secondary infection. See also Rubella.

Mechanical Advantage. *See* Machine.

Mechanical Equivalent of Heat. Symbol: J. The amount of mechanical energy equivalent to unit amount of thermal energy. It has the value 4.1855 joules per calorie.

Mechanics. Branch of Physics concerned with Motion and the Forces causing it. The field includes the study of the mechanical properties of matter, such as Density, elasticity, and Viscosity. Mechanics is divided into Statics, which deals with bodies at rest or in equilibrium, and Dynamics, which deals with bodies in motion. Isaac Newton, who derived three laws of motion and the law of universal Gravitation, was the founder of modern mechanics. For bodies moving at speeds close to that of light, Newtonian mechanics is superseded by the theory of Relativity, and for the study of very small objects, such as Elementary Particles, quantum mechanics is used.

Mechanized Warfare. Employment of modern mobile attack-and-defense tactics that depend upon machines, particularly armored motor vehicles, Tanks, and, increasingly, aircraft. Mechanized warfare was first used in World War I. It was of great importance in World War II, *e.g.*, in the German blitzkrieg against Poland and France. Gen. Erwin Rommel was a leading proponent and pratitioner of mechanized warfare, and the Allies' success with it (1944-45) in Europe under Gen. George Patton and others helped to end the war. Israeli offensives in 1956, 1967, and 1982 involved close coordination of motorized infantry with air forces. Helicopters proved to be a new and effective component of mechanized warfare during the Korean War and Vietnam War.

Mechanism. The complete set of changes occurring in a chemical reaction. In general, a reaction proceeds by a series of simple steps involving the producing of stable molecules, ions, free radicals, or other intermediates, some of which may have only a transient existence. A detailed description of the mechanism also involves specification of the transition states of the steps.

Medicine. Science and art of diagnosing, treating, and preventing disease. For centuries, because the origin of disease was unknown, its treatment was coupled with magic and superstition. The more scientific practice of medicine, however, began in ancient Asiatic civilizations. In Summer, the Laws of Hammurabi established the first known code of medical ethics. In China, the ancient practice of Acupuncture and ideas about the circulation of blood presuppose familiarity with anatomy, vascular systems, and the nervous system. The Greeks advanced medical knowledge in anatomy and physiology, diet, exercise, and other areas, and provided the Hippocratic oath, still used today. The Romans improved public health through their sophisticated sanitation facilities, and Galen provided a final synthesis of the medicine of the ancient world. After a period of decline during the Middle Ages, when medical knowledge was kept alive mainly by Arab and Jewish physicians, Vesalius proved that there were errors in Galen's work and again opened medicine to discovery In the 17th cent., William Harvey demonstrated the circulation of blood and the role of the heart as a pump. With the introduction of the compound microscope, minute forms of life were discovered and a major step was made toward diagnosing disease. In the 18th cent., Edward Jenner introduced the concept of Vaccination, and surgery was transformed into an experimental science. The beginnings of modern medicine data from the 19th cent., with the development of the germ theory of disease, the use of antiseptics and Anesthesia in surgery; and a revival of public-health measures and better sanitation. Medicine in the 20th cent. has been characterized by an increased undertanding of Immunity, the Endocrine System, and the importance of

Vitamins and nutrition; advances in Surgery, organ Transplantation, and diagnostic techniques; and the use of Drugs, especially Antibiotics. Advances have also been notable in the treatment of mental illness through both Psychotherapy, Tranquilizers). With growing Antidepressants; Psychiatry and the administration of drugs. With growing specialization and complex diagnostic and therapeutic technology, modern medicine faces problems in the allocation of both personnel and capital. Some authorities advocate an increased use of paramedical personnel. The cost of medical care in the U.S. is financed by various forms of private ocverage and government programs (e.g., Medicare for those over age 65, Medicaid for the indigent; public vaccination programs). With the cost of medical care continuing to rise, interest in some form of national health insurance persists in the U.S.

Medicine Man. Among American Indians, a tribal member, similar to a Shaman, credited with healing powers from supernatural sources. He cured bodily ailments and, as priest, could also cause pain and promote fertility and good hunting and fishing. He cast evil from the body with spells, sometimes supplemented by herbs or bloodletting.

Medium Frequency (MF). A frequency in the range 0.3 megahertz to 3 megahertz.

Meiosis. Process of nuclear division in a cell by which the Charomosomes (genetic material) are reduced to half their original number. Meiosis occurs only during formation of sex cells (Ovum and Sperm). An ordinary body cell contains two of each type of chromosome (diploid). Meiosis produces cells with one chromosome of each pair (haploid). In Fertilization, two haploid cells are united; the resulting zygote then contains a diploid number of chromosomes.

Melamine. A white crystalline solid, $C_3N_6H_6$, used together with formaldehyde in producing synthetic resins. M.pt. 354°C (decomposes); r.d. 1.57.

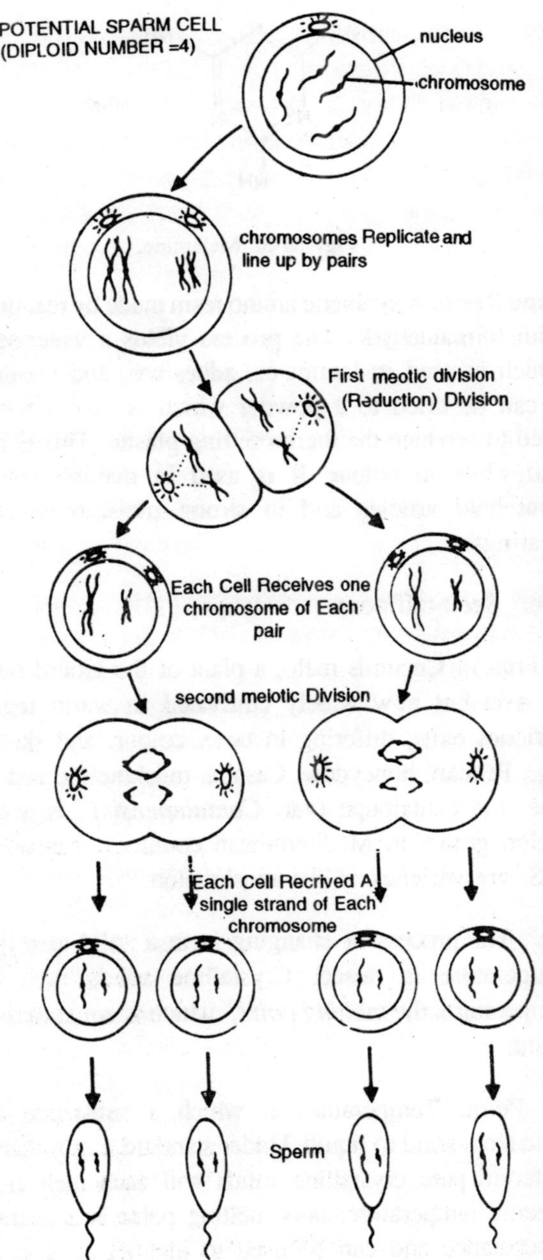

Fig. M-4. *Meiosis:* Formation of spermcell, through moiosis.

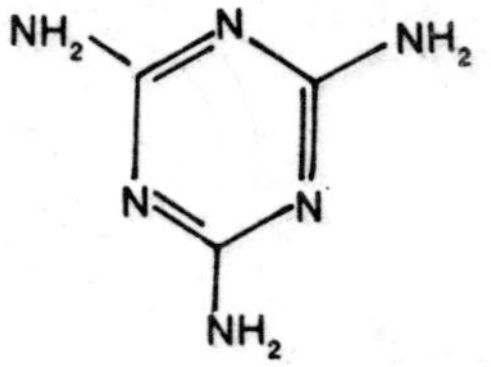

Fig. M-5. Melamine.

Melamine Resin. A synthetic amino resin made by reacting melamine with formaldehyde. The process yields a water-soluble syrup which is used in laminates, adhesives, and textile treatment. It can be dried to a powder which is mixed with filler and used to produce the thermosetting plastic. This is hard, strong, and white in colour. It is used in making many moulded household articles and in strong moisture-resistant surface coatings.

Melitose. *See* raffinose.

Melon. Fruit of Cucumis melo, a plant of the Gourd family, native to Asia but now widely cultivated in warm regions. Many varieties exist, differing in taste, colour, and skin texture — *e.g.*, Persian, honeydew, Casaba, muskmelon, and cantaloupe. The true cantaloupe (var. *Cantalupensis)* is a hard-shelled melon grown in Mediterranean countries; cantaloupes of the U.S. are varieties of the muskmelon.

Melting. The process of changing from a solid to a liquid as the temperature is raised. Crystalline solids melt at a fixed temperature, the *melting point,* which is equal to their freezing point.

Melting Point. Temperature at which a substance changes its state from solid to liquid. Under standard atmospheric pressure, different pure crystalline solids will each melt at a different specific temperature; thus melting point is a characteristic of a substance and can be used to identify it. The quantity of heat necessary to change 1 gram of any substance from solid to liquid at its melting point is known as its latent heat of fusion.

Mendel, Gregor Johann. 1822-84, Austrian monk noted for his experimental work on heredity. At the Augustinian monastery in Brno (1843-68) he conducted experiments, chiefly on garden peas and involving a controlled pollination technique and a careful statistical analysis of his results, that produced the first accurate and scientific explanation for hybridization. His findings, published in 1866, were ignored during his lifetime, but were rediscovered by three separate investigators in 1900. Mendel's conclusions have become the basic tenets of Genetics and a notable influence in plant and animal breeding. *Mendelism* is a system of heredity based on his conclusions. Briefly summarized, the Mendelian system states that an inherited characteristic is determined by the combination of two hereditary units (now called genes), one from each of the parental reproductive cells, or gametes.

Mendeleev, Dmitri Ivanovich. (mendela'ef), 1834-1907, Russian chemist. He formulated (1869) the periodic law and invented the Periodic Table, a system of classifying the elements that allowed him to predict properties of then-unknown elements. He was a professor (1868-90) at the Univ. of St. Petersburg, a government adviser on the development of the petroleum industry, and the director of the bureau of weights and measures.

Mendelevium (Md). Artificial radioactive element, detected in 1955 by A. Ghiorso and colleagues, who produced it one atom at a time by alpha-particle bombardment of einsteinium-253. Little is known about its properties.

Mendelevium. Symbol : Md. A transuranic actinide element obtained in trace amounts by bombarding einsteinium with alpha particles. The isotope produced, ^{256}Md, has a half-life of about 1.5 hours. A.N. 101. [After D. I. Mendeleev (1834-1907), Russian chemist who first formulated the periodic table.]

Meningitis or **Cerebrospinal Meningitis.** Acute inflammation of the membranes covering the Brain or spinal cord, or both. It can be caused by Bacteria, Viruses, Protozoa, Yeasts, or Fungi, usually introduced from elsewhere in the body. Symptoms include fever, headache, vomiting, neck and back rigidity, delirium, and convulsions. Examination of the cerebrospinal fluid by means of a spinal tap permits a specific diagnosis. Antibiotics have reduced mortality and decreased the indicence of such complications as brain damage and paralysis.

Menscus Lens. *See* lens.

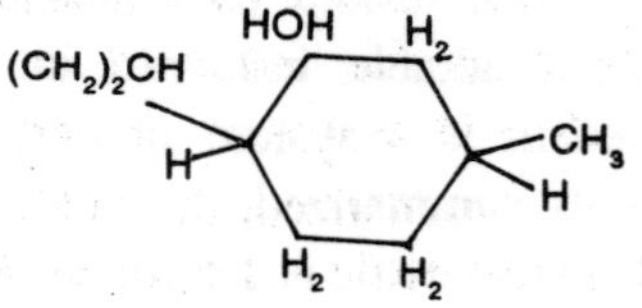

HOH H_2
$(CH_2)_2CH$
H CH_3
H
H_2 H_2

Fig. M-6. Menthol.

Menstruation. Periodic flow of bloody fluid which is part of the human female's reproductive cycle. In this cycle (menstrual cycle), the hormone Estrogen, secreted by the ovaries (egg-producing organs), first acts to thicken the lining (endometrium) of the uterus (the muscular organ that carries developing young). After 8 to 10 days the ovary releases an egg and begins to secrete the hormone Progesterone. If the egg is fertilized, it will embed itself in the thick uterine lining. If the egg is not fertilized, the secretion of progesterone declines, and the lowered level of progesterone causes the uterine lining to be sloughed off. The menstrual cycle, which is repeated approximately every 28 days, is the result of complex interactions between ovarian and pituitary hormones and the nervous system. Menstruation commences at puberty (about age 12) and ceases at the menopause (about age 45 to 50).

Mental Retardation. Subnormal mental development, manifested from birth or early childhood, that may be caused by a variety of conditions. Although mental retardation has traditionally been diagnosed in the U.S. by significantly lower than normal scores on Intelligence tests, other characteristics such as social maturity and ability to sustain personal and social independence are now used to evaluate achieve some language development and can be taught to perform manual tasks of moderate complexity.

Menthol. A white crystalline solid, $C_{10}H_{20}O$, with a strong minty taste. It is a terpene alchol, present in peppermint oil, used in flavourings and perfumes. M.pt. 42°C (*l*-); b.pt 212°C; r.d. 0.89.

Mecaptan. Any of a class of organic compounds of general formula RSH. Mercaptans are analogous to alcohols, with the oxygen atom replaced by sulphur.

Mercator Map Projection. A cylindrical Map Projection of the features of the surface of the earth that can be constructed only mathematically. The parallels of Latitude, which on the globe are equal distances apart, are drawn with increasing separation as their distance from the Equator increases in order to preserve shapes. However, the price paid for this is that areas are exaggerated with increasing distance from the equator. For instance, Greenland is shown with enormously exaggerated size, although its shape is preserved. The poles themselves cannot be shown on a Mercator projection. This type of projection gives an incorrect impression of the relative sizes of the world's countries.

Mercurous. *See* mercury.

Mercury (quick silver). Symbol: Hg. A heavy silvery liquid metallic element found in cinnabar (HgS), from which it is obtained by heating and condensing the vapour. It is used in mercury-vapour lamps, dental amalgams, thermometers, barometers, diffusion pumps, and other pieces of scientific apparatus. Mercury is a fairly reactive element. Like zinc, it combines with oxygen and other nonmentals: however, it is not attacked by nonoxidizing acids. It forms alloys (amalgams) with most other metals.

The element has two series of ionic compounds. Mercury II *(mercuric)* salts are the more stable, containing the Hg^{2+} ion. They can be reduced to mercury I *(mercurous)* salts, which contain not the Hg^{+} ion but the binuclear $^{+}Hg\text{-}Hg^{+}$ ion. Thus, mercurous fluoride is Hg_2F_2 (not HgF). Mercury also forms convalently bonded compounds, notably organometallic compounds. A.N. 80; A.W. 200.59; m.pt. 38.87° C; b.pt. 356.58°C; r.d. 13.35; valency 2.

Mercury (Hg) or **Quicksilver.** Metallic element, known to the ancient Chinese, Hindus, and Egyptians. silver-white and mirrorlike, it is the only common metal existing as a liquid at ordinary temperatures. Mercury is used in barometers, thermometers, electric switches, mercury-vapor lamps, and certain batteries; a mercury alloy, called an amalgam, is employed in dentistry. Mercury compounds have been used as insecticides, in rat poisons, and as disinfectants. Not easily discharged from the body, the metal is a cumulative poison; its ingestion in more than trace amounts in contaminated food or its absorption by the skin or mucous membranes results in *mercury poisoning,* which can cause skin disorders, hemorrhage, liver and kidney damage, and gastrointestinal disturbances. Workers in many industries have been affected, and mercury Pollution of rivers, lakes, and oceans, usually through the discharge of industrial wastes, has become a serious environmental problem . Most mercury pesticides have been withdrawn from the U.S. market, and in 1972 more than 90 nations approved an international ban on the dumping of mercury in the ocean, where the metal has tended to work its way into the food cycle of aquatic life and to reach dangerous levels in certain food fish.

Mercury. The closest planet to the sun in the solar system lying at a mean distance of 58 million kilometres from the sun. It is the smallest of the planets, comparable in size with the moon, and having an equatorial diameter of about 4870 km, a mass only one twentieth that of the earth, and a mean relative density of 5.2. Mercury is probably a rocky planet with little or no atmosphere—some traces of hydrogen and carbon dioxide have apparently been detected—and daytime temperatures that reach 400° C. Its orbital velocity averages 48 km per second and its period of revolution is 87.969 earth days. Its period of axial rotation, originally thought to equal its period of revolution, is about 58 or 59 days. Being the innermost of the planets, Mercury is very difficult to observe, and can only ever be seen as a morning or evening star; its maximum elongation is only 27°45' and it never appears high above the horizon.

Mercury. In astronomy nearest Planet to the sun, at a mean distance of 36.0 million mi (57.9 million km). It has a diameter of 3,031 mi (4,878 km), a cratered, lunarlike surface, and almost no atmosphere. Because its greatest Elongation is 28°, it can never be seen more than 2 hr after sunset or 2 hr before sunrise. The observed motion (43" each century) of Mercury's perihelion (closest point to the sun) is more than can be explained by planetary perturbations but is in nearly exact agreement with the prediction of the general theory of Relativity. Mercury has no known satellites. The only Space Probe to study Mercury was *Mariner 10,* which made three encounters in 1974-75.

Mercury Chloride. Either of two chlorides of mercury. Both are white powders. *Mercury H chloride* (mercuric chloride), $HgCl_2$, is obtained by direct combination of the elements. It is widely used in making other mercury compounds. M.pt 276° C: b.pt. 303° C; r.d. 5.4 (25° C). *Mercury I chloride* (mercurous chloride), Hg_2Cl_2, is prepared by heating mercury with mercury II chloride. It is used in medicine and pyrotechnics and as a fungicide. M.pt. 302° C; b.pt. 384° C; r.d. 7.0.

Mercury Fulminate. A grey crystalline powder, $Hg(CNO)_2$, made by treating mercury with nitric acid and ethanol. It explodes under slight friction or shock and is used in detonators. R.d. 4.4.

Mercury Oxide. Any of two basic oxides of mercury. *Mercury I oxide* (mercurous oxide) is a black powder, Hg_2O. *Mercury II oxide* (mercuric oxide) is a red or yellow powder, HgO, depending on the particle size. The red form is obtained by heating mercury I nitrate, while the finer yellow form is precipitated by adding an alkali to a solution of a mercury II salt. The compound is used in medicine and paint pigments.

Mercury Thermometer. A type of thermometer consisting of a small glass bulb containing mercury, connected to a fine sealed capillary tube. The temperature is meausred by the expansion of the mercury.

Meridian. Any great circle on the earth's surface passing through both geographical poles.

Mesitylene (1, 3, 5-trimethylbenzene). A colourless liquid, $C_6H_3(CH_3)_3$, extracted from coal tar or petroleum and used as an intermediate in the manufacture of dyes. M.pt. —52.7°C; r.d. 0.86.

Mesquite. Spiny trees or shrub (genus *Prosopis*) of the Pulse family, native to tropical and subtropical regions. The seed pods of *P. juliflora* contain an edible, sweet pulp that is used as forage and to make bread and a fermented drink; the durable wood is used for fence posts. Mesquite roots may penetrate 50 ft (15 m) into the ground for water, enabling the plant to grow in sites unsuited to most crops.

Meso Form. *See* optical isomerism.

Meson. Any member of a class of elementary particles characterized by a mass intermediate between those of the electron and the proton, an integral spin, and participation in strong interactions. Mesons are responsible for the forces between nucleons in the nucleus. They are all unstable particles, undergoing decay in a variety of ways. The particles can have positive and negative charge or can be electrically neutral. Positive and negative mesons are antiparticles of each other: their charge is the same magnitude as that of the electron. The three types of meson are the pion, the kaon, and η^- mesons. The muon (or mu-meson) was formerly classified as a meson but is in fact a lepton.

Meta. 1. Indicating that substituents in a benzene ring are attached to carbon atoms separated by one carbon atom: *i.e.* to the 1.3 positions.

2. Denoting the least hydrated acid of a series of acids formed from the same anhydride.

Metabolism. Sum of all living process in living systems. Two subcategories of metabolism are anabolism, the building up of organic molecules from simpler ones, and catabolism, the breaking down of complex substances, often accompanied by the release of energy. Thus the energy required for anabolism is obtained from catabolic reactions. Basal metabolism, the heat produced by an organism at test, represents the minimum amount of energy required to maintain life at normal body temperature.

Metal. 1. Any of a class of chemical elements that are typically lustrous malleable solids that conduct heat and electricity. There are some exceptions—mercury is a liquid and some metals are brittle. Chemically, metals are electropositive: they have basic oxides and hydroxides and tend to form ionic compounds containing positive ions. In fact, many common metals do not show this behaviour. Tin, for example, also forms covalent compounds, including SnH_4, and has an amphoteric oxide that dissolves in alkalis to form stannates.

2. An alloy of a metal with one or more other elements.

Metal. Chemical Element displaying certain properties, notably metallic luster, the capacity to lose electrons and form a positive Ion, and the ability to conduct heat and electricity, by which it is normally distinguished from a nonmetal. The metals comprise about two thirds of the known elements. Some elements, *e.g.*, arsenic and antimony, exhibit both metallic and nonmetallic properties, and are called metalloids. Metals fall into groups in the Periodic Table determined by similar arrangements of the orbital electrons and a consequent similarity in chemical properties. Such groups include the Alkali Metals (Group Ia in the periodic table), the Alkaline-Earth Metals (Group IIa), and the Rare-Earth Metals (Lanthanide and Actinide series). Most metals other than the alkali metals and the alkaline-earth metals are called transition metals. The oxidation states, or Valence, of the metal ions vary from +1 for the alkali metals to +7 for some transition metals. Chemically, the metals differ from the nonmetals in that they form positive ions and basic oxides and hydroxides. Upon exposure to moist air, a great many metals undergo corrosion, *i.e.*, enter into a chemical reaction, the oxygen of the atmosphere uniting with the metal to form the oxide of the metal, *e.g.*, rust on exposed iron.

Metaldehyde. A white crystalline polymer of acetaldehyde, $(CH_3CHO)_n$, where *n* is either 4 or 6. It is made by acid-catalysed polymerization below 0° C and is used in firelighters and slug killers. Sublimes at 112° C.

Metallic Bond. The type of chemical bond occurring in metallic crystals. In metallic bonding a regular lattice of positive ions is held together by a cloud of free electrons, which can move freely through the lattice.

Metallocene. A sandwich compound formed between a metal atom and two cyclopentadienyl ions, as in ferrocene.

Metalloid. A chemical element with properties characteristic both of metals and nonmetals. The term is often used for elements that are difficult to classify in one or other group. Germanium, for example, has a metallic appearance, semiconductivity, and many of the chemical properties of nonmetals. Arsenic has a metallic allotrope and two nonmetallic allotropes. Other metalloids include silicon, tellurium, and antimony.

Metallurgy. The study of metals, including industrial processes for smelting, refining, and working metals, and the study of the structure and properties of metals and alloys.

Metallurgy. Science extracting metals from their ores. The processes employed depend upon the chemical nature of the Ore to be treated and upon the properties of the Metal to be extracted. When an ore has a low percentage of the desired metal, a method of physical concentration, *e.g.*, the Flotation Process, must be used before the extraction process begins. Because almost all metals are found combined with other elements in nature, chemical reactions are required to set them free. These chemical processes are classified as *pyrometallurgy,* the use of heat for the treatment of an ore, *e.g.*, in Smelting and roasting; *electrometallurgy,* the preparation of certain active metals by Electrolysis; and *hydrometallurgy,* or leaching, the selective dissolution of metals from their ores. Modern metallurgical research is concerned with preparing radioactive metals, with obtaining metals economically from low-grade ores, with obtaining and refining rare metals hitherto not used, and with formulating Alloys.

Metamorphism. In geology, process of change in the structure, texture, or composition of Rocks caused by heat, deforming pressure, and/or hot, chemically active fluids. In general, metamorphic rock is coarser, denser, and less porous than the rock from which it was formed. The change in texture commonly results in a rearrangement of Mineral practices into a parallel alignment called foliation, probably the most characteristic property of metamorphic rocks; it is seen in Slate, Schist, and Gneiss. Local metamorphism is usually caused by the intrusion of a mass of igneous rock into older rock. Regional metamorphism accompanies mountain-building activity associated with large-scale crustal movements.

Metamorphosis. In zoology, a term used for a series of distinct stages in the development from egg to adult. For example, an insect such as the butterfly is active and wormlike (a caterpillar) in its larval stage. In its next stage (pupal) the insect is surrounded by an outer covering (a cocoon); it is outwardly inactive but undergoes many changes in internal organization. Eventually the insect emerges transformed into its adult stage (a butterfly). Metamorphosis is called complete when there is no suggestion of the adult in the larval stage. In incomplete metamorphosis, the successive larval stages resemble the adult, as in the Grasshopper. Many species change their habitat after undergoing complete metamorphosis *(e.g.,* the toad lives in water in its tadpole stage and on land in its adult stage).

Metaphosphoric Acid. *See* phosphoric acid.

Metaphysics. Branch of philosophy concerned with the ultimate nature of existence. Ontology (the study of the nature of being), cosmology, and philosophical theology are usually considered its main branches. The term comes from the metaphysical treatises of Aristotle, who presented the First Philosophy (as he called it) after the *Physics* [Gk. *metaphysic* = after physics]. Metaphysical systems in the history of philosophy have included Aristotelian Scholasticism and the rationalistic systems of the 17th cent. *(e.g.,* those of Descartes, Spinoza, and Leibniz). Kant, in the 18th cent., held scientific metaphysical speculation to be an impossibility but considered metaphysical questions of moral necessity. His work influenced that of Fichte, Schelling, and Hegel. Since the mid-19th cent. philosophy has generally denied validity to metaphysical thought.

Matastable. 1. *See* elquilibrium.

2. Designating an excited atom, molecule, ion, etc., with a relatively long lifetime.

Metathesis. *See* double decomposition.

Metavanadate. *See* vanadate.

Metavanadic Acid. *See* vanadate.

Meteor (shooting star). The spectacular streak of light caused by the trail of a small meteoroid as it burns up while passing through the earth's atmosphere. When the earth passes through the remains of a comet, there is a considerable increase in the number of meteors. One of the best showers, the *Perseids,* is visible every year at the beginning of August.

Meteor. Small piece of extraterrestrial matter that becomes visible as a "shooting star" or "falling star" when it enters the earth's atmosphere.–While still outside the atmosphere, it is called a meteoroid. As a meteor it is heated to incadescence through friction (due to collisions with air molecules) and usually disintegrates completely before reaching the earth; those meteors large enough to reach the ground are called Meteorites. A meteor of considerable duration and brightness is known as a fireball; a fireball that explodes in the air is called a bolide. The frequency of meteors increase when the earth, in its orbit, annually passes through a swarm of particles generated from the breakup of a comet. The meteors of such a *meteor shower* all appear to originate at a single point, or radiant, in the sky. Some of the better-known showers (named for the constellations in which their radiants are located) and their approximate dates are: Lyrids, Apr. 21; Perseids, Aug. 12; Orinoids, Oct 20; Taurids, Nov. 4; Leonids, Nov. 16, Geminids, Dec. 13.

Meteorite. A large meteoroid that enters the earth's atmosphere, becoming incandescent as a bolide (or fireball), and succeeds in reaching the ground. Usually of stone or metal, meteorites range in mass from a few grams to several thousand kilograms. The largest so far recorded was a meteorite found at Grootfontein, South Afirca, Weighing 65,000 kg.

Meteorite. Large meteor that survives the intense heat of atmospheric friction and reaches the earth's surface. Meteorites may have originated as fragments of asteroids. They are classified in these general categories. The siderites, or irons, are composed entirely of metal (chiefly nickel and iron). The aerolites, or stony meteorites, show a diversity of mineral elements including large percentages of silicon and magnesium oxides; the most abundant type of aerolite is the chondrite, so called because the metal embedded in it is in the form of grainlike lumps, or chondrules. The siderolites, or stony irons, which are rarer than the other types, are of both metal and stone in varying proportions. When a meteorite reaches the earth, the tremendous force of impact with the earth's surface causes great compression, heating, and partial vaporization of the outer part of the meteorite and of the materials in the ground; expansion of the gases thus formed and of steam produced from groundwater causes an explosion that shatters the meteorite and carves out a *meteorite crater* in the ground. One of the best-preserved craters is Meteor, or Barringer, Crater, near Winslow, Arizona, C.¾ mi ($1^1/_2$ km) in diameter and 600 ft (180 m) deep. The largest meteorite discovered, the 60-ton Hoba West, rests where it was found (in 1920), near Grootfontein, Namibia.

Meteoroid. Any of innumerable extraterrestrial pieces of matter found orbiting the sun and ranging in size from tiny particles to large—sometimes very large—objects.

Meteorology. Branch of science that deals with the Atmosphere of a planet, particularly that of the earth. Meteorology is based on the accurate scientific measurement of various atmospheric conditions with a wide assortment of instruments. Air temperature is measured with the Thermometer; air pressure with the Barometer; wind direction with the weather vane; wind speed with the anemometer; high-altitude air-pressure and wind information with the Weather Balloon; relative humidity with the Hygrometer; precipitation with the rain gauge; and cloud formations and weather fronts with both radar and high-altitude Weather Satellites. The meteorologist combines the data collected from many geographical locations into a weather map. On a typical map the various weather elements are shown by figures and symbols. Isobars are drawn to show areas of equal pressure, and Fronts and areas of precipitation are also indicated. Meteorologists analyze the data collected and illustrated on the weather map in order to predict, or forecast, the Weather for the next few hours and the next few days. Long-range weather forecasts, which are more general and less accurate, are also made for future periods of several months.

Methacrylic Acid (2-methylpropenoic acid). A colourless liquid, $CH_2:C(CH_3)COOH$, with an acrid odour, used in the manufacture of acrylic resins. M.pt. 16° C; b.pt. 163° C; r.d. 1.01.

Methadone. Synthetic Narcotic, similar in effect to Morphine, used primarily in the treatment of narcotic drug addiction. Given to addicts, it blocks the euphoric action of Heroin without itself causing euphoria and causes less severe and hazardous withdrawal symptoms than other narcotic drugs (although critics of methadone therapy point out that methadone patients are still addicts). Methadone is also used an Analgesic, especilally in patients who are terminally ill.

Methanal. *See* formaldehyde.

Methane (marsh gas). A colourless odourless tasteless flammable gas, CH_4, that is the chief component of most natural gas. It is the fiist member of the alkane series and is used as a fuel and a source of petrochemicals. M.pt. —182.5° C; b.pt. —161° C; r.d. 0.55 (air 1, 0° C).

Methane CH_4). Colourless, odorless, gaseous Hydrocarbon formed by the decay of plant and animal matter. It occurs naturally as the chief component of Natural Gas, as the firedamp of coal mines, and as the marsh gas released in swamps and marshes. Methane can also be made synthetically by various means. It is combustible and can form explosive mixtures with air. Used for fuel in the form of natural gas, methane is also an important starting material for making solvents and certain Freons.

Methanethiol. A colourless flammable liquid, CH_3SH, with a very powerful unpleasant odour, prepared by reacting methanol with hydrogen sulphide. It is used as a fuel additive and fungicide. M.pt. —121° C; b.pt. 6° C; r.d. 0.87.

Methanide. A carbide that yields methane on hydrolysis.

Methanoic acid. *See* formic acid.

Methanol (methyl alcohol, wood alcohol). A colourless poisonous flammable liquid, CH_3OH, usually manufactured by high-pressure catalytic synthesis from bydrogen and carbon monoxide. A typical alcohol, it is used as a solvent, antifreeze, and raw material for the manufacture of other chemicals. Methanol was formerly also called *carbinol.* M.pt. —98° C; b.pt. 64° C; r.d. 0.79.

Methanol, Methyl Alcohol, or **Wood Alcohol.** (CH_3OH), a colourless, flammable liquid and the simplest Alcohol. Methanol is a fatal poison. Small internal doses. prolonged exposure of the skin to the liquid, or continued inhalation of the vapor may cause blindness. It can be obtained from wood, but now is made synthetically from the direct combination of hydrogen and carbon monoxide gases. Methanol is used to make Formaldehyde, as a solvent, and as Antifreeze.

Method of Mixtures. A simple method of determining specific heat capacities and latent heats by mixing known amounts of substances (liquids or liquids and solids) at different temperatures and determining the final temperature of the mixture.

Methoxybenzene. *See* anisole.

Methoxyl Group. The organic group CH_3O.

Methyl acetate. A colourless flammable liquid ester. CH_3COOCH_3, with a fragrant odour, used as a solvent for paints and lacquers. M.pt. -98°C; b.pt. 54°C; r.d. 0.92.

Methyl Alcohol. *See* methanol.

Methylamine. A flammable colourless gaseous amine, CH_3NH_2, with a strong ammoniacal odour. It is prepared by reacting methanol and ammonia at high temperature with a catalyst and is an intermediate in preparing dyes, pharma-ceuticals, and insecticides. M.pt. -92° C; b.pt. -7° C; r.d. 0.69 (— 10.8°C).

Methylated spirits. Ethanol denatured by addition of about 9.5% methanol and 0.5% pyridine, with small amounts of blue dye. It is used as a solvent and fuel.

Methylation. A chemical reaction in which a methyl group is introduced into a molecule.

Methyl Bromide. *See* bromomethane.

Mythyl Cellulose. A Grey-white powder made by treating cellulose with an alkali followed by methylation. It swells in water forming a viscous colloidal solution and is used as an adhesive and a thickening and emulsifying agent.

Methyl Chloride. *See* shloromethane.

Methyl Cyanide. *See* acetonitrile.

Methylene Chloride. *See* dichloromethane.

Methylene Group. The divalent group H_2C =.

Methyl Ethyl Ketone. *See* butanone.

Methyl Group. The organic group CH_{3}.

Methyl Iodide. *See* iodomethane.

Methyl Methacrylate. A colourless volatile liquid ester, CH_2; $C(CH_3)COOH$. It polymerizes readily forming acrylic resins such as Perspex. M.pt. -48°C; b.pt. 101°C; r.d. 0.94.

Methyl Orange. An orange powder, $C_{14}H_{14}N_3NaO_3S$, used as an acid-base indicator. It changes colour in the pH range 3.1-4.4, being red below 3.1 and yellow above 4.4.

Methyl Red. A dark red powder, $C_{15}H_{15}N_3O_2$ used as an acid-base indicator. It changes colour in the pH range 4.4-6.0, being red below 4.4 and yellow above 6.0.

Methyl Salicylate. A colourless, yellow, or reddish oily ester, $C_6H_4(OH)(COOCH)_3$, with an odour of wintergreen. It is found in some essential oils and is used in perfumes and flavourings. M.pt. -8.3° C; b.pt. 222° C; r.d. 1.18.

Metre. Symbol : m. The basic SI unit of length, equal to the length of 1.650 763 73 wavelengths of the radiation from a transition between the $2p_{10}$ and $5d_5$ levels of a krypton-86 atom. The light is produced by a specified type of discharge lamp operating at the triple point of nitrogen (63.15 K).

The metre was originally defined, in 1791, as one ten-millionth of the length of the quadrant of the earth's meridian passing through Paris. This was redefined in 1927 as the distance between two marks on a platinum-iridium bar kept at the International Bureau of Weights and Measures at Sevres near Paris. The present definition was introduced in 1960.

Metre-Candle. *See* lux.

Metric System. System of weights and measures planned in France and adopted there in 1799. Now used by most of the technologically developed countries of the world, it is based

PREFIXES FOR BAISC METRIC UNIT

MULTIPLES

Prefix	*Abbreviation*	*Power of 10*	*Equivalent*
tera-	T	10^{12}	trillion
giga-	G	10^{9}	billion
mega-	M	10^{6}	million
kilo-	k	10^{3}	thousand
hectro-	h	10^{2}	hundred
deka-	da	10^{1}	ten
FRACTIONS			
deci-	d	10^{-1}	tenth part
centi-	c	10^{-2}	hundredth part
milli-	m	10^{-3}	thousandth part
micro-	µ	10^{-6}	millionth part
nano-	n	10^{-9}	billionth part
pico-	p	10^{-12}	trillionth part

on a unit of length called the meter (m) and a unit of mass called the kilogram (kg). The meter is now defined in terms of a reproducible, universally available atomic standard, being equal to 1,650,763.73 wavelengths of the redorange light given off by the krypton-86 isotope under certain conditions. The kilogram is defined as the mass of the International Prototype Kilogram, a platinum-iridium cylinder kept a Sevres, France, near Paris. Other metric units can be defined in terms of the meter and the kilogram. Fractions and multiples of the metric units are related to each other by powers of 10, allowing conversion from one unit to a multiple of it simply by shifting a decimal point. This avoids the lengthy arithmetical operations required by the English Units of Measurement. The prefixes in the accompanying table have been accepted for designating multiples and fractions of the meter, the gram (=1/1000 kilogram), and other units. Several other systems of units based on the metric system have been in wide use. The cgs system uses the centimeter (=1/100 meter) of length, the gram of mass, and the Second of time as its fundamental units; other cgs units are the dyne of Force and the erg of Work or energy. The mks system uses the meter of length, the kilogram of mass, and the second of time as its fundamental units; other mks units include the newton of foce, the joule of work or energy and the watt of Power. The units of the mks system are generally much larger and of a more practical size than the comparable units of the cgs system. Electric and Magnetic Units have been defined for both these systems. The International System of Units (officially called the Systeme International d'Unites, or Sl) is a system of units adopted by the 11th General Conference on Weights and Measures (1960). Its basic units of length, mass, and time are those of the mks system; other basic units are the Ampere of electric current, the kelvin of temperature (a degree of temperature measured on the Kelvin Temperature scale), the candela of luminous intensity, and the Mole, used to measure the amount of a substance present. All other units are derived from these basic units.

MeV. Million electron volts: 10^6 electronvolts.

MHD. *See* magnetohydrodynamics.

Mho. *See* siemens.

Mica. General term for a large group of hydrous aluminum and potassium silicate minerals, usually occurring in scales and sheets. The most important commercial micas are muscovite and phlogopite. Muscovite, the commoner variety, is usually colourless but may be red, yellow, green, brown, or grey; it is found most often in pegmatite dikes. Phlogopite ranges in colour from yellow to brown and occurs in crystalline limestones, dolomites, and serpentines. Sheet mica is used as an insulating material and in certain acoustic devices. Scrap and ground mica is used in wallpaper, fancy paint, ornamental tile, roofing, lubricating oil, and Christmas-tree snow.

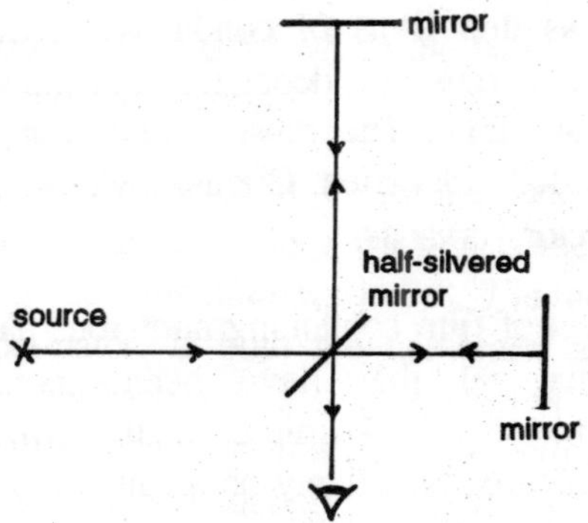

Fig. M-7. Michelson-Morley interferometer.

Michelson-Morley Experiment. An experiment designed to measure the velocity of the earth through the ether. The apparatus in the diagram was used. Light from a monochromatic source is split at a half-silvered mirror into two beams at right angles. The beams are reflected back along their path and recombined on reflection at the other side of the mirror, thus producing fringes as a result of interference.

If the earth is moving through, a stationary ether the light would have different velocities when measured in the direction of the earth's rotation and when measured perpendicular to this direction. In this case, rotation of the apparatus through 90° would cause a shift in position of the interference fringes. In fact, no change could be detected in any position or at any time. The experiment was one of the observations leading to the theory of relativity. [After Albert A. Michelson (1852-1931) and Edward Morley (1838-1923), American physicists.]

Microbalance. An extremely sensitive balance.

Microelectronics. Branch of Electronics devoted to the design and development of extremely small electronic devices that consume very little electric power. The simplest, but least effective, approach used is to make circuit elements, such as resistors, Capacitors, and Semiconductor devices, extremely small but discrete. In another approach, circuit elements fabricated as thin films of conductive, semiconductive, and insulating materials are deposited in sandwich form on an insulating substrate. The most advanced method is to form circuits, called Intergated Circuits, within and upon single semiconductor crystals.

Microfiche. Sheet of film containing numerous pages of printed or graphic material that have been greatly reduced by microphotography. Unlike microfilm, in which individual frames are reproduced consecutively on a roll of film, a single 4 X 6 in. (10 X 15 cm) microfiche card may contain hundreds of pages, providing faster access to a desired item. The image is magnified to approximately full size by a viewing machine or reader. Mocrofiche is increasingly used by libraries and businesses *(e.g.,* to store annual reports and catalogs).

Micrometer. Instrument used for measuring extremely small distances. In the micrometer caliper, the object to be measured is held between the two jaws of the instrument; the distance between the jaws is measured on a scale calibrated to the rotation of the finely threaded screw that moves one of the jaws. In astronomical and microscopic micrometers, the distance that a filament moves from one end to the other of the image of an object is read on a calibrated scale.

Mocron. Symbol: µm. A unit of length equal to 10^{-6} metre. In Sl units it is called the micrometre.

Microphone. Device (invented c. 1877) used in radio broadcasting, recording, and sound-amplifying systems to convert sound into electrical energy. Its basic component is a flexible diaphragm that responds to the pressure or particle velocity of sound waves. In a Capacitor, or condenser, mocrophone, used in high-quality sound systems, two parallel metal plates are given opposite electrical charges. One of the plates is attached to the diaphragm and moves in response to its vibrations, generating a varying voltage.

Microprocessor. Integrated Circuit that interprets and executes instructions from a Computer Program. When combined with other integrated circuits that provide storage for data and programs, often on a single Semiconductor base to form a chip, the microprocessor becomes the heart of a small Computer, or microcomputer. The evolution of the microprocessor has made possible the inexpensive handheld electronic Calculator, the digital wristswatch, and the Electronic Game. The microprocessor is also used to control consumer appliances, to regulate gasoline consumption in automobiles, and to monitor home and industrial alarm systems.

Microscope. Optical instrument used to increase the apparent size of an object. A magnifying glass, an ordinary double convex Lens having a short focal length, is a simple microscope. When an object is placed nearer such a lens than its principal focus, *i.e.,* within its focal length, an image is produced that is erect and larger than the object. The compound microscope, invented in the early 17th cent., consists essentially of two or more such lenses fixed in the two extremities of a hollow metal cylinder. This cylinder is mounted upright on a screw device, which permits it to be raised or lowered above the object until a clear image is formed. The lower lens (nearer to the object) is called the objective; the upper lens (nearer to the eye of the observer), the eyepiece. When an object is in focus, a real, inverted image is formed by the lower lens at a point inside the principal focus of the upper lens. This image serves as an "object" for the upper lens, which produces another image larger still (but virtual) and visible to the eye of the observer. The compound microscope is widely used in bacteriology, biology, and medicine in the examination of such extremely minute objects as bacteria, other unicellular organisms, and plant and animal cells and tissue. Technical advances making use of different forms of light and other forms of radiation have increased enormously the magnification and resolution of microscopes.

Microscope. An optical instrument for producing a magnified image of a small object. A *simple microscope* is a convex lens used as a magnifying glass. A *compound microscope* has a combination of lenses: an objective of short focal length to produce a real image that is viewed by the eyepiece.

Microscopic. 1. Visible only with the aid of a microscope.

2. Relating to the behaviour of individual atoms or molecules rather than matter in bulk. *Compare* macroscopic.

Microwave. Electromagnetic Radiation having a frequency range from 1,000 to 300,000 megahertz, corresponding to a wavelength range from 300 to 1 mm (about 12 to about 0.04 in.). Microwaves are used in Microwave Ovens, Radar, and communications links spanning moderate distances.

Microwave Oven. Cooking device that uses Microwaves to penetrate foods and rapidly cook them. The microwaves cause moisture molecules in the food to vibrate, a process that produces heat. Once used almost exclusively in fastfood restaurants, microwave ovens have become increasingly popular in home kitchens. Some microwave radiation has been found to leak from the ovens, however, and it is not yet known whether such low exposures might be harmful.

Microwave Background. An isotropic distribution of microwave radiation throughout the universe. The background seems to be black-body radiation corresponding to a temperature of 27 kelvin, and it may be a relic of the big bang that was the origin of the universe.

Microwaves. Electromagnetic radiation with wavelengths in the range 1 mm -0.3 m. Microwaves lie in the region between infrared radiation and radio waves.

Microwave Spectrum. An emission or absorption spectrum in the microwave region. Microwave radiation is absorbed by transitions between different rotational energy levels of molecules and gives information on the moments of inertia and dimensions of molecules.

Midnight Sun. Phenomenon in which the sun remains visible in the sky continuously for 24 hr or longer. It occurs in the polar regions because of the tilt of the equatorial plane to the plane of the ecliptic (the sun's apparent path through the sky). It occurs at the polar circles only at the solstice (summer for Arctic, winter for Antarctic), but as one approaches the pole it increases in occurrence up to a continuous six months (from vernal to autumnal equinox for the North Pole; the reverse for the South Pole).

Midwifery. Art of assisting at childbirth. The term *midwife* for centuries referred to a woman who was an overseer during the process of delivery. Professional schools of midwifery were established in Europe in the 16th cent. Midwives are still used widely in Europe and are experiencing an upsurge of popularity in the U.S.

Migration of Animals. Regular, periodic movements of animals in large numbers, usually away from and back to a place of origin. A round trip may take an entire lifetime or may be made more frequently, as on a seasonal basis. Seasonal migrations occur among many insects, birds, marine mammals, and large herbivorous mammals, *e.g.*, Wapiti, Caribou, and Moose. Such migrations provide more favorable conditions of temperature, food, or water and may involve a change of latitude, altitude, or both. The chief function is to supply a suitable breeding place. Migration may be initiated by physiological stimuli such as reproductive changes, external pressures such as drought, or a combination of both. Studies show that salmon depend on the olfactory sense to locate and return to the stream of their origin. Bats, whales, and seals use echolocation to navigate in the dark. Experiments in planetariums indicate that night-flying birds navigate at least in part by the stars. Day-flying birds orient themselves by the sun. A one-time, one-way wholesale migration out of an area prompted by explosive population increase is called an irruption. It is common among small rodents, notably Lemmings, and some species of birds and insects, *e.g.*, the so-called migratory locusts of North Africa.

Mil (thou). 1. One thousandth of an inch.

2. One thousandth of litre.

3. *See* Circular mil.

Mildew. Name for certain Fungi and the plant diseases they cause, and for the discolouration and disintegration of materials *(e.g.,)* leather, fabrics, and paper) caused by related fungi. The powdery mildews (class Ascomycetes) form a grey-white coating on plant tissues. The downy mildews (class Phycomycetes) form white, purplish, or grey patches; a downy mildew was the potato Blight that caused the Great Potato Famine (1845-49) in Ireland.

Mile. *See* Appendix.

Milk. Liquid secreted by the mammary glands of female mammals to feed their young. Cow's milk is most widely used by humans, but milk of such animals as the mare, goat, ewe, buffalo, camel, and yak is also consumed. An almost complete food, milk contains fats; proteins (mainly casein); salts; sugar (lactose); vitamins A,C, and D; some B vitamins; and minerals, chiefly calcium and phosphorus. The composition of milk varies with the species, breed, feed, and condition of the animal. Commercially produced milk commonly undergoes Pasteurization to check bacterial growth and Homogenization for uniformity. Dried (powdered) milk and concentrated milk have been in use since the mid 19th cent. Concentrated milk may be condensed (sweetened) or evaporated (unsweetened). Skim milk, valuable in fat-free diets, is low in vitamin A.

Milk of Magnesia. Common name for the chemical compound magnesium hydroxide [$Mg(OH)_2$]. The viscous, white, mildly alkaline mixture used as a medicinal antacid and laxative is a suspension of about 8% magnesium hydroxide in water.

Milk Sugar. *See* lactose.

Milky Way. The broad glowing band of faint stars, very conspicuous on a dark moonless night, that arches across the sky in a great circle. It is an effect caused by looking along the equatorial plane of our Galaxy, where the stars appear to be more densely packed. The *Milky-Way system* is another name for the Galaxy.

Milky Way. Large spiral Galaxy containing about 100 billion stars, including the sun. It is characterized by a central nucleus of closely packed stars, lying in the direction of the constellation Sagittarius, and a flat disk marked by spiral arms. Seen edgewise as a broad band of light arching across the night sky from horizon to horizon, the Milky Way passes through the constellations Sagittarius, Aquila, Cygnus, Perseus, Auriga, Orion, and Crux. The disk is c. 100,000 light-years in diameter and on the average 10,000 light-years thick (increasing up to 30,000 light-years at the nucleus). A thin halo of star Clusters surround the galaxy. The sun is c. 30,000 light-years from the nucleus and takes 200 million years to revolve once around the galaxy.

Millet. Common name for several plants of the Grass family cultivated mainly for cereals in the Old World and forage and hay in North America. The main varieties are foxtail, pearl, Barnyard, and proso (also called broomcorn and hog) millets. Proso millet is the chief cereal in parts of India,

Africa, and the USSR; the chief millet in the U.S. is foxtail millet.

Millikan, Robert Andrews. 1868-1953, American physicist and educator; b. Morrison, Ill. He taught (1896-1921) physics at the Univ. of Chicago and later (1921-45) was chairman of the executive council of the California Institute of Technology and director of the Norman Bridge Laboratory there. He received the 1923 Nobel Prize in physics for his measurement of the electron's charge and his work on the Photoelectric Effect. He also studied Cosmic Rays (which he named), X-rays, and physical and electric constants.

Millipede. Wormlike segmented Arthropod with two pairs of legs on each body segment except the first few and last. Most temperate species are small and dull in appearance, but tropical millipedes are often brightly coloured. In contrast to the carnivorous Centipedes, which they resemble, millipedes feed mostly on decaying vegetation.

Mimicry. In biology, the advantageous resemblance of one species to another, often unrelated species or to a feature of its own habitat. Mimicry serves to protect the mimic from predators or to deceive its prey *(e.g.,* ant-eating spiders that themselves resemble ants). Although most common among insects, mimicry occurs in both palnts and animals.

Mimic Thrush. Name for exclusively American Birds of the family Mimidae, allied to the Wrens and Thrushes and including the mockingbird, catbird, and thrashers. Mimic thrushes are most numerous in Mexico. They are slim, robin-sized birds with slender, downcurved bills, long tails, and strong legs suited to scratching through dead leaves for insects; they also eat berries and fruit. All are famous for their imitative vocal powers; the mockingbird, the preeminent North American songbird, may mimic some 30 calls in succession.

Mimosa. Tree, shrub, or herb (genus *Mimosa*) of the Pulse family, found mainly in the tropics. Mimosas usually have feathery foliage and rounded clusters of fragrant pink flowers atop the branches; they are grown as ornamentals. Best known is the sensitive plant (*M. pudica*), whose leaves fold up the collapse under stimuli such as touch, darkenss, or drought. The similar and related yellow-flowered Acacia is often sold as mimosa.

Mine. In warfare, a bomb placed in a fixed position, to be detonated by contact, magnetic proximity, or electrical impulse. Land mines, both antitank and antipersonnel, came into wide use in World War II; they were normally equipped with pressure sensors placed slightly above or below ground. Naval mines, known since the 16th cent., were first widely employed in World War I. The modern naval mine is often equipped with sonar or magnetic sensors and is laid on or anchored just below the surface of the sea.

Mineral. Natural inorganic substance having a characteristic and homogeneous chemical composition, definite physical properties, and , usually, a definite crystalline form. A few *(e.g.,* carbon, gold, iron, and silver) are elements, but most are chemical compounds. Rocks are combinations of minerals. Important physical properties of minerals include hardness, specific gravity, cleavage, fracture, luster, colour, transparency, heat conductivity, feel, magnetism, and electrical properties. Minerals originate by precipitation from solution, by the cooling and hardening of magmas, by the condensation of gases or gaseous action on country rock, and by metamorphism. They are of great economic importance in manufacturing; many are valued as Gems.

Mining Extraction of solid Mineral resources from the earth, including Ores (which contain commercially valuable amounts of metals), precious stones, building stones, and solid fuels. Surface mining, open-pit, or open-cut, mining, Strip Mining, and Quarrying are the most common mining methods that start from the earth's surface and maintain exposure to it. Under certain circumstances surface mining can become prohibitive, and underground mining is then considered. The objective of underground mining is to extract the ore below the surface of the earth safely and economically. Entry is through a tunnel or shaft, and the ore is mined in stopes, or rooms. Material left in place to support the ceiling is called a pillar and can sometimes be recovered afterward. A modern underground mine is a highly mechanized operation, using vehicles, rail haulage, and multiple drill units. To protect miners and their equipment, much attention is paid to mine safety, including proper ventilation and roof support. There are a number of other mining methods, including *solution mining,* in which the ore is brought into a liquid solution by a chemical or bacteria and pumped to the surface; *gopher mining,* an old-fashioned method in which small, narrow holes are driven to extract the ore; and *placer mining,* in which gravel, sand, or talus is removed from deposits by hand, hydraulic nozzles, or dredging.

Mink. Semiaquatic carnivorous Mammal (geneus *Mustela),* related to the Weasel and highly prized for its thick, lustrous, rich brown fur. Found in Europe and North America, it has a slender, arched body about 20 in. (51 cm) long and a bushy tail. Minks live near water, where they feed on Muskrats, fishes, frogs, and birds. The mink is widely bred on farms for the fur trade.

Minnow. Any of a large family (Cyprinidae) of freshwater Fish that includes Carp, bream, chub, dace, shiners, and Goldfish. Most minnows are small and drab, but a few species are brightly coloured. Minnows are important in freshwater aquatic life, feeding on insects, larvae, and Crustaceans and in turn serving as food for larger fishes. Superior hearing has given minnows the nickname "hearing aid" fish and accounts for their characteristic wariness.

Minoan Civilization. (Mino'en), a Bronze Age Aegean Civilization that flourished in Crete. It is divided into three periods. The early Minoan period (C. 3000-2200 B.C.) saw the rise of the culture from a neolithic state, with the tentative use of bronze and the appearance of hieroglyphic writing. The Middle Minoan period (C. 2200-1500 B.C.) was the high point: the great palaces appeared at Cnossus and Phaestus, culture thrived, and linear writing was used. The Late Minoan period (C. 1500-1000 B.C.) faded out in poverty and obscurity, and the cultural center passed to the Greek mainland.

Minor Axis. *See* ellipse.

Minority Carrier. The type of carrier responsible for transporting less than one half of the current in a semiconductor.

Minor planet. *See* asteroid.

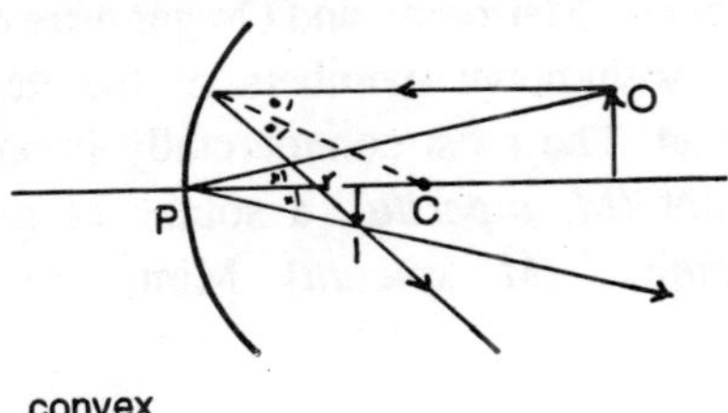

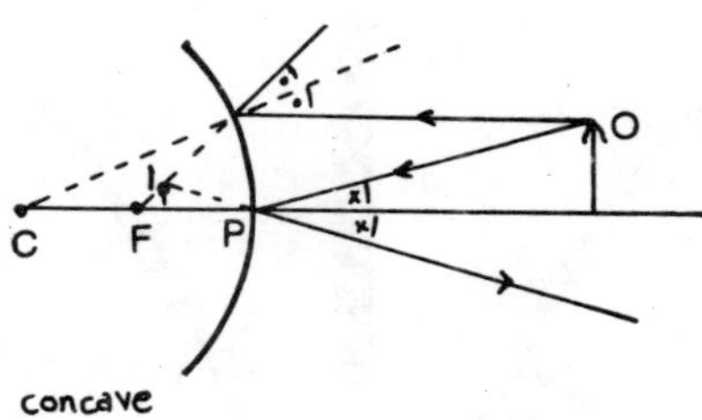

Fig. M-8. Ray diagram.

Mint. Place where legal coinage is manufactured. The name is derived from the temple of Juno Moneta [Latin, = *mint*] in Rome, where silver coins were made as early as 269 B.C.

Mints existed earlier elsewhere, as in Lydia and Greece. The first U.S. mint was established in Philadelphia in 1792. The U.S. Bureau of the Mint now maintains mints in that city, and in Denver and San Francisco.

Mint. In botany, common name for the Labiatae, or the Lamiaceae, a large family of chiefly annual or perennial herbs, distributed worldwide but most common in the Mediterranean region. The family is typified by square stems, paired opposite leaves, and white, red, blue, or purple flowers. The aromatic Essential Oils in the plants' foliage are used in perfumes, flavorings, and medicines. The true mints (genus *Mentha),* Sage, lavender, and Rosemary are important sources of essential oils; these and Basil, Thyme, Marjoram, and Oregano are common kitchen herbs. Other wellknown members of the family are Catnip and Horehound. The most commercially important true mints are peppermint *(M. piperita),* a source of menthol, and the milder spearminth *(M. spicata).* Mints are also grown as ornamentals.

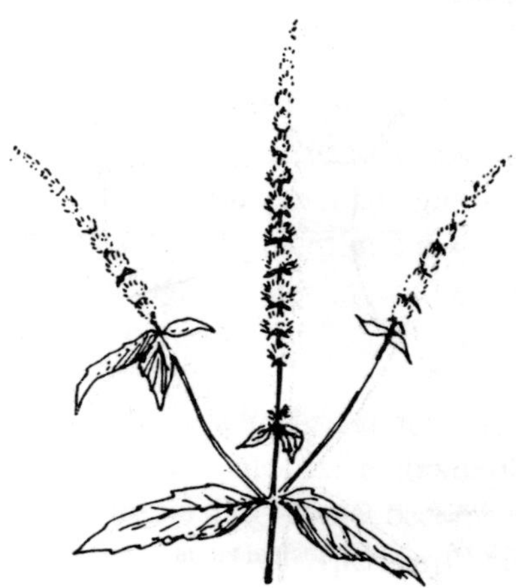

Fig. M-9. *Mint: Spearmint,* Mentha spicata, *a plant of the mint family.*

Mirror. A surface that reflects large amounts of light. A *plane* *mirror* forms a virtual image with the same size as the object. In a plane mirror the image has lateral inversion and is as far behind the mirror as the object is in front. Curved mirrors are usually spherical, being either *convex* or *concave* depending on whether the reflecting surface lies on the convex or concave side of the sphere. The *centre of curvature* of the mirror is simply the centre point of the sphere of which the reflecting surface forms a part. A line from this to the centre point of the mirror is the *optical axis* and the centre point itself is the *pole.* In a concave mirror, rays of light parallel to the axis are focused to a point on the axis: in a convex mirror they appear to diverge from a point on the axis. The point, in both cases, is the mirror's *principal focus* or *focal point.* The distance from the pole to this point is the *focal length.* Spherical mirrors form real or virtual images according to the equation $1/v + 1/u = 1/f = 1/r$, where u is the object distance, v the image distance, f the focal length, and r the radius of the sphere. All distance are measured from the pole, being taken as positive when in front of the mirror and negative when behind it. Spherical mirrors suffer from spherical aberration and in optical telescopes parabolic mirrors are used.

Mirror Image. An image of an object as it would appear in a plane mirror. If one object is the mirror image of the other, it is identical in all respects except that the two could not be superimposed. A left hand is the mirror image of a right hand.

Misch Metal. An alloy of cerium (about 25%) and various other lanthanide elements. It is a pyrophoric material, hence its use in lighter flints.

Miscible. Denoting two or more liquids that can be mixed together completely to form of homogeneous mixture.

Mispickel. *See* arsenopyrite.

Mist. A fine suspension of a liquid in a gas.

Mitscherlich's Law. The principle that substances that crystallize together in mixed crystals have similar chemical compositions. Perchlorate salts (such as $KClO_4$) are often isomorphous with the corresponding permanganate ($KMnO_4$). [After Eilhardt Mitscherlich (1794-1863), German chemist].

Mistletoe. Common name for the Loranthaceae, a family of chiefly tropical parasitic herbs and shrubs with leathery leaves and waxy white berries. Mistletoes, aerial hemiparasites with green leaves that carry out Photosynthesis, attach themselves to their hosts by modified roots called haustoria. They are widely associated with folklore and are used as Christmas decorations. The mistletoe commonly sold in the U.S. is *Phoradendron flavescens;* most popular in Europe is the "true" mistletoe (*Viscum album).*

Mite. Small, often microscopic Arachnid. Mites are often parasites of animals and plants and infest stored foodstuffs. Some burrow into the skin of mammals, causing Mange and scabies. Chiggers, which are the larvae of harvest mites, transmit the organism that causes scrub typhus. The larger *tick,* which is related to the mite, is usually a parasite of birds and mammals, embedding its entire head under the skin and sucking the host's blood. Ticks transmit many diseases, including Rocky Mountain Spotted Fever.

Mitosis. Process of nuclear division in a living cell by which the hereditary carriers, or Chromosomes, are exactly replicated, the two parts being distributed to identical daughter nuclei. In mitosis each cell formed receives chromosomes that are alike in composition and equal in number to the chromosomes of the parent cell. Mitotic division occurs in somatic (body) cells; in sex cells (Ovum and Sperm) Meiosis (halving of the number of chromosomes) also takes place.

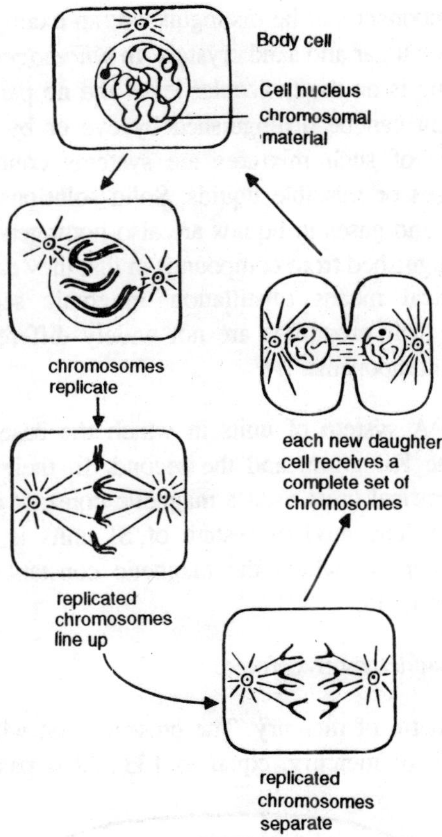

Fig. M-10. Mitosis in the body cell of an animal.

Mixed Crystal. A crystalline compound formed of a mixture of two or more simple compounds that have crystallized together in a homogeneous solid in definite proportions. Mixed crystals are solid solutions.

Mixture. A system containing two or more substances in which one substance is dispersed through the other and there is no chemical bonding between the different components. A *heterogeneous mixture* is one in which small amounts of each component can be distinguished: an example would be a mixture of sugar and sand crystals. In a *homogeneous mixture,* the mixing is on the molecular level and no particles of either component can be distinguished by eye or by a microscope. Examples of such mixtures are systems containing two or more gases or miscible liquids. Solid solutions and solutions of solids and gases in liquids are also homogeneous. Mixtures and distinguished from compounds in that they can be separated by physical means (distillation, magnetic separation, etc.) and that their properties are not widely different from those of their components.

MKS Units. A system of units in which the base units are the metre, the Kilogram, and the second. In their original form MKS electrical units used a magnetic constant of 10^{-7} henries per metre. The modern system of SI units is a rationalized MKS system in which the magnetic constant is $4\pi \times 10^{-7}$ henry per metre.

m.m.f. *See* magnetomotive force.

mmHg. Millimetre of mercury. The pressure that will support one millimetre of mercury, equal to 133.322 4 pascals.

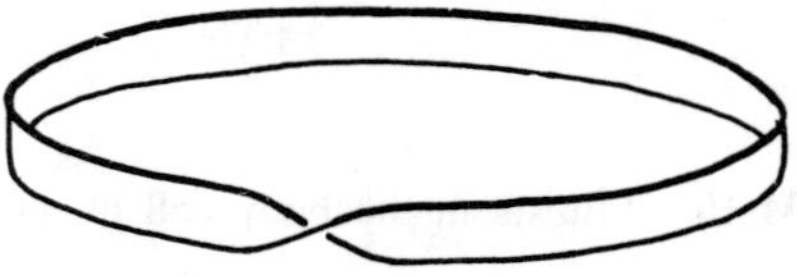

Fig. M-11. Mobius strip.

Mobius Strip. A surface made by twisting a band through 180° and joining the ends together. The Mobius strip is topologically interesting in that it is a single surface with a single bounding curve. [After August Ferdinand Mobius (1790-1868), German mathematician.]

Mode. 1. In music, any pattern or arrangement of the intervals of a Scale. In the Middle Ages eight modes each in the range of an octave (derived from ancient Greek theory), developed as the basis of Plainsong composition. They were grouped in four pairs, each pair containing an authentic mode and (at the interval of a fourth below that) a plagal mode. These modes were the basis of musical composition for 11 centuries. In the late 16th and early 17th centuries the series was limited to the major and minor modes in use today. The use of medieval modes by later composers such as Vaughan Williams is called modality in contrast to Tonality.

2. In the 13th cent., six rhythmical patterns in ternary meter that governed composition.

3. In 20th cent. music, any of four forms of the tone row, an arbitrary arrangement of the 12 equal chromatic tones of the diatonic scale of Western music.

Moderator. A substance used in fission reactors to slow down fast neutrons. Moderators are usually light materials, such as graphite and heavy water. The fast neutrons lose energy by colliding with nuclei of the moderator atoms and are then more likely to cause fission of heavy nuclei.

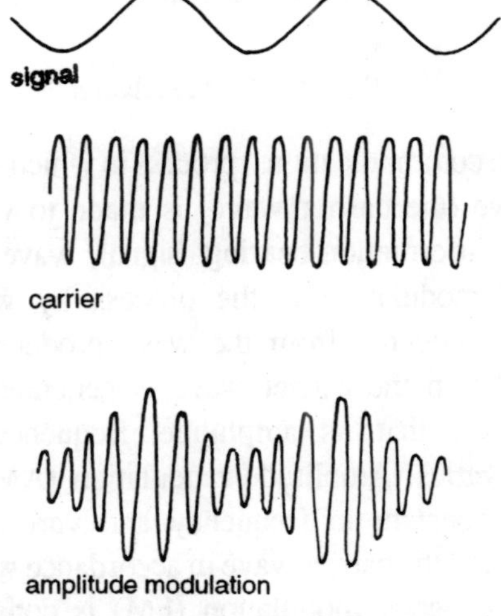

Fig. M-12. Ampitude modulation.

Modulation. The process of changing the waveform of one wave by combining it with another wave. Modulation is used in radio transmission to vary the characteristics of a high-frequency carrier wave by impressing the characteristic of the signal upon it. In *amplitude modulation*— the commonest method— the amplitude of the carrier wave is varied. *Frequency modulation* involves changing the frequency of the carrier wave so that it varies in proportion to the amplitude of the signal. *Phase modulation* involves changing the phase of the carrier wave so that its difference from the unmodulated wave is proportional to the signal amplitude.

amplitude modulation (AM)

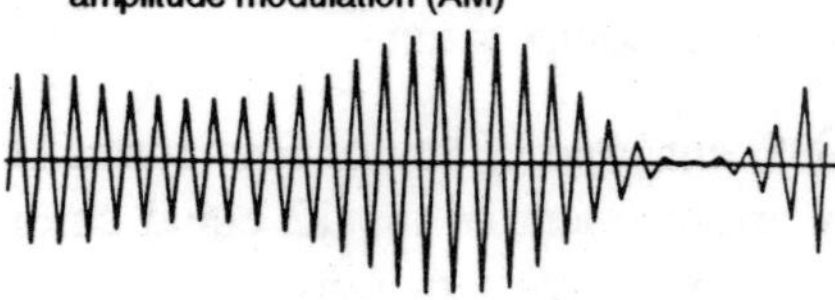

frequency modulation (FM)

Fig. M-13. Modulation.

Modulation. In communications, process in which some characteristic of a Wave (the carrier wave) is made to vary in accordance with an information-bearing signal wave (the modulating wave); demodulation is the process by which the original signal is recovered from the wave produced by modulation. In modulation the carrier wave is generated or processed in such a way that its amplitude, frequency, or some other property varies. Amplitude modulation (AM), widely used in radio, is constant in frequency and varies the intensity, or amplitude, of the carrier wave in accordance with the modulating signal. Frequency modulation (FM) is constant in amplitude and varies the frequency of the carrier wave in such a way that the change in frequency at any instant is proportional to another time varying signal. The principal application of FM is also in radio, where it offers increased noise immunity and greater sound fidelity at the expense of greatly increased bandwidth. In pulse modulation the carrier wave is a series of pulses that are all of the same amplitude and width and are all equally spaced. By controlling one of these three variables, a modulating wave may impress its information on the pulses. In pulse code modulation (PCM) it is the presence or absence of particular pulses in the carrier steam that constitutes the modulation.

Modulation. In music, shift in the Key center of a composition, a means of achieving variety in use since the late 15th cent. In modulating from one key to another, a chord common to both keys is used as a pivot chord. If there is no chord common to the two keys, the passage may move through several keys before the desired modualtion is effected.

Modulus (elastic modulus). The ratio of the stress applied to a body to the strain produced for a body obeying Hooke's law. The modulus of elasticity is a property of the material and there are several types depending on the type of applied stress. *Young's modulus* (symbol: E) is the ratio of tensile stress to tensile strain. The *bulk modulus* (symbol: K) is the ratio of the volume stress (or pressure) to the bulk strain. The *rigidity modulus* (symbol: G) is the ratio of the shear stress to the shear strain.

Mohs' Scale. A scale of hardness on which a mineral is compared with ten reference minerals given arbitrary numbers. The members of the scale, in ascending hardness, are : 1. talc, 2. gypsum, 3. calcite, 4. fluorite, 5. apatite, 6. orthoclase, 7. quartz, 8. topaz, 9. corundum, 10. diamond. Each member of the scale is capable of scratching a member with a lower number, and any other mineral can be assigned a number by carrying out a similar scratch test. For example, a mineral that is scratched by corundum and not by topaz has Mohs hardness 8-9. [After Friedrich Mohs (1773-1839), German mineralogist.]

Molality. The concentration of a solution in moles of solute per kilogram of solution *(i.e.,* 1 kilogram is the mass of solute plus solvent). *Compare* molarity.

Molar. 1. Indicating that a specified physical quantity is measured for unit amout of substance. For example, the *molar heat capacity* (C_m) is the heat capacity of unit amount of substance and is measured in joules per kelvin per mole. Molar quantities are properties of the substance or material rather than properties of objects or systems.

2. Denoting a solution that contains one mole of a specified substance per cubic decimetre of solution. A solution containing *x* moles per cubic decimetre is said to be *x molar*—written *x*M.

Molar Volume. Symbol: V_m. The volume occupied by one mole of a given substance under specified conditions. An ideal gas at standard temperature and pressure has a molar volume of 22.415 cubic decimetres and real gases aproximate to this value.

Mole. Any of the small, burrowing, insectivorous Mammals of the family Talpidae of the Northern Hemisphere. About 6 in. (15.2 cm) long, moles have pointed muzzles and powerful, clawed front feet for tunneling. Their eyes are covered with fur and they have no external ears, but their senses of hearing, smell, and touch are acute. They eat half their weight daily in worms, insects, and small animals. Moles are trapped as pests and for their fur.

Mole. Symbol : mol. The basic Sl unit of amount of substance equal to the amount of substance that contains the same number of entities as there are atoms in 0.012 kilogram of carbon-12. The entities may be atoms, molecules, ions, electrons, or similar elementary units. One mole of any substance contains *N* entities, where *N* is Avogadro's number ($6.022\ 52 \times 10^{23}$). One mole of a compound is equivalent to *M* grams, where *M* is the molecular weight.

Mole. In chemistry, a quantity of particles of any type equal to Avogadro's number (6.02252×10^{23}). One gram-atomic weight (or one gram-molecular weight)—the amount of an atomic (or molecular) substance whose weight in grams is numerically equal to the Atomic Weight (or Molecular Weight) of that substance—contains exactly one mole of atoms (or molecules). For example, one mole, or 12.011 grams, of carbon contains 6.02252×10^{23} carbon atoms, and one mole, or 180.16 grams, of glucose ($C_6H_{12}O_6$) contains the same number of glucose molecules.

Molecular Beam. A beam of atoms, ions, molecules, or free radicals in which all the particles in the beam are moving in the same direction and there are few intermolecular collisions. Molecular beams are formed by allowing gas or vapour to escape through a hole in an enclosure and collimating the beam with several apertures, using vacuum pumps to remove molecules that do not pass through the apertures. They are used in research into chemical reactions, spectroscopy, and surface studies.

Molecular crystal. *See* crystal.

Molecular Flow ***(Knudsen flow).*** Flow of a gas in which the mean free path of the gas molecules is large compared with the dimensions of the pipes or containers. Under these conditions, which occur at low pressures, the molecules make very few intermolecular collisions, most of their collisions being made with the walls of the container. The flow characteristics then depend on the molecular weight of the gas rather than its viscosity.

Molecular Orbital. An orbital for an electron in a milecule. In the molecular-orbital theory of valence, electrons are considered to move in the combined field of all the nuclei in the molecule. In atoms the electrons are spin-paired and occupy atomic orbitals. Similarly in molecules they occupy a number of molecular orbitals, which have definite energies and are visualized as regions in space. This approach to valence theory is distinguished from the valence-bond approach, which uses the idea of definite linkages between atoms. An idea of the possible types and shapes of molecular orbitals is gained by assuming that they result from overlap of the atomic orbitals. The hydrogen molecule, for instance, can be formed by overlap of the 1*s* orbitals, leading to a molecular orbital containing two electrons. The electrons in the orbital can be found anywhere in the molecule but there is a maximum of electron density mid-way between the atoms, and this holds the atoms together by electrostatic attraction. Two types of molecular orbital are distinguished: pi orbitals and sigma orbitals.

Molecular Weight. Symbol: *M* The ratio of the average mass per molecule of a compound to one twelfth of the mass of an atom of the isotope carbon-12. The molecular weight is the sum of the atomic weights of the atoms in the molecule.

Molecular Weight. Weight of a Molecule of a substance expressed in atomic mass units. The molecular weight is the sum of the atomic weights of the atoms making up the molecule.

Molecule. The smallest amount of a chemical compound that is capable of independent existence. In covalently bonded liquid or gaseous compounds the molecules are small groups of atoms held together by covalent bonds. Thus, a molecule of carbon dioxide, CO_2, has a carbon atom bound to two oxygen atoms. Discrete molecules can also be distinguished when such compounds from molecular crystals. In electrovalently bonded substances discrete molecules do not exist. A crystal of sodium chloride is simply a collection of sodium ions and chloride ions. In such compounds the molecule is considered to be the smallest group of atoms characteristic of the substance; in the case of sodium chloride it is NaCl. Similarly, a covalent crystal, such as boron nitride (BN), does not have distinguishable simple molecules because the covalent bonds extend throughout the whole crystal structure.

Molecule. Smallest particle of a Compound that has all the chemical properties of that compound. Molecules are made up of two or more Atoms, either of the same Element or of two or more different elements. Ionic compounds, such as common salt, are made up not of molecules but of ions arranged in a crystalline structure. Unlike ions, molecules carry no electrical charge. Molecules differ in size and Molecular Weight as well as in structure.

Mole Fraction. The ratio of the number of moles of a specified substance in a mixture to the total number of moles of mixture: *i.e.* if n_A moles of A are mixed with n_B of B, the mole fraction of A is $N_A/(n_A + n_B)$.

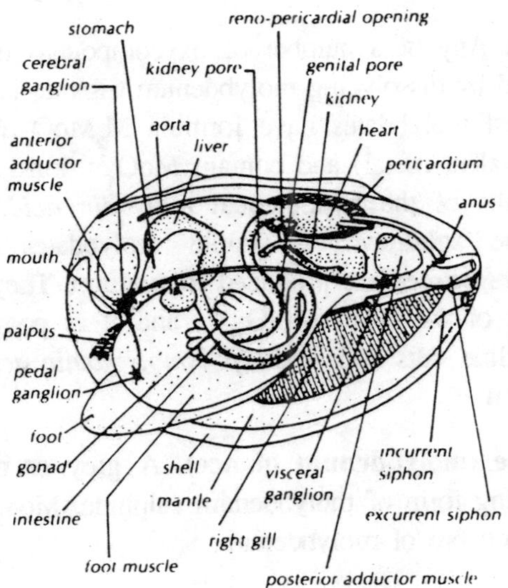

Fig. M-14. *Mollusk:* Internal anatomy of a clam, a mollusk of the class Pelecypoda.

Mollusk. Animal in the phylum Mollusca, the second largest invertebrate phylum. Mostly aquatic, mollusks have usually soft, unsegmented bodies enclosed in a shell; in some forms the shell is internal, and in a few it is absent. An organ called the mantle secretes the substance that forms the shell. A muscular foot under the body is used for locomotion. Certain mollusks, such as Clams, Oysters, and Scallops, are important food sources, and mollusk shells are highly valued by collectors. Mollusks include Gastropods, or univalves, Bivalves, Cephalopods (Octopuses and Squids), and Chitons.

Molting. Periodic shedding and renewal of the outer skin, exoskeleton, fur, or feathers of any animal. Most birds molt annually; development in young birds is marked by a succession of molts during the first year. Arthropods *(e.g.,* insects and crustaceans) molt their exoskeletons in order to grow; the process involves partial digestion of the old cuticle, emergence of the animal from the old covering, and hardening of the new cuticle. Amphibians and snakes molt a few times a year. Mammals change from heavy winter to light summer coats.

Molybdate. Any of a number of oxycompounds of molybdenum formed by dissolveing molybdenum trioxide in an alkali. The simplest molybdates have formula M_2MoO_4 (where M is a monovalent metal) and contain MoO_4^{2-} ions. Formally, they are salts of the hypothetical *molybdic acid,* H_2MoO_4. If alkaline solutions of simple molybdates are acidified, *polymolybdates* are produced in solution. They contain large anions of the type $(Mo_7O_{24})^{6-}$, and it is possible to obtain crystalline salts and acids *(polymolybdenic acids)* from such solution.

Molybdenite (molybdenum glance). A grey or black naturally occurring form of molybdenum sulphide, MoS_2, important as the main ore of molybdenum.

Molybdenum. Symbol: Mo. A very hard silvery white transition element found chiefly in molybdenite (MoS_2), from which it can be extracted by roasting to form the oxide and reduction of this with hydrogen. The main uses of molybdenum are in making alloy steels and other molybdenum compounds. The element is chemically unreactive, being inert to most acids and rendered passive by nitric acid. It is oxidized at red heat and it dissolves in molten alkalis. The chemistry of molybdenum is very complicated: it has a variable valency and forms numerous oxides and oxycompounds. A.N. 42; A.W. 95.94; m.pt. 2610°C; b.pt. 5560°C; r.d. 10.22; valency 2-6.

Molybdenum. (Mo), metallic element, recognized as a distinct element by Karl Scheele in 1778. Hard, malleable, ductile, silver-white, and high-melting, it is used in X-ray and electronic tubes, electric furnaces, and certain rocket and missile parts. It is a hardening agent in Steel alloys. Molybdenum disulfide is used as a lubricant in spacecraft and automobiles.

Moment. The turning effect of a force or system of forces about an axis. A single force has a moment equal to the product of the force and the perpendicular distance from the axis to the force's line of action. The moment of a system of forces is the algebraic sum of the individual moments. The moment of a force is the same as its torque.

Moment of Inertia. Symbol : *I*. The sum of the products of the masses of particles of a body and the squares of their distances from an axis of rotation. A single point mass m a distance r from an axis has a moment of inertia equal to mr^2. The moment of inertia of a system of such masses is the sum of these products: in the case of a rigid body an integral is used. Moment of inertia is used in place of mass in problems involving rotation. Thus, angular momentum is $I\omega$ and angular kinetic energy is $I\omega^2/2$, where ω is angular velocity.

Momentum. 1. Symbol : *p*. The product of a particle's mass and its velocity.

2. (angular momentum). Symbol : *L*. The product of the moment of inertia of a rotating body and its angular velocity.

The total momentum (linear or angular) is constant in any system. The principle is the *law of conservation of momentum.*

Momentum. In mechanics, the quantity of Motion of a body. The linear momentum of a body is the product of its mass and velocity. The angular momentum of a body rotating about a point is equal to the product of its mass, its angular velocity, and the square of the distance from the axis of rotation. Both linear and angular momentum of a body or system of bodies are conserved if no external force acts on it or them.

Monatomic. Composed of single atoms. Helium, argon, and other inert gases are monatomic.

Monazite. A yellowish-brown mineral consisting of a mixed phósphate of various lanthanide elements together with thorium silicate.

Monazite. Phosphate mineral [(Ce, La, Y, Th) PO_4], found in the form of sand in the U.S., Madagascar, Brazil, India, Sri Lanka, and Australia. Monazite is an important source of Cerium, Thorium, and other Rare-Earth Metals.

Mond Process. An industrial process for purifying nickel by heating the impure metal in carbon monoxide at 50°C. The volatile carbonyl is decomposed at 180°C. [After Ludwig Mond (1839-1909), British industrial chemist born in Germany.]

Mongoose. Small, carnivorous Mammal of the civet family, found in S Asia and Africa, with one species extending into S. Spain. Typical mongooses (genus *Herpestes)* are weasellime in appearance, with long, slender bodies and pointed faces. They range in length from 1½ to 3½ ft (45 to 106 cm). The Indian gray mongoose (*H. edwardsi)* is known for its ability to kill snakes, including Cobras. Mongooses are fierce hunters, and because of their destructiveness it is illegal to import them into the U.S., even for zoos.

Monkey-puzzle Tree. Evergreen tree (*Araucaria Araucana*) native to Chile, widely cultivated as an ornamental. The symmetrical branches have an unusual angularity and are completely covered by stiff, overlapping leaves. The monkey-puzzle tree and the related Norfolk Island pine (*A. excelsa*) and bunya-bunya (*A. bidwillii*) are good timber trees. Species of *Araucaria* form the dominant vegetation of the coniferous forests of Chile and S Brazil.

Monochromatic. Denoting light or other electromagnetic radiation that has only one wavelength; not polychromatic.

Monochromator. A device for producing monochromatic electromagnetic radiation, usually a spectrometer used to select one frequency of radiation from a polychromatic source.

Monoclinic. Crystal system.

Monohydrate. A solid compound with one molecule of water of crystallization per molecule of compound, as in sodium carbonate monohydrate, $Na_2\ CO_3.H_2O$.

Monohydric. Denoting an alcohol or phenol containing one hydroxyl group per molecule.

Monolayer. A unimolecular layer of atoms or molecules.

Monomer. A simple compound distinguished from a dimer, trimer, or polymer.

Mononucleosis, Infectious. Acute infectious disease of older children and young adults, occurring sporadically or in epidemic form. The causative organism is thought to be an airborne herbes virus. Symptoms include fever, enlarged spleen (in about half the cases), sore throat, and extreme fatigue. Hepatitis is common. Therapy includes bed rest and the treatment of symptoms.

Monorail. Railway system whose cars run on a single rail, either above it or suspended beneath it. Driving power is transmitted from the cars to the track by means of wheels that rotate horizontally, making contact with the rail between its upper and lower flanges. Short-run monorails have been built in Huston and Seattle, and, most notably, at Disneyland, in Anaheim, Calif.

Monosaccharide. A sugar that cannot be hydrolysed to simpler sugars.

Monosodium Glutamate. Sodium hydrogen glutamate.

Monotreme. Name for members of the primitive mammalian order Monotremata, found in Australia, Tasmania, and New Guinea. The only members are the Platypus and several species of Echidna. They are unique among Mammals in laying eggs instead of bearing live young. Certain skeletal features resemble those of Reptiles, from which the monotremes evolved. Adults are toothless; males have spurs connected to poison glands on their hind feet.

Monotropic. Denoting a substance that exists in only one crystalline form.

Monovalent (univalent). Having a valency of one.

Monsoon. Wind that changes direction with the seasons. Monsoons are the result of differing air pressures caused by the varied heating and cooling rates of continental land masses and oceans. Winter monsoons associated with India and Southeast Asia are generally dry; summer monsoons in those regions are extremely wet.

Montage. The art and technique of motion-picture editing in which contrasting shots or sequences are used impressionistically to affect emotional or intellectual responses.It was developed creatively after 1925 by Sergei Eisenstein.

Month. Time required for the Moon to orbit once around the earth. The *sidereal month,* or time needed for the moon to return to the same position relative to a fixed star, averages 27 days 7 hr 43 min 12 sec; the *synodic month,* or time needed for the moon to go through its complete cycle of Phases, averages 29 days 12 hr 44 min 3 sec. For the month's harmony with the solar calendar. Since ancient times certain lucky stones or birthstones have been connected with the months; they are often given as follows: *January* [from the god Janus]: Garnet; *February* [from Lat.,=expiatory, because of ancient rites]: Amethyst; *March* [from the god Mars]: Bloodstone or Aquamarine; *April*: Diamond or Sapphire; *May*: Agate or Emerald; *June* [from the gens *Junius*]: Pearl; *July* [From Julius Caesar]: Ruby or Onyx; *August* [from Augustus]: Carnelian or peridot; *September* [from Lat.,=seven; formerly the 7th month]: chrysolite or sapphire; *October* [eight]: Beryl, Tourmaline, or Opal; *November* [nine]: Topaz; *December* [ten]: ruby, Turquoise, or zircon.

Moon. 1. The earth's only natural satellite and, owing to its proximity, the brightest object in the night sky. Its size (diameter: 3476 kilometres; mass: 1/81 that of the earth; mean relative density: 3.34) may be compared to that of the planet Mercury. Though a solid body like the terrestrial planets, the moon seems to have no central core. Yet its chemical composition has recently been found to be not altogether dissimilar to that of the earth and chemical and physical data collected first by unmanned and later by manned spacecraft during the 1960s have led scientists to believe that the moon was formed either simultaneouly with or only shortly after the earth as part of the same planet-forming process. The moon orbits the earth once every 27.322 days (sidereal period of revolution) at a mean distance of 384 400 km (perigee: 363 000 km; apogee: 406 000 km). Its rotating is captured: it makes one rotation in the same time as it takes to make one orbit and therefore always keeps the same face turned towards us. However, because of an irregularity in the distribution together with the varied velocity of its orbit (faster at perigee and slower at apogee) more than half the moon's surface is visible over a period of time. This phenomenon is called *libration.*

In general, the moon is a cheerless place; rocky, barren, pitted with craters, it has high mountain ranges and huge flat smooth plains. When Galileo first observed them he called them "seas" and the Latin word for sea, *mare* (pl. *maria*) is still applied to these smooth expanses of dark basalt. Lunar craters range in size from tiny pits to huge depressed walled plains hundreds of kilometres across. The moon's smallness means that its gravitation and escape velocity are much lower than those of earth (about 16% and 20% respectively). As a result, any atmosphere that it may have once possessed has almost completely vanished. The fact that the moon's orbit is inclined at some 5° to the ecliptic ensures that lunar and solar eclipses are rare. Because of the gravitational attraction of the earth, the nodes of the lunar orbit (points of intersection with the ecliptic) move in a westerly direction, completing one revolution every 18.6 years (the *saros*). But the moon's gravity also has a crucial effect on the earth, since it causes the tides. One of the moon's most characteristic features is that because it reflects sunlight, it shares with the inferior planets the capacity to show phases. The cycle of phases, the synodic month or period, lasts 29.531 days.

2. Any natural satellite of a planet.

Moon. The single natural Satellite of the earth. The lunar orbit is elliptical, and the average distance of the moon from the earth is about 240,000 mi (385,000 km). The moon's orbital period around the earth, and also its rotation period, is 27.322 days. The true angular size of the moon's diameter is about $^{1}/_{2}$, which also happens to be the sun's apparent diameter. This coincidence makes possible total solar Eclipses. The moon's radius is about 1,080 mi (1,740 km); it has about 1/81 the mass of the earth and is $^{3}/_{5}$ as dense. The moon completely lacks both water and atmosphere. It has a rigid crust about 37 mi (60 km) thick, a mantle of denser rock extending down more than 500 mi (800 km), and possibly a small iron core (less than 370 mi/600 km in radius) surrounded by a partly molten zone. The lunar surface is divided into the densely cratered, mountainous highlands and the large, roughly circular, smooth-floored planis called maria.

Moose. Largest member (genus *Alces*) of the Deer family, found in N Eurasia and N North America. The Eurasian species (*A. alces*) is known in Europe as the ELK. The larger American moose is sometimes classed as a separate species (*A. americana*); it has a heavy, brown body with humped shoulders, long, lighter-coloured legs, a thick, almost trunk-like muzzle, and broad, flattened antlers in the male. Moose hunting is strictly regulated.

Mordant. A substance applied to a substrate to allow the attachment of a dye. The mordant is held on the fibres of the substrate and the dye forms a complex with it. Many mordants are inorganic compounds; the most widely used are chromium salts which are employed, together with certain acid dyes (*chrome dyes*), in colouring wool. The colour of the dye is affected by the nature of the mordant: alizarin, for instance, gives a red colour when mordanted with aluminium compounds and a dark purple with iron compounds.

Moraine. Rock and soil debris carried and finally deposited by a Glacier. A lateral moraine is the material that falls onto a glacier's edges from valley cliffs. A ground moraine is the debris deposited by a melting glacier. A terminal moraine is the debris left at the edge of a glacier's extreme forward movement. The great ice sheets of the Pleistocene epoch left terminal moraines stretching across North America and Europe.

Morphine. A white crystalline poisonous alkaloid, $C_{17}H_{18}NO_3.H_2O$, extracted from opium. Its soluble salts are used in medicine as narcotics.

Morphine. Highly addictive Narcotic derivative of Opium used for the relief of pain. Morphine suppresses anxiety and alters the perception of pain, thereby producing euphoria. It also impairs mental and physical performance, reduces sex and hunger drives, and induces apathy. Its use is strictly regulated.

Morse, Samuel Finley Breese. 1791-1872, American inventor and noted portraitist; b. Charlestown, Mass. After spending 12 years perfecting his own version of Andre Ampere's idea for an electric Telegraph, Morse demonstrated (1844) the practicability of his device to Congress and subsequently won world fame. Because many phases of the invention had been anticipated by others, his originality as the inventor of telegraphy has been questioned. Morse later experimented with submarine cable telegraphy.

MORSE CODE					
A	.-	J	.—	S	...
B	-...	K	-.-	T	-
C	-.-.	L	.-..	U..-	
D	-..	M	—	V	...-
E	.	N	-.	W	.—
F	..-.	O	—-	X	-..-
G	—.	P	.—.	Y	-.—
H	....	Q	—.-	Z	—..
I	..	R	.-.		
1	.——	5	.....	9	——.
2	..—	6	-....	0	——-
3	...—	7	—...	Period	.-.-.-
4	....-	8	——..	Comma	—..

Morse Code. (for Samuel Morse). The set of signals used on the Telegraph, or with a flash lamp for visible signaling. The unit of the code is the *dot,* representing a very brief depression of the telegraph key. The *dash* represents a depression three times as long as the dot. Different combinations of dots and dashes are used to code the alphabet, the numerals 1 to 9 and zero, the period, and the comma. American Morse differs considerably from International Morse.

Mortar. In warfare, a short-range weapon that fires a shell on a high trajectory. The name once applied to a heavy Artillery piece but lately has designated a much lighter, muzzle—loaded, smooth-bore Infantry weapon—consisting principally of a tube and a supporting bipod—that fires a fairly heavy projectile in a high arc.

Mosaic. Art of producing surface design by closely inlaying coloured pieces of marble, glass, tile, or semiprecious stone. In the Roman Empire, floors were decorated with mosaics made up of large marble slabs in contrasting colours or of small marble cubes (tesserae). Tessera floors varied from black-and-white geometrical patterns to large pictorial scenes. Glass mosaics were used in early Christian basilicas. The craft reached its height in the 6th cent. at Byzantium (later Constantinople), where the Hagia Sophia was decorated with gold mosaics. The use of gold and of colours produced by metallic oxides later reached the West. In the 5th and 6th cent. Ravenna became the center of Western mosaic art. A revival in Italy (11th—13th cent.) produced such mosaics as those in St. Mark's Church, Venice. Mosaic was also used in Russia, particularly in Kiev. The advent of Fresco decoration in 14th—cent. Italy caused mosaic art to decline. In the 19th cent., the Gothic revival produced modern attempts at mosaics, *e.g.,* those in Westminster Abbey. Mosaics are important in 20th cent. Mexico, where they continue a pre-Columbian tradition. Contemporary examples are also found in Europe, South America, and Israel.

Mosaic. 1. An irregular structure of small crystallites in a crystalline solid.

2. A layer of small particles of light-sensitive material deposited on an insulating support, used in a television camera for converting an optical image into an electrical image.

Moseley's Law. The principle that lines in the X-ray spectra of a set of elements have frequencies that are proportional to the squares of the elements' atomic numbers. The lines must all originate from the same type of transition. For example, the square root of the frequency for characteristic X-rays produced by transition of an electron from an *L* shell to a *K* shell, plotted against atomic number, gives a straight line. [After Henry Gwyn-Jeffreys Moseley (1887-1915), British physicist.]

Mosquito. small, long-legged, winged Insect (order Diptera), related to the Fly. Most females have piercing, sucking mouthparts and feed on the blood of mammals; males feed on plant juices. Eggs are usually laid in stagnant water and the larvae are aquatic. Many diseases, including Malaria, Yellow Fever, and human Encephalitis, are transmitted by certain species of mosquitores.

Moss. Small primitive plant (division Bryophyta, class Bryopsida) typified by tufted growth that is usually vertical. Although limited to moist habitats because they require water for Fertilization and lack a vascular system for absorbing water, mosses are extremely hardy and grow nearly everywhere. The green moss plant visible to the naked eye, seldom over 6 in. (15.2 cm) in height, is the Gametophyte generation, which gives rise to the sporophyte generation Mosses are important in soil formation, filling in surfaces lacking other vegetation, and providing food for certain animals. Sphagnum, or peat moss, is commercially valuable as the main constituent of Peat. Club Moss and Spanish Moss are unrelated to true moss.

Mossbauer Effect. An effect occurring in the gamma-ray emission from the nuclei of certain isotopes in solids. In the Mossbauer effect the gamma-ray photons have a sharply defined energy (normally a spread of energies would occur because of recoil of the nucleus).

The phenomenon is used in a spectroscopic technique for

investigating solids (*Mossbauer spectroscopy*). A gamma-ray source is mounted on a moving platform and a similar sample is placed nearby with a detector to measure the scattered gamma rays. The source is moved towards the sample at a varying speed, thus varying the gamma-ray frequency by the Doppler effect. A minimum in the scattered radiation indicates a resonance absorption of gamma rays by sample nuclei. The method is used for studying nuclear energy levels and for investigating chemical compounds. [After R.L. Mossbauer (b. 1922), German physicist.]

Moth. Any of a large group of Insects that, with the Butterflies, constitutes the order Lepidoptera. Moths have two pairs of wings that function as a single pair and are covered with dustlike scales. Wingspreads range from 1/6 in. (2 mm) to 10 in. (25 cm). Many moths have Protective Coloration matching their background. They are distinguished from butterflies by their stouter, usually hairy bodies and feathery antennae. Most moths are nocturnal and rest with wings outspread; butterflies usually fly by day and rest with wings upraised. Moths undergo a complete Metamorphosis, feeding on leaves or other plant material.

Mother Liquor. The liquid remaining after a substance has been crystallized from a solution.

Motion. In Mechanics, the change in position of one body with respect to another. The study of the motion of bodies is called Dynamics. The time rate of linear motion in a given direction by a body is its *velocity;* this rate is called the *speed* if the direction is unspecified. If during a time t a body travels over a distance s, then the *average speed* of that body is s/t. The change in velocity (in magnitude and/or direction) of a body with respect to time is its acceleration. The relationship between Force and motion was expressed by Isaac Newton in his three laws of motion: (1) a body at rest tends to remain at rest, or a body in motion tends to remain in motion at a constant speed in a straight line, unless acted on by an outside force; (2) the acceleration *a* of a mass *m* by a force *F* is directly proportional to the force and inversely proportional to the mass, or $a = F/m$; (3) for every action there is an equal and opposite reaction. The third law implies that the total Momentum of a system of bodies not acted on by an external force remains constant. Motion at speeds approaching that of light must be described by the theory of Relativity, and the motions of extremely small objects (atoms and elementary particles) are described by quantum mechanics.

Motion-picture Photography or **Cinematography.** Photographic arts and techniques involved in making Motion Pictures. The motion-picture Camera evolved from multi-image stop-action devices that recorded the parts of a continuous movement. D.W. Griffith gave the medium its first cohesive language of camera techniques; his innovations included the close-up, a device by which he heightened the emotional impact of his film. Silent pictures brought cinematographic art to its greatest heights: Expressive devices were developed for silent films, such as cutting (the reorganization of film footage by means of the removal of unwanted frames) and Montage (the creative cutting of images that, when juxtaposed, create a meaning absent from the single images, devised by the Russian director Sergei Eisenstein). The German directors Fritz Lang and F.W. Murnau evolved a highly subjective film style, expressive of psychic and emotional states, using distorted images presented by the camera as if seen from the principal character's vantage point. Von Sternberg combined spectacular sets with soft focus to create a sense of fantasy and mystery. Smooth sound synchronization was achieved in the 1930s. Orson Welles's *Citizen Kane* (1941) was a milestone among sound films, a showcase for numerous technical innovations copied and adapted throughout film's subsequent history. Colour processes, in existence since the 1920s, were perfected in the 1930s and 40s. In the 1950s and 60s various cinematographic gimmicks were developed, *e.g.*, three-dimensional photography, split-screen processes, **Aromarama** (in which audiences were bombarded with scents deemed appropriate to what they were watching), and several wide-screen processes, of which only CinemaScope was extensively used. Film techniques have been greatly refined in the past 50 years, largely through experimentation by gifted cameramen and their assistants. The foremost American cinema tographers include Karl Freund, Gregg Toland, Charles Rosher, James Wong Howe, Lee Garmes, Vilmos Zsigmond, Caleb Deshanel, and Gordon Willis. The French directors of the "new wave" of the 1960s (*e.g.*, Francois Truffaut and Jean-Luc Godard) evolved an influential intimate camera style. A choppy style termed *cinema verite*, characterized by frequent jump cutting, enjoyed a brief vogue in the 1960s in low-budget films such as John Cassavetes's *Shadows* (1960). Intensely personal, highly expressive cinematographic styles are the hallmark of the works of Alfred Hitchcock, Ingmar Bergman (working with the cameramen Gunnar Fischer and Sven Nykvist), Luis Bunuel, Ken Russell, Michelangelo Antonioni, Rainer Werner Fassbinder, Werner Herzog, and Lina Wertmuller.

Motion Pictures. Movie-making as an art and an industry, including its production techniques, its creative artists, and the distribution and displaying of its products. Experiments in photographing movement were made in the U.S. and Europe well before 1900. The first motion pictures made with a single camera were by E.J. Marey, a French physician, in the 1880s. In 1889 Thomas Edison developed the kinetograph, using rolls of coated celluloid film, and the Kinetoscope, for peep-show viewing. The Lumiere brothers, in France, created the Cinematography (1895). Projection machines were developed in the U.S. and first used in New York City in 1896. The first movie theater, a "nickelodeon," was built in Pittsburgh in 1905. Movies developed simultaneously as an art form and an industry. They had enormous immediate appeal, and were established as a medium for chronicling contemporary attitudes, fashions, and events. The camera was first used in a stationary position, then panned from side to side and moved close to or away from the subject. With the evolution of sound films in the late 1920s, language barriers forced national film industries to develop independently. In the U.S. a separation of motion-picture crafts had developed by 1908, and actors, producers, cinematographers, writers, editors, designers, and technicians worked interdependently, overseen and coordinated by a director. Hollywood, Calif., because the American movie capital after 1913. Films were at first sold outright to exhibitors and later distributed on a rental basis. By 1910 the "star system" had come into being. Directors became known for the individual character of their films and were as famous as their stars. During World War I the U.S. because dominant in the industry. In 1927 dialogue was successfully introduced in *The Jazz Singer*. Early colour experiments were achieved by hand-tinting each frame. In 1932 Technicolour, a three-colour process, was developed. The film industry in its heyday (1930-49) was managed by a number of omnipotent studios producing endless cycles of films in imitation of a few successful original types. In those great years Hollywood gave employment to a host of talented actors, *e.g.*, Ingrid Bergman, Humphrey Bogart, Joan Crawford, Bette Davis, Cary Grant, Katharine Hepburn, Spencer Tracy, and John Wayne. In the 1950s the overwhelming popularity of Television began to erode studio profits, necessitating technological innovations such as wide-screen processes, stereophonic sound systems, and three-dimensional cinematography (3-D). By 1956 studios were compelled to produce movies made expressly for television reruns. In the 1960s many film-makers began to work independently of the studio system, producing low-budget films departing from the glamorous, celebrity-packed works of earlier years. Costly and elaborate science fiction productions and horror films attained unprecedented popularity in the late 1970s and early 1980s. Among the great motion-picture directors are D.W. Griffith, Mack Sennett, John Ford, and Alfred Hitchcock in the U.S.; Jean Renoir, Jean-Luc Godard, and Francois Truffaut in France; Ingmar Bergman in Sweden; F.W. Murnau and Rainer Werner Fassbinder in Germany;

Lucino Visconti, Michelangelo Antonioni, and Federico Fellini in Italy; Sergei Eisenstein and Aleksandr Dovzhenko in the USSR; Satyajit Ray in India; and Kurosawa Akira in Japan.

Motor, Electric. Machine that converts electrical energy into mechanical energy. One type of electric motor consists of a conducting loop mounted on a nonconducting shaft. Current fed to carbon brushes enters the loop through two slip rings. A magnetic field around the loop, supplied by an iron-core field magnet, causes the loop to turn when current flows through it. In an alternating-current (AC) motor, the current flowing in the loop is synchronized to reverse direction at the moment when the plane of the loop is perpendicular to the magnetic field and there is no magnetic force exerted on the loop. Because the momentum of the loop carries it around until the current is again supplied, continuous motion results. In AC induction motors, the current passing through the loop does not come from external sources but is induced as the loop passes through a magnetic field. In direct-current (DC) motors, a split-ring commutator switches the direction of the current each half rotation to maintain the shaft's direction of motion. In any motor, the stationary parts constitute the stator, and the loop-carrying assembly is the rotor, or armature.

Mountain. High land mass projecting above its surroundings, usually of limited width at its summit. Some are isolated, but they usually occur in ranges. A group of ranges closely related in form, origin, and alignment is a mountain system; an elongated group of systems is a chain; and a complex of ranges, systems, and chains continental in extent is a cordillera, zone, or belt. Some mountains are remains of Plateaus dissected by erosion. Others are cones of Volcanoes or intrusions of igneous rock that form domes. Fault-block mountains occur where huge blocks of the earth's surface are raised relative to neighbouring blocks. All the great mountain chains are either Fold mountains or complex structures in which folding, faulting, and igneous activity have taken part. The ultimate cause of mountain building has been a source of controversy. The concept of Plate Tectonics, however, is the first reasonable unifying theory, hypothesizing that the earth's crust is broken into several plates that sideswipe each other or collide. Where they collide, compressional stresses are generated along the margin of the plate containing a continent, causing deformation and uplift of the continental shelf and continental rise, where accumulated sediments become complex folded and faulted mountain chains. Mountains have important effects on the climate, population, economics, and civilization of the regions where they occur. Major mountain ranges include the Alps, the Andes, the Caucasus, the Himalayas, the Pyrenees, and the Rocky Mountains. The highest elevation on earth above sea level is the peak of Mt. Everest.

Mountain Laurel. Evergree shrub (*Kalmia latifolia*) of the Heath family, native to E North America. Poisonous to live-stock, mountain laurel has leathery leaves and large clusters of spring-blooming pink or white flowers borne at the ends of the branches. It is the state flower of Connecticut and Pennsylvania. True Laurel is in a separate family.

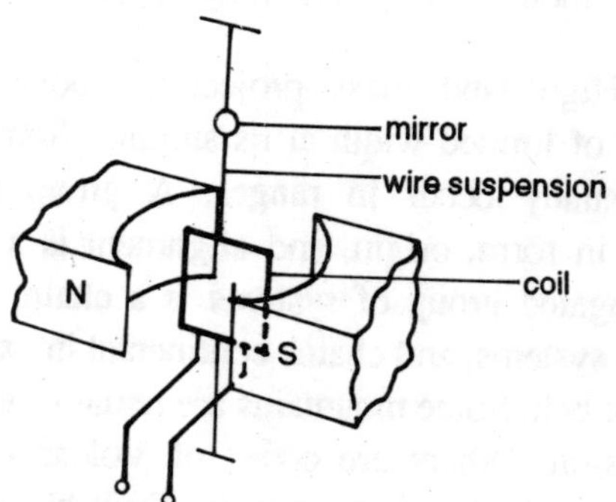

Fig. M-15. Moving-coil Instrument.

Moving-coil Instrument. An electrical measuring instrument consisting of a flat rectangular coil of wire suspended vertically between the curved pole pieces of a permanent magnet. If a current is passed through the coil the vertical sides experience opposite forces, causing the coil to turn. It is restrained by the torsion in a wire suspension or in a flat hair spring: the angle turned depends on the magnitude of the current. It is measured by attaching a small mirror to the suspension wire (in which case the instrument is a moving-coil galvanometer) or by use of a pointer attached to the coil. The pointer instrument is used as an ammeter or, by including a high resistance to reduce the current drawn from the circuit, as a voltmeter.

Moving-iron Instrument. An electrical measuring instrument consisting of a piece of soft iron pivoted so that it can move into a fixed coil of wire. If a current is passed through the coil, the iron is attracted and the piece of iron rotates. It is restrained by the torsion in a spring: the angle turned through depends on the magnitude of the current and is measured by movement of the pointer. The attraction is independent of the direction of the current and the instrument can be used as an ammeter for both alternating and direct currents. If a high resistance is connected in series with the coil, so that little current is drawn from the circuit, it can be used as a voltmeter.

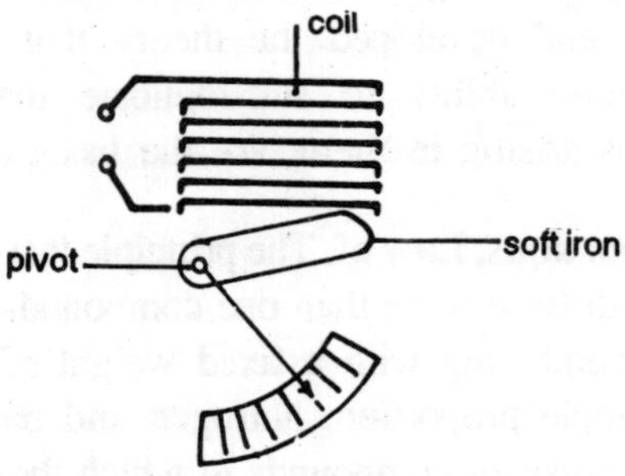

Fig. M-16. Moving-iron Instrument.

Mulberry. Common name for the Moraceae, a family of deciduous or evergreen trees and shrubs, often climbing, mostly of pantropical distribution and typified by milky sap. Several genera bear edible fruit, *e.g., Morus* (Mulberries), *Ficus*

(Figs), and *Artocarpus,* which includes the breadfruit. Both the white (*M. alba*) and the red (*M. rubra*) mulberries are cultivated in North America. Mulberry fruits are tender and juicy and resemble blackberries; the fruit of *M. rubra* is used to make wine. Silkworms feed on Mulberry leaves. The Osage orange (*Maclura pomifera*) is a hardy tree native to the south central U.S. and is a source of a durable wood and a dye. The breadfruit. (*A. ultilis*) is a cultivated, staple food plant in the Pacific tropics and West Indies; its wood, fiber, and latex are also utilized.

Mule. Sterile, hybrid offspring of a male donkey and a female Horse, bred as a work animal. It has long ears, slender legs, small hooves, and a loud bray. Cautious and temperamental, mules are slower but more surefooted than horses and have great powers of endurance. They have been used as pack and draft animals since prehistoric times.

Muller, Hermann Joseph. 1890-1967, American geneticist and educator; b. N.Y.C. He was awarded the 1946 Nobel Prize in physiology or medicine for discovering a technique of artificially inducing Mutations by means of X-rays. He also proposed and developed the theory that genes, because of their unique ability to self-replicate themselves and any alterations arising in them, are the basis of life.

Multiple Proportions, Law of. The principle that when two elements combine to form more than one compound, the weights of one element combining with a fixed weight of the other element are in simple proportion. Nitrogen and oxygen, for example form a number of compounds in which the weights of oxygen combining with 14 g of nitrogen are in the ratio 8 : 16 : 24 : 32 : 48 = 1/2 : 1 : 3/2: 2 : 5/2, corresponding to the oxides N_2O, NO, N_2O_3, NO_2, N_2O_5. The law is one of three laws of chemical combination.

Multiple Sclerosis. Chronic degenerative disease of the central Nervous System in which patches of the myelin sheath around nerve fibers are lost. The cause is unknown. Symptoms include disturbances in vision, speech, balance, and coordination, as well as numbness and tremors. The onset of the disease generally occurs between ages 20 and 40. Although it may result in severe disability, its course varies widely, with symptoms appearing at irregular intervals for years. There is no specific treatment, although drugs can sometimes reduce the severity of the symptoms.

Mumetal (R). An alloy containing nickel (about 75%), iron (about 18%), and copper and chromium. It has a high magnetic permeability and is used in applications requiring a low hysteresis loss and in shielding electrical equipment from magnetic fields.

Mmmy. Human or animal body preserved by embalming or by natural conditions. The word refers primarily to ancient burials found in Egypt. Mummies, embalmed and tightly wrapped, were preserved for over 5,000 years in the dry air of Upper Egypt, making it possible to determine fairly accurately how the great pharaohs appeared in life. Mummification seems to have been performed to prepare the body for reunification with the soul in an afterlife; royal figures, their retinue, and even food were preserved. Similar practices occurred in other parts of the world, *e.g.,* among the Incas. Natural mummification, caused by certain soil and climatic conditions, is seen in bodies found in Danish peat bogs dating from 300 B.C. to A.D. 300.

Mumps or Epidemic Parotitis. Acute contagious viral disease whose symptoms include pain and swelling of the salivary glands, pain on swallowing, and fever. Mumps usually affects children between ages five and fifteen and rarely lasts more than three days. In adults it is often more severe, especially in males, who may experience such complications as pain and swelling of the testes and, infrequently, sterility. Vaccination can prevent the disease in individuals over age one.

Muon. An elementary particle having a positive or negative charge and a mass equal to 206.77 times the mass of the electron. The muon was formerly called the *mu-meson* but is now considered to be a lepton.

Muriate. A chloride of a metal; a salt of muriatic acid. Thus, muriate of potash is potassium chloride, KCl.

Mustard Gas. A colourless oily liquid, $S(CH_2CH_2Cl)_2$, with a pungent odour, made by reacting ethylene with sulphur chloride. It has been used as a war gas. M.pt. 14°C; b.pt. 217°C; r.d. 1.27.

Muscle. Contractile Tissue that effects the movement of the body. Muscle tissue is classified according to its structure and function. Striated, or skeletal, muscle is under conscious control, effecting purposeful movements of limbs and other body parts. It is called striated because microscopic examination reveals alternating bands of light and dark. Smooth muscle, which lines most hollow organs, is involuntary, *i.e.*, regulated by the autonomic nervous system. It produces movement within internal organs such as those of the Digestive System. Cardiac muscle is straited like skeletal muscle but, like smooth muscle, is controlled involuntarily. It is found only in the Heart, where it forms that organ's thick walls. Contraction, thought to be a similar process in all types of muscle, involves the proteins actin and myosin and their arrangement within the muscle tissue.

Muscular Dystrophy. Any of several inherited diseases characterized by progressive weakness and wasting of muscles, believed due to an abnormality in the muscle tissue itself. The most common form, Duchenne, affects boys, beginning with leg weakness before age 3 and progressing rapidly, with death often occurring before age 30. Another form involves primarily facial and shoulder muscles and affects both sexes, usually from adolescence. There is no known treatment.

Mushroom. Fungus characterized by spore-bearing gills on the underside of an umbrella or cone-shaped cap. The term *mushroom* is properly restricted to the plant's above-ground portion, which is the reproductive organ. Once a delicacy for the elite, edible mushrooms are now grown commercially, especially strains of the meadow mushroom *(Agaricus campestris)*. Although mushrooms contain some protein and minerals, they are largely water and hence of limited nutritive value. Inedible, or poisonous, species are often popularly referred to as toadstools; one of the best-known poisonous mushrooms is the death angel (genus *Amanita)*.

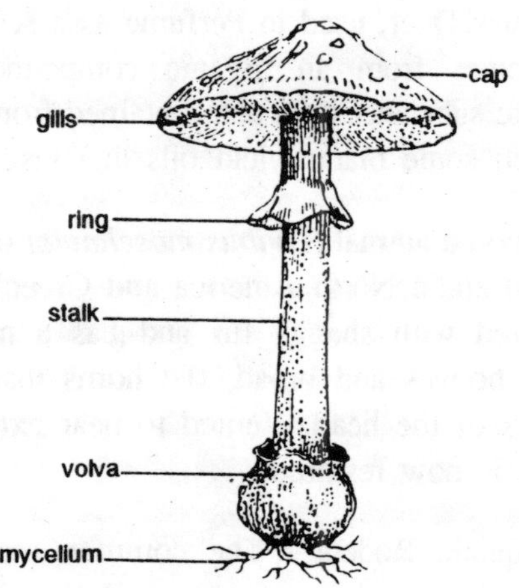

Fig. M-17. *The poisonous mushroom* Amanita.

Musical Notation. Symbols used to make a written record of musical sounds. Boethius applied the first 15 letters of the alphabet to notes in use at the end of the Roman period. By the 6th cent. notation of Gregorian chant was by means of neumes, thought to derive from Greek symbols for pitch; they indicated groupings of sounds to remind a singer of a melody already learned by ear. By the end of the 12th cent. The Bendictine monk Guido d'Arezzo had perfected the staff, placing letters on certain lines to indicate their pitch. These letters evolved into the clef signs used today. In the 15th cent. the shape of notes became round and time signatures replaced coloration to indicate note value. The key signature developed early, although sharps were not used until the 17th cent. The five-line staff (with ledger lines used to extend the range) became standard in the 16th cent. Expression signs and Italian phrases to indicate tempo and dynamics came into use in the 17th cent. Notation for Electronic Music is still not standardized but generally combines traditional symbols with specially adapted rhythm and pitch notation. On notation of lute and keyboard music.

Musk. Odorous substance secreted by an abdominal gland in the male musk Deer, used in Perfume as a scent and fixative. Its odor comes from an organic compound called muscone. Musklike substances are also obtained from the Muskrat and the civet; some plants yield oils that resemble musk.

Musk Ox. Hoofed animal *(Ovibus moschatus)* of the Cattle family; found in arctic North America and Greenland. The musk ox is covered with shaggy fur and has a musky odor. It has splayed hooves and broad, flat horns that downcurve along the sides of the head. Hunted to near extinction in the 19th cent., it is now restored.

Muskrat. Aquatic Rodent. The common muskrats (genus *Ondatra)* are found in marshes, quiet streams, and ponds in most of North America north of Mexico. They have partially webbed feet, a long tail, and shiny brown outer fur (much used commercially) with a dense undercoat. Muskrats are 10 to 14 in. (25 to 36 cm) and weigh 2 to 3 lb (0.9 to 1.4 kg). Muskrats are closely related to Voles.

Mussel. Edible Bivalve mollusk, abundant in cooler seas. Mussels form extensive, crowded beds, anchoring themselves to pilings or rocks by a secretion of strong threads known as the byssus. The dark-shelled burrowing freshwater mussel is a source of Pearls and Mother-of-Pearl.

Mustard. Common name for the Gruciferae, a large family chiefly of herbs of north temperature regions, typified by flowers with four petals arranged diagonally ("cruciform") and alternating with the four sepals. The Cruciferae, often rich in Sulfur compounds and Vitamin C, include many important food and condiment plants, *e.g.,* Rape, rutabaga, Turnip, mustard, numerous Cabbage varieties, Watercress, Horseradish, and Radish. The herbs called mustard are species of *Brassica* native to Europe and W Asia. Black *(B. nigra)* and white *B. alba)* mustard are cultivated for their seeds, which are ground and used as a condiment, usually mixed to a paste with vinegar or oil. Mustards are also grown as salad plants and for greens.

Mutarotation. A change with time in the optical rotation produced by a solution. The phenomenon is seen in freshly prepared optically active solutions of some sugars, which change as a result of change of the sugar into another optically active form. Freshly prepared glucose solutions change as a result of conversion of glucose into another isomer of different activity.

Mutation. In biology, a sudden change in a Gene, or unit of hereditary material, that results in a new inheritable characteristic. In higher animals and many higher plants a mutation may be transmitted to future generations only if it occurs in germ, or sex cell, tissue; body cell mutations cannot be inherited. Changes within the chemical structure of single genes may be induced by exposure to radiation, temperature extremes, and certain chemicals. The term *mutation* may also be used to include losses or rearrangements of segments of Chromosomes, the long strands of genes. Drugs such as colchicine double the normal number of chromosomes in a cell by interfering with cell division. Mutation, which can establish new traits in a population, is important in Evolution.

Mutual inductrance. *See* electromagnetic induction.

Myasthenia Gravis. Chronic disorder of the muscles, characterized by weakness and a tendency to tire easily. Most commonly found in young adults, the disease is caused by a loss of muscle receptors for chemicals that induce muscle contraction. The muscles of the head and neck are most frequently involved, those of the trunk and extremities less frequently. Symptoms typically advance irregularly, with varying intensity and severity, but there is a gradual worsening over a period of years, with the major danger resulting from respiratory paralysis or infection. Drugs that improve neuromuscular function are used to treat the symptoms of the disease.

Myna, or **Mynah.** An Asiatic Starling, a bird found chiefly in India and Sri Lanka and known for its power of mimicry. The hill myna *(Gracula religiosa),* C. 12-15 in. (30-38 cm), is the best known. Glossy black with yellow head wattles, it is a forest dweller and lives mostly on fruit. When trained, it is a better mimic than the Parrot.

Myopia. A defect in the eye in which parallel rays of light are focused to a point in front of the retina when the eye is at rest so that distant objects cannot be accommodated. Near objects can be seen. In myopia the distance between the front and back of the eye is too long and it can be corrected by concave spectacle lenses. It is sometimes called *short sight* or *near sight*.

Mythology. Collective myths of a people and scientific study of such myths. Myths are traditional stories occurring in a timeless past and involving supernatural elements. Products of prerational cultures, myths express and explain such serious concerns as the creation of the universe and of humanity, the evolution of society, and the cycle of agricultural fertility. Myths are differentiated from folktales *(e.g.,* Cinderella and tales from *The Arabian Nights)* by being more serious, less entertaining, more supernatural, and less rational and logical. Legends and sagas, by contrast with myths, are historical or quasi-historical in nature. Many theories have been advanced to explain myths. The Greeks' explanation of their own mythology was most fully developed in Stoicism, which reduced the gods to moral principles and natural elements. Such allegorical interpretations continued into the 18th cent. Theologians have tended to view myths *(e.g.,* the blood myth or the myth of a golden age) as foreshadowings or corruptions of Scripture. Modern investigations of mythology began with the 19th cent. philologist Max Muller, who saw myths as having evolved from linguistic corruptions. Anthropological explanations have also abounded. Sir James Frazer in his *Golden Bough* (1890) proposed that all myths were originally connected with the idea of fertility in nature, with the birth, death, and resurrection of vegetation as a constantly recurring motif. Bronislaw Malinowski considered myths to be validations of established social patterns. Among influential psychologists, Sigmund Freud related the unconscious myth and dream, while Carl Jung believed that all peoples unconsciously formed the same mythic symbols. In the 20th cent. Mircea Eliade believes that myths serve to return their adherents to the time of the original creative act, and Claude Levi-Strauss contends that myths should be interpreted structurally. Important mythologies include the Greek, largely codified and preserved in the works of Homer and Hesiod; the Roman, primarily derived from the Greek, and, with it, the best known; the Norse, which is less anthropomorphic, the Indian (Vedic), which tends to be abstract and otherworldly; the Egyptian, which is closely related to religious ritual; and the Mesopotamian, which exhibits a strong concern with the relationship between life and death, Mythology has enriched literature since the time of Aeschylus and has been used by some of the major English poets, *e.g.,* Milton, Shelley, Keats. Some great literary figures, *e.g.,* William Blake, James Joyce, Franz Kafka, W.B. Yeats, T.S. Eliot, and Wallace Stevens, have constructed symbolic personal myths by reshaping old mythological materials.

N

Nadir. The point opposite the zenith on the celestial sphere.

Nagasaki. City (1980 pop. 447,091), capital of Nagasaki prefecture, W Kyushu, Japan. Nagasaki's port was the first to receive (16th cent.) Western trade, and remains one of Japan's leading ports. Shipbuilding is the chief industry. During World War II, Nagasaki was the target of the second Atomic Bomb ever detonated (Aug. 9, 1945) on a populated area. About 75,000 people were killed or wouded, and more than one third of the city was devastated.

Napalm. Incendiary material used in bombs and flame throwers. Developed during World War II, napalm is a mixture of gasoline (sometimes mixed with other petroleum fuels) and a thickening agent. The thickener turns the mixture into a dense jelly that flows under pressure, as when shot from a flame thrower, and sticks to a target as it burns. Earlier Soap thickeners have been replaced by polystyrene and similar polymers.

Naperian Logarithm. Logarithm, [After John Napier (1550-1617), Scottish mathematician.]

Naphthalene. A white flammable crystalline solid, $C_{10}H_8$, with an odour of moth balls. It is the most abundant constituent of coal tar, from which it is extracted by fractional distillation. Naphthalene is an aromatic hydrocarbon and is used in the manufacture of dyes and synthetic resins and as a moth repellent. M.pt. 80.2°C; b.pt. 218°C; r.d. 1.14.

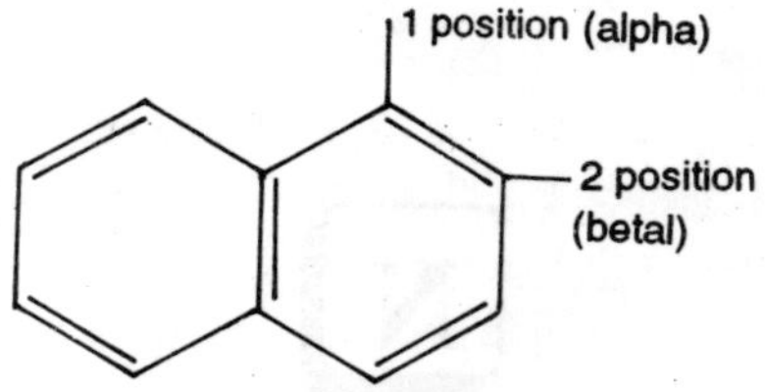

1 position (alpha)
2 position (betal)

Fig. N-1. Naphthalene.

Naphthol. Any one of two isomers of hydroxynaphthalene, $C_{10}H_7OH$, both of which are white powders obtained by fusing the corresponding sodium naphthalene sulphonate with caustic soda. The more important is *beta-naphthol* (2-hydroxynaphthalene) which is used in antioxidants for rubber, dyes, and pharmaceuticals. M.pt. 122°C; b.pt. 285°C; r.d. 1.2.

Napier, John. 1550-1617, Scottish mathematician, the inventor of Logarithms. He also introduced the decimal point in writing numbers. His *Rabdologiae* (1617) gives various methods for abbreviating arithmetical calculations, including a method of multiplication using a system of numbered rods called Napier's rods, or Napier's bones. Napier was also known as an outspoken exponent of the protestant cause.

Narcissus. Showy-blossomed plant (genus *Narcissus*) of the Amaryllis family, native chiefly to the Orient and the Mediterranean region but now widely distributed. The genus includes the yellow daffodil (*N. pseudo-narcissus*), with a long, trumpet-shaped central corona; the yellow jonquil (*N. jonquilla*), with a short corona; and the narcissus, any of several usually white-flowered species, *e.g.*, the poet's narcissus (*N. poetica)*, with a red rim on the corona. The biblical Rose of Sharon may have been a narcissus.

Narcotic. Group of drugs with potent analgesic effects, associated with alteration of mood and behaviour. The chief narcotic drugs are Opium, Codeine, Morphine, and the morphine derivative Heroin. Narcotics are thought to act by mimicking and/or enhancing the activity of Endorphins, proteins produced by the brain and believed to modulate pain and other nervous system functions. Narcotics are valuable in numbing the senses, alleviating pain, inducing sleep, and relieving diarrhea. Common side effects include nausea, vomiting, and allergic reactions. In large doses, narcotics can cause respiratory depression, Coma, and death. All narcotics are addictive; synthetic narcotics such as meperidine and Methadone tend to be less addicting and possess fewer side effects, but they are also less potent.

Nascent. Denoting a gas that is formed in a reaction mixture and consequently has a higher activity. Nascent hydrogen, for example, can be produced *in situ* by adding zinc and hydrochloric acid to the reaction medium. The hydrogen produced often reduces compounds that are not reduced by "normal" hydrogen. The high activity of the nascent hydrogen can be ascribed to the production of hydrogen atoms or to catalysis involving intermediates formed during the reaction.

Natural Gas. Natural mixture of flammable gases found issuing from the ground or obtained from specially driven wells. Largely a mixture of Hydrocarbons, natural gas is usually 80 to 95% Methane. The composition varies in different localities, and minor components may include carbon dioxide, nitrogen, hydrogen, carbon monoxide, and helium. Often found with Petroleum, natural gas also occurs apart from it in sand, sandstone, and limestone deposits. Natural gas began to be used as an illuminant and a fuel on a large scale in the late 19th cent., when pipelines were built to provide it to large industrial cities. *Liquified natural gas* (LNG) is natural gas that has been cooled and pressurized to liquify it for convenience in shipping and storage.

Naturalism. A position that attempts to explain all phenomena by means of strictly natural (as opposed to supernatural) categories. Generally considered the opposite of Idealism, naturalism looks for causes and takes little account of reasons. It is often equated with Materialism, Positivism, and Empiricism. Some naturalists (*e.g.*, Comte, Nietzsche, and Marx) have professed Atheism, while others (*e.g.*, Aristotle, Spinoza, and William James) have accepted some form of a deity. Later thinkers such as Whitehead have sought to unify the scientific viewpoint with the concept of an all-encompassing reality.

Natural Law. Theory that some laws are fundamental to human nature and discoverable by human reason without reference to man-made, or positive, law, which is conditioned by history and subject to continuous change. Roman Law, drawing on theories of Greek Stoicism, recognized a common cause regulating human conduct; this was the basis for the later development by Grotius of the theory of international law. St. Thomas Aquinas, Spinoza, and Leibniz all interpreted natural law as the basis of ethics and morality. J.J. Rousseau regarded it as the basis of democratic principles. The influence of natural law declined greatly in the 19th cent. under the impact of Positivism, Empiricism, and Materialism, but regained importance in the 20th cent. as a necessary opposition to totalitarian theory.

Natural Selection. Important mechanism in Charles Darwin's theory of Evolution. As a result of various factors in the environment (*e.g.*, temperature and the quantity of food and water available) and the geometrically increasing over-production of plants and animals that results from the process of reproduction, a struggle for existence arises. In this struggle, according to Darwin, those organisms better adapted to the environment (that is, those having favorable differences or variations) survive and reproduce, while those least fitted do not. Favorable variations among members of the same species are thus transmitted to the survivors' offspring and spread to the entire species over successive generations. Natural selection suggests that the origin and diversification of species results from the gradual accumulation of individual modifications. Artificial selection, the selection by humans of individuals best suited for a specific purpose, is common in plant and animal breeding.

Nautilus. Mollusk with a spirally coiled shell consisting of a series of chambers; as the nautilus grows, it builds larger chambers, sealing off the old ones. A Cephalopod, the animal lives in the largest and newest chamber, breathing by means of gills and feeding on crabs and other animals it catches with long, slender tentacles. Nautiluses are found in deep waters of the S. Pacific and Indian oceans. The Paper Nautilus, which is not a true nautilus, is related to the Octopus.

Navigation. Science and technology of finding the position, and directing the course, of vessels and aircraft. In ancient times navigation was based on observing landmarks along the coast and the positions of the sun and the stars. A tremendous advance took place with the introduction (C. 12th cent.) of the Compass into Europe. Instruments used to find latitude in medieval times included the Astrolabe, the cross-staff, and the quadrant. The problem of finding the longitude, however, was not satisfactorily solved until the 18th cent. inventions of the Chronometer and the Sextant and the appearance (1765) of the British *Nautical Almanac.* The next great revolution in navigation occurred in the 20th cent., when radio signals came into wide use. The development of Radar, Loran, and radio direction-finding during World War II and, subsequently, of Navigation Satellites caused fundamental changes in navigational practice.

Navigation Satellite. Artificial Satellite designed expressly to aid navigation at sea and in the air. Two major navigational satellite systems have been launched into orbit, both by the U.S. In the Transit system (first launch, 1960), a navigator determines a ship's position by measuring the Doppler shift in radio signals from a Transit satellite passing overhead. In the Navstar Global Positioning system (GPS), which will eventually replace the Transit system, each satellite will broadcast time and position messages continuously.

Neanderthal Man. Type of early man, existing 100,000-40,000 years ago, generally assigned to the species *Homo sapiens.* Fossil remains were first found (1856) in Neanderthal, W. Germany. His middle Paleolithic Culture comprised stone tools, fire, burial, and cave shelters. The so-called classic Neanderthal *(e.g.,* Rhodesian man) had a large, thick skull, a sloping forehead, a chinless jaw, and a brain somewhat larger than that of modern man; he stood slightly over 5 ft (152 cm). There is no evidence to indicate whether he became extinct, or evolved into or interbred with a more modern type such as Cro-Magnon Man.

Nearsightedness. Or Myopia, defect of vision in which far objects appear blurred, but near objects are seen clearly. Because the eyeball is too long or the eye's refractive power too strong, the image is focused in front of the retina of the Eye rather than upon it. Eyeglasses with concave lenses can compensate for the refractive error.

Nebula. Immense body of highly rarified gas and dust in the interstellar spaces of galaxies. A diffuse nebula, such as the Crab Nebula, is irregular in shape and ranges up to 100 light-years in diameter. A bright emission nebula, composed primarily of hydrogen gas ionized by nearby hot blue-white stars, radiates its own light; a bright reflection nebula, located near cooler stars, reflects the starlight. A dark nebula which neither emits nor reflects light because it is too distant from any star, appears as an empty patch in a field of stars or as a dark cloud obscuring part of a bright nebula in the background. A planetary nebula consists of a well-denfined shell of gaseous material that glows from the radiation emitted by the central hot star it surrounds. The shell, measuring about 20,000 Astronomical Units in diameter, is slowly expanding, indicating that it was expelled in a nova or Supernova explosion.

Nebula. Any of numerous clouds of intersteller gas and dust. *Galactic nebulae* occur within the Milky Way and fall into a number of classifications. *Bright-emission nebulae* emit light and other forms of radiation, *reflective nebulae* shine by the light of nearby stars, and *dark nebulae* absorb the light of stars lying beyond them. Because of their misty appearance galaxies were originally referred to as *extragalactic nebulae,* a name still found in modern catalogues. *Planetary nebulae* are also found, consisting of a very hot central star surrounded by an expanding ring or envelope of gas and particles.

Nebular Hypothesis. Any of a group of theories concerning the origin of the solar system, all of which assume that the sun and planets have been formed by condensation of a nebula.

Nectarine. Name for a tree (*Prunus persica nectarina*) of the Rose family and for its fruit, a smooth-skinned variety of the Peach. In appearance, culture, and care the nectarine tree is almost identical to the peach tree. Occasionally a nectarine tree will produce peaches, and a peach tree, nectarines.

Needlepoint. Type of embroidery worked on a cotton or linen mesh with a blunt needle. The mesh, called a canvas, may be printed with a pattern, or original patterns may be created. Wool, cotton, or silk yarns are used, their weight varying with the fineness of the canvas. Among the many needlepoint stitches are tent stitch, Florentine stitch (also called bargello or flame stitch), and cross stitch. Needlepoint worked on a canvas with more than 16 holes per inch is called petit point; with 8 to 16 holes, gros point; and with fewer than 8, quick point. Needlepoint was also known as canvas work; after wool yarn and patterns from Berlin became popular in the 19th cent. it was called Berlin wool work.

Neel Temperature. Antiferromagnetism.

Negatron. A negative electron: *i.e.* an electron as opposed to a positron.

Nematode. Any of a large class of Roundworms. Nematodes live in the water or soil. Many species, such as roundworms and hookworms, are parasites of plants and animals, including humans.

Neodymium (Nd). Metallic element, discovered in 1885 by C.A. von Welsbach. A lustrous, silver-yellow Rare-Earth Metal in the Lanthanide Series, it is present in Monazite and bastnasite. Neodymium is used in the manufacture of certain lasers. Its oxide is used in colouring eyeglasses and in an alloy in cigarette-lighter flints.

Neodymium. Symbol: Nd. A silvery metallic element belonging to the lanthanide series. A.N. 60; A.W. 144.24; m.pt. 1010°C; b.pt. 3967°C; r.d. 6.9; valency 3.

Neon. Symbol: Ne. A colourless odourless tasteless nonflammable gaseous element present in the atmosphère (0.0012%) from whiich it is obtained as a by-product in the liquefaction of air. Neon is used in fluorescent lamps. It is one of the inert gases. A.N. 10; A.W. 20.283; b.pt.—248.6°C; m.pt. —245.92°C; r.d. 0.694 (air=1).

Neon (Ne). Gaseous element, discovered in 1898 by William Ramsay and M.W. Travers. A colourless, odourless, and tasteless Inert Gas, it emits a bright-red glow when conducting electricity in a tube. Neon is used in advertising signs. Lasers, Geiger counters, Particle Detectors, and high-intensity beacons. Liquid neon is a cryogenic refrigerant.

Neopentane (2,2-dimethylpropane). A colourless gas or volatile liquid, $C(CH_3)_4$, found in small amounts in natural gas. It is isomeric with pentane and isopentane. M.pt. -16.6°C; b.pt. 9.5°C; r.d. 0.59.

Neopentyl Group. Pentyl Group.

Neoprene. A synthetic rubber made by polymerizing chloroprene. It is resistant to heat and solvents and is also used in cements and paints.

Neptune. The eighth planet in order of succession from the sun; the outermost of the giant planets. It orbits the sun once every 164.79 years at a distance of 4.497 million kilometres. Nepture is similar to Uranus, having a diameter of 48 400 km, a mass 17.2 times that of the earth, and a relative density of 1.77. Its rotation period is almost 16 hours and its temperature is about — 205°C. As with Uranus, the atmosphere is dense and comprises mainly methane and hydrogen. Neptune has two satellites *Triton* and *Nereid.*

Neptune. In astronomy, 8th Planet from the sun, at a mean distance of 2.7941 billion mi (4.4966 billion km). It has an equatorial diameter of c. 30,760 mi (c. 49,500 km) and an atmosphere composed of hydrogen, helium, methane, and ammonia. Neptune was the first planet to be discovered on the basis of theoretical calculations; such calculations, based on the observed irregularities in the motion of the planet Uranus, were made independently by John Couch Adams and Urbain Leverrier. The German astronomer Johann Galle discovered Neptune on Sept. 23, 1846, within 1° of the position predicted and sent to him by Leverrier. Neptune has two known natural satellites: *Triton* (discovered 1846) has an estimated diameter of 1,100 to 1,600 mi (1,800 to 2,600 km) and travels in an unusual retrograde orbit; Nereid (discovered 1949) has an estimated diameter of 145 to 290 mi (235 to 470 km). The *Voyager* 2 Space Probe is expected to encounter Neptune in 1989.

Neptunium (Np). Radioactive element, discovered in 1940 by E.M. McMillan and P.H. Abelson by neutron bombardment of uranium. It is a silvery metal in the Actinide series and is the first Transuranium Element. Neptunium is found in very small quantities in nature in association with uranium ores. The neptunium-237 isotope has a half-life of 2 million years.

Neptunium. Symbol: Np. A silvery metallic actinide element: the first synthetic transuranic element to be produced. The isotope

^{237}Np is made in nuclear reactors in the production of plutonium. A.N. 93; m.pt. 640°C; valency 3, 4, 5, or 6.

Nernst, Walther Hermann. 1864-1941, German physicist and chemist. A founder of modern physical chemistry, he won the 1920 Nobel Prize in chemistry for his work in Thermodynamics. He established what is often called the third law of thermodynamics, which deals with the behaviour of matter at temperatures approaching absolute zero. He also did work in electrochemistry, electroacoustics, and astrophysics.

Nervous System. Network of specialized tissue that controls actions and reactions of the body, enabling it to adjust to its environment. In general the system functions by receiving signals from all parts of the body, relaying them to the Brain and Spinal Cord, and then sending appropriate return signals to muscles and body organs. Virtually all multicellular animals have at least a rudimentary nervous system; in vertebrates the system is most complex. The basic unit of the nervous system is the nerve cell (neuron). Of the billions of neurons in humans, half are in the brain. The neuron consists of a cell body, containing the cell nucleus; dendrites, branchlike extensions that receive incoming signals; and the axon, the long cell extension that carries signals long distances. A neuron works by receiving chemical signals—some excitatory, some inhibitory-through its dendrites and sending electrical impulses along its axon. Chemical transmitters released at the terminal fibers of the axon diffuse across a junction called the synapse and bind to dendrites of recipient neurons. Dendrites and axons are called nerve fibers; a nerve is a bundle of nerve fibers. The nervous system has two divisions: the central nervous system and peripheral nervous system. The central nervous system, consisting of the brain and spinal cord, receives impulses from sensory (afferent) nerve fibers delivering impulses from receptors in the skin and organs; it returns impulses via motor (efferent) fibers to terminals in muscles and glands. Peripheral nerves mediate these pathways. The peripheral nervous system comprises cranial nerves, controlling face and neck; spinal nerves, radiating to other parts of the body; and autonomic nerves, which form a subsidiary system regulating the iris of the eye and muscles of heart, glands, lungs, stomach, and other visceral organs.

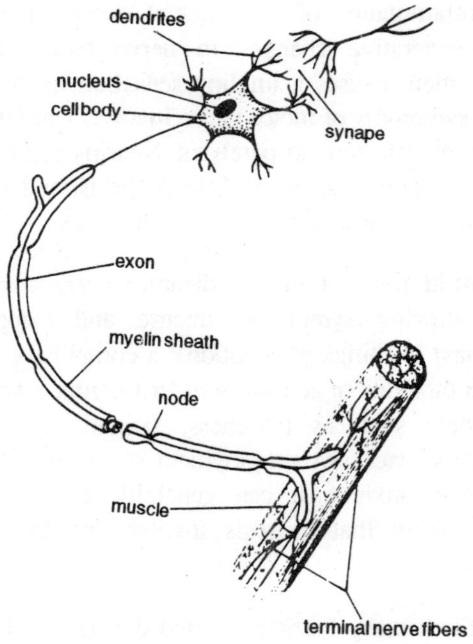

Fig. N-2. Nerve cell (neuron).

Nessler's Reagent. A colourless alkaline solution of mercury II iodide in potassium iodide used to detect traces of ammonia, with which it gives a brown coloration.

Nettle. Common name for the family Urticaceae, fibrous herbs, small shrubs, and trees found chiefly in the tropics and subtropics. Several species are covered with small stringing hairs that on contact emit formic acid, a skin irritant. Stinging nettles in the U.S. include species of *Urtica,* widely distributed, and *Laportea canadensis,* a characteristic plant of eastern forests.

Neuralgia. Acute, throbbing pain along a peripheral sensory nerve, commonly in the area of the facial, or trigeminal, nerve. Its causes include Shingles, infections, and extreme cold. Unlike Neuritis, neuralgia does not involve degeneration of the nerve tissue.

Neuritis. Inflammation of a peripheral nerve, often accompanied by degenerative changes in nerve tissue. Sensory nerve involvement causes a tingling sensation or loss of sensation, while symptoms of motor nerve involvement range from slight loss of muscle tone to paralysis. Neuritis commonly occurs in Diabetes, Shingles, rheumatoid Arthritis, and other disorders. Treatment varies, depending on the cause.

Neurosis. Mild form of mental disorder characterized by any of the following symptoms: intense and inappropriate fears (Phobias) of things or situations; a compulsive need to pursue certain thoughts or actions in order to reduce Anxiety; physical symptoms such as tenseness, fatigue, and Psychosomatic Disorders; excessive employment of Defense Mechanisms to overcome anxiety. It can generally be differentiated from Psychosis in that neurosis involves no loss of a sens of reality.

Neutrino. Elementary Particle emitted during the decay of certain other particles. It was first postulated in 1930 by Wolfgang Pauli in order to maintain the law of conservation of energy during beta decay. Further studies showed that the neutrino was also necessary to maintain the conservation laws of momentum and spin. The neutrino was not detected directly until 1956. The neutrinos associated with the electron and with the muon are distinct; each has its own antiparticle. Neutrinos are stable; they are created and destroyed only by particle decays involving the weak nuclear Force, or weak interaction. Neutrinos have little or no mass.

Neutrino. An elementary particle with zero rest mass, a velocity equal to that of light, and a spin of one half. The existence of the neutrino was originally postulated to explain energy conservation in beta decay. There are two types: one produced in decay processes that yield positrons and the other produced in decays of some mesons. The antiparticles of these two types of neutrino are also called neutrinos.

Neutron. An elementary particle with zero charge and a rest mass of 1.674 92 x 10^{-27} kg, which is similar (but not identical) to that of the proton. Together with protons, neutrons are present in the nuclei of all atoms except the hydrogen atom. When free, they decay to protons and electrons with a mean-life of 932 seconds.

Neutron. Uncharged Elementary Particle, discovered by James Chadwick in 1932, of slightly greater mass than the Proton. The stable isotopes of all elements except hydrogen and helium contain within the nucleus a number of neutrons equal to or greater than the number of protons. The preponderance of neutrons becomes more marked for very heavy nuclei. A neutron bound within the nucleus may be stable. A nucleus with an excess of neutrons, however, is radioactive; the extra neutrons (as well as any free neutrons not bound within a nucleus) convert by beta decay into a proton, an electron, and an antineutrino. The neutron and the proton are regarded by physicists as two aspects, or states, of a single entity, the nucleon. The antineutron, the neutron's antiparticle was discovered in 1956.

Neutron Excess. Isotopic number.

Neutron Star. A type of star, less than 20 kilometres in diameter, that is postulated to result from the gravitational collapse of the matter in a star after the source of nuclear energy has been exhausted. As the matter becomes compressed it reaches a degenerate state in which its nuclei and electrons are packed together. At densities above about 10^7 kg m^{-3}, protons and electrons may fuse together to give neutrons. A very fast rotational spin is imparted to the new superdense star and energized particles are sent out in a highly directional beam. In theory, a neutron star could form a black hole if it were of sufficient size.

Neutron Star. Extremely small, extremely dense star comparable to the sun in mass but only a few miles in radius. According to current theories, the core of a neutron star is composed of Elementary Particles and is surrounded by a fluid composed primarily of Neutrons squeezed in close contact. The fluid in turn is encased in a rigid, extremely dense crust a few hundred miles thick. The only observational evidence of the existence of neutron stars to date is provided by Pulsars, radio sources that fluctuate in intensity in a manner indicating that they might be rotating neutron stars. Neutron stars are believed to be the dead stellar corpses remaining after the Supernova explosion of an intermediate-mass star.

Newcomen, Thomas. 1663-1729, English inventor of an early atmospheric Steam Engine (c.1711) used to pump water. It was an improvement over Thomas Savery's engine (patented 1698).

Newlands' Law of Octaves. The observation made by the British chemist John Newlands (1837-98) that if the elements are arranged in order of their atomic weight, then every eighth element is chemically similar. He compared this to the intervals of the musical scale.

Newton. The SI unit of force, equal to the force that will give a mass of one kilogram an acceleration of one metre per second per second. [After Sir Isaac Newton (1647-1727), English physicist.]

Newtonian Fluid. A fluid that obeys Newton's law of viscosity.

Newtonian Mechanics. The system of mechanics based on Newton's laws of motion.

Newtonian Telescope. A type of astronomical telescope in which light is reflected from the primary mirror onto a small plane mirror on the axis.

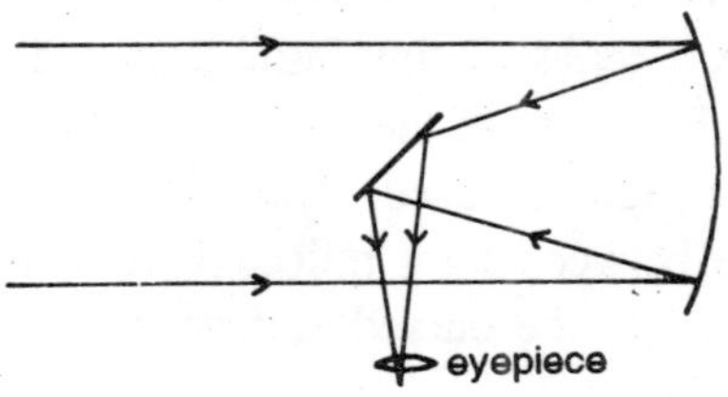

Fig. N-3. Newtonian telescope.

Newton's Law of Cooling. The principle that the rate at which a body loses heat is proportional to the difference in temperature between the body and its surroundings. Strictly, it only applies to forced convection.

Newton's Law of Gravitation. Gravitation.

Newton's Laws of Motion. Three laws formulated by Newton to describe the motion of bodies.

1. Everybody will continue in a state of rest or of uniform motion in a straight line unless it is compelled to change that state by an external impressed force.

2. When a force is applied the rate of change of momentum is proportional to the force and takes place in the direction of action of the force: *i.e.* $F = d(mv)/dt$, where F is the magnitude of the force, m the body's mass, and v its velocity. Usually m is constant and the equation becomes $F = ma$, where a is the acceleration of the body.

3. Every action on a body produces an equal and opposite reaction.

The laws are the basis of *Newtonian mechanics,* a system of mechanics applicable at velocities that are much less than the velicity of light.

Newton's Rings. A interference effect observed when a plano-convex lens of large radius is placed on a flat glass plate and illuminated from above by monochromatic light. When observed from above, a series of concentric light and dark rings is seen centred on the centre of the lens. Coloured rings are produced when white light is used. The phenomenon is caused by interference between light reflected from the glass plate and light reflected at the curved surface of the lens.

Newton, Sir Isaac. 1642-1727, English mathematician and natural philosopher (physicist); considered by many the greatest scientist of all time. He was Lucasian professor of mathematics (1669-1701) at Cambridge Univ. Between 1664 and 1666 he discovered the law of universal Gravitation, began to develop the Calculus, and discovered that white light is composed of every colour in the Spectrum. In his monumental *Philosophiae naturalis principia mathematica [Mathematical principles of natural Philosophy]* (1687), he showed how his principle of universal gravitation explained both the motions of heavenly bodies and the falling of bodies on earth. The *Principia* covers Dynamics (including Newtons's thre laws of Motion), Fluid Mechanics, the motions of the planets and their satellites, the motions of the comets, and the phenomena of Tides. Newton's theory that Light is composed of particles—elaborated in his *Opticks* (1704)—dominated optics until the 19th cent., when it was replaced by the wave theory of light; the two theories were combined in the modern Quantum Theory. Newton also built (1668) the first reflecting Telescope, anticipated the calculus of variations, and devoted much energy towards alchemy, theology, and history, particularly problems of chronology. He was president of the Royal Society from 1703 until his death.

Nickel. Symbol: Ni. A silvery-white malleable and ductile transition element found in pentlandite ((Fe,Ni)S), millerite (NiS), and garnierite $(Ni,Mg)_6(OH)_6Si_4O_{11}.H_2O)$. Nickel is also present in many meteorites. The element is obtained by roasting the sulphide to the oxide, which is then reduced or refined dby electrolysis or by the Mond process. It is ferromagnetic and is used in some ferromagnetic alloys, such as Mumetal, as well as alloy steels and protective electroplated coatings. Nickel is also an efficient hydrogenation catalyst, used for hardening oils in making margarine. The element is resistant to corrosion, dissolves in dilute nonoxidizing acids, and is rendered passive by nitric acid. The normal valency is 2: many ionic nickel II (*nickelous*) salts exist as well as numerous complexes. A few complexes (*nickelic*) are formed in higher oxidation states. A.N. 28; A.W. 58.71; m.pt. 2468°C; b.pt. 4927°C; r.d. 8.57; valency 2, 3, or 5.

Nickel (Ni). Metallic element, discovered in 1751 by A.F. Cronstedt. It is a silver-white, hard, malleable, ductile, and lustrous metal whose chief use is in the preparation of alloys, to which it brings strength, ductility, and resistance to corrosion and heat. Many stainless Steels contain nickel. Nickel's chief ores are garnierite, pentlandite and pyrrhotite. Nickel is present in most Meteorites. Trace amounts are found in plants and animals.

Nickel Carbonyl. A poisonous flammable yellow liquid, $Ni(CO)_4$, made by passing carbon monoxide over finely divided nickel. It is used as a catalyst and is also the intermediate in the Mond process for purifying nickel. M.pt. -25°C; b.pt. 43°C; r.d. 1.32.

Nickel Oxide. Any of two basic oxides of nickel made by heating the corresponding nitrate. *Nickel II oxide* (nickelous oxide) is a green powder, NiO. R.d. 6.6. *Nickel III oxide* (nickelic oxide, nickel peroxide, nickel sesquioxide) is a grey-black powder used in nife cells. R.d. 4.8.

Nickel Peroxide. Nickel oxide.

Nickel Sesquioxide. Nickel oxide.

Nicol Prism. A prism made by cutting two pieces of calcite in particular directions and cementing them together with Canada balsam. The arrangement is designed to transmit the ordinary ray and reflect the extraoridinary ray at the interface between the two pieces. Nicol prisms are used for producing and analysing plane-polarized light. [After william Nicol (1768-1851), British physicist.]

Nicotine. A colourless poisonous hygroscopic oily alkaloid. $C_5H_4NC_4H\text{-}_7NCH_3$, extracted from tabacco. It is used as a horticultural insecticide.

Nife Cell. A type of accumulator consisting of a nickel positive plate and an iron negative plate dipping into a solution of sodium hydroxide.

Nightingale. Migratory Old World Bird of the Thrush family, celebrated for the song the male sings during the breeding season. The common nightingale of England and W Europe (*Luscinia megarhynchos*), reddish-brown above and grayish-white below, winters in Africa.

Nightshade. Common name for the family Solanaceae, herbs, shrubs, and trees of warm regions. Many are climbing or creeping types. Rank-smelling foliage typifies many species; the odor is due to the presence of various Alkaloids, *e.g.*, scopolamine, nicotine, and atropine. The chief drug plants of the family are Belladonna, Mandrake, Jimson weed, and Tobacco. The family also includes important food plants *e.g.*, Potato, Tomato, red Pepper, and Eggplant, and ornamentals, *e.g.*, Petunia. The name *nightshade* is commonly restricted to members of the genus *Solanum*, typified by white or purplish star-shaped flowers and orange berriers. Among the better-known species are Jerusalem cherry (*S. pseudocapsicum*), a house plant popular for its scarlet berries, and the Bittersweet, or woody nightshade.

Nimbus. The luminous disk, circle, or other indication of light around the head of a sacred personage. Employed in Buddhist and other Oriental art, it was used by the ancient Greeks and Romans to designate gods and heroes. The device appeared in Christian art in the 5th cent. In Christian Iconography, the nimbus may have a number of geometric shapes. *Halo* is a nontechnical term for *nimbus*.

Ninhydrin. A white crystalline solid, $C_9H_6O_4$, used as a test reagent for detecting free amino and carboxyl groups in proteins, with which it gives a blue colour.

Niobium. Symbol: Nb. A soft ductile silvery-white metallic element, developing a bluish caste in air, found in columbite-tantalite $(Fe,Mn)(Nb, Ta)_2O_6$ and pyrochlore $(NaCaNb_2O_6F)$. Niobium is separated from tantalum by fractional crystallization of its complexes. It is extracted by reducing the oxide with calcium or electrolysis of the fused complex fluoride. The metal is superconducting at low temperatures. Its main uses are in alloy steels and vacuum getters.

Niobium (Nb). Metallic element, discovered in 1801 by Charles Hatchett. Called columbium by metallurgists, it is a rare, soft, malleable, ductile, gray-white metal that is used in high-temperature-resistant alloys and special stainless steels.

Chemically, it is not attacked by acids, dissolves in fused alkalis, and reacts with oxygen and other nonmetals at high temperatures. In its compounds it exhibits several valencies, generally behaving like a nonmetal in forming volatile covalent compounds, such as halides and oxyhalides. Niobium was formerly called *columbium* in the U.S. A.N. 41: A.W. 92.906; m.pt. 2468°C; b.pt. 4927°C; r.d. 8.57; valency 2, 3, or 5.

Nitration. A chemical reaction in which a nitro group ($-NO_2$) is introduced into a molecule.

Nitre. Potassium nitrate.

Nitric Acid (aqua Fortis). A colourless or yellowish fuming corrosive liquid, HNO_3, obtained by the catalytic oxidation of ammonia. It is a strong oxidizing agent, which attacks almost all metals. Nitric acid is used in manufacturing ammonium nitrate for use in fertilizers and explosives. M.pt. -41°C; b.pt. 83°C; r.d. 1.5.

Nitric Acid. Chemical compound (HNO_3), colourless, highly corrosive, poisonous liquid that gives off choking fumes in moist air. It is miscible with water in all proportions. Commercially, it is usually available in solutions of 52% to 68% nitric acid in water. Solutions containing over 86% nitric acid are commonly called fuming nitric acid. Nitric acid is a strong oxidizing agent. It reacts with metals, oxides, and hydroxides, forming nitrate salts.

Nitric Oxide. Nitrogen oxide.

Nitrile. Any of a class of organic compounds containing the group -CN.

Nitrobenzene. A poisonous pale yellow oily liquid, $C_6H_5NO_2$, with an almond-like odour. It is prepared by nitrating benzene and used in the manufacture of aniline. M.pt. 6°C; b.pt. 211°C; r.d. 1.2.

Nitrocellulose (gun cotton). A cotton or pulp-like polymer and variable composition made by treating cellulose with a mixture of nitric and sulphuric acids. It is highly flammable and is used in explosives and solid rocket propellants. Strictly speaking it is a nitric acid ester of cellulose, not a nitro compound, and the more correct name is *cellulose nitrate.*

Nitro Compound. A compound with the formula RNO_2, where R is usually an aryl group.

Nitrogen. Symbol: N. A colourless odourless tasteless nonflammable gas forming 78% of the air by volume. Pure nitrogen is obtained by liquefying air. The element is used in the manufacture of ammonia and other nitrogen compounds and as an inert atmosphere in welding, forging, and similar operations. The element is diatomic, N_2, and relatively inert: it reacts with hydrogen under the influence of a catalyst, forms nitrogen oxides in electric sparks, and combines with magnesium, lithium, and calcium to produce nitrides. Nitrogen compounds are essential to plant growth. A.N. 7; A.W. 14.0067; m.pt. -209.86°C; b.pt. -195.8°C; r.d. 0.967 (air = 1).

Nitrogen (n). Gaseous element, discovered by Daniel Rutherford in 1772. Nitrogen is a colourless, odourless, tasteless, diatomic gas that is relatively inactive chemically; it occupies about 78% (by volume) of dry air. Its cheif importance lies in its compounds, which include Nitrous Oxide, Nitric Acid, Ammonia, many Explosives, Cyanides, Fertilizers, and Proteins. Nitrogen is present in the Protoplasm of all living matter; it and its compounds are necessary for the continuation of life.

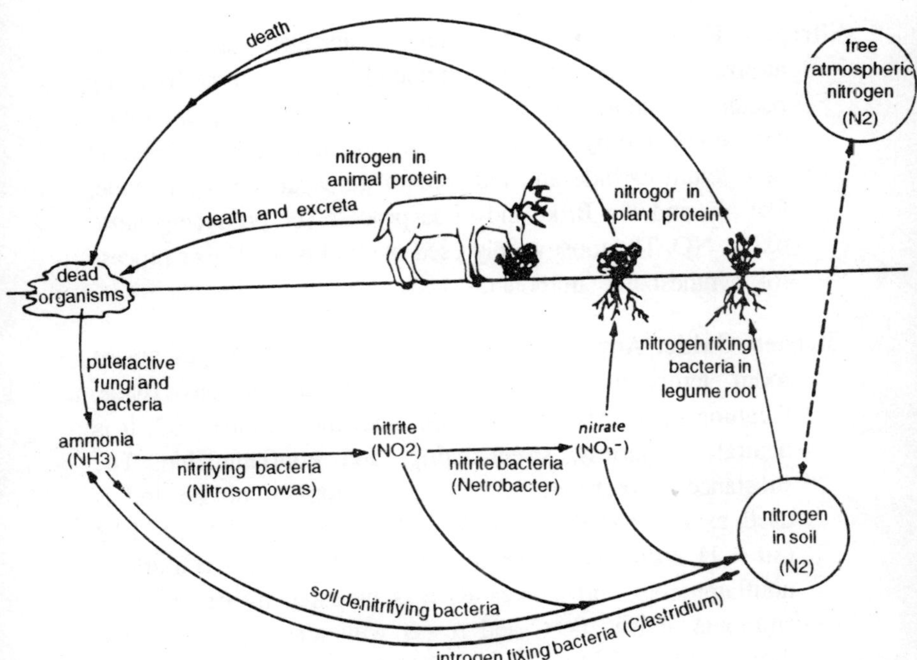

Fig. N-4. Nitrogen cycle.

Nitrogen Cycle. The continuous flow of nitrogen through the Biosphere by the processes of nitrogen fixation, ammonification (decay), nitrification, and denitrification. Nitrogen is an essential constituent of all protoplasm. To enter living systems, however, it must first be "fixed" (combined with oxygen or hydrogen) into compounds that plants can utilize, such as nitrates or ammonia. Most fixation is performed by certain bacteria living in the soil or in nodules in the roots of leguminous plants. Plants elaborate the fixed nitrogen into plant protein; then animals consume the plants and convert plan protein into animal protein. Organic nitrogen is returned to the soil as ammonia when animal remains and wastes decay. Then nitrifying bacteria oxidize the ammonia to nitrites and the nitrites to nitrates, which can, like ammonia, be taken up by plants. Still other soil microorganisms can reduce ammonia nitrates to molecular nitrogen.

Nitrogen Fixation. The conversion of nitrogen gas from the atmosphere into nitrogen compounds. Nitrogen is fixed by bacteria in the roots of some plants. Industrial nitrogen fixation can be effected by several processes including the combination of calcium carbide and nitrogen to give calcium cyanamide, $CaCN_2$, and the Birkeland-Eyde process for producing nitric oxide, NO. The most widely used method is the Haber process for synthesizing ammonia.

Nitrogen Oxide. Any of seven oxides of nitrogen all gaseous at room temperature. *Dinitrogen monoxide* (nitrous oxide, laughing gas) is the lowest oxide, with the formula N_2O. It is neutral, colourless, sweet-tasting, and nonflammable. The substance is prepared by heating ammonium nitrate and is used as an anaesthetic. M.pt. -91°C; b.pt. -88°C; r.d. 1.52 (air = 1). *Nitrogen monoxide* (nitric oxide), NO, is colourless, nonflammable, and poisonous. It is prepared by oxidation of ammonia above 500°C and reacts with air to give nitrogen dioxide. M.pt. -164°C; b.pt. -152°C; r.d. 1.3. *Nitrogen dioxide* (nitrogen peroxide) is a brown acidic gas, NO_2, which on cooling is converted to a colourless gas, N_2O_4, *dinitrogen tetroxide*. It is prepared by oxidation of nitrogen monoxide and used as a nitrating and oxidizing agent. M.pt. -11.2°; b.pt. 21°C; r.d. 1.4 (20°C). *Dinitrogen trioxide,* N_2O_3, *dinitrogen pentoxide,* N_2O_5, and *nitrogen trioxide,* No_3, are the other less stable oxides of nitrogen.

Nitrogen Peroxide. Nitrogen oxide.

Nitroglycerine. A pale yellow thick explosive liquid, $C_3H_5(NO_3)_3$, used in dynamite and other explosives.

Nitroglycerin ($C_3H_5N_3O_9$). Colourless oily, liquid Explosive. An unstable compound that decomposes violently when heated or jarred, nitroglycerin is made less sensitive to shock when mixed with an absorbent material to form Dynamite. Nitroglycerin is also a component of smokeless powder and is used in medicine for relief from the symptoms of angina pectoris. It was first produced commercially by Alfred Nobel.

Nitrous Oxide. Chemical compound (N_2O), colourless gas with a sweetish taste and odor. Although it does not burn, it supports combustion because it decomposes into oxygen and nitrogen when heated. a major use is in dental anesthesia. It is often called laughing gas because it produces euphoria and mirth when inhaled in small amounts. It is also used in making certain canned pressurized foods, *e.g.* instant whipped cream.

Nitro Group. The monovalent group $-NO_2$, characteristic of nitro compounds.

NMR. Nuclear magnetic resonance.

Nobel, Alfred Bernhard. 1833-96, Swedish chemist and inventor. He was involved, with his family, in the development and manufacture of explosives, and his invention of Dynamite, a mixture of nitroglycerine and inert filler, greatly improved the safety of explosives. Inclined toward pacifism and concerned about the potential uses of the explosives he had invented, he established a fund to provide annual awards, called Nobel Prizes, in the sciences, literature, and the promotion of international peace.

Nobelium. Symbol: No. A transuranic actinide element made by bombarding curium with carbon nuclei. The isotope produced, ^{254}No, has a half-life of about 3 seconds. The name *nobelium* was proposed by earlier workers claiming the production of a longer-lived isotope with a half-life of about 10 minutes, A.N. 102.

Nobelium (No). Radioactive element, first produced artificially in 1958 by A. Ghiorso, T. Sikkeland, J.R. Walton, and Glenn T. Seaborg by carbon-ion bombardment of curium. It is a Transuranium Element in the Actinide Series. Seven isotopes are known.

Nobel Prize. Award, est. and endowed by the will of Alfred Nobel, given annually for outstanding achievement in one of five fields. By the terms of Nobel's will, the physics and chemistry prizes are judged by the Royal Swedish Academy of Sciences; the physiology or medicine prize, by Sweden's Royal Caroline Medico-Chirurgical Institute; the literature prize, by the Swedish Academy; and the peace prize, by a committee of the Norwegian parliament. Each recipient is presented a gold medal and a sum that by 1980 was about $200,000. These five awards were first given in 1901. A sixth, related award, the *Nobel Memorial Prize in Economic Science,* was established and endowed in 1968 by Sveriges Riksbank, the Swedish national bank, and first awarded in 1969. It is judged by the Royal Swedish Academy of Sciences.

Node. 1. A point of zero displacement of a standing wave.

2. Either of two points at which the orbit of a celestial body crosses a reference plane. The south-to-north crossing occurs at the ascending node; the north-to-south crossing at the descending node. Perturbations due to other bodies cause the nodes to move along the reference plane.

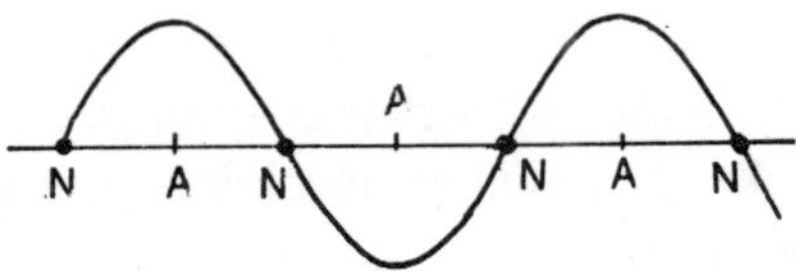

Fig. N-5. Nodes and antinodes.

Noise. Unwanted random disturbances in an electrical circuit.

Noise Pollution. Sounds produced by human commercial and industrial activities at levels and/or frequencies harmful to health or welfare. Transportation vehicles *(e.g.,* trains) and construction equipment *(e.g.,* pneumatic drills) are particular offenders in this regard. Federal estimates indicate that more than one in three Americans is affected by noise pollution. Apart from hearing loss, excessive noise can cause lack of sleep, irritability, heartburn, indigestion, ulcers, and high blood pressure. Noise-induced stress can create severe tension in daily living and may contribute to mental illness. The Noise Control Act of 1972 empowers the Environmental Protection Agency to determine the noise limits required to protect health and to set noise emission standards.

Non-Euclidean Geometry. Any logical self-consistent mathematical system concerning shape, based on postulates that do not include the parallel postulate of Euclid. Non-Euclidean geometry is used in the theory of relativity.

Non-Euclidean Geometry. Branch of Geometry in which the fifth postulate of Euclidean Geometry is replaced by one of two alternative postulates. Euclid's fifth postulate states that one and only one line parallel to a given line can be drawn through a point external to the line. The first alternative, which allows two parallels through any external point, leads to the hyperbolic geometry developed independently by Nikolai

Lobachevsky (1826) and Janos Bolyai (1832). The second, which allows no parallels through any external point, leads to the elliptic geometry developed by G.F.B. Riemann (1854). The results of these two types of non-Euclidean geometry are identical with those of Euclidean geometry in every respect except for the propositions involving parallel lines, either explicitly or implicitly.

Nonmetal. A chemical element that is not a metal. Nonmetals are poor electrical and thermal conductors (although graphite does conduct electricity) and are not generally lustrous or malleable when solid. They include hydrogen, the inert gases, the halogens, oxygen, sulphur, nitrogen, and carbon. Chemically, such elements are electronegative, forming either convalent compounds or negative ions in ionic compounds. Their oxides and hydroxides are usually acidic.

Non-Newtonian Fluid. A fluid that does not obey Newton's law of viscosity: *i.e.* the viscosity is not constant but depends on the rate of shear. Fluids that *depends on the rate of shear. Fluids that* consist of a mixture of different phases show non-Newtonian behaviour: usually the viscosity decreases as the rate of shear increases (thixotropy), although in some suspensions the fluid becomes more viscous.

Nonoxidizing Acid. An acid that does not act as an oxidizing agent. The term is used for compounds such as hydrochloric acid and dilute sulphuric acid, which dissolve metals that are higher than hydrogen in the electrochemical series but do not attack lower metals such as copper. In contrast, oxidizing acids do dissolve such metals.

Nonpolar. Denoting a substance whose molecules do not have a permanent dipole moment. Carbon tetrachloride is an example of a nonpolar compound. Nonpolar solvents are not capable of dissolving ionic solids or other polar compounds. They do dissolve other nonpolar substances: carbon tetrachloride, for example, is a good solvent for many organic compounds.

Nonreducing Sugar. Sugar.

Norepinephrine or **Noradrenaline.** The major transmitter substance of the postganglionic sympathetic nervous system. An impulse reaching the end of a nerve cell stimulates the cell to secrete norepinephrine, which diffuses to a receptor site on a target organ; this role parallels that of Acetylcholine in the rest of the nervous system. Norepinephrine also acts as a Hormone, with effects sometimes opposite to those of Epinephrine.

Normal. 1. Denoting a solution containing one gram-equivalent per litre. A solution containing x gram-equivalents per litre is said to be *x-normal,* written xN.

2. Denoting an isomer with a straight chain of atoms. Normal hexane, for example, has the formula $CH_3CH_2CH_2CH_2CH_2CH_3$, and is distinguished from isomers with branched chains. The abbreviation n- is used, as in *n*-hexane.

3. A line or plane perpendicular to a given line or surface. At any point, the normal to a curved line or surface is perpendicular to the tangent at that point.

Normality. The concentration of a solution in gram-equivalents per litre of solution. A two-normal (2N) solution for example has a normality of 2.

Normal Temperature and Pressure (N.T.P.). Standard temperature and pressure.

North Pole. Northern end of the earth's axis, lat. 90°N, long. 0°, distinguished from the north Magnetic Pole. It was first reached by Robert E. Peary in 1909.

Norway Saltpetre. Ammonium nitrate.

Nose. Organ of breathing and smell. The external nose consists of bone and cartilage. The hollow internal nose, above the roof of the mouth, is divided by the septum (wall) into two nasal cavities extending from the nostrils to the Pharynx. The cavities are lined with a mucous membrane, which is convered with fine hairs that help to filter dust and impurities from the air before it reaches the lungs; the air is also moistened and warmed in its passage. High in each nasal cavity is a small tract of mucous membrane containing olfactory cells. Hairlike fibers in these nerve cells, responding to various odors, send impulses along the olfactory nerve to the brain and thus produce the sense of smell.

Nova. A star that undergoes an explosion, increasing in Iuminosity by up to 100 000 times its original value.

N-P-N Transistor. Transistor.

N.T.P. Standard temperature and pressure.

N-Type. Semiconductor.

Nuclear Energy. The energy stored in the nucleus of an Atom and released through fission, fusion, or Radioactivity. In these processes a small amount of mass, equal to the difference in mass before and after the reaction, is converted to energy according to the relationship $E = mc^2$, where E is energy, m mass, and c the speed of light. In fission processes, a fissionable nucleus absorbs a neutron, becomes unstable, and splits into two nearly equal nuclei. In fusion processes, two nuclei combine to form a single, heavier nucleus. Fission occurs for very heavy nuclei, while fusion occurs for the lightest nuclei. Nuclear fission was discovered in 1938 by Otto Hahn and Fritz Strassman, and was explained in 1939 by the Lise Meitner and Otto Frisch. Fission energy can be obtained by bombarding the fissionable isotope Uranium-235 with slow neutrons in order to split it. Because this reaction releases an average of 2.5 neutrons, a chain reaction is possible, provided at least one neutron per fission is captured by another nucleus and causes a second fission. In an Atomic Bomb the number of neutrons producting additional fission is greater than 1, and the reaction increases rapidly to an explosion. In a Nuclear Reactor, where the chain reaction is controlled, the number must be exactly 1 in order to maintain a steady reaction rate. Uranium-233 and Plutonium-239 can also be used but must be produced artificially. Moreover, the fuel for fusion reactors, deuterium, is readily available in large amounts. Temperatures greater than 1,000,000°C are required to initiate a fusion, or thermonuclear, reaction. In the Hydrogen Bomb such temperatures are provided by the detonation of a fission bomb. Sustained, controlled fusion reactions, however, require the containment of the nuclear fuel at extremely high temerpature long enough to allow the reactions to take place. At these temperatures the fuel is a Plasma, and magnetic fields have been used in attempts to contain this plasma. To produce fusion energy, scientists have also used high-powered laser beams aimed at tiny pellets of fission fuel. Once practical controlled fusion is achieved, it will have great advantages over fission as a source of energy.

Nuclear Fission. Fission.

Nuclear Fusion. Fusion.

Nuclear Magnetic Resonance (NMR). A method of investigating the spins of atomic nuclei. In the presence of a strong magnetic field the magnetic moment of a nucleus precesses around the field direction. Only certain orientations are possible, leading to the existence of a number of discrete energy states for the nucleus. Radiofrequency radiation is supplied to the sample by passing a high-frequency current through a coil and the radiation is detected by a second coil. If the magnetic field is changed slowly, thus changing the difference in energy between levels, radiation is absorbed at certain values of the field when the frequency of the radiation corresponds to the difference between two energy levels. A graph of detector response against magnetic field is an *NMR spectrum* of the sample. The technique is extensively used in chemistry for studying molecules—the energy levels of the nucleus are affected by the electrons surrounding the nucleus and the frequency at which the nucleus absorbs radiation depends on its position in the molecule. A modification of the method described can be used for the accurate measurement of magnetic fields.

Nuclear Magnetic Resonance (NMR). In medicine, a diagnostic technique using radio waves and magnets. Used for many years in chemistry and physics to analyze samples of solids and liquids, as well as tissues removed from the body, nuclear magnetic resonance became a diagnostic tool early in the 1980s for detecting and analyzing changes in body structure and function. The patient is placed in the field of an electromagnet, which causes the nuclei of certain atoms in the body (especially those of hydrogen) to align magnetically. The patient is then subjected to radio waves, which cause the aligned nuclei to "flip"; when the radio waves are withdrawn the nuclei return to their original positions, emitting radio waves that are then detected by a receiver and analyzed by computer. Able to pass through bones and record changes in blood and body tissues. NMR is expected to aid in the early diagnosis of such diseases as cancer, multiple sclerosis, and heart disease. In its earliest use as a diagnostic tool, NMR has been considered to be without risk to the patient.

Nuclear Physics. The branch of physics concerned with the study of the nuclei of atoms.

Nuclear Physics. Study of the components, structure, and behaviour of the nucleus of the Atom. It is especially concerned with the nature of matter and with Nuclear Energy. The subject is commonly divided into three fields: low-energy nuclear physics, the study of Radioactivity; medium-energy nuclear physics, the study of the force between nuclear particles; and high-energy, or particle, physics, the study of the transformations among subatomic particles in reactions produced in a Particle Accelerator.

Nuclear Reactor. Device for producing Nuclear Energy by controlled nuclear reactions. It can be used for either research or power production. The reactor is so constructed that the fission of atomic nuclei produces a self-sustaining nuclear chain reaction, in which the produced neutrons are able to split other nuclei: A fission reactor consists basically of (1) a fuel, usually uranium or plutonium, enclosed in shielding, (2) a moderator—a substance such as graphite, beryllium, or heavy water—that slows down the neutrons so that they may be more easily captured by the fissionable atoms; and (3) a cooling system that extracts the heat energy produced. The fuel is sometimes enriched—*i.e.,* its concentration of fissionable isotopes is artificially increased—to increase the frequency of neutron capture. The breeder reactor is a special type of reactor that produces more fissionable atoms than it consumes by using surplus neutrons to transmute certain nonfissionable atoms into fissionable atoms. The design of fusion reactors is still in an experimental stage because of the problems involved in containing the plasma fuel and attaining the high temperatures needed to initiate the reaction.

Nuclear Weapon. A bomb in which the explosion is the result of uncontrolled nuclear fission or fusion. In a *fission bomb (atomic* or *A-bomb)* two mases of uranium or plutonium, each below the critical mass, are brought together to form a mass greater than the critical mass. The resulting chain reaction occurs with enormous production of energy. In a *fusion bomb (hydrogen* or *H-bomb)* a fission bomb is used to increase the temperature of a hydrogen-containing material to the point at which nuclear fusion occurs.

Nucleic Acid. Organic substance, found in all living cells, in which the hereditary information is stored and from which it can be transferred Nucleic acid molecules are long chains that generally occur in combination with proteins. The two cheif types are DNA (deoxyribonucleic acid), found mainly in cell nuclei, and RNA (ribonucleic acid), found mostly in cytoplasm. Each nucleic acid chain is composed of subunits called nucleotides, each containing a sugar, a phosphate group,

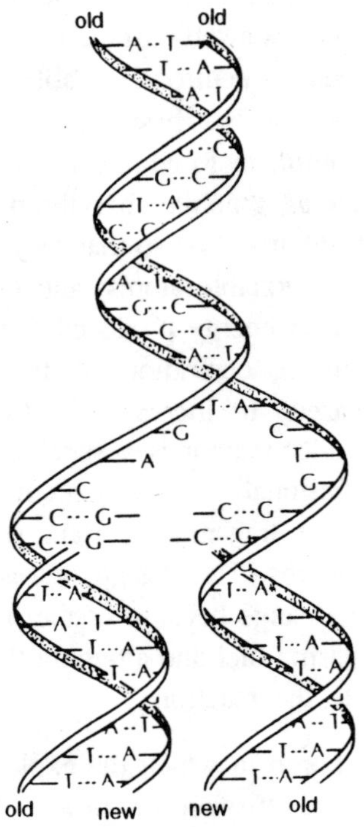

Fig. N-6. *Nucleic acid:* Reokucatuib if stabds if DNA.

and one of four bases: adenine (symbolized A), guanine (G), cytosine (C), and thymine (T). RNA contains the sugar ribose instead of deoxyribose and the base uracil (U) instead of thymine. The specific sequences of nucleotides constitute the cell's genetic information: Each three-nucleotide DNA sequence specifies one particular amino acid. The long sequences of DNA nucleotides thus correspond to the sequences of amino acids in the cell's proteins. In order to be expressed as protein, the genetic information is carried to the protein-synthesizing machinery of the cell, usually in the cell cytoplasm. Forms of RNA mediate this process. DNA not only provides information but also specifies its own exact replication: The cell replicates its DNA by making a complementary copy of its exact nucleotide sequence: T for every A, C for every G, G for every C, A for every T. Although the triplet nucleotide code seems to be universal, the actual sequences of the nucleotides vary according to the species and individual.

Nucleon. A particle that is a constituent of an atomic nucleus; either a proton or a neutron.

Nucleonics. The technology of nuclear physics and its applications.

Nucleon Number. Symbol: *(N)*, The number of nucleons in the nucleus of an atom.

Nucleophilic. Having or involving an affinity for positive electric charge. A nucleophilic reagent (or *nucleophile*) is one that acts by attacking a positive region of a molecule. Examples are negative ions such as Cl- and CN-. Nucleophilic reactions are ones involving such a reagent. Thus, in a *nucleophilic substitution* one atom in a molecule is displaced by a nucleophile; this occurs in displacement reactions involving alkyl halides. In *nucleophilic addition* the initial attack is by a nucleophile: this occurs in addition to the carbonyl group of aldehydes and ketones. *Compare* electrophilic.

Nucleus. The positively charged part of the atom about which the electrons orbit. The nucleus is composed of neutrons and protons (nucleons) held together by strong interactions. The number of protons determines the element and the number of neutrons determines the particular isotope of the element. The principal theoretical models of the nucleus are the liquid drop model and the shell model.

Null Method. a method of measurement in which a zero reading is obtained by balancing one quantity against another, as in the Wheatstone bridge or the Lummer-Brodhun photometer.

Number. Entity describing the magnitude or position of a mathematical object or extensions of these concepts. Cardinal numbers describe the size of a collection of objects; ordinal numbers refer to position relative to an ordering, such as first, second, third, etc. Both types can be generalized to infinite collections. The finite cardinal and ordinal numbers, represented by the Numerals 1, 2, 3,..., are called the *natural numbers*. The *integers* are the natural numbers with their negatives and zero adjoined. The ratios *a/b*, where *a* and *b* are integers and $b \neq 0$, constitute the *rational numbers*, which may be also be represented by repeating decimals, *e.g.*, $^1/_2$ = 0.500..., 2/3 = 0.666.... The *real numbers* are all numbers representable by an infinite decimal expansion, which may be repeating or nonrepeating; they are in a one-to-one correspondence with the points on a straight line. Real numbers that have nonrepeating decimal expansions, *i.e.*, that cannot be represented by any ratio of integers, are called *irrational*. The Pythagoreans knew in the 6th cent. B.C. that $\sqrt{2}$ was irrational. The number $\sqrt{2}$ is also an example of an algebraic number, *i.e.*, it is the Root of a Polynomial equation, in this case $x^2 - 2 = 0$. Numbers that are not algebraic are called *transcendental; e* and π (pi) are examples. The *imaginary numbers* were invented to deal with equations, such as $x^2 + 2 = 0$, that have no real roots. The basic imaginary unit is $i = \sqrt{-1}$. Imaginary numbers take the form *yi*, where *y* is a real number, *e.g.*, $\sqrt{-2} = \sqrt{(-2)}i$. Numbers of the form $x + yi$, where *x* and *y* are real *(e.g.*, $8 + 7i$), are called *complex numbers*. The complex numbers are in a one-to-one correspondence with the points on a plane, with one axis defining the real parts of the numbers and another axis defining the imaginary parts.

Number Theory. Branch of mathematics concerned with the properties of the integers (the Numbers 0, 1, —1, 2, —2,...). Modern number theory made its first great advances through the work of Leonhard Euler, Carl Gauss, and Pierre de Fermat. Much of the focus is on the analysis of prime numbers, *i.e.*, those integer *p* greater than 1 tht are divisible only by 1 and *p*; the first few primes are 2, 3, 5, 7, 11, 13, 17, and 19. The fundamental theorem of arithmetic asserts that any positive integer *a* is a product of primes that are unique except for the order in which they are listed. For example, the number 20 is uniquely the product 2 . 2. 5. This theorem was known to the Greek mathematician Euclid, who also proved that there are an infinite number of primes.

Numeral. Symbol denoting Number. The Arabic numerals (which apparently originated in India) are 1, 2, 3, 4,....; the Roman numerals are I, II, III, IV,.... Both types probably derive from counting on the fingers. The word *digit*, used for the ten numerals 0, 1, 2,...., 9, is from the Ltin for *finger*. Some languages show traces of reckoning by units of 20, using the toes as well as the fingers. The numeral for zero in the Arabic system is much more recent than the other numerals.

Numeration. In mathematics, process of designating Numbers according to a particular system. In any system of numeration a base number is specified, and groupings are then made by powers of the base number. The most widely used system of numeration is the Decimal System, which uses base 10. In the decimal system the Numeral 302 means $(3 \times 10^2) + (0 \times 10^1) + (2 \times 10^0)$, or $300 + 0 + 2$. The binary system, used in most computers, has a base of 2 and only two digits, 0 and 1. The binary numeral 302 means $(3 \times 2^2) + (0 \times 2^1) + (2 \times 2^0)$, *i.e.* $12 + 0 + 2$, or 14, in the decimal system. The decimal numeral 14 and the binary numeral 302 thus represent the same number. The ancient Babylonians used a system of base 60, which survives in our smaller divisions of time and angle, *i.e.*, minutes and seconds.

Numerator. The number placed above the line in a vulgar fraction. In 3/4, 3 is the numerator and 4 the denominator.

Nursing. Profession dealing with prevention of illness and the care and rehabilitation of the sick, encompassing the physical and emotional well-being of the patient as a whole. Nursing includes individualized care on a continuing basis, the carrying out of medical regimes, coordination of necessary interdisciplinary services, and health counseling. Until the middle of the 19th cent. nurses were trained in hospitals to provide beside care. The first school designed primarily to train nurses—rather than provide nursing service for the hospital—was established (1860) in London by Florence Nightingale. The profession in recent years has expanded to include the specialized service of nurse practitioners, physician's assistants, nurse midwives, and nurse anesthetists. Although traditionally most nurses have been women, the number of males in the profession has increased. Training includes classroom and clinical experience, as well as knowledge of automated equipment, artificial organs, and computers.

Nut. In botany, a dry, one-seeded, usually oily Fruit. True nuts include the acorn, chestnut, and hazelnut. The term *nut* also refers to any seed or fruit with a hard, brittle covering around an edible kernel, *e.g.*, the peanut pod (a Legume), and almond (a drupe fruit). Others that are not ture nuts are the Cashew, Coconut, Litchi, Pistachio, and Walnut. Nuts are a valuable food and are often cultivated in nuts orchards.

Nutation. Slight wobbling motion of the earth's axis, superimposed on that which produces the Precession of the Equinoxes. Caused by the difference in gravitational attraction exerted by the sun and the moon, it has an 18.6-yr period.

Nutmeg. Evergreen tree (*Myristica fragrans*) native to the Moluccas. Its fruit is the source of two spices: whole or ground nutmeg, from the seed; and mace, from the fibrous seed covering that separates the seed from the husk.

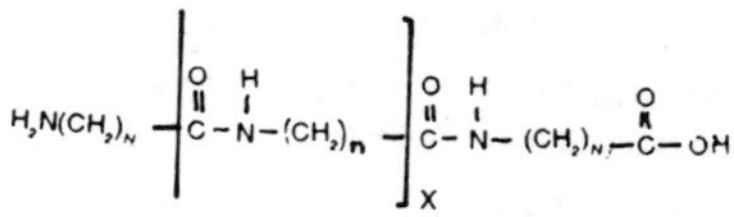

$$H_2N(CH_2)_N - \left[\overset{O}{\overset{\|}{C}} - \overset{H}{\overset{|}{N}} - (CH_2)_n \right]_X - \overset{O}{\overset{\|}{C}} - \overset{H}{\overset{|}{N}} - (CH_2)_N - \overset{O}{\overset{\|}{C}} - OH$$

Fig. N-7. Nylon polymer.

Nylon. Any of a class of synthetic polymeric materials in which the monomers are linked by -CO.NH- groups. Nylon is a polyamide made either by copolymerization of a diamine with a dicarboxylic acid or by polymerization of a compound containing both amine and carboxyl groups in its molecules. It is a translucent creamy-white material used in many moulded and machined articles and parts and in clothing fibres, bristles, rope, etc.

Oak. Tree or shrub (genus *Quercus)* of the Beech family, found in north temperate zones and Polynesia; the more southerly species are usually evergreen. Oaks are cultivated for ornament and are a major source of hardwood lumber. Their durable, attractively grained Wood is valued for shipbuilding, construction, flooring, furniture, barrels, and veneer. The black oaks *(e.g.,* scarlet, pin, willow, live, and shingle oaks) are typified by leaves with sharp-tipped lobes and by acorns that mature in two years. White oaks *(e.g.,* the white, post, cork, and holly oaks) have smooth-lobed leaves and acorns that mature in one year. *Q. alba,* the white oak, is the most valuable timber tree of the genus; the cork oak *(Q. suber)* supplies Cork. Acorns, the fruit of oak trees, are a source of food, tannin, oil, and hog feed. Poison oak belongs to the Sumac family.

Oasis. Fertile area in a desert where there is enough moisture for vegetation. The water may originate in springs or collect in mountain hollows. Oases vary considerably in size, ranging from a pond with a group of date palms to oasis cities with extended agricultural cultivation.

Oats. Cereal plants (genus *Avena)* of the Grass family. Most species are annuals growing in moist temperate regions. Oats are valued chiefly as a pasturage and hay crop and for crop rotation; less than 5% of the oats grown commercially is for human consumption The common cultivated species, *A. sativa,* is native to Eurasia.

Oberon. *See* Uranus.

Objective (object lens). The lens or lens system that is nearest the object in a telescope, microscope, or other optical instrument. It forms an image that is viewed through the eyepiece.

Object lens. *See* objective.

Oblate. *See* ellipsoid.

Observatory, Astronomical. Scientific facility especially equipped to detect and record astronomical phenomena. Early civilizations established primitive observatories to regulate the calendar and predict the changes of season. Later observatories were established to compile accurate star charts and an annual ephemeris that would be of use to navigators in determining longitude at sea. Early instruments used included the armillary sphere, the Astrolabe, the quadrant, and the Sextant. The 17th cent. invention of the Telescope permitted not only more accurate measurement of the positions and motions of celestial bodes, but also analysis of the physical nature of the bodies. The 19th cent. development of dry-plate photography, which permitted long exposure times, offered a much more sensitive method of recording images than the drawings made from visual observations by earlier observers. The spectroscopic study of starlight by various instruments has provided information on the temperature and chemical composition of stars, stellar motions, and magnetic fields. Optical observatories are generally located at high altitudes in sparsely populated areas to minimize adverse seeing conditions caused by weather disturbances, air turbulence, air glow, and any source of extraneous illumination. Astronomical observations have been recently extended to wavelengths outside the visible spectrum *(e.g.,* Gamma Radiation, radio waves, Ultraviolet Radiation, and X-Rays) by, among other means, the use of artificial Satellites equipped with telescopes.

Obsidian. Volcanic glass, commonly black, but also red or brown, formed by Lava that has cooled too quickly for crystals to form. Chemically it is rich in silica and similar to Granite.

Obsidian is used extensively by primitive peoples to make knives, arrowheads, and other weapons and tools.

Obtuse. Denoting an angle that is greater than 90° and less than 180°.

Occlusion. The trapping of one substance in small pockets in crystals of another substance. The occluded material may be solid, liquid, or gas—small amounts of liquid are often occluded when crystals are formed from solution. The term is also used for the absorption of a gas by a solid, in which case the gas molecules or atoms occupy interstitial positions in the lattice and the substance formed is homogeneous.

Occultation. The process in which one celestial object cuts off the light or radio emission from another by moving between it and the earth.

Occupational Disease. Illness resulting from the conditions or environment of employment. Among environmental causes are unusual dampness, causing diseases of the respiratory tract, skin, or muscles and joints; changes in atmospheric pressure, causing Decompression Sickness; and exposure to infrarted or ultraviolet rays or radioactive substances. Widespread use of material used in nuclear power production has led to heightened awareness of Radiation Sickness. The most common of the dust-related disorders are the lung diseases caused by Silica, Asbestos, iron ore, and other metals to which miners and others are exposed.

Ochre. Any of various naturally occurring forms of iron III oxide, Fe_2O_3, used as red or yellow pigments. *Burnt ochre,* which has a brownish-red colour, is a pigment made by calcining natural yellow ochre.

Octadecanoic Acid. *See* stearic acid.

Octagon. A polygon that has eight sides.

Octahedron. A polyhedron with eight faces. A *regular octahedron* has congruent equilaternal triangles as its faces.

Octahydrate. A solid compound with eight molecules of water of crystallization per molecule of compound, as in Iron II fluoride octahydrate, $FeF_2.8H_2O$.

Octal non-tation. A system of representing numbers using the base 8, thus using eight integers, 0-7. For example, the number represent by 125 in decimal notation is $(1 \times 8^2) + (7 \times 8) + 5$, or 175 in octal notation. Each digit in an octal number can be represented by a three digit binary number; *e.g.* 7 is 111, 5 is 101, etc. For this reason octal is used in computers, each digit corresponding to one byte.

Octane. Any of eighteen isomeric alkanes with formula C_8H_{18}. The octanes are colourless liquids obtained from petroleum: the most important is *isooctane,* $(CH_3)_2CCH_2C(CH_3)_2$, which is used as a standard for the octane rating of fuels.

Octane Number. A quality rating for Gasoline indicating the ability of the fuel to resist premature detonation and to burn evenly when exposed to heat and pressure in an Internal Combustion Engine. Premature detonation, indicated by knocking and pinging noises, wastes fuel and may cause engine damage. The octane number can be increased by varying the relative amounts of the different Hydrocarbons that make up the gasoline or by additives, *e.g.,* tetraethyl lead. Federal regulations in the U.S. require commercial gasoline pumps to indicate the octane number, usually about 90 for regular grade gasoline and between 95 and 96 for premium grade. Since the early 1970s most Automobiles have been built to operate on low octane gasoline with little or no lead added.

Octane rating. A number measuring the ability of petrol to burn in an internal-combustion engine without knocking. It is the volume percentage of isooctane (C_8H_{18}) in a mixture of isooctane and (normal) heptane (C_7H_{16}) with the same properties as the fuel when the two are compared under standard conditions. A similar measure, the cetane rating, is used for diesel fuel.

Octant. A segment of a circle equal in area to one eighth of the area of the circle: *i.e.,* having an angle of 45°.

Octave. An interval between two notes such that the frequencies of the two are in the ratio 2:1. An octave covers eight notes of a scale in music.

Octet. A group of eight electrons in the shell of an atom. This configuration has a special stability: it is found in the inert-gas atoms (except helium) and in atoms in compounds.

Octopus. Marine mollusk, with a pouch-shaped body and eight muscular arms or tentacles; a Cephalopod. It seizes its prey with the sucker-bearing arms and paralyzes it with a poisonous secretion. Octopus species range in size from only 2 in. (5 cm) in the North Atlantic to 30 ft (9 m) in the Pacific. Octopuses can change colour, from pinkish to brown, and eject a dark "ink" from a special sac when disturbed. They are used for food in many parts of the world.

Odd-even Nucleus. An atomic nucleus that has an odd number of protons and an even number of neutrons.

Odd-odd Nucleus. An atomic nucleus that has odd numbers of neutrons and of protons.

Oersted. A CGS-elecromagnetic unit of magnetic field strength equal to the field strength that produces a force of one dyne on unit magnetic pole. It is equal to $10^3/4\pi$ amperes per metre. [After Hans Christian Oersted (1777-1851), Danish physicist.]

Ohm. Symbol: ΩThe SI unit of electrical resistance, equal to the resistance between two points of a conductor when a potential difference of one volt between the points produces a current flow of one ampere. In 1948 this definition replaced the *international ohm,* which was defined as the resistance at 0°C of a column of mercury of length 106.300 centimetres, mass 14.4521 grams, and uniform cross section 1 Ω_{int} = 1.000 49 Ω. [After Georg Simon Ohm (1787-1854), German physicist.]

Ohm. Symbol Ω, unit of electrical Resistance, defined as the resistance to the flow of a steady electric current offered by a column of mercury 14.4521 grams in mass, with a length of 1,06300 m, and with an invariant cross-sectional area, when at a temperature of 0° C.

Ohmic. Denoting a resistance or other electrical device that obeys Ohm's law.

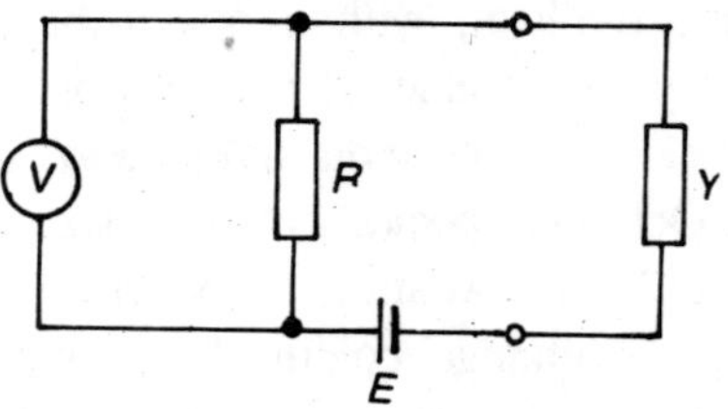

Fig. O-1. Ohmmeter circuit.

Ohmmeter. An instrument for measuring electrical resistance, usually a modified voltmeter containing a small dry battery (e.m.f. E). In the circuit Y is the unknown resistance connected to the meter: the voltage reading is $RE/(R + Y)$. The resistance R is adjusted so that the reading is full scale when $Y = 0$ and other values of Y give lower deflections. The scale is calibrated so that Y is read directly.

Ohmmeter. Instrument used to measure, in Ohms, the electric Resistance of a conductor. It is usually included in a single package with a Voltmeter and often an Ammeter. In normal usage, the ohmmeter operates by using the voltmeter to measure a voltage drop, then converting this reading into a corresponding resistance reading through Ohm's Law.

Ohm's Law. The principle that at constant temperature the current *(I)* flowing through a conductor is directly proportional to the potential difference *(V)* applied across it. The constant of proportionality is the resistance (R) of the material : *i.e.* $V = IR$.

Ohm's Law. Law stating that the electric current *i* flowing through a given Resistance *r* is equal to the applied voltage *v* divided by the resistance, or $i = v/r$. In alternating-current (AC) circuits, where Inductances and Capacitances may also be present, the law must be amended to $i = v/z$, where *z* is Impedance. The law was formulated by the German physicist *Georg Simon Ohm,* 1787-1854.

Oil-immersion. A method of increasing the resolving power of a microscope by placing a few drops of cedar-wood oil on the sample slide and lowering the objective so that it is immersed in the oil. The oil has the same refractive index as glass and serves to direct more light from the object into the lens.

Oil of Vitriol. *See* sulphuric acid.

Olbers' Paradox. An astronomical paradox concerning the appearance of the night sky. The amount of light reaching earth from distant stars is inversely proportional to the square of the star's distance. On the other hand, the number of stars within a given solid angle is directly proportional to the square of the distance. It follows that the amount of light reaching the earth from stars is the same from all distances and that, if the universe contains an infinite number of uniformerly distributed stars, the night sky should be uniformely bright. It is thought that this is not the case because the recession of distant galaxies results in a lower brightness. [After Heinrich Wilhelm Olbers (1758-1840). German astronomer.]

Oleate. Any salt or ester of oleic acid.

Olefine (olefin). *See* alkene.

Oleic Acid (cis-9-octadecenoic acid). A yellow oily liquid, $CH_3(CH_2)_7CH{:}CH(CH_2)_7COOH$, found, in the form of its glycerides, in natural fats and oils. It is used in the manufacture of soaps, ointments, polishes, and lubricants. M.pt. 13°C; b.pt. 286°C; (100 mmHg); r.d. 0.9.

Oleoresin. *See balsam.*

Oleum. See sulphuric acid.

Omega-minus. Symbol: Ω. An elecmentary particle with negative charge and a mass 3276 times that of the electron. It is classified as a hyperon.

Onion. Plant (genus *Allium*) of the Lily family of the same genus as the chive (*A. schoenoprasum*), garlic (*A. sativum*), leek (*A. porrum*), and shallot (*A. ascalonium*). Believed native to SW Asia, these plants are typified by an edible bulb composed of sugar-rich food-storage leaves that are also the source of a pungent oil. Their long, tubular, above-ground leaves are also eaten. The onion (*A. cepa*) is a cultivated biennial with many varieties; it is no longer found in the wild form. Common varieties included the red onion, the yellow onion, the white onion, and the large, delicately-flavored Bermuda and Spanish onions. The more pungent garlic, a pernnial, has a bulb consisting of small bubils called cloves. The perennial shallot has clusters of small, onionlike bulbs; the biennial leek has a single small bulb. The chive, found wild in Italy and Greeve, is a perennial whose leaves are the disirable portion. *Scallion* is a popular term for any edible *Allium* species with a reduced bulb, especially the leek and shallot.

Onium Ion. A positive ion formed by addition of a proton to a neutral molecule as in the formation of an ammonium, phosphonium, or oxonium ion.

Onyx. Variety of Chalcedony, similar to Agate but with parallel, regular bands. Black-and-white specimens are used for cameos. Sardonyx has alternate layers of onyx and carnelian, or sard.

Oolite. Sedimentary rock composed of small concretions, usually of calcium carbonate, containing a nucleus and clearly defined concentric shells. In Britain, oolitic Limestone is characteristic of the Jurassic geologic period.

Opacity. The reciprocal of the transmittance of a medium.

Opal. Hydrous silica mineral (SiO_2 . nH_2O), formed a low temperatures from silica-bearing water, that can occur in cavities and fissures of any rock type. Gem opal has rich iridescence and a remarkable play of colours, usually in red, green, and blue. Most precious opals come from South Australia; other sources included Mexico (fire opal) and part of the U.S.

Opaque. *See* translucent.

Open-chain. Denoting a chemical compound with a straight or branched chain of atoms: *i.e.* a compound that does not contain a ring.

Open-hearth Process (siemens-Martin process). A process for making steel by heating a mixture of pig iron and iron oxide with producer gas on an open health.

Operator. A mathematical symbol representing a particular operation. Thus, the differential operator d/dx represents differentiation with respect to the variable x.

Ophthalmology. Branch of medicine specializing in the function and diseases of the eye. It is concerned with prevention of Blindness; treatment of disorders, *e.g.* Glaucoma; and surgery, including Cataract removal and corneal Transplantation. *Optometry* is concerned with correcting visual abnormalities *(e.g.,* Nearsightedness) and the prescription of corrective lenses, including Contact Lenses.

Opium. Dried milky juice of unripe seedpods of the opium poppy

(*Papavera somniferum*). The chief constituents of opium are the alkaloids Codeine, papaverine, noscapine, and Morphine, from which Heroin is synthesized. Opium is grown worldwide; despite international laws and agreement to control its use, an illicit opium traffic persists.

Opossum. Name for several Marsupials of the family Didelphidae, native to Central and South America, with one species in the U.S. Mostly arboreal and nocturnal animals, opossums have long noses, naked ears, prehensile tails, and black-and-white fur. They eat small animals, eggs, insects, and fruit. When frightened they collapse as if dead. Opossums are hunted as pests as well as for food and sport.

Opposition. The position in which two celestial objects, in parctice the sun and any of the outer planets, have a difference of celestial longitude of 180°. In this position, the outer planet in question is best seen, crossing the observer's meridian at midnight. The moon is at opposition when it becomes full.

Optical Activity. The property of certain substances of rotating the plane of polarization of plane-polarized light. The effect is observed in passing a polarized beam through crystals, liquids, solutions, or vapour. It is caused by asymmetry of the molecules: a solution of one optical isomer will rotate the light in one direction and the other optical isomer causes equal rotation in the opposite sense.

The extent to which the plane of polarization is rotated depends on the wavelength of the light, the path length, and the concentration (of a solution). The optical activity of a substance is expressed by its *specific rotation* $[\alpha]^{t}_{\lambda}\ \alpha = a\ V/lm$, where α is the rotation, *V* the volume, *m* the mass of dissolved or pure substance, and *l* the path length. In the notation, *t* is the temperature and λ the wavelength used (often D is used to indicate the D line of sodium). Optical rotation is measured in a polarimeter.

A substance that rotates the plane in a clockwise sense (looking towards the oncoming light) is *dextrorotatory;* one that causes anticlockwise rotation is *laevorotatory*. Often prefixes *d*- [or (+)) or *l*- or (-)] are used in the names of optically active isomers. These should not be confused with the *D*-and *L*- prefixed to the namesof sugars and other natural compounds. *D*- compounds are named as having similar configurations to an arbitarily chosen isomer of glyceraldehyde and *L*- compounds resemble the other isomer. Thus *D*- and *L*- refer to the configuration of the molecules, not the optical rotation, and it is possible for a *D*- isomer to be laevorotatory. *DL*- and *dl*- denote recemates.

Optical Axis. The axis of symmetry of a lens. mirror, or other optical system.

Optical Bench. A table, track, etc., on which sources, lenses, mirrors, and other optical components can be mounted and moved: used in experiments in optics.

Optical centre. *See* lens.

Optical Character Recognition (OCR). A method of feeding information into a computer in the form of conventional printed characters, which are identified by the input device (the *optical character reader)*. A special typeface is usually necessary.

Optical Crown. Any of various types of optical glass with low dispersion of light. *Compare* optical flint.

optical Flat. A flat surface on which the irregularities are not greater than the wavelength of light. Optically flat surfaces are necessary in many optical devices are experiments.

Optical Flint. Any of various types of optical glass with high dispersion of light *Compare* optical crown.

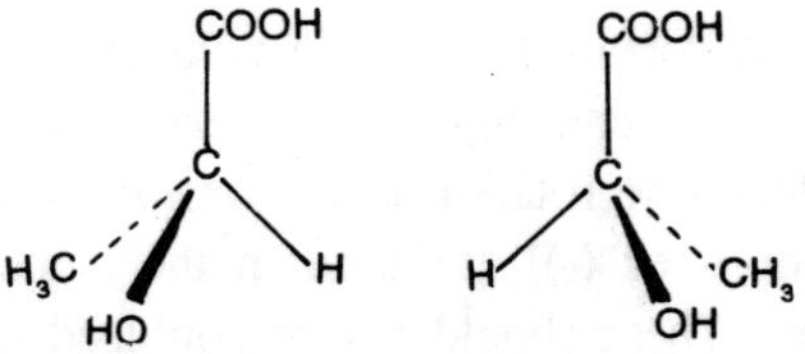

Fig. O-2. Lactic acid.

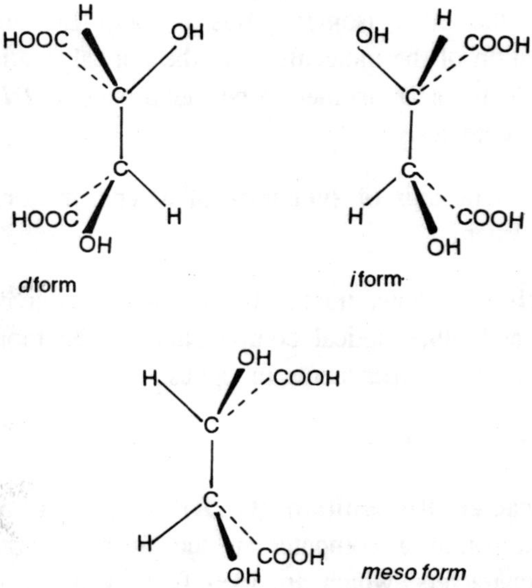

d form *i* form

meso form

Fig. O-3. Tartaric acid

Optical Isomerism (enantiomorphism). A type of stereoisomerism in which a compound can have two structures, one of which is ak mirror image of the other. Optical isomerism can occur in many substances, particularly those that contain an *asymmetric carbon atom, i.e.* a carbon atom to which is attached four different groups arranged tetrahedrally. This occurs in lactic acid, in which the central carbon atom is asymmetric and one structure is a mirror image of the other. The two forms are said to be *optical isomers* or *enantiomers.* They have very similar physical properties but differ in their optical activity.

Tartaric acid is more complicated because it has two asymmetric carbon atoms in its molecules (see fig. 2). In addition to the *l*- and *d*- forms it has a third form —the *meso* form—which is not a mirror image of either. Stereoisomers that are not mirror images are called *diastereoisomers*. The *meso* form of tartaric acid is optically inactive because one half of the molecule is a mirror image of the other.

Optical Pyrometer. *See* pyrometer.

Optical Rotatory Dispersion. A technique in which the optical activity of a substance in investigated as a function of the wavelength of light. A graph of specific rotation against wavelength gives information on the molecular structure.

Optical Sensing. In general, any method by which information that occurs as variations in the intensity, or some other property, of light is translated into an electric signal. This is usually accomplished by the use of various photoelectric devices. Optical sensing is used in various pattern-recognition systems, *e.g.*, in military reconnaissance and astronomical observation, and in photographic development, to enhance detail and contrast.

Optic Axis. The direction in a birefringent crystal in which no double refraction occurs.

Optics. The study of light and optical instruments. *Geometrical optics* is the use of ray diagrams and the laws of reflection and refraction to determine the properties of lenses and mirrors. *Physical optics* is the study of the wave nature of light, and includes such topics as diffraction, interference, and polarization.

Optics. Scientific study of Light. Physical optics is concerned with the genesis, nature, and properties of light; physiological optics with the part light plays in vision and geometrical optics with the geometry involved in the Reflection and Refraction of light as encountered in the study of the Mirror and the Lens.

Orange. Tree (genus *Citrus*) of the Rue family, native to China and Indochina, and its fruit. A Citrus Fruit, the orange is rich in Vitamin C. Among the commercially important species are the sweet, or common, orange *(C sinensis),* which furnishes varieties such as the navel and Valencia; thje sour, or Seville, orange (*C. aurantium*), used as an understock on which to bud sweet orange varieties and in marmalade; and *C. reticulata*, which includes the mandarin orange, the tangerine, and the hardy Satsuma varieties. Orange hybridize readily. The citrange is a cross between two varieties of orange; the tangelo is produced by crossing a tangerine and a grapefruit. Orange may be artificially coloured before marketing. They are eaten fresh, made into juice, or used in preserves and confections. Essential oils from orange rind, flowers, and leaves are used in perfumes. The orange blossom is Florida's state flower.

Orangutan. Ape (*Pongo pygmaeus*) found in swampy coastal forests of Borneo and Sumatra. With their extremely long arms and short, bowed legs, orangutans are highly specialized for arboreal life and rarely descend to the ground. An adult male is about $4^1/_2$ ft (1.4 m) tall and weighs about 150 lb (68 kg); the body is covered with reddish fur.

Orbit. Path is space described by a smaller body revolving around a second, larger body where the motion of the orbiting body is dominated by their mutual gravitational attraction. The size and shape of an orbit are specified by (1) the semimajor axis (a length equal to half the greater diameter of the orbit) and (2) the eccentricity (the distance of the larger body from the center of the orbit divided by the length of the orbit's semimajor axis). The position of the oribit in space is determined by three factor: (3) the inclination, or tilt, of the orbital plane to the reference plane (the Ecliptic for sun-orbiting bodies; a planet's Equator for natural and artificial satellites); (4) the longitude of the ascending Node (measured from the vernal Equinox to the point where the smaller body cuts the reference plane moving south to north); and (5) the argument of pericenter (measured from the ascending node in the direction of motion to the point at which the two bodies are closest). These five quantities, plus the time of pericenter passage, are called orbital elements. The gravitational attractions of bodies other than the larger body causes perturbations in the smaller body's motions that can make the orbit shift, or precess, in space or cause the smaller body to wobble slightly.

Orbit. 1. The path of one celestial body around another.

2. The path of an electron around the nucleus of an atom, often visualized as a circle or ellipse.

Orbital. A wave function in an atom or molecule.

Orchid. Name for the Orchidaceae, a large family of perennial herbs distributed worldwide, but most abundant in tropical and subtropical forests. The family includes many Epiphytes and Saprophytes. Orchid flowers have three petals and three sepals, the central one modified and specialized to secrete insect-attracting nectar. The diverse flower forms are apparently complicated adaptations for Pollination by specific insects. Orchids are highly prized ornamentals. Many are native to North America, and most are bog plants or flowers of moist woodlands and meadows, *e.g.*, the pink-blossomed lady's slipper, or moccasin flower (*Cyripedium acaule*). A species of the tropical American genus *Vanilla* is the source of natural Vanilla flavoring.

Order. The power to which a concentration is raised in an expression for the rate of a chemical reaction. Thus, if the rate of reaction between A and B is proportional to $[A][B]^2$, the reaction said to be first order in A and second order in B. The overall order is three. The order of reaction is distinct from the molecularity.

Order of magnitude. An approximate magnitude to within a factor of ten. For instance, values of 0.01 and 0.03 are the same order of magnitude: 10 and 1000 differ by two orders of magnitude.

Ordinate. The *y* coordinate of a point on a two-dimensional Cartesian graph: *i.e.*, the distance of the point from the *x* axis, measured along the *y* axis from the origin. *Compare* abscissa.

Ore. A mineral used as a source of a metal or other solid element.

Ore. Metal-bearing Mineral mass that can be profitably mined. Nearly all rocks contain some metallic minerals, but often the concentration of metal is too low to justify Mining. Ores often occur in veins in rock, varying in thickness from fractions of an inch to several hundred feet. Minerals with no commercial value, called gangue minerals, are usually found mixed with the ore in the vein. Recovering minerals from their ores is one area in the field of Metallurgy.

Oregano. Name for several herbs using in cooking. *Organum vulgare,* of the Mint family, is the usual source of the spice sold as oregano in the U.S. and Mediterranean countries. Its flavor is similar to that of Marjoram but slightly less sweet. Other *Origanum* species are also sold as oregano. A related herb, *Coleus amboinicius,* is known as oregano in Mexico and the Philippines.

Organ. Musical wind instrument in which sound is produced by one or more sets of pipes, each producing a single pitch by means of a mechanically or electrically controlled wind supply. Several keyboards (manuals) are played with the hands. Projecting knobs (stops) to the sides of the keyboard operate wooden sliders that pass under the mounths of a rank of pipes to "stop" a particular rank. The pedals of the organ are like another keyboard, played with the feet. The prevailing organ for several centuries from the 3d cent. B.C. was the Greek *hydraulos.* Organs in the Middle Ages already had several ranks of *diapason* pipes, their timbre characteristic only of the organ. The 15th cent. added stops, including those imitative of other instruments, *e.g.*, flute. Organ building reached a peak in the German Baroque, then declined. The 19th cent. obscured diapason tone by adding stops imitative of orchestral sound and by the use of crescendo. While the early 20th cent. developed electrification of the mechnacial parts of the organ, Albert Schweitzer and others led a movement back to baroque ideals.

Organic. Relating to carbon compounds and their chemistry. The term was originally applied to compounds produced by living matter but now applies to any carbon compound with the exception of simple compounds such as oxides, carbon disulphide, cyanides, cyanates, and carbonates, which are considered to be inorganic. The study and preparation of organic compounds is the province of *organic chemistry*

Organic Chemistry. Branch of Chemistry dealing with Carbon compounds. Of all the elements, carbon forms the greatest number of different compounds; moreover, compounds that contain carbon are about 100 times more numerous than those that do not. Compounds containing only carbon and Hydrogen are called Hydrocarbons. Organic compounds containing Nitrogen are of great importance to Biochemistry. Organic chemistry is of importance to the petrochemical, pharmaceutical, and textile industries; in textile a prime concern is the synthesis of new organic molecules and Polymers.

Organic Farming. Farming practices that exclude or avoid synthetic Pesticides, Fertilizers, growth regulators, and feed additives. Organic farmers prefer biological pest control, manure, and practices such as Rotation of Crops to supply plant nutrients and control pests. Interest in organic farming has been fostered by awareness of the dangers caused by Insecticides and the excessive use of chemical fertilizers.

Organo-. Prefix indicating that compounds contain organic groups. *Organomercury* compounds, for example, are organometallic compounds of mercury.

Organometallic. Denoting an organic metal compound. Organometallic compounds have direct bonds between the metal atom and carbon atom and include alkyls, aryls, Grignard reagents, and sandwich compounds. Metal carbonates, cyanides, cynates, and carbides are not classed as organometallic.

Origin. The point of intersection (0, 0) of the coordinate axes in Cartesian coordinates; the point from which distances are measured.

Orpiment. Arsenic trisulphide, As_2S_3, found naturally as a yellow solid and also synthesized for use as a pigment.

Ortho. 1. Indicating that substituents in a benzene ring are in adjacent (1, 2) positions.

2. Denoting the most hydrated acid of a series of acids formed from the same anhydride.

Orthochromatic Film. Photographic film that is sensitive to green light in addition to blue. *Compare* panchromatic film.

Orthogonal. Indicating that lines, planes, curves, or other surfaces cross at right angles; perpendicular.

Orthohydrogen. The form of molecular hydrogen in which the atomic nuclei have parallel spins. It exists in equilibrium with *parahydrogen*, in which the spins are antiparallel. At room temperature the ratio of orthohydrogen to parahydrogen is about 3:1 and as the temperature is lowered the equilibrium proportion of parahydrogen increases. Pure parahydrogen can be obtained by using certain catalysts at liquid-air temperature. Chemically the two forms are the same although they differ slightly in their physical properties.

Oscillator. An electronic circuit for producing an oscillating signal.

Oscilloscope. Device based on a Cathode-Ray Tube used to produce a visual display of electrical signals. Typically the horizontal position of the illuminated point is controlled by the value of the independent variable (often time), while the vertical position is controlled by the dependent variable. A third signal is often used to control the brightness of the point.

Osmium (Os). Metallic element, discovered by Smithson Tennant in 1804. It is a very hard and dense, brittle, lustrous, bluish-white metal found in platinum ores. Its tetroxide is used as a stain in microscopy, in fingerprint detection, and as a catalyst. Osmium alloy are used in fountain-pen points, phonograph needles, and instrument bearings.

Osmium. Symbol: Os. A hard brittle lustrous bluish-white transition element found associated with platinum. It is used as a catalyst and in making platinum alloys—osmium-iridium alloys are used in pen nibs, compass bearings, and other applications requiring a hard corrosion-resistant material. Chemically, osmium resembles iridium. It forms a large number of complexes in a range of oxidation states. A.N. 76; A.W. 190.2; m.pt. 3045°C; b.pt. 5027°C; r.d. 22.57; valency 1-8.

Osmium Tetroxide. A yellow highly poisonous crystalline solid, OsO_4 with a pungent odour. It is prepared by heating osmium in air and is used as an oxidizing agent and catalyst. M.pt. 40° C; b.pt. 130°C; r.d. 4.9.

Osmometer. Any apparatus for measuring osmotic pressure.

Osmosis. Preferential flow of certain substances in solution through a semipermeable membrane. If the membrane separates a solution from a pure solvent. The solvent will flow throgh the membrane into the solution. Osmosis can be stopped by applying a pressure to the solution — the value that prevents osmosis occurring is the *osmotic pressure* (symbol: Π). Osmosis is a colligative property, depending on the number of molecules present. In dilute solutions the osmotic pressure of a solution is the pressure that the solute would exert if it were a gas occupying the same volume: *i.e.* $\Pi V - RT$, where V is the volume of solution, R is the gas constant, and T is the thermodynamic temperature.

Osmosis. Spontaneous transfer of a liquid solvent through a semipermeable membrane that does not allow dissolved solids (solutes) to pass. Osmosis refers only to the transfer of solvent; transfer of solute is called Dialysis. In either case, the direction of transfer is from the area of higher Concentration of the material transferred to the area of lower concentration. If a vessel is speared into two compartments by a semipermeable membrane, if both compartments are filled to the same level with a solvent, and if solute is added to one side, then osmosis will occur, and the level of the liquid on the side containing the solute will rise. If an external pressure is exerted on the side containing the solute, the transfer of solvent can be stopped. The minimum pressure to stop solvent transfer is called the osmotic pressure. Osmosis pays an important role in the control of the flow of liquids in and out of a living Cell.

Osprey. Bird of prey related to the Hawk and the New World Vulture, found near water in much of the world. The American osprey or fish hawk (*Pandion haliaetus*) has a wingspan of 5 to 6 ft (152 to 183 cm) and feeds solely on live fish.

Osteomyelitis. Acute or chronic infection of the bone and bone marrow characterized by pain, high fever, and an Abscess at the site of infection. The infection may be caused by variety of microorganisms and reaches the bone through an open wound or fracture, or through the bloodstream. Treatment includes Antibiotics and sometimes surgery.

Osteopathy. Practice of therapy based on manipulation of bones and muscles. Osteopaths maintain that the normal body produces forces necessary to flight disease and that most ailments are due to the misalignment of bones and other faulty conditions of the muscle tissue and cartilage. Osteopathy was founded in the U.S. in 1874 by Andrew Taylor Still.

Ostwald's Dilution Law. The expression $K = \alpha^2/(1 - \alpha)V$, for dissociation of an acid in solution. K is a constant (the equilibrium constant), V the volume of water, and α the degree of dissociation. If α is small, the law can be written $\alpha = (KV)^{1/2}$. [After Wilhelm Ostwald (1853-1932), German chemistist.]

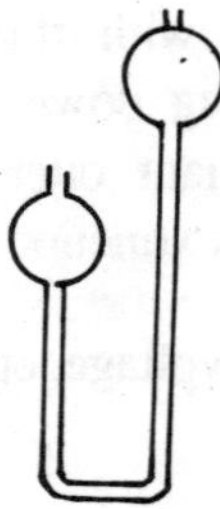

Fig. O-4. Ostwald viscometer.

Ostwald Viscometer. An apparatus for measuring the viscosity of liquids by comparing their rate of flow with that of a standard substance. The liquid is held in the upper bulb and the time *(t)* taken for it to flow to the lower bulb is measured. The density *(p)* is also found. The experiment is repeated with an equal volume of standard substance with viscosity η_o and density ρ_o. If this flows in a time t_o, the required viscosity η is given by $\eta/\eta_o = \rho t_o/\rho_o t$.

Ostwald, Wilhelm. 1853-1932. German physical chemist and natural philospher; b. Lativa. He won the 1909 Nobel Prize in chemistry for his work on catalysis, equibrium, and rates of reaction. He also studied colour and originated the *Ostwald process* for preparing nitric acid. In this process, ammonia mixed with air is heated and led over a catalyst (plantinum). It reacts with oxygen to form nitric oxide, which is then oxidized to nitrogen dioxide; this in turn reacts with water to form nitric acid.

Otis, Elisha Graves. 1811-61. American inventor; b. Halifax, Vt. From his invention (1852) of an automatic safety device to prevent the fall of hoisting machinery he developed the first passenger Elevator (1857), a basic in the development of the Skyscraper.

Otter. Several aquatic, carnivorous Mammals of the Weasel family, found on all continents except Australia. Common river otter of Eurasia and the Americas (genus *Lutra*) are slender, streamlined, and agile, with thick, brown fur. Otters are social and playful, sliding down mudbanks and snowbanks. The South American giant otter and the sea otter of the Pacific, hunted to near extinction, are now protected.

Output. 1. The signal, current, voltage, etc., delivered by an electrical circuit or device.

2. The processed data delivered by a computer. The term also refers to the part of the computer system that converts the data into usable form. Commonly this is some form of printer for producing conventional printed characters but may also be a device for producing punched cards, magnetic tape, or a visual display.

Overtone. A partial of higher frequency than the fundamental in a musical note.

Ovum. In biology, specialized female plant or animal sex cell, also called the egg. In higher animals the ovum differs from the Sperm (male sex cell) in that it is larger and nonmotile. Its nucleus contains the Chromosomes, which bear the herediatary material of the parent. Ova are produced in the ovary of the female. They undergo a maturation process that includes Meiosis, by which the number of their chromosomes is reduced by half. The union of mature sperm and ovum, with each bearing half the normal number of chromosomes, results in a single cell (the zygote) with a full number of chromosomes. The zygote eventually becomes a mature individual. The pattern is similar in plants that reproduce sexually. The term *egg* also refers to a complex structure, such as a bird's egg, in which the ovum is swollen with yolk (food material) and the rest of the egg is secreted around the ovum. Development from an unfertilized ovum is called Parthenogenesis.

Owl. Nocturnal Bird of prey found worldwide, belonging to the families Tytonidae and Strigidae. Owls resemble shortnecked Hawks, except that their eyes are directed forward and surrounded by disks of radiating feathers. Their eyes are specially adapted to seeing in partial darkness, and most sleep during the day. Owls' soft, fluffy plumage makes them almost noiseless in flight. They feed on rodents, frogs, insects and small birds. Many usurp the deserted nests of other birds; others live in burrow.

Oxalate. Any salt or ester of oxalic acid.

Oxalic Acid (ethanedioic acid). A colourless poisonous crystalline solid, HOOC.COOH, present in many plants. It is prepared by passing carbon monoxide into sodium hydroxide solution under, pressure and is used as a reducing agent and metal cleaner. M.pt. 101°C; r.d. 1.6.

Oxalis or **Wood Sorrel.** Plant (genus *Oxalis*), usually with cloverlike leaves that respond to darkenss by folding back their leaflets. Most cultivated forms are tropical herbs. Wild North American species include the white wood sorrel (*O. acetosella*), a plant identified as the Shamrock. True sorrel is a plant of the Buckwheat family.

Oxidant. An oxidizing agent, particularly one that supplies oxygen. The term is often used for the liquid oxygen, hydrogen peroxide, or similar substance that oxidizes the fuel in a rocket.

Oxidation. A process in which oxygen is combined with a substance or hydrogen is removed from a compound. Thus, the reaction

$C + O_2 = CO_2$ is an oxidation of carbon and $H_2S + O = H_2O + S$, where O is supplied by an oxidizing agent, is an oxidation of hydrogen sulphide. The term has been extended to any reaction in which electrons are lost, as in the oxidation of ferrous ions to ferric ions: $Fe^{2+} - e = Fe^{3+}$. *Compare* reduction.

Oxidation and Reduction. Complementary chemical reactions characterized by the loss of gain, respectively, of one or more electrons by an atom or molecule. When an atom or a molecule combines, or forms a chemical bond, with oxygen, it tends to give up electrons to the oxygen. Similarly, when it loses oxygen, it tends to gain electrons. Oxidation is defined as any reaction involving a loss of electrons, and reduction as any reaction involving the gain of electrons. The two processes, oxidation and reduction, occur simultaneously and in chemically equivalent quantities; the number of electrons lost by one substance is equaled by the number of electrons gained by another substance. The substance losing electrons (undergoing oxidation) is said to be an electron donor, or a reductant. Conversely, the substance gaining electrons (undergoing reduction) is said to be an electron acceptor, or an oxidant. Common reductants (substances readily oxidized) are the active Metals, Carbon, Carbon Monoxide, Hydrogen, hydrogen sulfide, and sulfurous acid. Common oxidants (substances readily reduced) include the Halogens, Nitric Acid, Oxygen, Ozone, potassium permanganate, potassium dichromate, and concentrated Sulfuric Acid.

Oxidation State or Number. A measure of the electronic state of an atom in a particular compound, equal to the number of electrons it has more than or less than the number in the free atom. In ionic compounds the oxidation states are simply the charges of the ions, with due regard to sign. Thus, in calcium chloride, $CaCl_2$, the calcium has an oxidation state of +2 (Ca^{2+}) and the chlorine has a state of -1 (Cl^-). In covalently bonded compounds the electrons are assigned to the more electronegative of the pair of atoms. Thus in ammonia, NH_3, the nitrogen atom is in a -3 oxidation state and the hydrogen is in a + 1 state (as if the compound was ($3H^+N^{3-}$). A pure element has an oxidation state of 0. Coordinate bonds are considered to have no effect on the oxidation state. Thus the compound $Ni(CO)_4$ has nickel in the 0 oxidation state. The complex ion $[Cu(NH_3)_4]^{2+}$ contains copper in its + 2 oxidation state. It is considered to be a cupric ion (Cu^{2+}) with four ammonia molecules coordinated to it. The use of oxidation numbers is the basis of a system of nomenclature for inorganic compounds. Thus ferrous oxide, FeO, is called iron (II) oxide and ferric oxide, Fe_2O_3, is iron (III) oxide.

Oxide. Any compound of oxygen with another element. Oxides can be classified into two main types. Covalent oxides have simple covalently bonded molecules. Examples are carbon dioxide, sulphur trioxide, and nitrogen dioxide. Most covalent oxides are acidic, although some, such as carbon monoxide, are neutral. Ionic oxides are metal compounds containing the O^{2-} ion. They are all basic compounds: some are amphoteric. In addition, some metals can form *peroxides,* which contain the O_2^{2-} ion, *superoxides,* which contain the O_2^- ion, and *ozonides,* which contain the O_3^- ion.

Oxidizing Acid. An acid that acts as an oxidizing agent. The term is used for compounds, such as concentrated nitric acid, which dissolve metals that are lower than hydrogen in the electro-chemical series. Copper, for example, is dissolved according to the equations: $Cu + 2HNO_3 = CuO + H_2O + 2NO_2$; $CuO + 2HNO_3 = Cu(NO_3)_2 + 2H_2O$. In contrast. nonoxidizing acids do not dissolve such metals.

Oxidizing Agent. A compound that causes oxidation, either by supplying oxygen, abstracting hydrogen, or accepting electrons. Many compounds act as oxidizing agents in particular circumstances but the term is also used for compounds that are particularly effective oxidizing agents. Examples are oxygen, hydrogen peroxide, ferric salts, and concentrated nitric and sulphuric acids.

Oxonium Ion. A positive ion of formula R_3O^+, where R is either hydrogen or an organic group. The simplest example is the hydroxonium ion, H_3O^+.

Oxy. Designating a compound that contains oxygen.

Oxygen. Symbol: O. A colourless odourless tasteless gaseous element forming 28% of the atmosphere by volume; the most abundant element in the earth's crust (49.2% by weight). Oxygen is essential for combustion and respiration of plants and animals. It is obtained by liquefying air and used in blast furnaces, welding, respirators, and the manufacture of organic chemicals. The element is diatomic, O_2, and highly reactive, combining with most other elements. It is one of the chalconide elements. A highly active triatomic allotrope, ozone, also exists. A.N. 8; A.W. 15.9994; m.pt. -218.4° C; b. pt. -183.0°C; r.d. 1.105 (air = 1).

Oxygen (O). Gaseous element, first isolated (c.1773-74) independently by Joseph Priestley and Karl Scheele. A colourless, odourless, tasteless gas, it is the most abundant element on earth, constituting about half of the surface material. It makes up about 90% of water, two-thirds of the human body, and 20% by volume of air. Normal atmospheric oxygen is a diatomic gas (O_2). Ozone is highly reactive triatomic (O_3) allotrope of oxygen. Oxygen forms compounds with almost all of the elements except the inert gases. The common reaction in which it unites with another substance is called oxidation. The burning of substances in air is rapid oxidation, or combustion. The Respiration of plants and animals is a form of oxidation essential to the liberation of the energy stored in such food materials as carbohydrates and fats. Chief industrial uses are in Steel production (*e.g.*, in the Bessemer Process) and the oxyacetylene torch; (in medicine it is used to treat respiratory diseases. Liquid oxygen is used in rocket fuel systems.

Oyster. Bivalve mollusk found in beds in the shallow, warm waters of all oceans. Except in the free-swimming larval stage, oysters spend their lives attached to substrates of rocks, shells, or roots by means of a cementlike secretion. The pearl oyster, from which the Pearl is obtained, is a large (12 in./ 30.5 cm) tropical species. The edible American, or common, oyster is harvested commercially in artificial beds on the U.S. coast.

Ozone. A pale blue gaseous allotrope of oxygen, O_3, prepared by passing an electrical discharge through oxygen. Ozone is a poisonous unstable compound, a powerful oxidizing agent, used for water purification. It reacts with unsaturated organic compounds to give ozonides. M.pt. -192°C; b.pt. -112°C.

Ozone. Triatomic Oxygen (O_3). Pure ozone is an unstable, faintly bluish gas with a characteristic fresh, penetrating odour. It is the most chemically active forms of oxygen. Ozone is formed in the ozone layer of the stratosphere by the action of solar ultraviolet light on oxygen; this layer plays an important role in preventing most ultraviolet and other high-energy radiation, which is harmful to life, from penetrating to the earth's surface. Some environmental scientists fear that certain man-made pollutants, *e.g.*, nitric oxide (NO), may cause a drastic depletion of stratospheric ozone. Ozone is also formed when an electric discharge passes through air; it is produced commercially by passing dry air between two electrodes connected to an alternating high voltage. Ozone is used as a disinfectant and decontaminant for air and water.

Ozonide. 1. A type of compound prepared by addition of ozone to double bonds. Ozonides contain a three-membered ring of two carbon atoms and an oxygen atom in their molecules and are unstable, often explosive, substances. The process of forming ozonides *(ozonolysis)* is used as a tool in organic chemistry to detect the positions of double bonds in milecules.

2. An inorganic compound containing the O_3^- ion. An example is potassium ozonide, KO_3.

Ozonizer. An apparatus for producing ozone by maintaining an elecrical discharge in a stream of oxygen.

Ozonosphere (ozone layer). A layer containing ozone in the earth's atmosphere. It lies between heights of 15 and 30 kilometres and absorbs the sun's higher-energy ultraviolet radiation.

P

Pacemaker, Artificial. Device used to stimulate a rhythmic heartbeat for means of electrical impulses. Implanted in the body when the heart's own electrical conduction system does not function normally, the battery-powered device emits impulses that trigger heart-muscle contraction at a rate present or controlled by an external remote switch.

Packing Fraction. The difference between the atomic mass of an isotope and its mass number, divided by its mass number: *i.e.* $f = (M - A)/A$. The fractions are sometimes multiplied by 10^4. They give an indication of the stability of atomic nuclei, being proportional to the energy binding the nucleons together.

Pair Production. Production of an elementary particle and its antiparticle from a photon in the field of an atomic nucleus. A gamma-ray photon, for example, can be converted into an electron and a positron. The total mass produced is equivalent to the energy lost by the photon according to the law of conservation of mass-energy.

Palaeomagnetism. The study and measurement of the magnetization of rocks, used to investigate the way the earth's field has changed with time. The method is most useful for studying igneous rocks, which are formed at high temperatures and contain iron compounds. As they cool the rocks are magnetized by the earth's field and their intensity and direction of magnetization gives an indication of the intensity and direction of the field at the time the rocks were formed. Such studies show that the earth's polarity has changed direction many times in the past. The technique can also be used as a method of dating rocks.

Palladium. Symbol: Pd. A steel-white soft ductile transition element occurring in certain copper and nickel ores. The metal is used in jewellery and as a hydrogenation catalyst. It is fairly unreactive and does not tarnish in air, although an oxide forms at high temperatures. It dissolves slowly in cold nitric acid and hot hydrochloric and sulphuric acids and dissolves quickly in fused alkalis. Palladium is also capable of absorbing (occluding) up to 900 times its own volume of hydrogen. Unlike nickel it has very few simple ionic salts containing cations. Most of its compounds are complexes, being mainly palladium II (*Palladous*) compounds with a few palladium IV (*Palladic*) compounds. It also forms some complexes in other oxidation states. A.N. 46; A.W. 106.4; m.pt. 1552°C; b.pt. 3140°C; r.d. 12.02; valency 2, 3, or 4.

Palladium (Pd). Metallic element, discovered in 1803 by William Wollaston. It is lustrous, corrosion-resistant, silver-white metal with a great ability to absorb hydrogen. Major uses are in alloys used in low-current electrical contracts, jeweliry, and dentistry.

Palm. Common name for the Palmae, or the Arecaceae, a large family of chiefly tropical trees, shrubs, and vines. Most species are trees, typified by a crown of compound leaves (fronds) terminating a tall, woody, unbranched stem. The fruits. covered with a tough, fleshy, fibrous, or leathery outer layer, usually contain a large amount of stored food. Important economically, especially in the tropics, palms *e.g.*, Coconut and Date palms, provide food and other products. Important palm fibers and raffia and Rattan. *Palm oil* is fat pressed from the fruit, principally that of the coconut palm, African oil palm (genus *Elaeis*), and some other species. The oils are widely used in soap, candles, lubricants, margarine, fuel, and other products.

Palmitic Acid (hexadecanoic acid). A white crystalline solid, $CH_3CCH_2)_{14}$ COOH, occurring in the form of its glyceride, in many oils and fats, from which it may be extracted by hydrolysis. One of the more common fatty acids, palmitic acid is used in soaps and lubricants and as a food additive. M.pt. 63°C; b.pt. 351°C; r.d. 0.85.

Palomar Observatory. Astronomical observatory located at an altitude of 5,500 ft. (1,680 m) on Palomar Mt., NE of San Diego, Calif., and operated by the California Institute of Technology. Its primary instrument is the 200-in. (500-cm) Hale reflector, which was the largest in the world at the time of its completion (1948). Its 48-in. (120-cm) Schmidst camera telescope was used to prepare a monumental photographic atlas of about three fourths of the entire sky.

Panchatantra. Anonymous collection of Sanskrit animal fables, probably compiled before A.D. 500. Derived from Buddhist sources and intended for instruction of sons of royalty, the prose fables are interspersed with aphoristic verse.

Panchromatic Film. Photographic film that is sensitive to all the colours of the visible spectrum. *Compare* orthochromatic film.

Pancreas. Glandular organ of the Digestive System that secretes digestive enzymes and hormones. In humans, the pancreas is a yellowish organ that lies crosswise beneath the stomach and is connected to the small intestine at the duodenum. It produces trypsin, amylase, and lipase—enzymes essential to the digestion of proteins, carbohydrates, and fats respectively. Small groups of cells in the pancreas, called islets of Langerhans, secrete two hormones, Insulin and glucagon, which regulate blood-sugar levels.

Panda. Two nocturnal Asian Mammals, the red panda (*Ailurus fulgens*), and giant panda (*Ailuropoda melanoleuca*). The red panda, or lesser panda, resembles a Racoon and is found in the Himalayas and the mountains of W China and N Burma. The giant panda resembles a Bear, although it is anatomically more like a racoon; it lives in the high bamboo forests of central China. Its body is mostly white, with black limbs, ears, and eye patches; adults weight from 200 to 300 lb (90 to 140 kg).

Panther. Name commonly applied to black Leopards, a colour variant of the leopard, not a distinct species. The generic name *Panthera* refers to all big roaring cats.

Paper. Thin, flat sheet usually made from plant fiber. It was probably invented c.105 in China, where it was made from a mixture of bark and hemp. The Moors introduced the papermaking process into Spain c.1150, and by the 15th cent., when Printing developed in Europe, paper mills had spread throughout the Continent. The basic papermaking process exploits the ability of plant cell fibers to bond together when a pulp made from the fibers is spread on a screen and dried. Today, paper is made principally from wood pulp combined with pulps from waste paper or, for fine grades of paper, with fibers from cotton rags. For newsprint, tissue, and other inexpensive papers, the pulp is prepared mechanically, by griding the wood. Chemical pulp is made by boiling a mixture of wood chips with either soda, sulfite, or sulfate, a process that removes legnin. The pulp is poured onto a wire screen, where the water drains away and the fibers begin to mat. The paper layer then passes through a series of rollers that dry, press, and smooth it, and add various finishes. Writing papers contain a water-resistant substance such as rosin to prevent the spreading of ink. Kraft paper, made chemically with sulfate, is used for bags and wrapping papers because of its strength.

Paper nautilus. Mollusk, closely related to the octopus, with a rounded body, eight tentacles, and no fins. It is named for the delicate, papery shell, actually an egg case, that surrounds the egg-brooding female. A Cephalopod found in most warm waters, it is not a true Nautilus.

Papyrus. Plant (*Cyperus papyrus*) of the Sedge family, now almost extinct in Egypt but universally used there in antiquity. The roots were used as fuel; the pith was eaten. The stem was used for sandals, boats, twine, mats, and cloth, and, most notably, in a paperlike writing material.

Para. Indicating that substituents in a benzene ring are attached to carbon atoms on opposite sides of the ring: *i.e.* to the 1, 4 positions. See structural isomerism.

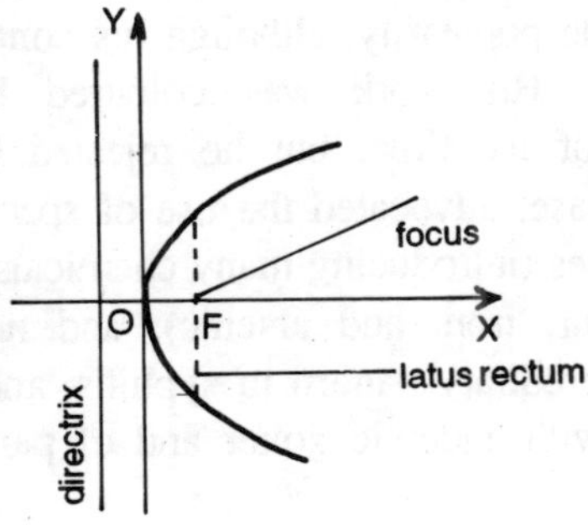

Fig. P-1. Parabola.

Parabola. A conic with an eccentricity equal to 1, It has an equation of the type $y^2 = 4mx$, where (m, o) is the focus and y = -m is the directrix. In this form the vertex of the parabola is at the origin and the axis is the x axis. A line drawn from the focus to any point on the parabola makes an angle with the normal at that point equal to the angle made by a line parallel to the x axis, a property known as the *reflection property* of the parabola.

Parabolic. Having the shape of a parabola or a paraboloid.

Paraboloid. A surface that has parabolic cross sections. There are two types. A *hyperbolic paraboloid* $(x^2/a^2 + y^2/b^2 = 2cz)$ has parabolic cross sections in two planes and a hyperbolic cross section in the third. An *elliptic paraboloid* $(x^2/a^2 - y^2/b^2 = 2cz)$ has an elliptical cross section in its third plane. A *paraboloid of revolution* is a special type of elliptic paraboloid produced by rotating a parabola about its x axis. This shape is used in devices for producing a parallel beam of radiation from a point source or for focusing a parallel beam to a point, as a result of the reflection property of the parabola. Parabolic reflectors are used in searchlights, electric fires, radar antenna radio telescopes, etc.

Paracelsus, Philippus Aureolus. 1493?-1541, Swiss physician and alchemist, originally named Theophrastus Bombastus von Hohenheim. Learned in alchemy, chemistry, and metallurgy, he gained wide popularity, although his contemporaries often opposed him. His work was coloured by the fantastic philosophies of the time, but he rejected Galen's humoral theory of disease; advocated the use of specific remedies for specific diseases (introducing many chemicals, *e.g.*, laudanum, mercury, sulfur, iron, and arsenic); and noted relationship such as the hereditary pattern in syphilis, and the association of cretinism with endemic goiter and of paralysis with head injuries.

Parachute. Umbrellalike device designed to retard the descent of a falling body by creating drag as it passes through air. A parachute is usually constructed from a flexible material; when extended, it takes the form of an umbrella from which a series of cords converge downward to a harness strapped to the user. It can be folded into a small package and thus can be easily carried aboard an aircraft or strapped onto a person's body. The rate of descent for a human-carrying parachute is about 18 ft/sec (5.5 m/sec). The first successful parachute descent from a great height was made in 1797 by the French aeronaut Jacques Garnerin, who dropped 3,000 ft (920 m) from a balloon. Parachutes are used as escape systems for persons abroad aircraft unable to land safely, as braking devices for rockets, space vehicles, airplanes, and high speed surface a vehicles, and as a means to land airborne military units and their equipment from transport planes. Parachute jumping for sport is known as skydiving.

Paracyanogen. A white polymeric substance formed by heating cyanogen.

Paraffin.1. *See* alkane.

2. (kerosene). A mixture of hydrocarbons boiling in the range 150-300°C obtained by distillation of petroleum. It is used as a fuel, especially for domestic heaters.

Paraffin. White, semitranslucent, odourless, tasteless, water-insoluble, waxy solid. Though relatively inert, it burns readily in air. A mixture of Hydrocarbons obtained from Petroleum during refining, paraffin is used in candles and for coating paper.

Paraformaldehyde. A white solid polymer of formaldehyde, $(HCHO)_n$, where *n* is 6 or more. It is made by dehydrating aqueous formaldehyde and is used as a disinfectant. M.pt. 120-170°C.

Parahydrogen. *See* Orthohydrogen.

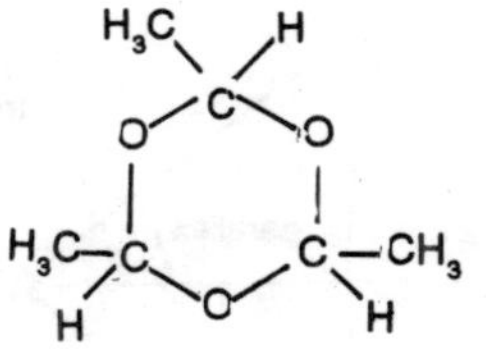

Fig. P-2. Paraldehyde.

Parakeet or **Parrakeet.** Name for a widespread group of small Parrots, with generally green plumage, native to the Indo-Malayan region and popular as cage birds. The budgerigar, also called the shell, zebra, or grass, parakeet (*Melopsittacus undulatus*), is the best known of the true parakeets. The extinct Carolina parakeet was the only parrot native to the U.S.

Paraldehyde. A colourless cyclic polymer of acetaldehyde, $C_6H_{12}O_3$, made by treating acetaldehyde with sulphuric acid. It is less reactive than acetaldehyde and is used as a substitute for it. M.pt. 12°C; b.pt. 124°C; r.d. 0.99.

Parallax. 1. The apparent change in the separation between two objects when they are viewed from different positions. If two objects are in line with the eye they appear to move apart as the observer moves to the left or right. "No parallax" only occurs when both objects occupy the same position: this is used as a test for finding the position of the image in simple experiments in optics. Parallax errors can occur in measuring instruments when the pointer is a small distance in front of the scale. In many instruments a mirror is placed behind the pointer, which is read from an angle at which it is in line with its reflection—the error is then eliminated.

2. (Annual Parallax). The maximum angle subtended at a given star by the earth's mean orbital radius. The parsec is a unit of distance based on parallax.

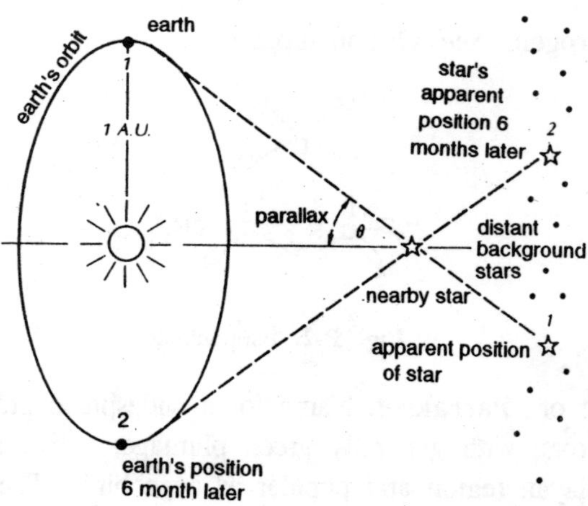

Fig. P-3. *Parallax*: A star's trigonometric parallax θ is a measure of its apparent motion against the background of more distant stars, a motion resulting from the earth's orbit around the sun. To obtain the maximum apparent motion of the star, measurements are made at two points six months apart.

Parallax. Any alteration in the relative apparent positions of objects produced by a shift in the position of the observer. Stellar parallax is the apparent displacement of a nearby star against the background of more distant stars resulting from the motion of the earth in its orbit around the sun; formally, the parallax of a star is the angle at the star that is subtended by the mean distance (1 Astronomical Unit) between the earth and the sun. A star's distance d in Parsec is thus the reciprocal of its parallax p in seconds of arc (or $d = 1/p$). Friedrich Bessel measured (1838) the first stellar parallax (0.3 seconds of arc for the star 61 Cygni). Geocentric parallax, used to determine the distances of solarsystem objects, is measured similarly; the diameter of the earth, rather than that of its orbit, however, is used as the baseline.

Parallel. 1. Denoting lines or planes that are equally distant from each other at all points.

2. (In parallel). Denoting electrical circuit elements joined so that each element is connected across the same two points of the circuit. If resistances R_1, R_2, R_3,... are in parallel, the total resistance R is given by: $(1/R) = (1/R_1) + (1/R_2) + ...$ Capacitances in parallel have a total capacitance equal to their sum: *i.e.* $C = C_1 + C_2 + C_3 + ...$ *Compare* series.

Parallelepiped. A solid whose faces are all parallelograms.

Parallelogram. A quadrilateral that has both pairs of opposite sides parallel. The area of a quadrilateral is the length of the base multiplied by the perpendicular distance between the base and the opposite side.

Parallelogram rule. A method of adding two vectors by representing them by lines forming adjacent sides of a parallelogram, so that the vector sum (the *resultant*) is represented by the diagonal of the parallelogram. For example, in a parallelogram OACB, *OC* is the resultant of *OA* and *OB* and its length and direction can be found by solving the triangle OAC. A physical example of vector addition would be two forces acting simultaneously on a body, for which the resultant force could be found by a *parallelogram of forces*. The method can be extended to adding more than two vectors by simply placing them end to end to form a polygon. If vectors are represented in three dimensions in the form $a = ia_x + Jay + ka_r$, then addition of vectors follows the rule: $a + b = i\,(a_x + b_x) + j\,(a_y + b_y) + k\,(a_z + b_z)$.

Paramagnetism. A type of magnetic behaviour in which the material has a fairly low positive susceptibility that is inversely proportional to temperature. A paramagnetic sample will tend to move towards an applied magnetic field. The effect is caused by the spins of unpaired electrons in the atoms or molecules, which give the atoms a magnetic moment. The effect of paramagnetism always overwhelms the diamagnetic behaviour of the solid. Measurement of paramagnetic susceptibility is one method of finding the number of unpaired electrons in compounds of transition metals.

Paramecium. Microscopic, one-celled, slipper-shaped animal (genus *Paramecium*), in the phylum Protozoa. Paramecia are among the most complex single-celled organisms. Found in fresh water, they swim rapidly, usually in a cork-screw fashion, by means of coordinated, wavelike beats of their many short, hairlike projections, called cilia. Paramecia feed on smaller organisms, such as bacteria.

Parameter. An element of a mathematical expression that may, by varying, cause different values to be assigned to the whole. For example, if the equation of a circle is of the form $y^2 = r^2 - x^2$, then x and y are variables and r is a parameter, determining which circle centred on the origin is described.

Paranoia. Condition characterized by persistent and logically reasoned false beliefs, or delusions, often of persecution or grandeur (the belief that one has an important mission). The mildly paranoid person may lead a relatively normal life, since most aspects of general behaviour are unaffected by the delusional beliefs. In paranoid Schizophrenia, however, there is a deterioration in judgement and perception of reality as the patient develops delusional systems that control all aspects of behaviour.

Paraquat. A yellow water-soluble highly poisonous solid, $C_9H_{20}N_2(SO_4)_4$, widely used as a herbicide.

Parent. *See* daughter.

Parity. Symbol: *P*. A property of elementary particles depending on the symmetry of their wave function with respect to changes in sign of the coordinated. The wave function will generally be a mathematical function of coordinates x, y and z. It these are replaced by $-x$, $-y$, and $-z$ the wave function either remains the same, in which case the particle has parity +1, or changes sign, in which case it has parity –1. In any system of particles interacting by strong or electromagnetic interactions the principle of *conservation of parity* applies: *i.e.* the total parity is unchanged in the interaction.

This does not occur in process that involve weak interactions. In beta decay it is found that the emitted electrons always have a spin in the opposite direction to their direction of motion—a violation of the principle of parity invariance.

Para rubber tree. Large tree (*Hevea brasiliensis*) of the Spurge family, native to tropical South America and the source of the greatest amount and finest quality of natural Rubber. They yellow or white latex from which rubber is made occurs in vessels in the bark layers. Trees, which may grow over 100 ft (30 m) high, are trapped for the latex.

Parasite. Organism that obtains nourishment from another living organism (the host). The host, which may or may not be harmed, never benefits from the parasite. Many parasites have more than one host and most cannot survive apart from their host. Parasites include bacteria (*e.g.*, those causing Tuberculosis), invertebrates such as worms (*e.g.*, Tapeworm), and vertebrates (*e.g.*, the Cuckoo, which lays its eggs in the nests of other birds).

Parathyroid glands. Four small endocrine bodies behind the Thyroid Gland that govern calcium and phosphorus metabolism. A low calcium ion concentration in the blood causes these glands to produce parathormone, which increase calcium absorption in the intestines and kidneys and takes calcium salts from the bones. The hormone also decreases the concentration of phosphate ions, which form a relatively insoluble salt with calcium. High parathormone levels can lead to bone degeneration; low levels lead to calcium deficiency (typified by muscle spasms and, eventually, convulsions and psychiatric symptoms).

Parchment. Untanned animals skins, especially of sheep, calf, or goat, prepared for use as a writing material. The skins were soaked, scraped, and stretched, then rubbed with chalk and pumice. More durable than the papyrus used in ancient times, parchment was the chief writing material in Europe until the advent of printing brought Paper into wide use. Vellum is a fine grade of parchment.

Parkinson's disease or **Parkinsonism.** Degenerative brain disorder initially characterized by trembling lips and hands and muscular rigidity, later producing body tremors, a shuffling gait, and eventually possible incapacity. The disease may occur as the result of influenza Encephalitis, carbon monoxide poisoning, or drugs, but in most cases the cause is unknown. Symptoms, which usually appear after age 40, are treated with the drugs L-Dopa (combined with carbidopa to reduce the side effects) and amantadine. Emotions may be affected and mental capacity impaired, but assessment of these is difficult because depression often accompanies the disease. Parkinsonism is named for the English surgeon James Parkinson, who first described it in 1817.

Parrot. Common name for brilliantly coloured Birds of the pantropical order Psittaciformes. Parrots have large heads, short necks, and strong feet with two front and two back toes (for climbing and grasping). Parrots range from the $3^1/_2$ in.

(8.7 cm) pygmy parrot of the South Pacific tc the 40 in. (100 cm) Amazon parrot of South America. Species include the Parakeets, cockatoos, cockateels, and macaws. Parrots are long-lived, and many are popular as cage birds; some species can learn to mimic speech.

Parsec. A unit of distance used in astronomy, equal to the distance corresponding to a parallax of one second of arc. It is equivalent to 3.26 light-years or 3.086 X 10^{16} metres.

Parsley. Mediterranean aromatic herb (*Petroselinum crispum*) of the Carrot family. It has been cultivated since Roman antiquity for its foliage, used as a seasoning and garnish.

Parsons, Sir Charles Algernon. 1854-1931, English engineer and inventor of a revolutionary steam Turbine. His first turbines drove electric generators. In 1897 he constructed the *Turbinia,* the first turbine-propelled vessel; its amazing speed led to the construction of many turbine-powered warships for the British navy.

Parthenogenesis. In zoology, reproduction in which an unfertilized egg develops into a new individual. It is common in lower animals, especially insects, *e.g.*, the Aphid. In many social insects, *e.g.,* the honeybee, unfertilized eggs produce male drone and fertilized eggs produce female workers and queens. Artificial parthenogenesis has been achieved by scientists for most major groups of animals, although it usually results in abnormal development. In plants, the phenomenon (called parthenocarpy) is rare.

Partial. A tone that is an integral multiple of the fundamental frequency in a musical note. The partials of a note are the same as its harmonics. The fundamental is the fist partial. Higher frequencies are also called overtones.

Partial derivative. A derivative taken with respect to only one variable, other variables being treated as constants. The symbol

∂ is used. For example, if $y = 3x^2z^3$, $\partial y/\partial x = 6xz^3$ and $\partial y/\partial z = 9x^2z^2$.

Partial pressure. The pressure that a component of a mixture of gases would exert if it were present alone in the same volume as the mixture.

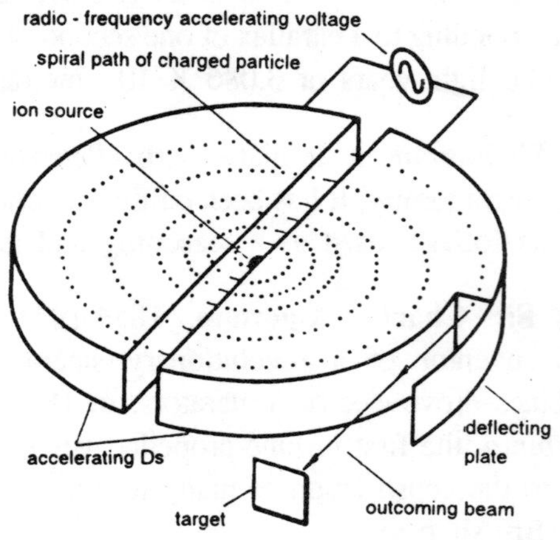

Fig. P-4. *Particle accelerator:* In a cyclotron, as the charged particle move faster, they spiral out to the edge of the D-shaped sections (Ds).

Particle accelerator. Device used to produce beams of energetic charged particles and to direct them against various targets for studies of the structure of the atomic nucleus and of the forces holding it together. Accelerator also have applications in medicine and industry, most notably in the production of radioisotopes. The first stage of any accelerator is an Ion source to produce the charged particles from a neutral gas. The charged particles are accelerated by electric fields. In linear accelerators, which are the most powerful and efficient electron accelerators, the particle path is straight line. The early linear accelerator used large static electric charges, which produced an electric field along the length of an evacuated tube to accelerate the particles. Present linear accelerators use electromagnetic waves to accomplish the acceleration. To reach high energies without prohibitively long paths, E.O.

Lawrence designed the cyclotron, in which a cylindrical magnet bends the particle beam into a circular path in a hollow circular metal box that a split in half to form two *D*-shaped sections. A radio-frequency electric field is applied across the gap, accelerating the particle each time it crosses the gap. In the synchrocyclotron, used to accelerate protons, the frequency of the accelerating electric field steadily decreases to match the decreasing angular velocity of the proton caused by the increase of its mass at relativistic velocities, *i.e.,* those close to the speed of light. In the synchrotron, a ring of magnets surrounding a doughnut-shaped vacuum tank produces a magnetic field that rises in step with the proton velocities, thus keeping the radius of their paths constant; this design eliminates the need for a center section of the magnet, allowing construction of rings with diameters measured in miles.

Particle detector. In physics, one of several devices for detecting, measuring, and analyzing particles and other forms or radiation entering it. For example, in the ionization chamber (consisting basically of a sealed chamber containing a gas and two electrodes between which a voltage is maintained by an external circuit) and the Geiger counter (consisting commonly of a gas-filled metal cylinder that acts as one electrode, and a needle or thin, taut wire along the axis of the cylinder that acts as the other electrode), the ionizing radiation is measured by changes in the external circuit; these changes are caused by a current resulting from the increase in charges. In other devices, ionization is used to make visible the track of the charged particle causing the ionization: the bubble chamber is filled with liquid hydrogen or some other liquefied gas; the cloud chamber is filled with a supersaturated vapor; and the spark chamber contains a high-pressure gas that fills the gaps between a stack of metal plates or wire grids that are maintained with high voltage between alternate layers. In the scintillation counter, radiation is detected and measured by means of tiny, visible flashes produced by the radiation when it strikes a sensitive substance known as a phosphor.

Particle physics. The branch of physics concerned with the properties, interactions, and structure of elementary particles.

Partridge. Name for henlike Birds of various families. The true partridges of the Old World belong to the Pheasant family. The gray partridge (*Perdix perdix*) of Europe, about 1 ft (30 cm) long, has been successfully introduced in parts of the U.S. The name *partridge* is also applied to the ruffed Grouse, the Bobwhite, and the plumed Quail.

Partition Coefficient. The ratio of the concentrations of a given substance in two different phases of a system. Thus, if a substance is added to a system of two immiscible liquids the ratio of its concentration in one to its concentration in the other is its partition coefficient for that system. The value depends on the temperature.

Parton. An ultimate fundamental particle postulated as a basic unit of other elementary particles. In the simplest theories the partons are quarks.

Pascal. symbol: Pa. The SI unit of pressure, equal to a pressure of one newton per square metre. [After Blaise Pascal (1623-62), French scientist.]

Pascal, Blaise. 1623-62, French scientist and religious philosopher. A mathematical prodigy, Pascal founded the modern theory of probability, discovered the properties of the cycloid and contributed to the advance of differential calculus. In physics his experiments in the equilibrium of fluids led to the invention of the hydraulic press. As a young man Pascal came under Jansenist influence, and after a profound religious experience in 1654 he entered the convent at Port-Royal, thereafter devoting his attention primarily to religious writing. His best-know works are *Provincial Letters* (1656), a defense of the Jansenists; and the posthumously published *Pensees* (1670), which preach the necessity of mystic faith in understanding the universe.

Passive. 1. Denoting a solid substance that is chemically unreactive as a result of the formation of a thin protective layer on its surface. Aluminium, for instance, is highly reactive and quickly forms a transparent oxide, which coats the surface and prevents further attack. Similarly, iron is rendered passive by concentrated nitric acid, again by the formation of an oxide.

2. Denoting an electronic component that does not amplify the signal. Resistors, capacitors, and inductors are passive components.

Pasta or Macaroni. Shaped and dried dough prepared from semolina or farina wheat flour and water, associated especially with Italian cuisine. Often mixed with eggs or egg solids, the dough is cut into many shapes, such as shells and tubes (maçaroni) and ribbons (*e.g.*, spaghetti). Long known in Asia, similar flour and rice pastes are believed to have been introduced into Europe during the 13th cent. Mongol invasions.

Pasteur, Louis (pastur). 1822-95, French chemist. Noted for his studies of fermentation and Bacteria, he disproved the theory of spontaneous generation and advanced the germ theory of infection. Of great economic importance are the process of Pasteurization, which he developed; his studies of silkworm disease and chicken cholera; and his discovery of Anthrax and Rabies vaccines. In 1888 the Pasteur Inst. was founded in Paris, with Pasteur as director, to provide a teaching and research center on virulent and contagious diseases.

Pasteurization. Treatment of food with heat to destroy disease-causing and other undesirable organisms. The process was developed (but not discovered) by Louis Pasteur in the 1860s. Modern pasteurization standards for milk require temperatures of about 145°F (63°C) for 30 min. followed by rapid cooling. The harmless lactic acid bacteria survive the process and can sour warm milk.

Patent. In law, governmental grant of the exclusive privilege of making, using, and selling and authorizing others to make, use, and sell an invention. The term derives from the medieval letters patent, public letters granting monopolistic control of useful goods to an individual. The first U.S. patent legislation was enacted in 1790, and the U.S. Patent Office was created in 1836. U.S. law allows the patenting of any original and useful device or process; the patent is valid for 17 years. Although there have been many independent inventors in the U.S., most important patents today are the property of large corporations.

Patronite. A naturally occurring sulphide of vandium, VS_4, used as a vandium ore.

Pauli exclusion principle. The principle that no two electrons in an atom can have the same set of quantum numbers. [After Wolfgang Pauli (1900-58), Austrian physicist.]

Pelican. Large aquatic Bird of warm regions, related to the Cormorants and gannets. Pelicans are long-necked birds with large, flat bills; they store fish in a deep, expandable pouch below the lower mandible. The white pelican (*pelecanus onocrotalus*) of North America ranges from the NW U.S. to the Gulf and Florida coasts. It is about 5 ft (152.5 cm) long with a wingspread of 8 to 10 ft (244 to 300.5 cm).

Peltier Effect. A thermoelectric effect in which an electric current passed through a junction between two different solids causes heat into be produced or absorbed, depending on the direction of the current. The phenomenon is the converse of the Seeback effect. Semi-conductor junction can be used as cooling elements. [After Jean Peltier 1785-1845), French scientist.]

Pen. Pointed implement used in writing or drawing to apply Ink or a similar coloured fluid to any surface, such as paper. Various pens have been used since ancient times, including reeds, styluses, quills, metal fountain pens, ballpoint pens, and felt- and fiber-tipped pens.

Penodulum. 1. A body suspended from a point above its centre of gravity and allowed to swing freely under the action of gravity. The *simple pendulum* consists of a weight hung from a fixed support by a string of negligible weight. For small displacements the bob undergoes simple harmonic motion with a period given by $2\Pi(l/g)^{½}$, where l is the length and g the acceleration of free fali. A *compound pendulum* is any rigid body swinging on a pivot. Its period is $2\Pi[(k^2 + h^2)^{½}/hg]^{½}$, where h is the distance between the pivot and the centre of gravity and k is the radius of gyration. Pendulums can be used for measuring g, the acceleration of free fall, and are used in prospecting for oil and minerals. Kater's pendulum is an example of acompound pendulum designed for determining g. Many other specialized types of pendulum exist, including compensated pendulums for regulating clocks, the ballistic pendulum for measuring velocities, and Foucalt's pendulum for demonstrating the rotation of the earth.

2. Any of various analogous devices in which a mass is suspended on a spring or torsion wire and allowed to rotate with a backwards and forwards twisting action. Such devices are called *torsion pendulums.*

Pendulum. A mass suspended from a fixed point so that it can swing in an arc. The length of a pendulum is the distance from the point of suspension to the center of gravity of the mass. The period T, or time for one complete swing, depends on the length l of the pendulum and on the acceleration g of gravity at the pendulum's location, according to a formula derived by Christian Huygens: $T = 2\Pi \sqrt{l/g}$. Huygens introduced (1673) the use of the pendulum to regulate the speed of clocks.

Pelican. Large aquatic Bird of warm regions, related to the Cormorants and gannets. Pelicans are long-necked birds with large, flat bills; they store fish in a deep, expandable pouch below the lower mandible. The white pelican (*Pelecanus onocrotalus*) of North America ranges from the NW U.S. to the Gulf and Florida coats. It is about 5 ft (152.5 cm) long with wingspread of 8 to 10 ft (244 to 300.5 cm).

Penicillin. Any of a class of related antibiotics produced by the *Penicillium* mould or made synthetically.

Penicillin. Any of a group of Antibiotics obtained from the molds of the genera *Aspergillis* and *Penicillium* and the first to be used successfully to treat bacterial infections in man. Penicillin acts to inhibit cell wall formation in many gram-positive bacteria. Although not active against most gram-negative bacteria, the tubercle bacillus, protozoans, fungi, or viruses, penicillin remains important in antibiotic therapy. Use of penicillin is limited by its tendency to induce allergic reactions and because of the development of microorganisms that are resistant to its actions. Synthetic derivatives of penicillin include ampicillin, methicillin, and oxacillin. Erythromycin and other antibiotics have become important in treating infections that are resistant to penicillin.

Pentaerythritol. A white crystalline powder, $C(CH_2OH)_4$, prepared by reacting acetaldehyde with excess formaldehyde. It is used in making synthetic resins. M.pt. 262°C; r.d. 1.4.

Pentagon. A polygon that has five sides.

Pentahydrate. A solid compound with five molecules of water of crystallization per molecule of compound, as in copper sulphate pentahydrate, $CuSO_4 . 5H_2O$.

Pentane. A colourless flammable liquid straight-chain alkane, C_5H_{12}, obtained from petroleum. Pentane is also called *n-pentane* to distinguish it from isopentane and neopentane. M.pt. -129.7°C; b.pt. 36.1°C; r.d. 0.63.

Pentanoic Acid (valeric acid). A colourless liquid. $CH_3(CH_2)_3)COOH$, with a penetrating odour, made by oxidation of pentanol. It is used as an intermediate for perfumes. M.pt. -34°C; b.pt. 185°C; r.d. 0.9.

Pentavalent (quinquevalent). Having a valency of five.

Pentode. A thermionic valve with five electrodes.It is a tetrode modified by the inclusion of a third grid, the *suppressor grid,* between the anode and the screen grid to reduce the effects of secondary emission from the anode. It is held at a negative potential.

Pentose. A simple sugar that has five carbon atoms in its molecules.

Pentyl Group. A monovalent group C_5H_{11}-, derived from pentane by removing one atom of hydrogen. The term is often used for the group with a straight chain of atoms: *i.e.* for the *n-pentyl group.* Other isomers exist including the *tert-pentyl group.* $CH_3CH_2C(CH_3)_2$-, the *isopentyl group,* $(CH_3)_2CHCH_2CH_2$-, and the *neopentyl group,* $(CH_3)_3CCH_2$. The pentyl group was formerly called the *amyl group.*

Penumbra. 1. A region of partial shadow surrounding the umbra.

2. The outer, lighter portion of a sunspot.

Peony. Popular flowering plant (genus *Paeonia*) of the buttercup family. Herbaceous penonies (most varieties of *P. lactiflora*) are hardy, bushy perennials that die back each year. The large, usually spring-blooming, single or double flowers are generally shades of red, pink, or white. Tree peonies (*P. suffroticosa*) have a somewhat woody base and are usually taller, with more abundant and larger blossoms.

Pepper. Name for the fruits of several plants used as condiments or in medicine. *Black pepper (Piper nigrum),* the true pepper, is economically the most important species of the pantropical pepper family (Piperaceae). a perennial climbing shrub native to Java, it bears pea-sized berries, the "peppercorns" of commerce Black pepper, sold whole or ground, is the whole fruit; white pepper, made by removing the dark, outer hull, has a milder, less pungent flavor. Other *Piper* species of value include the betel pepper (*P. betle),* whose leaves are a principal ingredient in Betel. The *red peppers,* native to warm temperate and tropical regions of the Americas, are various species of *Capsicum* (of the Nightshade family). The hot varieties include cayenne pepper, whose dried, ground fruit is sold as a spice, and chili-pepper, sold similarly as a powder or in a chili sauce. Paprika (the Hungarian word for red pepper) is a ground spice from a less pungent variety. The pimiento, or Spanish pepper, is a mild type; its small fruit is used as a condiment and for stuffing olives. The common garden, or bell, pepper has larger, also mild fruits; they are used as vegetables and in salads. Bell peppers are also known as green peppers because they are most often marketed while still unripe.

Pepsin. Digestive enzyme in the Stomach that degrades protein. Most active in the acidic medium normally present in the stomach, pepsin breaks down proteins into their readily absorbed components, *i.e.,* Peptides and Amino Acids.

Peptide. A compound formed by two or more amino acids linked together. The amino group ($-NH_2$) of one acid can be thought of as reacting with the carboxyl group (-COOH) of another to give the group -NH-CO-. This is known as the *peptide linkage.*

Peptide or **Polypeptide.** Biochemial formed by the linkage of up to about 50 Amino Acids to form an unbranched chain. Longer chains are called proteins. The amino acids are coupled by a peptide bond, a special linkage in which the nitrogen atom of one amino acid binds to the carboxyl atom of another. Many peptides, such as the hormones vasopressin and ACTH, have physiological or antibacterial activity.

Peptization. Dispersion of a substance to form a colloid by addition of a chemical to stabilize the sol.

Perboric Acid. A hypothetical acid, HBO_3, known in the form of *perborate* salts.

Perch. Symmetrical freshwater Fishes of the family Percidae, related to sunfishes and sea Basses. The yellow perch, *Perca flavescens,* is a popular game and food fish, abundant in lakes and streams, where it feeds on insects, Crayfish, and small fish, and grown to an average length of 1 ft (30 cm) and weight of 1 lb (.5 kg). The family includes the walleye or walleyed pike (*Stizostedion vitreum*), the sauger or sand pike (*S. canadense*), and the brilliantly coloured American darters, found E of the rockeies.

Perchlorate. Any salt or ester of perchloric acid.

Perchloric Acid. A colourless hygroscopic explosive liquid, $HClO_4$, made by distilling potassium perchlorate with 96% sulphuric acid under reduced pressure. It is a strong oxidizing agent, used in analytical chemistry, electroplating of metals, and explosives. M.pt. -112°C; b.pt. 16°C (8 mmHg); r.d. 1.8.

Percussion Instrument. Any instrument that produces a musical sound when struck with an implement such as a mallet, a stick, a disk, or with the hand. They are used in Orchestras and Bands most often to emphasize rhythm. The most common type is the Drum, technically known as a membranophone, which consists of a membrane stretched over a frame to be struck with sticks or with the hand. Idiophones are percussion instruments in which the vibrating agent is the solid substance of the instrument itself. These include the triangle, a steel rod bent into an angle and struck with a straight rod; sticks clicked against each other; castanets, commonly consisting of two joined pieces of wood or ivory snapped together between the palm and the fingers; cymbals, a pair of concave metal plates that are struck together; the gong, a disk struck with a mallet or drumstick; the celesta, a high-pitched instrument consisting of steel bars fastened over wood resonators and struck by hammers operated from a keyboard; the glockenspiel, a curved frame enclosing metal bars that are struck with a hammer; the xylophone, an instrument with gradated wooden slats that are struck with small mallets, and sometimes equipped with tubular resonators (in which case it is called a marimba), and the Bell. In general, percussion instruments are not turned by construction; pitch, tone, and volume depend on the skill of the player.

Perennial. Plant that normally lives for more than two growing seasons, as contrasted to an Annual or Biennial. In horticulture the term is usually restricted to hardy herbaceous plants, such as Iris, Peony, and Tulip, all of which die down to the ground each year and survive the winter on food stored in specialized underground stems.

Perfect Gas. Ideal gas.

Perfume. Aroma produced by essential oils of plants and by synthetic aromatics. The burning of incense in religious rites of ancient China, Palestine, and Egypt led gradually to the personal use of perfume, widespread in ancient Greece and Rome. During the Middle Ages Crusaders brought knowledge of perfumery to Europe from the East. After 1500 Paris was the major center of perfume-making. Since the early 19th cent. chemists have produced thousands of synthetic scents. In the 20th cent. perfumes for men and women, most a blend of natural and synthetic scents, were produced by prominent fashion designers.

Periclase. A natural form of magnesium oxide, MgO.

Perigee. The point at which the moon or any other orbiting satellite is closest to the earth.

Perihelion. The point of closest approach to the sun attained by a planet, asteroid, comet, or other orbiting body. *Compare* aphelion.

Perimeter. The distance around a plane figure. Thus, the perimeter of a polygon is the sum of the lengths of its sides and the perimeter of a circle is its circumference.

Period. The constant interval between identical states of a system whose properties vary periodically. For a system in which a quantity varies with time, as in a wave motion or simple harmonic motion, the period is the time taken to complete a cycle. The period of a crystal lattice in a particular direction is the distance between similar lattice points in that direction.

Periodate. Any salt of periodic acid.

Periodic Acid. A white crystalline acid, HIO_4, made by the action of perchloric acid on iodine. Sublimes at 110°C.

Periodic Table. Chart that reflects the periodic recurrence of chemical and physical properties of the Elements when the elements are arranged in order of increasing Atomic Number. The periodic table was devised by Dmitri Mendeleev and revised by Henry Moseley. It is divided into vertical columns, or groups, numbered from I to VIII, with a final column numbered 0. Each group is divided into two categories, or families, one called the a series (the representative, or main group, elements) the other the b series (the Transition Elements, or subgroup elements). All the elements in a group have the same number of Valence electrons and have similar chemical properties. The horizontal rows of the table are called periods. The elements of a particular period have the same number of electron shells; the number of electrons in these shells, which equals the element's atomic number, increases from left to right within each period. In each period the lighter Metals appear on the left, the heavier metals in the center, and the nonmetals on the right. Elements on the borderline between metals and nonmetals are called metalloids. Elements in group la are called the Alakali Metals; in group lla, the Alkaline-Earth Metals; in group VIIa, the Halogens; and in group 0, the Inert Gases.

Periodontitis or **Pyorrhea.** Inflammation and degeneration of the gums and other tissues surrounding the teeth. Symptoms are bleeding gums, followed by the receding of gums from the

Teeth, loosening of the teeth, and resorption of the bone supporting the teeth. Causes include poor nutrition, plaque, and poor oral hygiene.

Periodic Table. A tabular arrangement of the elements in order of increasing atomic number such that similarities are displayed between groups of elements. Early attempts at classifying the elements include Dobereiner's triads and Newlands's law of octaves. In the modern form of the table, the columns are called *groups*. All the elements in a group have similar electron configurations and a consequent similarity in chemical properties. Thus, in group IA the alkali metals all have a single electron in their outer shell and are electropositive metals. There is a tendency for electropositive behaviour to increase down the group because the atoms are larger. Thus potassium has a lower ionization potential than sodium and is more reactive. This behaviour is very marked in group IV, where there is a transition from the typical nonmetal, carbon, to the metal, lead. The rows across the table are called *periods*. Those from lithium to neon and sodium to argon are *short periods,* the others are *long periods.* Across a period there is a change from electropositive metallic behaviour to electronegative nonmetallic behaviour. The table also shows the transition series and the lanthanides and actinides.

Periscope. 1. An apparatus for viewing objects when there is no direct line of sight to the eye. Essentially, it contains a mirror at 45° to the incident light, thus reflecting the light at right angles along the periscope tube onto a second mirror that reflects the light out of the apparatus, so that the reflected light is parallel to the original incident light bat is displaced. Total internal reflection by right-angled prisms is sometimes used instead of reflection by mirrors.

2. Instrument to permit the viewing of an object either out of one's direct line of vision or concealed by some intervening body. The image is received in a mirror and reflected through a tube with lenses to a mirror visible to the viewer. Submarine periscopes, with tubes up to 30 ft (9.1 m) long, can be rotated to permit a scan of the entire horizon.

Peritonitis. Acute or chronic inflammation of the peritoneum, the membrane lining the abdomen and surrounding internal organs. It is caused by invasion of bacteria or foreign matter following rupture of an internal organ, by infection from elsewhere in the body, by penetrating injury to the abdominal wall, or by accidental pollution during surgery. Treatment includes Antibiotic therapy.

Periwinkle. Mollusk with a conical spiral shell, a variety of Snail. Periwinkles are marine Gastropods that feed on algae and seaweed and are found at the water's edge.

Permanent Gas. A gas that cannot be liquefied by pressure alone: *i.e.* a gas with a critical temperature below normal temperature, so that some cooling is necessary for the gas to be liquefied.

Permanent Hardness. Hard Water.

Permanent Magnetism. Magnetism.

Permanganate. A salt containing the ion MnO_4. Permanganates are powerful oxidizing agents derived from the hypothetical *permanganic acid,* $HMnO_4$. The permanganate ion is an intense purple colour. In strongly basic solutions it changes to the manganate ion.

Permanganic Acid. Permanganate.

Permeability. Relative permeability.

Permittivity. Relative permittivity.

Permutation. An ordered selection of a number of entities from a set of entities. The number of permutations of r elements from a set of n is denoted ^nP_r and is given by $n!/(n - r)!$ A permutation differs from a combination in that the order is considered. There are six permutations of two letters from the letter A,B, and C: namely AB, BA, AC, CA, BC, and CB. There are only three combinations because pairs like AB and BA are the same combination of letters.

Permutations and Combinations. The study of techniques for counting arrangements and choices of objects; such techniques are often used in Probability problems. A permutation of a set is a way in which the elements of the set can be arranged or ordered. In general, the number of permutations of n things taken r at a time is given by $P(n, r) = n/(n - r)$, where the symbol $n!$, denoting the product of the integers from *1* to n, is called the n factorial. A combination is a choice of different elements from a larger set, without regard to order. In general, the number of combinations of n things taken r at a time is $C(n,r) = n!r!(n - r)!$

Peroxide. Any inorganic compound containing the O_2^{2-} ion. Peroxides can be hydrolysed to hydrogen peroxide.

Peroxide. Chemical compound containing two oxygen atoms, each of which is bonded to the other and to a radical or some element other than oxygen; *e.g.*, in hydrogen peroxide (H_2O_2) the atoms are joined together in the chainlike structure H–O–O–H. Peroxides are unstable, releasing oxygen when heated, and are powerful oxidizing agents. Peroxides may be formed directly by the reaction of an element or compound with oxygen.

Peroxodisulphuric Acid (persulphuric acid). An acid, $HOSO_2OOSO_2OH$, produced in solution by the electrolysis of potassium sulphate in the dilute sulphuric acid at high current density.

Peroxosulphuric Acid (Caro's acid). A white crystalline strongly oxidizing acid, $HOSO_2OOH$, formed by the action of hydrogen peroxide on sulphuric acid.

Perpendicular. Indicating that a line or plane is at right angles to another line or plane: a line or plane that is at right angles to another. A line is perpendicular to a curve or surface if it is perpendicular to its tangent line or plane.

Perpetual Motion. Continuous motion without any supply of energy. A machine or device that, once set in motion, would continue moving for ever would only be possible in the absence of forces due to friction, viscosity, etc. It would be impossible to construct such a machine that performed useful work as its operation would contravene the first law of thermodynamics.

Perpetual-motion Machine. A machine, considered impossible to build, that would be able to operate continuously and supply useful work without needing a continuous supply of heat or fuel. A perpetual-motion machine of the first kind, which would produce more Energy in the form of work than is supplied to it in the form of heat, violates the first law of Thermodynamics. A perpetual-motion machine of the second kind, which would continuously supply work without a flow of heat from a warmer body to a cooler body, violates the second law of thermodynamics.

Personal Computer. A small but powerful Computer primarily used in a home or office and not connected to a larger computer. Personal computers evolved after the development of the Microprocessor made possible the hobby-computer movement of the late 1970s, when some computers were built from components or kits. They became popular during the early 1980s, when the first low-cost, fully assembled units were mass-marketed. The typical personal computer consists of a video display, keyboard, logic unit, and external storage device.

Perspex. Polymethyl methacrylate.

Persulphate. Any salt of persulphuric acid.

Persulphuric Acid. Peroxodisulphuric acid.

Pesticide. Biological, physical or chemical agent used to kill plants or animals considered harmful to human being.

Petrochemicals. Chemicals manufactured or obtained from petroleum or natural gas.

Petrol. A mixture of octane, heptane, and other hydrocarbons used as a fuel in internal-combustion engines. It is obtained from petroleum and usually contains additives to prevent knocking, retard corrosion, etc.

Petrolatum (petroleum jelly). A translucent semisolid amorphous mixture of hydrocarbons obtained from petroleum. It is used in medical dressings and as a lubricant.

Petroleum. A mixture of hydrocarbons and other organic compounds formed originally from marine sediments. The crude oil is refined by distillation to give petrol, paraffin oil, lubricating oils, waxes, and tar. Petroleum is the raw material for the production of many organic chemicals.

Petroleum or **Crude Oil.** Oily, flammable liquid that occurs naturally in deposits, usually beneath the surface of the earth. The exact composition varies according to locality, but it is chiefly a mixture of Hydrocarbons. Petroleum is a fossil fuel thought to have been formed over millions of years from incompletely decayed plant and animal remains buried under thick layers of rock. Drilling for oil is a complex, often risky process. Scientific methods are used to locate promising sites for wells, some of which must be dug several miles deep to reach the deposit. Many wells are now drilled offshore from platforms standing on the ocean bed. Usually the crude oil in a new well comes to the surface under its own pressure. Later it has to be pumped or forced up with injected water, gas, or air. Pipelines or tankers transport it to refineries, where it is separated into fractions, *i.e.,* the portions of the crude oil that vaporize between certain defined limits of temperature. Fractions are obtained by a refining process called fractional

Distillation, in which crude oil is heated and sent into a tower. The vapors of the different fractions condense on collectors at different heights in the tower. The separated fractions are then drawn from the collectors and further processed into various petroleum products. Generally the fractions are vaporized in the following order: dissolved Natural Gas, Gasoline, naphtha, Kerosene, diesel fuel, heating oils, and finally tars. Lighter fractions, especially gasoline, are in greatest demand and their yield can be increased by breaking down heavier hydrocarbons in a process called cracking. The leading producers of petroleum in 1980 were the USSR, Saudi Arabia, the U.S., Iraq, Venezuela, China, Nigeria, Mexico, Libya, and the United Arab emirates. The largest reserves are in the Middle East. Modern industrial civilization depends heavily on petroleum for motive power, fuel, lubrication, and a variety of synthetic products, *e.g.*, Dyes, drugs, and Plastics. The widespread burning of petroleum as fuel has resulted in serious problems of air Pollution, and oil spilled from tankers and offshore wells has damaged oceans and coastlines. Unless the need for oil is reduced, conservationists may be unable to prevent the development of oil deposits whose exploitation poses threats to the environment.

Petroleum Ether (ligroine, benzine). A colourless flammable liquid, a mixture of hydrocarbons incorrectly named an ether. It is prepared by distillation of petroleum and used as a laboratory solvent.

Petrology. Branch of Geology concerned with the origin, composition, structure, and properties of Rocks, as well as the laboratory simulation of rock-forming processes.

Pewter. An alloy of tin (about 80%) and lead with small amounts of antimony (up to 3%).

Pewter. Ductile, silver-white alloy consisting mainly of tin. The addition of lead imparts a bluish tinge and increased malleability.

Other metals such as antimony, copper, bismuth, and zinc may also be added. Pewter is shaped by casting, hammering, or lathe spinning on a mold. Ornamentation is usually simple. Pewter was early used in the Far East, and Roman pieces are extant. In England during the Middle Ages pewter was the chief tableware, later being supplanted by China. It was made in America from c.1700. The craft had virtually died out by 1850 but was revived in the 20th cent. in reproductions and pieces of modern design.

PH. 1. A measure of the acidity or alkalinity of a solution, equal to the logarithm to base 10 of the reciprocal of the concentration of hydrogen ions: *i.e.* $pH = \log_{10}(1[H^+])$. Neutral solutions have a pH of 7, acid solutions a pH less than 7, and alkaline solutions a pH greater than 7.

2. Range of numbers expressing the relative acidity or alkalinity of a solution. The pH value is the negative common Logarithm of the hydrogen-ion Concentration in a solution, expressed in Moles per litre of solution. A neutral solution, *i.e.,* one that is neither acidic nor alkaline, such as pure water, has a concentration of 10^{-7} moles per litre; its pH is thus 7. Acidic solutions have pH values ranging with decreasing acidity from 0 to nearly 7; alkaline or basic solutions have a pH ranging with increasing alkalinity from just beyond 7 to 14.

Pharmacology. Study of the changes produced in living animals by drugs, chemical substances used to treat and diagnose disease. It is closely related to other scientific disciplines, particularly Biochemistry and Physiology. Areas of pharmacologic research include mechanisms of drug action, the use of drugs in treating disease, and drug-induced side effects.

Pharmacopoeia. Authoritative publication designating the properties, actions, uses, dosages, and standards of strength and purity of Drugs. The Nuremberg pharmacopoeia, published in Germany in 1546, was the first work of its kind. Drug standards in the U.S. were not established until 1906, when the *United States Pharmacopoeia* (USP; first published 1820) and the *National Formulary* (NF; est. 1888) were recognized by the federal government. Published every five years and supplemented as needed, the USP and NF were combined in 1980.

Pharmacy. Science of compounding and dispensing medication; also, an establishment used for such purposes. Modern pharmaceutical practice includes the dispensing, identification, selection, and analysis of Drugs. Pharmacy began to develop as a profession separate from medicine in the 18th cent., and in 1821 the first U.S. school of pharmacy was established in Philadelphia.

Pharynx. Section of the Digestive System between the mouth and esophagus. In humans the pharynx is a cone-shaped tube, continuous with both the mouth and nasal passages at its upper end and esophagus at its lower end. It connects with the Ears via the eustachian tubes and with the Larynx by an opening covered by the epiglottis during swallowing.

Phase. 1. The stage that a periodically varying system has reached in its cycle at a given time measured from a specified reference point. Two alternating currents, waves, etc., of the same frequency are said to be *in phase* if they have their maximum and minimum values at the same time. Otherwise they are *out of phase.* A varying quantity can be represented by a line (vector) rotating about a point and the difference in phase of two such quantities can be expressed by the angle between their rotating vectors. This is the *phase angle:* it is zero when the two systems are in phase.

2. Any of the apparent changes in shape that the moon cyclically passes through on its orbit around the earth. As more and then less of the moon's face is illuminated by the sun it is seen to grow from a slender crescent to a fully rounded disc and back again during the course of its synodic period. When the moon lies between the sun and the earth (inferior conjunction) it cannot be seen at all, except possibly as a black disc during a total solar eclipse. The phase is referred to as *new.* With increasing elongation a crescent appears, popularly but not astronomically also called *new moon.* At quadrature, half the moon is illuminated, this being *first quarter.* As more of the surface becomes visible the moon reaches a stage where it is called *gibbous.* It continues onward until it reaches superior conjunction, with the earth lying between the sun and the moon. The whole of the lunar disc is now illuminated and the phase is called *full.* As the elongation between the sun and the moon decreases the phases are repeated in reverse order, the phase at quadrature this time being called *last quarter,* until new moon is once again reached.

3. Ahomogeneous part of a system, divided from other parts of the system by definite boundaries. This, a mixture of ice crystals in water contains two phases as does a suspension of solid particles in a gas or an alloy consisting of small inclusions of one metal in a matrix of another. A solution of salt in water, a solid solution, and a mixture of gases are all single phases.

Phase. In astronomy, the measure of how much of the illuminated surface of a planet or natural satellite can be seen from a point at a distance from that body. The phase depends on the overlap of the half of the surface that is seen by the observer and the half that is illuminated by the sun. an inferior planet, whose orbit lies inside the earth's shows all the phases that the moon shows; a superior planet, whose orbit lies outside earth's, is always gibbous or full.

Phase Modulation. Modulation.

Phase Rule. The principle that at equilibrium $P + F = C + 2$, where P is the number of phases in the system, C the number of components, and F the number of degrees of freedom.

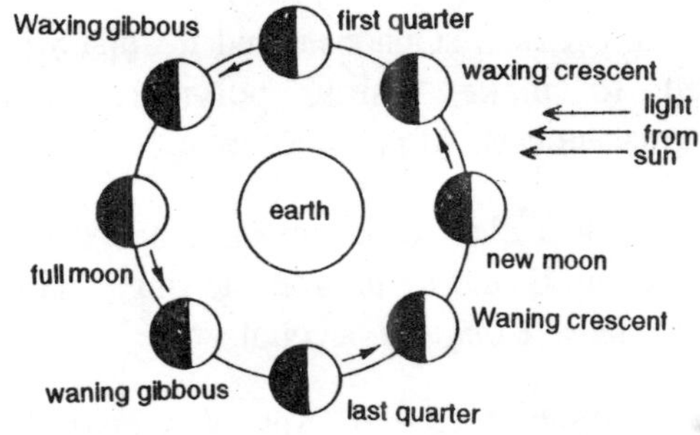

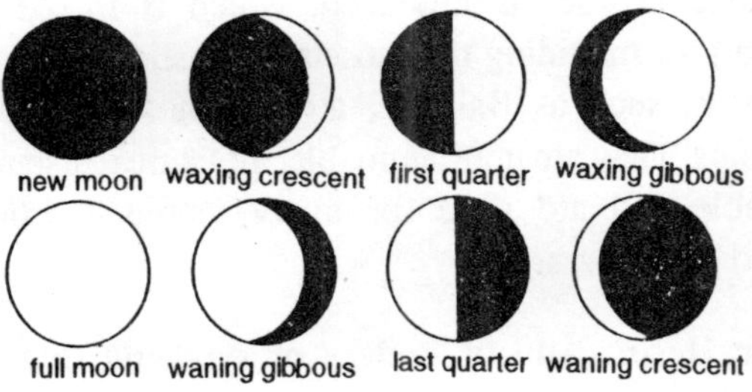

Fig. P-5. Phases of the moon as seen from the earth.

Pheasant. Name for some henlike birds of the family Phasianidae, related to the Grouse and including the Old World Partridges, the Peacock, some domestic and jungle fowl, and the true pheasants (genus *Phasianus*). Pheasants are typified by wattled heads and long tails, and by the brilliant plumage and elaborate courtship displays of the male. All are indigenous to Asia. The hybrid ring-necked pheasant, *P. colchicus,* is a common game bird in the U.S. where the ruffed grouse is also called *pheasant*.

Phenetole (phenyl ethyl ether). A colourless liquid, $C_6H_5OC_2H_5$, B.pt. 172°C; r.d. 0.97.

Phenol. 1. (*carbolic acid*). A white hygroscopic crystalline solid, C_6H_5OH, which turns pink if impure or in light. An important aromatic hydroxy compound, it is manufactured by preparing sodium benzene sulphonate and treating with acid. Phenol is used to make resins, polymers, weed-killers, and pharmaceuticals. M.pt. 42°C; b.pt. 182°C; r.d. 1.1.

2. Any of a class of compounds that contain a hydroxyl (-OH) group bound to an aromatic ring. The phenols are acidic compounds. *Compare* alcohol.

Phenol-formaldehyde Resin. Any of a class of synthetic resins made by condensation of formaldehyde with phenol or with a substituted phenol. The reaction is catalysed by acid or alkali and a powder is produced, which is mixed with filler and used for moulding thermosetting plastics. Phenol-formaldehyde resins, such as Bakelite, are the most widely used phenolic resins. They are nonflammable, moisture resistant, and thermally stable and are used for many moulded articles, laminates, and adhesives.

Phenolic Resin. Any of a class of synthetic thermosetting resins made by copolymerizing phenol, or a substituted phenol, with an aldehyde such as formaldehyde or furfural. The phenolformaldehyde resins are the most widely used phenolics.

Phenolphthalein. A colourless crystalline compound. $C_{20}H_{14}O_4$, used as an indicator. It is colourless in acid solutions and red in alkaline solutions, changing in the pH range 8.3-10.4.

Phenyl Group. The monovalent group C_6H_5- derived from benzene.

Phenyl Hydrazine. A yellow oil or crystalline solid, $C_6H_5NHNH_2$. It forms phenylhydrazones with aldehydes and ketones. M.pt. 190°C; b.pt. 243°C; r.d. 1.1.

Phenylhydrazone. Any of a class of compounds with the general formula $R_2C = NNH.C_6H_5$, where *r* is an alkyl or aryl group or a hydrogen atom. The phenylhydrazones, which are crystalline solid, are made by reaction between phenylhydrazine and an aldehyde or ketone.

Phenyl Methyl Ketone. Acetophenone.

Pheromone. Any chemical substance secreted by members of an animal species that alters the behaviour of other members of the same species. Sex-attractant pheromones are widespread, particularly among insects. Other pheromones act as signals for alarm and defense, territory and trail-marking, and social regulation and recognition.

Philosopher's Stone. Alchemy.

Philosopher's Wool. Zinc oxide.

Phlogiston. A substance formerly supposed to be present in all flammable substances and to be released during combustion, leaving ash. The phlogiston theory was disproved in the eighteenth century by Lavoisier.

Phlox. Name for plants (genus *Phlox*) and the family Polemoniaceae, consisting of herbs, shrubs, and vines. Found chiefly in the W U.S., the family includes many popular wild and garden flowers, especially the genera *Phlox, Polemonium* (called Jacob's ladder), and *Gilia*. Most phlox are perennial, but some common garden varieties are annual hybrids.

Phobos. Mars.

Phoenix. Fabulous bird of ancient legend. When it reached 500 years of age it burned itself on a pyre from whose ashes another phoenix arose. It commonly appears in literature as a symbol of death and resurrection.

Phon. A unit of loudness of sound defined with reference to a standard tone of frequency 1000 hertz. The intensity of this standard is varied until it is judged to be as loud as the sound to be measured. The loudness of the sound is then *n* phons, where the reference is *n* decibels above a standard datum level.

Phonon. A quantized lattice vibration in a solid.

Phosgene (carbonyl chloride). A colourless poisonous gas, $COCl_2$, with an odour of freshly mown hay, made by passing carbon monoxide and chlorine over activated carbon. It is used as a war gas and as a reagent in the manufacture of many organic chemicals. M.pt. -118°C; b.pt. 8°C.

Phosphate. Any salt or ester of a phosphoric acid. Phosphates include *orthophosphates.* M_3PO_4, *diphosphates (pyrophosphates),* M_4PO_7, and various *polyphosphates.* M is a monovalent metal.

Phosphide. A compound of phosphorus with a more electropositive element.

Phosphine. A colourless poisonous spontaneously flammable gas, PH_3, with the odour of rotting fish. It is prepared by dropping potassium hydroxide on to phosphorus and is used as a doping agent for semiconductors. M.pt. –133°C; b.pt. –85°C; r.d. 1.2.

Phosphite. Any salt or ester of a phosphorous acid. The phosphites include the *orthophosphites,* M_2HPO_3, the *diphosphites (pyrophospites)* $M_2H_2P_2O_5$, and the *hypophosphites,* MH_2PO_2. M is a monovalent metal.

Phosphonium Ion. The monovalent ion PH_4^+ produced by protonation of phosphine. Phosphonium salts contain these ions and are analogous to ammonium compounds.

Phosphor. Any substance that exhibits fluorescence or phosphorescence.

Phosphorescence. Luminescence that persists after the source of excitation has been removed. It is distinguished from fluorescence—in which emission ceases immediately. In scientific usage the term can either denote a luminescence that persists long enough to be recognized with the eye, or can denote one that lasts longer than 10^{-8} seconds. Phosphorescence was originally the "cold light" emitted from slowly oxidizing phosphorus. In general non-scientific usage the term is often used for cold light of the type emitted seaweed or other organic matter.

Phosphorescence. Luminescence produced by certain substances after absorbing radiant energy or other types of energy. Phosphorescence is distinguished from Fluorescence in that it continues even after the radiation causing it has ceased. The luminescence is caused by electrons that are excited by the radiation and trapped in potential troughs, from which they are freed by the thermal motion within the crystal. As they fall back to a lower energy level, they emit energy in the form of light.

Phosphoric Acid. Any of a number of oxyacids of phosphorus. The term is often applied to *orthophosphoric acid,* H_3PO_4, which is a syrupy colourless liquid made by the action of sulphuric acid on a phosphate. It is used in the manufacture of fertilizers and soaps. M.pt. 42.35°C; r.d. 1.8. *Diphosphoric acid (pyrophosphoric acid),* $H_4P_2O_7$, is a glassy solid produced by heating orthophosphoric acid. It is the first member of a series of *polyphosphoric acids* with the general formula $H_{n+2}P_nO_{3n+1}$. The series includes *metaphosphoric acid,* a highly polymeric deliquescent glassy solid, $(HOP_3)_x$, which is used as a dehydrating agent.

Phosphorous Acid. A white crystalline hydroscopic dibasic acid, H_3PO_3, obtained by the action of water on phosphorous III oxide. M.pt. 73.6°C; r.d. 1.7. This compound is also called *orthophosphorous acid.* Other acids include *diphosphorous acid (pyro-phosphorous acid*), $H_4P_2O_5$, which is dibasic; *hypophosphorous acid,* H_3PO_2, which is monobasic; and the polymeric metaphosphorous acid, HPO_2.

Phosphorus. Symbol: P. A nonmetallic element occurring chiefly in phosphate rocks. It exists in three allotropic forms that differ in chemical activity. White phosphorus is manufactured by reducing phosphates with carbon and silica in an electric furnace. It is the most reactive form of phosphorus, burning spontaneously in air, and is insoluble in water and soluble in organic solvents. It contains P_4 molecules, in which the four atoms are arranged at the corners of a tetrahedron. When heated at about 400°C for a few hours it converts into red phosphorus, a more stable allotrope of complex structure. Black phosphorus is obtained by heating white phosphorus at 200-300°C using a mercury catalyst. It is stable in air. The element is used in the manufacture of phosphoric acid, alloys, and semiconductors. Red phosphorus is used in matches. Phosphorus forms covalently bonded compounds with a valency of three, as in PCl_3, and with a valency of 5, as in PCl_5. It also forms binary phosphides with many other elements. A.N. 15; A.W. 30.9738; m.pt. 44.1°C (white); b.pt. 280°C (white); red phosphorus sublimes at 416°C; r.d. 1.82 (white), 2.20 (red), 2.25-2.69 (black).

Phosphorus (P). Nonmetallic element, discovered c.1674 by Hennig Brand. It is an extremely poisonous, yellow to white, waxy, solid substance. Because phosphorus ignites spontaneously when exposed to air, it is stored underwater. Its major source is the mineral Apatite found in phosphate rocks. The principal use of phosphorus is in compounds in fertilizers, detergents, insecticides, soft drinks, toxic nerve gases, pharmaceuticals, and dentifrices. Phosphorus compounds are essential in the diet. Phosphorus is a component of Adenosine Triphosphate (ATP), a fundamental energy source in living things, and of calcium phosphate, the principal material in bones and teeth.

Phosphorus Chloride. Either of two chlorides of phosphorus made by reaction between the elements. *Phosphorus trichloride,* PCl_3, is a colourless fuming liquid. M.pt. –112°C; b.pt. 76°C; r.d. 1.6. *Phosphorus pentachloride,* PCl_5, is a yellow volatile solid. Sublimes at 160°C; r.d. 3.6. Both compounds are used as chlorinating agents.

Phosphorus Oxide. Any of three oxides of phosphorus, P_4O_6, P_4O_8, and P_4O_{10}. The most important is *phosphorus pentoxide,* P_4O_{10}, which is a white hygroscopic powder made by burning phosphorus in dry air. It is the anhydride of phosphoric acid and is used as a dehydrating agent and catalyst. The formula is often written O_2O_5.

Phosphorus Oxychloride (phosphoryl chloride). A colourless fuming liquid, $POCl_3$, with a pungent odour. It is made by reacting phosphorus pentoxide and chlorine, and is used to manufacture organic phosphates and as a chlorinating agent. M.pt. 125°C; b.pt. 107°C; r.d. 1.7.

Phosphoryl Chloride. Phosphorus oxychloride.

Phot. A unit of intensity of illumination equal to one lumen per square centimetre.

Photocathode. A cathode from which electrons are emitted as a result of the photoelectric effect.

Photocell (photoelectric cell). Any device for converting light or other electromagnetic radiation directly into an electric current. Photocells depend for their action on the photoelectric effect, photoconductivity, or the photovoltaic effect.

Photochemistry. The branch of chemistry concerned with chemical reactions produced by light or ultraviolet radiation.

Photochromism. Change of colour occurring when certain substances are exposed to light. Sometimes the change is reversible. The effect is displayed by some dyes.

Photoconductivity. The increase of the electrical conductivity of certain solids, usually semiconductors, as a result of the impact of photons. It occurs when the photons have sufficient energy to increase the energy of an electron in a filled band, so that it is promoted into the conduction band.

Photocopying. Processes that use various chemical, electrical, or photographic techniques to copy printed or pictorial matter. Most familiar is *xerography,* an electrostatic process that utilizes the attractive force of electric charges to transfer an image to a charged plate. Light, reflecting off the white areas of the object to be copied, erases the charge on the corresponding areas of the plate. a plastic ink powder, called toner, sticks to the charged areas of the plate and transfers an image of the original to paper. *Thermography* user infrared rays on heat-sensitive paper to transfer an image. Several processes use cameras to make copies of an original. The *Photostat* can reduce or enlarge copied material photographically. *Microfilming* generates copies from 1/12 to 1/100 the size of the original, and advances in microphotography allow even greater miniaturization in the *microfiche,* where extremely small microphotographs are printed side by side on a card made of film. In the *transfer process,* chemically coated paper is placed in contact with the original, exposed to light, and developed. The blueprint and the whiteprint, used to reproduce the drawings of architects and engineers, are both examples of transfer processes.

Photodisintegration. Disintegration of an atomic nucleus when it is hit by a gamma-ray photon. The photon may displace a nucleon or may cause fission (*photofission*) of the nucleus.

Photoelectric cell or **Photocell.** Device whose electrical characteristics (*e.g.*, current, voltage, or resistance) vary when light is incident upon it. Common photoelectric cells consist of two electrodes separated by a light-sensitive Semi-Conductor material. A battery or other voltage source connected to the two electrode sets up a current even in the absence of light; when light strikes the semiconductor section of the photocell, the current increases in proportion to the light intensity. Photocells can be used to operate Switches, Relays, door openers, and intrusion alarms.

Photoelectric Effect. The ejection of electrons from a solid as a result of irradiation by light or other electromagnetic radiation. The number of electrons emitted depends on the intensity of the light and not on its frequency: the energy of the electrons is proportional to the frequency. This is explained by the theory that each electron is ejected by a single photon of radiation with energy hv, where h is the Planck constant and v the frequency of the radiation. The energy E of the electrons is given by the *Einstein photoelectric equation:* $E = hv - \phi$, where ϕ is the work function of the solid, *i.e.* the minimum energy required to remove an electron. Some metals, such as caesium and rubidium, have a low enough work function for electrons to be ejected by visible light but most solids only emit in the ultraviolet region of the spectrum.

Photoelectric Effect. The emission of electrons by substances, especially metals, when light falls on their surfaces. The effect was discovered by Heinrich Hertz in 1887 and explained by Albert Einstein in 1905. According to Einstein's theory, light is composed of discrete particles of energy, or quanta, called Photons. When the photons with enough energy strike the material, they liberate electrons that have a maximal kinetic energy equal to the energy of the photons less the work function (the energy required to free the electrons from a particular material).

Photoelectron. A electron emitted by the photoelectric effect or by photoionization.

Photoelectron Spectroscopy. A form of electron spectroscopy in which the sample, usually a gas or solid, is irradiated with a beam of monochromatic ultraviolet radiation or X-rays. Electrons are ejected by photoemission or the photoelectric effect and measurement of their energy gives information on the ionization potentials of molecules and on the energy levels in the sample.

Photoemission. The emission of electrons by the photoelectric effect or by photoionization.

Photofission. Photodisintegration.

Photography, Still. Science and art of making permanent images on light-sensitive materials. Photography's basic principles, processes, and materials were discovered independently and virtually simultaneously by a diverse group of individuals early in the 19th cent. Johan Heinrich Schulze had discovered in 1727 that silver nitrate darkens upon exposure to light. Using Schulze's research, Thomas Wedgwood and Sir Humphry Davy created the first photogram. The French physicist Joseph Nicephore Niepce made the first paper negative (1816) and the first known photograph, on metal (1827). He formed a partnership with a painter, Louis Jacques Daguere, who in 1839 announced his method for making a direct positive image on a silver plate—the daguerrotype. The English scientist William Henry Fox Talbot developed a paper negative (the calotype) from which an infinite number of paper positives could be printed. Sir John Herschel discovered (1819) a suitable photographic fixing agent for paper images and is credited with giving the new medium its name. In 1851 Frederick Scott Archer developed the collodion process, or "set plate" technique, which resulted in a negative image with the fine detail of the daguerrotype and was infinitely reproducible. Soon the photograph was considered incontestable proof of an event, experience, or state of being. Among the chief photographers of the 19th cent. were the explorer W.H. Jackson; Roger Fenton, who documented the Crimean War; Mathew B. Brady and his photographic corps, who photographed the American Civil War; the painter Thomas Eakings; and Eadweard Muybridge, who devised a means of making stop-action photographs. As accessory lenses were perfected, the moon and the microcosm became accessible. The development of the halftone process in 1881 made photographic reproduction possible in books and newspapers. In 1888 Geoıge Eastman introduced roll film and the simple Kodak box Camera, and photography became available to all. The medium's legitimacy as an art form was challenged by artists and critics and upheld by Alfred Stieglitz, founder of the Photo-Secession movement. Initially photography imitated painting, but soon a "straight" aesthetic was widely adopted whereby the image was left free of manipulation. In the early 20th cent. the documentary power of photography was exemplified in the works of Jacob Riis. Paul Strand combined documentary concerns with a lean, modernist vision. Eugene Atget and Henri Cartier-Bresson developed intensely personal styles. Technical advances, including smaller cameras, faster films, and portable lighting, granted photojournalists unprecedented versatility, and the popular picture magazines (*e.g., Life*) provided a vast audience for documentary work. The Great Depression and World War II were major sources of important documentary photography, as evidenced in the work of Margaret Bourke-White, Walker Evans, Dorothea Lange, and W. Eugene Smith. Colour photography also emerged as an important force, as in the nature photographs of Eliot Porter and the microphotographs of roman Vishniac. Although many photographers have adhered to the straight aesthetic, e.g., Ansel Adams, Diane Arbus, and Edward Weston, younger photographers have felt little inhibition against manipulation of the image, often using such devices as handwork, collage, and multiple images.

Photoionization. Ionization of atoms or molecules by light or other electromagnetic radiation. The process can be thought of as a collision of a photon, with energy hv where h is the Planck constant and v the frequency, with an atom. If the photon energy exceeds the ionization potential an ion is formed: $M + hv = M^+ + e$.

Photoluminescence. Luminescence occurring as the result of irradiation by electromagnetic radiation. Incident radiation is absorbed and produces atoms or molecules in excited states. They then decay, either directly back to the ground state or via an intermediate excited state, with the emission of light or other radiation. The wavelength of the emitted radiation is always longer than that of the absorbed radiation. Photoluminescence is the mechanism causing the brightness of fluorescent paints and materials. It is also used in detergent whiteners, which are compounds that absorb ultraviolet radiation and then emit blue light over a long period of time—thus giving a blue cast to a white fabric and counteracting any yellowing.

Photolysis. The dissociation of a chemical compound into other compounds, atoms, and free radicals by irradiation with electromagnetic radiation, usually light or ulraviolet radiation.

Photometer. An instrument for measuring luminous intensity, usually by comparing a light source with a standard source. Typical photometers are the Lummer-Brodhun, flicker, and grease-spot photometers.

Photometry. The measurement of intensities of light sources and of illumination by use of photometers.

Photometry. Branch of physics dealing with the measurement of the intensity of light sources. Instruments used for such measurements are called photometers, most types are based on the comparison of the light source to be measured with a light source of known intensity. The modern unit, adopted in 1948, for the measurement of light intensity is the candela (cd); it is equal to 1/60 of the intensity of one square centimeter of a Blackbody radiator at the temperature at which platinum solidifies (2046°K).

Photomultiplier. A device for detecting photons, consisting of an electron multiplier with a photocathode on the front. Photons hitting this electrode produce electrons by the photoelectric effect and these are amplified and detected by the electron multiplier.

Photon. A quantum of electromagnetic radiation. The photon is usually thought of as an elementary particle with zero rest mass and an energy *hv*, where *h* is the Planck constant and *v* the frequency of radiation.

Photon or **Light Quantum.** The particle composing light and other forms of electromagnetic radiation. The Photoelectric Effect and Blackbody radiation can be explained only by assuming that light energy is transferred in discrete packets, or photons, and that the energy of each photon is equal to the frequency of the light multiplied by Planck's constant *h*. Light imparts energy to a charged particle when one of its photons collides with the particle.

Photopic Vision. Vision by the eye when the cones of the retina are the main receptors. This is the type of vision occurring at normal luminance levels. The cones can distinguish different colours. *Compare* scotopic vision.

Photosynthesis. Process in which green plants use the energy of sunlight to manufacture carbohydrates from carbon dioxide and water in the presence of Chlorophyll. The chlorophyll molecule is uniquely capable of converting active light energy into a latent for (glucose) that is stored in food. The initial phase of the process requires direct light; water (H_2O) is broken down into oxygen (which is released as a gas) and hydrogen. Hydrogen and the carbon and oxygen of carbon dioxide (CO_2) are then converted into a series of increasingly complex compounds that result finally in a stable organic compound, glucose ($C_6H_{12}O_6$), and water. The simplified equation for the overall reaction is $6CO_2 + 12H_2O + \text{energy} \rightarrow C_6H_{112}O_6 + 6O_2 + 6H_2O$. The oxygen released as a byproduct is atmospheric oxygen, vital to respiration in plants and animals. Photosynthesis, in general, is the reverse of Respiration, in which carbohydrates are broken down to release energy.

Photosphere. The surface region of the sun.

Photovoltaic Cell. Semiconductor diode that converts light to electric current. When light strikes the exposed active surface, it knocks electrons loose from their sites in the cyrstal. Some of the electrons have sufficient energy to cross the Diode junction and pass through an external circuit. Because the current and voltage obtained from these devices are small, they are usually connected in large series-parallel arrays. Practical photovoltaic cells are currently about 10 to 15% efficient. Although cells constructed from indium phosphide and gallium arsenide are, in principle, more efficient, silicon-based cells are generally less costly. Solar photovoltaic cells have long been used to provide electric power for spacecraft. Recent developments, still in progress, have driven costs down to the point where they are being use more and more as terrestrial energy sources.

Photovoltaic Effect. The production of an e.m.f. between two layers of different materials when the surface is irradiated by light or other electromagnetic radiation. The phenomenon is shown by a thin layer of copper oxide on copper and by gold on selenium.

Phthalic Acid. A colourless crystalline dibasic carboxylic acid, $C_6H_4(COOH)_2$. The carboxyl groups are in *ortho positions.* M.pt. 207°C: r.d. 1.6.

Phthalic Anhydride. A white crystalline compound, $C_8H_4O_3$, made by dehydrating phthalic acid or by oxidation of naphthalene. It is used as a drying agent and a raw material for making dyes. M.pt. 131°C; sublimes at 285°C; r.d. 1.5.

Physical Change. Any change that does not involve the production of different chemical compounds.

Physical Chemistry. The branch of chemistry concerned with such topics as the physical properties of chemical compounds and general investigations of their structure, chemical bonding, and mechanism and rates of reaction.

Physical Chemistry. Branch of science that combines the principles and methods of Physics and Chemistry. It provides a fundamental theoretical and experimental basis for all of chemistry, including organic, inorganic, and analytical chemistry. Important topics are chemical equilibrium, Electrochemistry, molecular structure, Molecular Weights, reaction rates, Solutions, and States of Matter.

Physical Optics. Optics.

Physical Therapy or **Physiotherapy.** Treatment of disorders of the Muscles, Bones, or Joints resulting from injury or disease of the muscles or nerves. Treatment, by a trained physiotherapist, includes the use of such agents as water (*e.g.*, whirlpool baths), manual and electronic massage, heat, and exercise to stimulate nerves, prevent muscular atrophy, and train new muscles to compensate for damaged ones.

Physics. Branch of science traditionally defined as the study of Matter, Energy, and the relation between them. Physics today may be loosely divided into classical physics and modern physics. Classical physics includes the traditional branches that were recognized and fairly well developed before the beginning of the 20th cent.: Mechanics (the study of Motion and the Forces that cause it), Acoustics (the study of Sound), Optics (the study of Light), Thermodynamics (the study of the relationships between Heat and other forms of energy), and Electricity and Magnetism. Most of classical physics is concerned with matter and energy on the normal scale of observation. By contrast, much of modern physics is concerned with the behaviour of matter and energy under extreme conditions or on the very small scale. On the very small scale, and for rapidly-moving objects, ordinary, commonsense notions of space, time, matter, and energy are no longer valid, and two chief theories of modern, physics present a different picture of these concepts from that presented by classical physics. The Quantum Theory is concerned with the discrete, rather than the continuous, nature of many phenomena at the atomic and subatomic level, and with the complementary aspects of particles and waves in the description of such phenomena. The theory of Relativity is concerned with the description of phenomena that take place in a frame of reference that is in motion with respect to an observer.

Physics. The study of matter and energy without reference to chemical changes occurring. "Traditional" physics includes such topics as heat, light, sound, mechanics, electricity, and magnetism. "Modern" physics includes atomic, nuclear, and particle physics, relativity, quantum mechanics, and astrophysics.

Physiology. Study of the normal functioning of animals and plants, and of activities that maintain and transmit life. It is usually accompanied by the study of structure, the two being intimately related. Physiology considers basic activities such as Metabolism and special functions within cells, tissues, and organs.

Pi. In mathematics, the ratio of the circumference of a Circle to its diameter, its symbol is Π. The ratio is the same for all circles and is approximately 3.1416. The Number Π is irrational and transcendental. An early value was the Greek approximation 3-1/7; by the mid-20th cent. a computer had calculated Π to 100,000 decimal places.

Physisorption. Adsorption in which the adsorbed substance is held by van der Waals forces.

Pi (Π). The ratio of the circumference of a circle to its radius: 3.141 592 653 ...

Picrate. Any salt or ester of picric acid.

Picric Acid (2, 4, 6-trinitrophenol). A poisonous yellow explosive crystalline solid, $C_6H_2(NO_2)OH$, used in explosives and the manufacture of dyes. M.pt. 122°C; r.d. 1.8.

Piezoelectric Effect. An effect observed in certain crystals whereby they develop a potential difference across a pair of opposite faces when subjected to a stress. The phenomenon is observed in Rochelle salt and quartz and is used in transducers. *Compare* electrostriction.

Piezoelectric Effect. Voltage produced between surfaces of a solid Dielectric when a mechanical stress is applied to it. This effect is exhibited by certain crystals, *e.g.,* quartz and Rochelle salt, and ceramic materials. When a voltage is applied across certain surfaces of a solid exhibiting the piezoelectric effect, the solid undergoes a mechanical distortion. Piezoelectric materials are used in Transducers, *e.g.,* phonograph cartridges, microphones, and strain gauges, which produce an electrical output from a mechanical input, and in earphones and ultrasonic radiators, which produce a mechanical output from an electrical input.

Pig, Iron. Impure iron obtained by smelting iron in a blast furnance: used for making steel.

Pigeon. Land Bird of the family Columbidae, cosmopolitan in temperate and tropical regions, characterized by a stout body, small head, and thick plumage. The names *dove* and *pigeon* are interchangeable, although the former generally refers to smaller birds. The rock dove (*Columba livia*) of temperate W eurasia is the wild progenitor of the common street and domestic pigeons. The most common American wild pigeon is the small, brown mourning dove, or turtle-dove (*Zenaidura carolinensis*), similar to the once-abundant passenger pigeon, which became extinct in 1914.

Pigment. 1. An insoluble colourant. Pigments are used in dispersion, as in paints, or can be melted together with the substance to be coloured, as in many synthetic fibres and plastics. They may be organic compounds or insoluble inorganic solids.

2. Substance that imparts colour to other materials. Most paint pigments are metallic compounds, but organic compounds are also used. Some metallic pigments occur naturally, *e.g.*, the oxides the produce the brilliant colouring of rocks and soil in the W U.S. Plants and animals also contain pigments. Chlorophyll (green) and carotene (yellow) produce bright colours in plants. Blood receives its red colour from Hemoglobin, and various pigments colour human skin.

Pike. Freshwater Fishes of the family Esocidae, found in Europe, Asia, and Northern America. The pike, muskellunge, and pickerel are long, thin fishes with spineless dorsal fins, large anal fins, and long, narrow jaws with formidable teeth. The muskellunge is the largest of the genus *Esox,* from 2 to 7 ft long (61 to 213.5 cm) and weighing 10 to 20 lb (4.5 to 9 kg). Carnivorous and solitary except while spawning, it eats fish, frogs, snakes, young aquatic mammals, and water fowl. The great northern pike (*Esox lucius*), called jackfish in Canada, is thought to consume one fifth its weight (10-35 lb/4.5-16 kg) daily. The pickerels are smaller members of the family. Pikes are strong fighters and valued as game and food. The walleyed pike is a Perch.

Pincushion Distortion. *See* distortion.

Pine. Common name for the family Pinaceae, resinous woody trees chiefly of north temperate regions, with needlelike, usually evergreen leaves. The Pinaceae reproduce by means of cones rather than flowers, and have winged seeds suitable for wind distribution. The family is the largest and most important of the Conifers, providing naval stores (pitch, turpentine, and rosin), paper pulp, and more lumber than any other family. The family's genera include the Fir, Larch, Spruce, Hemlock, Cedar, Douglas Fir, and true pines. True pines (genus *Pinus*) can be indentified by the leaf arrangement: in each species a specific number of needles (one to five) is contained in a sheath. The ponderosa, or Western yellow, pine (*P. ponderosa*) is second only to the Douglas fir as a commercial timber tree in North America; the Scotch pine

(*P. sylvestris*), ranging from Scotland to Siberia, is one of the most valuable European timber trees. The white pine (*P. strobus*) of E North America has straight-grained softwood with little resin; it is used for interior trim and cabi-network. A major source of naval stores is the longleaf, or Southern yellow, pine (*P. palustris*); its highly resinous wood is also used for heavy construction and paper pulp. Several Mediterranean and American pines yield edible seeds, called pine nuts.

Pineal Body or **Pineal Gland.** Pea-sized organ situated in the Brain. Generally regarded as an endocrine gland even though no pineal Hormone has been isolated in humans, it may exert some influence on sexual development by secreting a substance called melatonin.

Pineapple. Fruit of a spiny herbaceous plant (*Ananas comosus*) of the Bromeliad family. Native to South America, the pineapple plant is widely cultivated in tropical regions; Hawaii supplies the major portion of the world's canned pineapple. The fruit, whose spiny skin is yellowish brown when ripe, is sweet and juicy; it is topped by a distinctive rosette of green leaves.

Pink. Common name for some members of Caryophyllaceae, a family of small herbs chiefly of north temperate zones, typified by swollen stem nodes and notched, or "pinked", petals ranging in colour from white to red and purple. The family includes several ornamentals and many wildflowers and weeds. Ornamental pinks include the fragrant flowers of the genus *Dianthus,* among which are the many varieties of carnation and sweet William. Baby's breath (*Gypsophila paniculata*) is an unusual pink in being a bushy plant; it is often used by florists as a bouquet filler.

Pion (pi-meson). A type of meson having either zero charge and a mass 264.2 times that of the electron, or a positive or negative charge and a mass 273.2 times that of the electron.

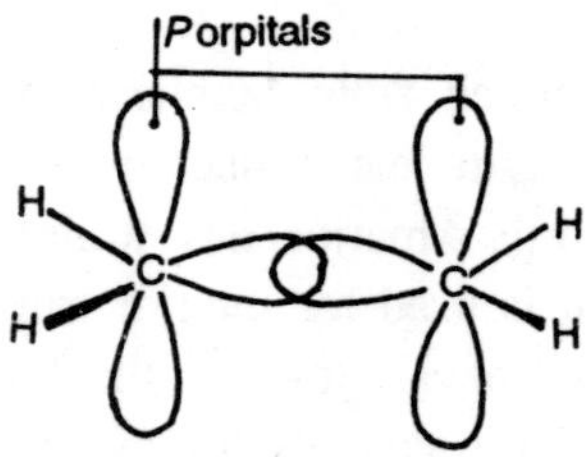

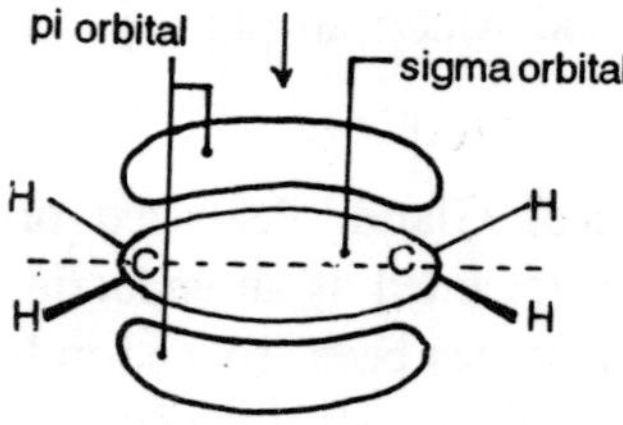

Fig. P-6. Molecular orbitals in ethylene.

Pi Orbital (Π-orbital). A type of molecular orbital in which there are two regions on each side of a line between the atoms. Pi orbitals form one of the bonds in a double bond between atoms. In ethylene, for example, each carbon atom is sp^2 hybridized and two of these hybrid orbitals form a sigma orbital between the carbon atoms. The remaining *p* orbitals, one on each carbon atom, overlap laterally and form a pi orbital, which has lobes each side of the sigma orbital. It is occupied by a pair of electrons with opposite spins. The double bound between the carbon atoms thus consists of a sigma bond and *pi bond.*

Piperazine. A colourless crystalline heterocyclic solid, $C_4H_{10}N_2$, used as a corrosion inhibitor and insecticide. M.pt. 104°C; b.pt. 145°C.

Pipette. A graduated glass tube for measuring out a known fixed quantity of a liquid.

Piranha or **Caribe.** Predatory freshwater fish of the family Characidae, found in E and central South America, especially in the Amazon. The piranha (genus *Serrasalmus*) has powerful jaws and razor-sharp triangular teeth and is capable of killing cattle and human beings. The largest piranha is c.2 ft (60 cm) long.

Pistachio. Tree or shrub (genus *Pistacia*) of the Sumac family. The pistachio nut of commerce is obtained from *P. vera,* native to Turkey, Syria, and Palestine; the trade supply comes primarily from Iran, Syria, Turkey, Greece, Italy, and, in the U.S., California. The nut is a greenish seed that is eaten salted or used in confections.

Pitch. In music, the position of a tone in the musical scale, today designated by a letter name and determined by the frequency of vibration of the source of the tone. An international conference held in 1939 set a standard for A above middle C of 440 cycles per second.

Pitch. The frequency of a sound as judged by comparison with the frequency of a pure tone.

Pitchblende. A heavy black radioactive mineral consisting mainly of uranium oxides. It is the most important ore of uranium.

Pitchblende. Dark, lustrous, massive variety (largely UO_2) of the mineral uraninite, a source of Radium and Uranium. Pitchblende and uraninite occur as primary constituents of quartz veins and with other metals, chiefly in the Great Lakes region in Canada, the Colorado Plateau in the U.S., Australia, Czechoslovakia, South Africa, and Zaire.

Pitcher Plant. Any of several insectivorous plants with leaves adapted as "pitchers" for trapping insects. Lured by nectar and the plant's coloration, the insects drown in the rain-water solution contained in the pitcher and are digested by plant Enzymes and, perhaps, Bacteria. There are three families: the American family, Sarraceniaceae, including the common pitcher plant (*Sarracenia purpurea*), found in the bogs of E North America; the Old World tropical family, consisting of the genus *Nepenthes,* found chiefly in Borneo; and the Australian pitcher plant (*Cephalotus follicularis*), the only species in its family.

Pituitary Gland. Small, oval endocrine gland that lies at the base of the Brain. It is called the master gland because the other endocrine glands depend on its secretions for stimulation. The pituitary has two distinct lobes, anterior and posterior. The anterior lobe secretes at least six hormones: growth hormone, which stimulates overall body growth; ACTH (adrenocorticotropic hormone), which controls steroid hormone secretion by the adrenal cortex; thyrotropic hormone, which stimulates the activity of the Thyroid Gland; and three gonadotropic hormones, which control growth and reproductive activity of the gonads. The posterior lobe secretes antidiuretic hormone, which causes water retention by the kidneys, and oxytocin, which stimulates the Mammary Glands to release milk and also causes uterine contractions. An overactive pituitary during childhood can cause gigantism; during adulthood, it can cause acomegaly. Dwarfism results from pituitary deficiency in childhood.

Pit Viper. Poisonous Snake of the family Crotalidae. Like the Old World true Vipers, pit vipers have long, hollow, erectile fangs. In addition, they have special heat-receiving organs, or pits, that help them sense warm-blooded animals, an ability especially useful at night, when many of them hunt. Pit vipers include the Rattlesnake, Copperhead, and Water Moccasin.

PK. The logarithm to base 10 of the reciprocal of the equilibrium constant of a given chemical reaction: *i.e.* $pK = \log_{10} (1/K)$. It is usually used as an indication of the strengths of acids, K being the dissociation constant. Weak acids have high pK values.

Placebo. Insert chemical substance used instead of a Drug Placebo preparations contain no medicine but are given for their positive psychological effects. They are also used as controls in scientific research to assure unbiased, statistically reliable results. In so-called double blind experiments, neither the doctor nor the patient knows whether a placebo or medication has been administered.

Plague. A general term used for any contagious epidemic disease, but usually used to refer specifically to bubonic plague, or the Black Death, an acute infectious disease caused by the bacterium *Pasteurella pestis (Yersinia pestis)*, transmitted to humans by fleas from infected rats. Symptoms include high fever; chills; prostration; enlarged, painful lymph nodes (buboes), particularly in the groin; and, in its black form, hemorrhages that turn black. Invasion of the lungs by the bacterium causes a rapidly fatal form of the disease (pneumonic plague), which can be transmitted from one person to another via droplets. epidemics have occurred throughout history, the best known being the Black Death that swept Europe and parts of Asia in the 14th cent; killing as much as three quarters of the population in less than 20 years. The disease is still prevalent in some areas of the world, but such Antibiotics as tetracycline and streptomycin have greatly reduced the mortality rate.

Plaid. A long shawl or blanketlime outer wrap of woolen cloth, usually patterned in checks or tartan figures. Now a feature of the Scottish Highland costume, it was once worn all over Scotland and N England by men and women. A tartan plaid has crossbars of three or more colours in designs signifying the various Highland clans. *Plaid* may also refer to any fabric patterned like the traditional plaid.

Planarian. Any of several groups of turbellarians—free-living, primarily carnivorous Flatworms—with a three-branched digestive cavity. Most are freshwater forms, but marine and terrestrial planarians exist. White, gray, brown, black, or sometimes transparent, planarians range in size from 1/8 to 1 in. (0.32 to 2.54 cm), although some tropical forms are as big as 2 ft (60 cm). Some species can regenerate severed parts of the body, in some cases even producing entire individuals from small pieces.

Planck Constant. Symbol: *h*. The universal constant 6.626 196 X 10^{-34} joule-second. A vibration of frequency *f* has energy quantized in units of *hf*. [After Max Planck (1858-1947), German physicist.]

Planck, Max, 1858-1947. German physicist. From his hypothesis (1900) that atoms emit and absorb energy only in discrete bundles (quanta) instead of continuously, as assumed in classical physics, the Quantum Theory was developed. Planck received the 1918 Nobel Prize in physics for his work on Black Body radiation. He was professor (1889-1928) at the Univ. of Berlin and president (1930-35) of the Kaiser Wilhelm Society for the Advancement of Science, Berlin. Planck's constant is named for him.

Planet. Any of the nonluminous bodies that orbit the sun. Strictly the term is applied to the group of major bodies of which nine are known. In order of succession outwards from the sun they are: Mercury, Venus, the earth, Mars, Jupiter, Saturn, Uranus, Neptune, and Pluto. Technically, the asteroids are also planets. The planets in the solar system fall into two main categories: namely, the inner or terrestrial planets, Mercury, Venus, the earth, and Mars, and the outer or giant planets, Jupiter, Saturn, Uranus, and Neptune. The terrestrial planets are relatively small and complicated in chemical composition, the earth being the largest: the giant planets are much larger in size and possess atmospheres of simple composition, usually mathane and ammonia, with cores of highly condensed frozen hydrogen. The outermost planet, Pluto, is probably comparable to the earth in size but similar to the giant planets in composition.

Plane. In mathematics, flat surface of infinte extent but no thickness. A plane is dermined by (1) three points not in a straight line; (2) a straight line and a point not on the line; (3) two intersecting lines; or (4) two parallel lines.

Planet. Any of the nine relatively large, nonluminous bodies—Mercury, Venus, Earth, Mars, Jupiter, Saturn, Uranus, Neptune, and Pluto—that revolve around the sun. By extension, any similar body discovered revolving around another star would be called a planet. The Asteroids are sometimes called minor planets. The major planets are classified either as inferior, with an orbit between the sun and the orbit of the earth (Mercury and Venus), or as superior, with an orbit beyond that of the earth (Mars, Jupiter, Saturn, Uranus, Neptune, and Pluto). The terrestrial planets—Mercury, Venus, Earth, Mars, and Pluto—resemble the earth in size, chemical composition, and density. The Jovian planets—Jupiter, Saturn, Uranus, and Neptune—are much larger in size and have thick, gaseous atmospheres and low densities. The rapid rotation of the latter planets results in polar flattening of 2-10%, giving them an elliptical appearance.

Planetarium. Optical device used to project a representation of the heavens onto a domed ceiling; the term also designates the building that houses such a device. As the axis of the device moves, beams of light are emitted through lenses and travel in predetermined paths on the ceiling. The juxta-position of lights reproduces a panorama of the sky at a particular time as it might be seen under optimum conditions. The motions of the celestial bodies—typically the fixed stars, the sun, moon, and planets, and various nebulae—are accurately represented, although they can be compressed into much shorter time periods.

Plane Tree, Sycamore, or **Buttonwood.** Deciduous tree (genus *Platanus*) indigenous to northern temperate regions. The dry, seedlike fruits are compressed into a hard, brown ball, which, when ripe, separates into windborne, downy tufts. The genus includes the American sycamore (*P. occidentalis*) and the Oriental plane (*P. orientalis*), both used for their wood. The London plane (*P. acerifolia*) is much used as an ornamental shade tree in cities.

Plankton. Very small to microscopic plants and animals that have little or no power of locomotion and drift or float in surface waters. Plankton is found worldwide in fresh and salt water. The plant forms include Diatoms and dinoflagellates; planktonic animals include protozonas, small Crustanceans, Jellyfish, Comb Jellies, and Fish eggs and larvae. In the ocean, plankton is the source of food, either directly or indirectly through the food chain. for all marine animals.

Plant. An organism of the plant kingdom (Planta) as opposed to one of the Animal kingdom (Animalia). A plant may be microscopic in size and simple in structure *(e.g.,* Algae) or a many-celled, complex system *(e.g.,* a Tree). Plants differ from animals in that, with few exceptions, they possess Chlorophyll, are fixed in one place, have no nervous system or sensory organs, and have rigid, supporting cell walls containing Cellulose. In addition, most plants grow continually and have no maximum size or characteristic form in the adult stage. Green plants, *i.e.,* those with chlorophyll, manufacture their own food (Glucose, a sugar) and give off oxygen in the process of Photosynthesis, thus representing the primary source of food for animals and providing oxygen for the earth's atmosphere. Some plants *(e.g.* Fungi) to not make their own food, and certain unicellular forms *(e.g., Euglena)* are motile and can either make or ingest food. The scientific study of plants is Botany.

Plantain. Annual or perennial weed (genus *Plantago*), of wide distribution. Many species are lawn pests, and the pollen is often a hay fever irritant. *P. psyllium,* or fleawort, is cultivated in Spain and France for its mucilaginous seed coats, which are used as a laxative. A tropical plant related to the banana is also called plantain.

Plasma. A mixture of ions and electrons. The plasma in an arc, spark, or other electrical discharge is composed of simple ions an delecrons together with unionized gas. The plasma occurring in thermonuclear reactions, as in the sun, is highly ionized and consists of a mixture of atomic nuclei and electrons.

Plasma. In physics, a fully ionized gas containing approximately equal numbers of positive and negative Ions. A plasma is an electric conductor and a affected by magnetic fields. The study of plasmas, called plasma physics, is important in efforts to produce a controlled thermonuclear reaction. In nature, plasmas occur in the interior of stars and in interstellar gas, making plasma a form of matter in the universe.

Plaster of Paris. A hemihydrate of calcium sulphate, $CaSO_4 \cdot {}^1/{}_2H_2O$, made by partial dehydration of gypsum by heating. When mixed with water it sets to a hard mass.

2. Any of various synthetic materials that contain or consist of polymers and are moulded at some stage in the manufacturing process. *Thermosetting* plastics are polymeric material that harden on heating to give a rigid product that cannot then be softened. *Thermoplastic* materials soften when heated and harden again when cooled. The terms *plastic* and *resin* are often used synonymously, although in strict technical usage *resin* is used for the original polymer whereas *plastic* refers to the final material, which may contain pigments, fillers, plasticizer, stabilizer, and antioxidant.

Plastic. Any synthetic organic material that can be molded under heat and pressure into a shape that is retained after the heat and pressure are removed. There are two basic types of plastic: thermosetting, which cannot be resoftened after being subjected to heat and pressures; and thermoplastic, which can be repeatedly softened and reshaped by heat and pressure. Plastic are made up chiefly of a binder consisting of long chainlike molecules called Polymers. Binders can be natural materials, *e.g.,* Cellulose, or (more commonly) synthetic Resins, *e.g.,* Bakelite. The permanence of thermosetting plastics is due to the heat and pressure—nduced cross-linking reactions the polymers undergo. Thermoplastics can be reshaped because their linear or branched polymers can slide plast one another when heat and pressure are applied. Adding plasticizers and fillers to the binder improves a wide range of properties, *e.g.*, hardness, elasticity, and resistance to heat, cold, or acid. Adding Pigments imparts colour. Plastic products are commonly made from plastic powders. In compression molding, heat and pressure are applied directly to the powder in the mold cavity. Alternatively, the powder can be plasticized by outside heating and then poured into molds to harden (transfer molding); be dissolved in a heating chamber and then forced by a plunger into cold molds to set (injection molding); or be extruded through a die in continuous form to be cut into lengths or coiled (extrusion molding). The first important plastic, Celluloid, has been largely replaced by a wide variety of plastics known by such trade names as Plexiglas, Lucite, Polaroid, and Cellophane. New uses continue to be found and include contact lenses, machine gears, and artificial body parts. The widespread use of plastics has led to environmental problems. Because plastic products do not decay, large amounts accumulate as waste. Disposal is difficult because they melt when burned, clogging incinerators and often emitting harmful fumes, *e.g.*, the hydrogen chloride gas given off by Polyvinyl Chloride.

Plasticizer. A material mixed with synthetic resin to improve the flexibility or other properties of the plastic. Camphor, for example, is used in celluloid.

Plastic surgery. Surgical repair of congenital or acquired deformities and the restoration of contour to improve the appearance of tissue defects, *e.g.*, disfigurements resulting from accidents such as severe burns or from removal of extensive skin cancers. It is also used to restore vital movement and function of destroyed tissues. The technique was first developed after World War I and is now also used for cosmetic purposes, *e.g.*, to eliminate wrinkles.

Plateau. Elevated, more or less level portion of the earth's surface bounded on at least one side by steep slopes. Plateaus are formed by successive lava flows, upward-folding earth movements, or the erosion of adjacent lands. Notable plateaus include the Colorado and Columbia Plateaus in the U.S. and the Deccan in India.

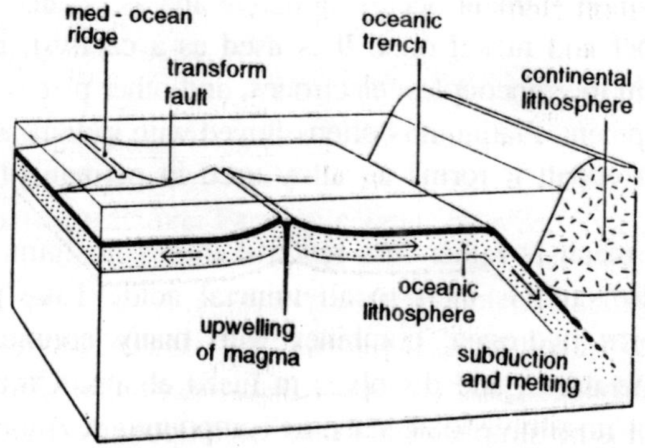

Fig. P-7. *Plate tectonics:* Schematic model of the oceanic crust, showing the three types of plate boundaries.

Plate Tectonics. Modern theory of Continental Drift that has revolutionized geologists' understanding of Earth history. It holds that the earth's crust is divided into contiguous, moving plates that carry the embedded Continents. Plate boundaries are marked by lines of Earthquake and volcanic activity. One kind of boundary is at the mid-ocean ridges, where tensional forces open rifts, allowing new crustal material to well up from the earth's mantle and become welded to the trailing edges of the plates. When a continent straddles such a rift it is split apart, forming a new ocean area *(e.g.,* the Red Sea and the Gulf of California). The ocean trenches mark subduction zones, where plate edges drive steeply into the mantle and are reabsorbed. A third boundary type occurs where two plates slide past each other in a shearing manner along great transform Faults *(e.g.,* the San Andreas Fault in California). Mountain ranges form where two plates carrying continents collide (*e.g.*, the Himalayas), or where ocean crust is subducted along a continental margin (*e.g.*, the Andes). Geologists believe that c.200 million years ago there was a supercontinent, Pangaea, which subsequent plate movements have split and resplit into the continents and islands we recognize today.

Platinum. Symbol: Pt. A silvery-white malleable and ductile transition element occurring native and associated with certain copper and nickel ores. It is used as a catalyst, in jewellery, and in thermocouples, electrodes, and other pieces of scientific equipment. Platinum is often alloyed with iridium and rhodium. With cobalt it forms an alloy used in permanent magnets.

The metal is rather less reactive than palladium: it does not oxidize and is inert to all mineral acids. Like palladium it absorbs hydrogen, combines with many nonmetals at high temperatures, and dissolves in fused alkalis. Unlike nickel, it forms no simple ionic salts: its compounds are mainly platinum II (*platinous*) and platinum IV (*platinic*) complexes. A.N. 78; A.W. 195.09; m.pt. 1772°C; b.pt. 3820°C; r.d. 21.45; valency 1, 2, 3, 4 or 6.

Platinum Chloride. Either of two chlorides of platinum. *Platinum IV chloride* (platinic chloride), $PtCl_4$, is a brown solid when anhydrous, made by evaporating a solution of platinum in aqua regia. It also exists as *chloroplatinic acid,* $H_2PtCl_6.6H_2O$. Platinum IV chloride is used in platinum plating and photography. R.d. 4.3 *Platinum II chloride* (platinous chloride) is a green-grey powder, $PtCl_2$, made by heating platinum with chlorine. It is used to make platinum salts. R.d. 5.9.

Platinum Metals. The group of related transition metals ruthenium, osmium, rhodium, iridium, palladium, and platinum.

Plato. 427?-347 B.C., Greek philosopher. In 407 B.C. he became a pupil and friend of Socrates. After living for a time at the Syracuse court, Plato founded (c.387 B.C.) near Athens the most influential school of the ancient world, the Academy, where he taught until his death. His most famous pupil there was Aristotle. Plato's extant work is in the form of epistles and dialogues, divided according to the probable order of composition. The early, or Socratic, dialogues, *e.g.*, the *Apology, Meno,* and *Gorgias,* present Socrates in conversations that illustrate his major ideas—the unity of virtue and knowledge and of virtue and happiness. They also contain Plato's moving account of the last days and death of Socrates. Plato's goal in dialogues of the middle years, *e.g.*, the *Republic, Phaedo, Symposium,* and *Timaeus,* was to show the rational relationship between the soul, the state, and the cosmos. The later dialogues, *e.g.*, the *Laws* and *Parmenides,* contain treatises on law, mathematics, technical philosophic problems, and natural science. Plato regarded the rational soul as immortal, and he believed in a world-soul and a Demiurge, the creator of the physical world. He argued for the independent reality of Ideas, or Forms, as the immutable archetypes of all temporal phenomena and as the only guarantee of ethical standards and of objective scientific knowledge. Virtue consists in the harmony of the human soul with the universe of Ideas, which assure order, intelligence, and pattern to a world in constant flux. Supreme among them is the Idea of the Good, analogous to the sun in the physical world. Only the philosopher, who understands the harmony of all parts of the universe with the Idea of the Good, is capable of ruling the just state. In Plato's various dialogues he touched upon virtually every problem that has occupied subsequent philosophers; his teachings have been among the most influential in the history of Western civilization, and his works are counted among the world's finest literature.

Pleochroism. The phenomenon in which certain crystals have different colours when viewed from different directions.

Pleurisy. Inflammation of the pleura, the membrane that covers the Lungs and lines the chest cavity. It is sometimes accompanied by fluid (effusion) that fills the chest cavity; when the fluid is infected, the condition is known as empyema. Dry pleurisy usually occurs with bacterial infections, whereas pleurisy with effusion is associated with chronic lung conditions, such as Tuberculosis. Treatment is directed at the underlying cause.

Plum. Name for many species of trees (genus *Prunus*) of the Rose family, and their fruits. Numerous varieties and hybrids exist. The name *damson* is applied to several varieties of the common garden plum (*P. domestica*)having small leaves and small, oval, usually tart fruits. Greengage and prune plums are also varieties of *P. domestica*. Many varieties are used as ornamentals; these usually have red or purple foliage and double pink, white, or lilac flowers. Prunes are dried plums.

Pluto. The ninth planet in order of succession outwards from the sun and the furthest of all the planets currently known in the .entire solar system. Lying at a mean distance of 5907 million kilometres from the sun, Pluto orbits it at an average velocity of less than 5 km per second, taking 248.5 years to complete one revolution. Pluto is one of the most obscure objects in the night sky. No reliable facts are available for the planet, but it seems certain that it is radically different from the other outer planets. Its diameter is probably about 5900 km (less than half that of earth) and its mass must be assumed to be of the order of one tenth to one third that of earth. It is thought that the temperature of the sunlit side of Pluto probably does not rise much above -214°C; as a result many astronomers believe that there can be no atmosphere except possibly for traces of helium and neon. Pluto is therefore believed to be a very small planet and it has been suggested that it was originally a satellite of Neptune that escaped. In fact at a perihelion of 4505 million km Pluto's highly eccentric orbit falls within that of Neptune. Off all the planets, Pluto's path is the most deviant, having an eccentricity of 0.25.

Pluto. In astronomy, 9th and usually most distant Planet from the sun, at a mean distance of 3.67 billion mi (5.90 billion km).

Because of the high eccentricity (0.250) of its elliptical orbit, Pluto occasionally *(e.g.,* between 1979 and 1999) comes closer than the planet Neptune to the sun. Discovered in 1930 by Clyde Tombaugh, Pluto has an estimated diameter of 1,500 to 2,400 mi (2,400 to 3,800 km) and is thought to have a rocky, silicate core and a thin atmosphere containing methane. Its one known satellite, *Charon,* was discovered on June 22, 1978, by the American astronomer James Christy. It has a diameter estimated to be about a third that of Pluto.

Plutonium. Symbol: Pu. A silvery metallic transuranic actinide element made in nuclear reactor by neutron bombardment of uranium. It is extensively used in nuclear reactor and nuclear weapons. The fissile isotope, ^{239}Pu, has a half-life of 24 360 years. A.N. 94; mp.pt 639.5°C; b.pt. 3235°C; r.d. 19.84; valency 3, 4, 5, or 6.

Plutonium (Pu). Radioactive element, first produced artificially by Glenn Seaborg and colleagues in 1940 by deuteron bombardment of uranium oxide. It is a silver-gray Transuranium Element in the Actinide Series. Plutonium is a fission fuel for Nuclear Reactors and weapons. It is an extremely dangerous poison, collecting in bones and altering the production of white blood cells.

Pneumatic. Operating by air pressure.

Pneumoconiosis. Any of a group of chronic Lung diseases caused by the inhalation of dust particles. Commonly known as black lung and primarily found among coal miners, sandblaster, and metal grinders, it may be caused by the inhalation of Silica (silicosis), Asbestos (asbestosis), Iron filings (siderosis), Coal dust, or other mineral or metal dust. Particles collect in the lungs and become sites for the development of fibrous tissue that replaces elastic lung tissue, often resulting in decreased lung function. Symptoms include shortness of breath, wheezing, cough, and susceptibility to Pheumonia, Tuberculosis, and other respiratory infections.

Pneumonia. Acute infection of one or both lungs that can be caused by a bacterium, usually the pneumococcus bacterium, or by a virus. Symptoms include high fever, pain in the chest, difficulty in breathing, coughing, and sputum. Viral pneumonia is generally milder than the bacterial form. Antibiotics are used to treat bacterial pneumonia and have greatly reduced the mortality rate of the disease.

P-n-p transistor. *See* Transistor.

Podiatry. Science concerned with disorders, diseases, and deformities of the feet. Podiatrists treat such common conditions as bunions, corns and calluses, and ingrown toenails; perform minor surgery; and prescribe medicines and orthopedic devices. A practitioner in the. U.S. must be licensed; training is similar to that for medicine in most respects, except that it is largely limited to a single area of the body.

Point-contact Transistor. *See* transistor.

Point Defect. *See* defect.

Pointer. Large Sporting Dog; shoulder height, 23-26 in. (58.4-66.4 cm); weight, 50-60 lb (22.7-27.2 kg). Its short coat is usually white, with liver, black, yellow, or orange markings. Developed in England over 300 years ago, it hunts game birds by scent. It stands rigidly poised with its nose facing the game, directing the hunter to it.

Poise. Symbol: P. A CGS unit of dynamic viscosity equal to the dynamic viscosity of a fluid for which a tangential force of one dyne per square centimetre is necessary to create a velocity gradient of one centimetre per second per centimetre. [After Jean Poiseuille (1799-1869), French physicist.]

Poiseuille's Equation. The equation $Q = \Pi(p_1 - p_2)r^4/8\eta l$, giving the rate of fluid flow Q through a circular pipe of radius r and length l. The dynamic viscosity is η and $(p_1 - p_2)$ is the pressure difference between the ends of the tube. The equation can be used in a method of determining viscosity by measuring the rate of flow under steady conditions.

Poison. A substance that destroys the activity of a catalyst.

Poison. Any chemical that produces a harmful effect on a living organism. Almost any substance can act as a poison if it enters the body in sufficiently large quantities or in an abnormal way; *e.g.*, water inhaled into the lungs becomes an asphyzial poison. The severity of a poison is determined by the nature of the poison itself, the concentration and amount ingested, the route of entry, the length of exposure, and the age, size and health of the victim. Common poisonous substances include Arsenic, Cyanide, Strychnine, Lead, Mercury, acids, venoms, Herbicides, and Curare and other drugs.

Poison Gas. Any of various gases sometimes used in Chemical Warfare or riot control because of their poisonous or corrosive nature. These gases may be roughly grouped according to the portal of entry into the body and their physiological effects. Vesicants (blister gases, *e.g.*, mustard gas) produce blisters on all body surfaces; lacrimators (tear gas) cause severe eye irritation; sternutators (vomiting gases) cause nausea; nerve gases inhibit proper nerve function; and lung irritants cause pulmonary edema. World War I marked the first effective use of poison gas and the introduction of gas masks. The use of poison gas has been limited since World War I by fear of retribution, although the military powers have continued to develop new gases.

Poison Hemlock. Lethally poisonous herb (*Conium maculatum*) of the Carrot family, native to the Old World but now naturalized in parts of the U.S. It has cluster of small white flowers and a purple-mottled stem. The poisonous principle (the Alkaloid coniine) causes paralysis, convulsions, and eventual death. The plant was used in ancient Greece to execute criminals; a famous example was Socrates.

Poison Ivy, Poison Oak, and **Poison Sumac.** Woody vines and trailing or erect shrubs (genus *Toxicodendron*, although sometimes considered geuns *Rhus*) of the Sumac Family, native to North America. The names poison ivy and *poison oak,* often used interchangeably, describe several species, the most common being *T. radicans,* which has three smooth leaflets. Poison human (*T. vernix*) is larger plant. Both species have whitish, berrylike fruits and red autumn foilage. Urushiol,an irritant present in almost all parts of the plant, causes a skin eruption that may vary from itching inflammations to watery blisters.

Poisson's Ratio. Symbol: v. The ratio of the lateral strain in a body to the longitudinal strain. Thus if a rod of material of length l and diameter d undergoes tensile elongation to produce an increase in length Δl and a decrease in diameter Δd, Poisson's ratio is $(\Delta d/d)$ / $(\Delta l/l)$. If the volume does not change, the ratio is 0.5. [After Simeon Denis Poisson (1781-1840), French mathematician.]

Polar. Denoting a substance whose molecules have a dipole moment. Water, for example, is a polar compounds because of unequal sharing of electrons between the oxygen and hydrogen atoms.

Polar solvents are capable of dissolving ionic solids because parts of the energy required to separate the ions in the crystal is recovered in solvation of the ions. They are not usually capable of dissolving nonpolar compounds.

Polar Bond. *See* electrovalent bond.

Polar Coordinates. Coordinates used to locate the position of a point by its distance from a fixed point (the *pole*) and its angular displacement form a line. In two dimensions the coordinates are (r,θ), r being the *radius vector* and θ the *azimuth* or *amplitude* of the point. In three dimensions *spherical coordinates* can be used, in which the point is located by a radius vector, a *colatitude* θ measured from a vertical line, and a *longitude* ϕ measured from a horizontal line.

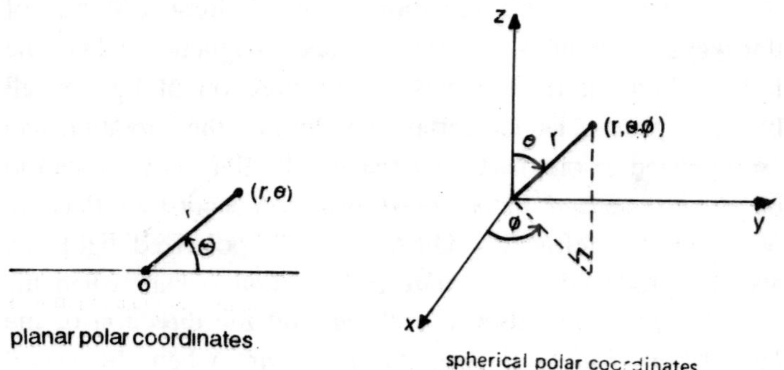

Fig. P-8. Polar coordinates.

Polarimeter. A instrument for determining the specific rotation of optically active substances. It consists of a *polarizer,* which is a Nicol prism or other device to polarize light, a transparent cell to contain the sample, and an *analyser,* similar to the polarizer but capable of being rotated and fitted with an angular scale. A beam of monochromatic light passes through the polarizer, cell, and analyser and is observed through an eyepiece. The analyser is rotated until the light has maximum intensity—the angular rotation can then be read on the scale.

Polariscope. Any instrument for viewing objects in polarized light. Strains in glass or transparent plastics are observed as coloured interference lines in the material.

Polarized Light. Light in which the vibration of the electric or magnetic field is confined to one plane. Ordinary light consists of a mixture of waves vibrating in all directions perpendicular to its line of propagation. Polarized light can be obtained to a varying extent (depending on the angle of incidence) by reflection. It can also be obtained by double refraction in certain crystals, such as calcite. These crystals have the property of refracting unpoloarized light in two different directions, the ordinary ray and the extraordinary ray; both are polarized in directions perpendicular to each other.

Polarization. 1. Restriction of the direction of vibration in light or other electromagnetic radiation. Normal light consists of transverse vibrations of electric and magnetic fields: the fields vibrate at right angles to the direction of light in all possible planes. Under certain conditions, the vibration can be restricted to one particular plane—the light is then said to be *plane-polarized.* Plane polarization can occur by reflection, or by double refraction. Other types of polarized light can also be produced. *Cicurcular polarization* occurs when the electric vector describes a circle around the direction of the light beam. *Elliptical polarization* occurs when the vector describes an ellipse.

2. Separation of charges in a dielectric material. When an electric field is applied across an insulator current does not flow through it, but the molecules are polarized so that they act as dipoles. In this way charge is stored in a capacitor.

3. A reduction in the e.m.f. of a cell as a result of the formation of a layer of small bubbles of hydrogen on the cathode. These decrease the effective area of the electrode and also produce a back e.m.f. The effect is reduced by use of a depolarizer.

Polarography. An analytical technique for identifying and determining the concentrations of ions by electrolysis of a solution. The cathode is a dropping-mercury electrode: *i.e.* one in which mercury is continually forming at and dropping away from the end of a vertical glass tube. In this way the electrode is kept clean and has a low surface area. A slowly increasing voltage is applied between this electrode and an anode and different positive ions in the solution are discharged at different values of the voltage. A graph of current against voltage (a *polarogram*) shows a series of steps, each corresponding to a particular type of ion.

Polaroid. ® A type of film containing many minute doubly refracting crystals aligned with their axes paralle. The film polarizes transmitted light and is used in sunglasses and other devices to reduce glare.

Polaroid Camera. Trade name for a single-step camera invented (1947 for black and white; 1963 for colour) by Edwin H. Land. In the Land process, the camera holds both negative film and printing paper, as well as a pod of developer that is spread between film and paper when the film is exposed. A finished print is produced within seconds.

Poliomyelitis. Acute viral infection which, in its severe form, invades the nervous system and causes paralysis. In its mild form the disease produces mild symptoms (*e.g.*, low-grade fever, malaise), or none. Also known as polio and infantile paralysis, it is found worldwide, occurring mainly in children. The Salk vaccine (killed-virus by injection) and the Sabin vaccine (live-virus vaccine taken orally) have greatly reduced the incidence of polio, nearly eradicating it from developed nations.

Pollen. Minute, usually yellow grains, borne on the another sac at the tip of the stamen (the male reproductive organ of the Flower) or in the male cone of a conifer. Pollen granis are the male Gametophyte generation of seed plants. They are formed by the division, or Meiosis, of pollen mother cells and contain half the number of chromosomes of the parent plant. See also Pollination.

Pollination. Transfer of Pollen from the male reproductive organ (the stamen of a Flower or staminate cone of a Conifer) to the female reproductive organ (pistil or pistillate cone) of the same or another flower or cone. The most common agents of pollination are flying insects (for most flowering plants) and wind (for most trees and all Grasses and conifers). The devices that operate to ensure cross-pol-lination and prevent self-pollination are highly varied and intricate. They include different maturation times for the pollen and eggs of the same plant; separate staminate and pistillate flowers on the same or separate plants; chemical properties that make the pollen and eggs of the same plant sterile to each other; and special mechanisms and arrangements that prevent the pollinating agent from transferring a flower's pollen to its stigma. Pollination should not be confused with Fertilization, which it may precede by some time—a full season in some conifers.

Polonium. A radioactive element: one of the decay products of radium. Polonium is one of the chalconides. A.N. 84; A.W. 210; mp.t. 225°C; b.pt. 950°C.

Polyamide. Any polymer in which the monomers are linked by amide groups, -CO.NH-. The various forms of nylon are the most common examples.

Polyatomic. Having molecules composed of several atoms. The term is often used to distinguish compounds from diatomic and monatomic compounds.

Polycarbonate Resin. Any of a class of synthetic resins that are polyesters of carbonic acids and dihydric phenols. Usually, polycarbonate resin is made by reacting phosgene and bisphenol. It is a strong transparent thermoplastic material used for many moulded articles.

Polychromatic. Denoting light or other electromagnetic radiation that has a mixture of wave lengths; not monochromatic.

Polycyclic. Denoting a chemical compound that contains two or more rings of atoms in its molecular structure.

Polyene. An organic compound containing three or more double bonds in its molecules.

Polyester Resin. Any of a class of synthetic resins made by reacting polyhydric alcohols with polybasic acids or their anhydride. Saturated polyesters can be made by condensation of simple diols, such as ethane diol, with unsaturated acids, such as phthalic acid. Compounds of this type are widely used in synthetic fibres like Terylene and Dacron. If unsaturated acids, such as maleic and fumaric acids, are used, then the polyester itself is unsaturated. This resin can then be further polymerized to a thermosetting plastic; often styrene is added at this stage and the material is reinforced with glass fibre. The result is very strong, light, and hard wearing making it suitable for use in car bodies, boat hulls, and similar applications. *Alkyd* (or *glyptal*) resins are made from Phthalic acid and glycerol, often with inclusion of styrene. They are used in paints and as moulding materials.

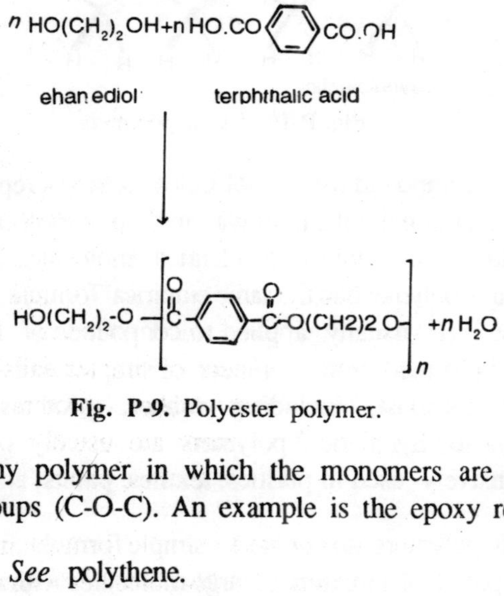

$$n\ HO(CH_2)_2OH + n\ HO.CO\text{-}C_6H_4\text{-}CO.OH$$

ehanediol terphthalic acid

$$\downarrow$$

$$HO(CH_2)_2\text{-}O\left[\overset{O}{\overset{\|}{C}}\text{-}C_6H_4\text{-}\overset{O}{\overset{\|}{C}}\text{-}O(CH2)2\,O\right]_n + n\,H_2O$$

Fig. P-9. Polyester polymer.

Polyether. Any polymer in which the monomers are linked by ether groups (C-O-C). An example is the epoxy resins.

Polyethylene. *See* polythene.

Polygon. Any plane figure with three or more sides.

Polyhedron. A geometric solid whose plane faces are polygons. For any simple polyhedron, the sum of the number of vertices and faces less the number of edges is equal to two. This is known as *Euler's theorem*. A *regular* polyhedron is one that has all its faces congruent. Only five types of polyhedron may be regular, namely the tetrahedron, hexahedron, octahedron, dodecahedron, and icosahedron.

Polyhydric. Denoting an alcohol that contains two or more hydroxyl groups in its molecules.

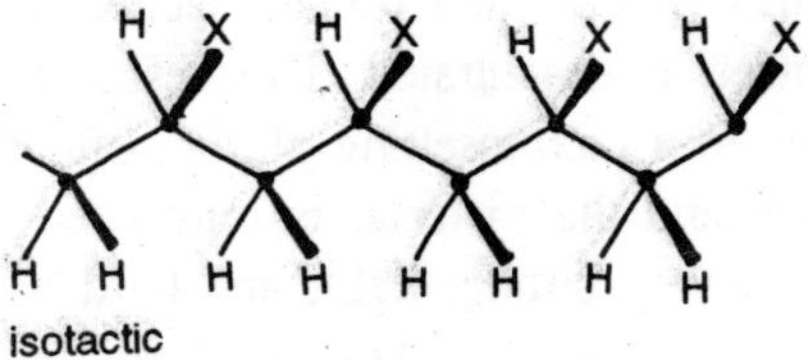

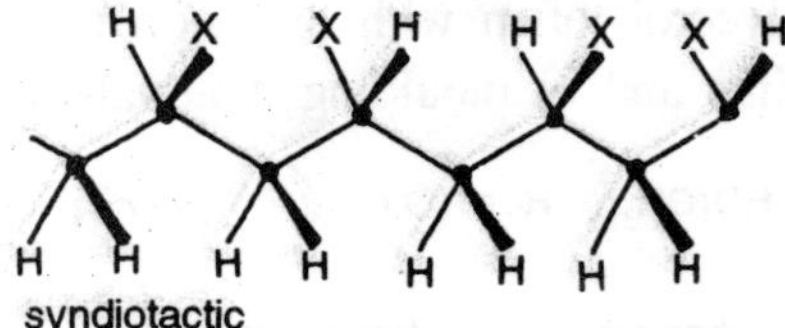

Fig. P-10. Tactic polymers.

Polymer. A compound whose molecules consist of repeated molecular units. Originally the term was applied to molecules of any size produced by addition of identical molecules (*monomers*), so that the polymer had the same empirical formula as the monomer. Now, it is usually applied to conpounds of high molecular weight formed from numbers of simpler units. Many natural polymers exist, including rubber, proteins, starch, and cellulose. Synthetic polymers are usually organic and are extensively used in plastics, textiles, paints, and adhesives.

Most polymers do not have a simple formula. Instead, they are composed of a mixture of large molecules with a distribution of molecular weights. The properties of polymeric materials also depend on the structure of the molecules, which is classified in various ways. The simplest molecules have straight chains: more complicated polymers contain side chains. Polymers are also characterized by the arrangement of monomers in the chain. *Homopolymers* are formed of the same type of monomer. A polymer consisting of two or more different units is a *copolymer*, in which the types of monomer may have a random order or may have some type of regular structure. A *block copolymer* has short chains of one monomer alternating with chains of another. A *graft copolymer* has a main chain of one type of monomer with side chains of the other. If the groups attached to atoms along the chain of a polymer have a regular arrangement in space the polymer is said to be tactic; otherwise it is *atactic*. Atactic polymers are of two types. In *isotactic* polymers the same configuration is repeated at each point along the chain. *Syndiotactic* polymers have asymmetric atoms in their chain and the arrangement is alternately inverted along the chain. Tactic polymers are stereospecific.

The properties of polymeric materials depend not only on the structure of chains but also on the shape of the chains and the way they are arranged in the bulk material. Thus, if the chains are tangled together or otherwise orientated at random the material will be amorphous. If they are parallel the polymer will have some degree of crystallinity. In some materials polymer chains are linked together by side chains, leading to an extended three-dimensional structure, which again may be random or regular.

Polymer. Chemical compound with high molecular weight consisting of a number of structural units linked together by covalent bonds. The simple molecules that may become structural units are themselves called monomers. A structural unit is a group having two or more bonding sites. In a linear polymer, the monomers are connected in a chain arrangement and thus need only have two bonding sites. When the monomers have three bonding sites, a nonlinear, or branched, polymer results. Naturally occuring polymers include Cellulose, Proteins, natural Rubber, and Silk; those synthesized in the laboratory have led to such commercially important products as Plastics, synthetic fibers, and synthetic rubber.

Polymerization. A chemical reaction leading to the formation of a polymer. Synthetic polymers may be produced by two types of process. In *addition polymerization* the monomers simply add together with no other compound formed. This happens for monomers that contain double bonds, as in the production of polythene, or rings, as in the production of epoxy resins. In *condensation polymerization* water, an alcohol, or other simple molecule is eliminated. This occurs when the monomer contains two reactive groups, as in the production of nylon. Addition and condensation copolymers can also be made (*copolymerization*) by reaction of two or more different compounds.

Polymethylmethacrylate. A colourless transparent thermoplastic material made by polymerization of methyl methacrylate. It is widely used as a substitute for glass. Polymethylmethacrylate is produced under the trade name *Perspex.*

Polymolybdate. Molybdate.

Polymolybdenic Acid. Molybdate.

Polymorphism. Existence of more than one solid (crystalline) form of the same chemical substance. If two forms exist the phenomenon is dimorphism.

Polynomial. A mathematical expression of the form $A_o + A_1x + A_2x^2 + A_3x^3 + ...$

Polynomial. Mathematical expression containing terms of one or more variables or constants that are connected by addition or subtraction. No variable can appear as a divisor or have fractional exponents. In one unknown, the general form of a polynomial is $a_ox^n + a_2x^{n-1} + a_1x^{n-2} + ... + a_{n-1}x + a_n$ where n is a positive integer and $a_o, a_1, a_2...., a_n$ are any numbers. The *degree* of a polynomial in one variable is the highest power of the variable. A polynomial of degree 2, *i.e*, $ax^2 + bx + c$ (where *a, b,* and *c* are any numbers), is called a *quadratic*.

Polyp and Medusa. Names for the two body forms of animals (Coelenterates) of the phylum Cnidaria—one, the polyp or hydroid, is stationary; the other, the medusa, is free-swimming. Both consist of a cylindrical body with a mouth surrounded by stinging tentacles at one end. The polyp is elongated and attaches to the ocean bottom with the mouth and tentacles pointing upward; the medusa is rounded and swims with the mouth down and tentacles dangling. Some cnidarians are always polyps, some are always medusae, and some take both medusa and polyp forms during their life cycles. Common polyps are the Sea Anemone and Hydra; a common medusa is the Jellyfish. In many species, such as Coral, the polyps form colonies. The Portuguese man-of-war is an elaborate floating colony of polyplike and medusalike individuals.

Polypeptide. A chain of amino acids held together by peptide linkages. Polypeptides are found in proteins.

Polyphony. Music whose texture is formed by the harmonic interweaving of several independent melodic lines through the use of Counterpoint. Contrasting terms are *homophony,* wherein one part dominates while the other form a chordal accompaniment; and *monophony,* wherein there is a single melodic line (as in Plainsong). Polyphony developed by the late 9th cent. and culminated in the great age of polyphony in the 15th and 16th cent. Although polyphony was overshadowed by Harmony in the Baroque period and by homophony in the classical and romantic periods, in the 20th cent. there has been renewed interest in polyphonic aspects of musical texture and structure.

Polyphosphate. Phosphate.

Polyphosphoric Acid. Phosphoric acid.

Polypropylene. A thermoplastic material made by polymerization of propylene. It is similar to high-density polythene but is stronger, lighter, and more rigid.

Polypropylene. Lightweight Plastic, a Polymer of propylene. It is less dense than water and resists moisture, oils, and solvents. It is used to make packaging material, textiles, luggage, ropes that float, and, because of its high melting point (250°F/121°C), objects that must be sterilized.

Polysaccharide. A sugar or other carbohydrate with molecules containing more than one monosaccharide unit.

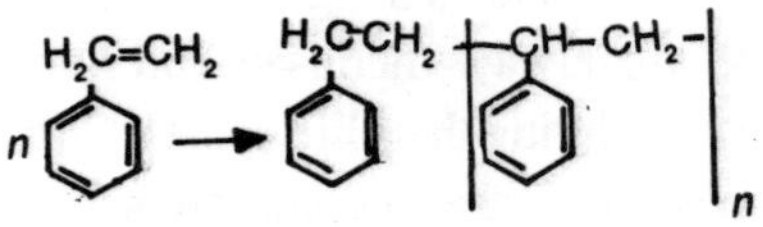

$$n\,H_2C{=}CH{-}C_6H_5 \longrightarrow H_2C{-}CH_2{-}C_6H_5 \quad \left[{-}CH(C_6H_5){-}CH_2{-}\right]_n$$

Fig. P-11. Polystyrene polymer.

Polystyrene. A clear colourless thermoplastic material made by polymerization of styrene. It is a clear rather brittle flammable substance used in electrical insulation and in making moulded articles. *Expanded polystyrene* is a light rigid foam extensively used in packing and in thermal insulation.

Polysulphide. Sulphide.

Polyurethanes. Large group of Plastics that occur in a wide variety of form. As a flexible foam, it is used for cushions and carpet backings. As a rigid foam, it can be molded into furniture or used as insulation. Some polyurethanes are highly elastic, *e.g.*, Lycra, a fiber used in stretch clothing; others form hard protective coatings.

Polytetrafluoroethylene (PTFE). A synthetic material produced by polymerization of tetrafluoroethylene. It has several useful properties, notably its ability to withstand temperatures up to above 400°C and its low coefficient of friction (equal to that of wet ice). PTFE is used in coating nonstick cooking utensils and in gaskets, bearings, and electrical insulation. It is sold under the trade names *Teflon* and *Fluon*.

Polythene (polyethylene). A white translucent waxy thermoplastic material made by polymerizing ethylene at high pressure. Polythene is made in the gas phase and is produced in two forms. Low-density polythene is a soft material made at high pressures. High-density polythene is made at lower pressures with the use of solid catalysts. It is more crystalline and more rigid than the low-density form and softens at a higher temperature. Both types are widely used: the high-density material being used in large numbers of moulded articles and the low-density form being also used for flexible tubes and sheets.

Polytungstate. Tungstate.

Polytungstic Acid. Tungstate.

Polyurethane. Any of a class of synthetic resins formed by copolymerization of diisocyanates and polyhydric alcohols. All contain the urethane group, –NH.CO.O-. Many types of thermosetting and thermoplastic polyurethanes exist. They are used in fibres, moulded articles, adhesives, rubbers, paints, lacquers, and foams.

Polyvanadate. Vanadate.

Polyvanadic Acid. Vanadate.

Polyvinyl Acetate (PVA). A vinyl resin or plastic produced by polymerization of vinyl acetate. It is a soft material used in emulsion paints and adhesives.

Polyvinyl Chloride (PVC). A vinyl resin or plastic produced by polymerization of vinyl chloride. The plastic is used in two forms; rigid PVC, which is moulded into many products, and flexible PVC, produced by addition of a plasticizer, which has an immense range of applications in moulded articles, sheets, tubes, electrical insulation, clothing, etc. PVC is tough, nonflammable, resistant to moisture, and a good electrical insulator. After polythene, it is the second most widely used synthetic plastic.

Polyvinyl Chloride (PVC). Thermoplastic that is a Polymer of vinyl chloride. By adding plasticizers, hard PVC Resins can be made into a flexible, elastic Plastic, used as an electrical insulator and as a coating for paper and cloth in making fabric for upholstery and raincoats.

Polywater. A supposedly polymeric form of water made by condensing water in very narrow glass or quartz tubes. It is now known to be impure water, containing ions from the glass or quartz.

Pomegranate. Deciduous, thorny shrub or small tree (*Punica granatum*) native to semitropical Asia and grown in the Mediterranean region. It has long been cultivated as an ornamental and for its fruit. The roughly apple-sized fruit contains many seeds, each within a fleshy red seed coat enclosed in a tough yellowish-to-deep-red rind. It is eaten fresh or used for grenadine syrup.

Pomeranian. Sturdy Toy Dog; shoulder height, c. 6 in. (15.3 cm); weight, 3-7 lb (1.4-3.2 kg). Its abundant double coat forms a ruff on the neck. Descended from the larger sledge dogs of Iceland and Lapland, it became popular when Queen Victoria obtained one in the late 19th cent.

Pompano. Marine Fish of the family Carangidae, mackerel-like fishes abundant in warm seas around the world. Swift swimmers with deeply forked tails, most are valued as food fish. The family includes the amberfishes, leather jacks, cavallas or jacks, moonfishes, casabes, and pompanos. The best-known leather jack is the pilot (fish) (*Naucrates ductor),* which follows ships and sharks, feeding on discarded scraps. The amberfishes (genus *Seriola*) include the California yellowtail, weighing up to 40 lb (18 kg), and the amberjack, common off the Florida coast and weighing up to 100 lb (45 kg). The warm-water common pompano, found from the Carolinas to Texas, reaches 18 in. (45 cm) in length and is considered among the most delicious of food fish.

Pony Express. Horseback relay mail service (1860-61) from Saint Joseph, Mo., to Sacramento, Calif., extending c. 2,000 mi (3,200 km) Relay riders traveled the distance in about eight days. After a transcontinental telegraph line was completed in Oct., 1861, the pony express was ended.

Poodle. Popular dog, divided into three types, which differ in size: the standard, a Nonsporting Dog, shoulder height, over 15 in. (38.1 cm), weight, 40-55 lb (18.1-24.9 kg); the miniature, also a nonsporting dog, shoulder height, 10-15 in. (25.4-38.1 cm), weight, 14-16 lb (6.4-7.3 kg); and the toy, a Toy Dog, shoulder height, up to 10 in. (25.4 cm), weight, c. 6 lb (2.7 kg). The dense coat, of any solid colour, is usually clipped in a variety of styles. The breed probably originated in Germany, but is associated with France, where it was used as a waterfowl retriever.

Poppy. Name for some members of Papaveraceae, a family composed chiefly of herbs of the Northern Hemisphere, typified by a milky or coloured sap. Many are cultivated for their brilliantly coloured but short-lived blossoms. The true poppy genus, *Papaver,* includes the Oriental poppy (*P. orientale*), bearing a large scarlet flower with a purplish-black base; the corn poppy (*P. rhoeas*); and Iceland poppy (*P. nudicaule*). The opium poppy (*P. somniferum*) is the most important species economically. The sap of its unripe seed pods is the source of the narcotics Opium, Morphine, Codeine, and Heroin. Its seed (poppyseed) is not narcotic and is used as birdseed and in baking; oil from the seed is also used in cooking and for paints, soaps, and varnishes. Other genera in the family are also called poppies, *e.g.,* the celandine poppy (*Stylophorum diphyllum*) and the California poppy (*Eschscholtzia californica*), the state flower of California.

Populations, Stellar. Two distributions of star types. Population I stars are recently formed stars that lie mostly on the main sequence of the Hertzsprung-Russell Diagram. They are located in the interstellar dust of the spiral arms of galaxies and in galactic, or open, star Clusters. The older and highly evolved population II stars are formed early in the history of a galaxy from pure hydrogen with a possible admixture of primordial helium. Population II stars are found in the central bulges of galaxies and in globular clusters in the halo surrounding galaxies.

Population Type. *See* Star.

Porcelain. White, hard, nonporous, translucent pottery. First made by the Chinese during the T'ang period (618-906) to withstand the great heat of their kilns, traditional porcelain was hard paste (a combination of kaolin, a white clay that melts at high temperature; and petuntse, a feldspar mineral that forms a glassy cement to bind the vessel permanently). It was exported to the Islamic world and highly prized. During the Yuan period (1280-1368) blue-and-white ware was produced with cobalt blue from the middle East, and from the 14th to 17th cent. other colours were used as well. Most European porcelain is soft paste (clay combined with an artificial compound, e.*g*., ground glass) and is not as strong as Chinese porcelain. In Europe, porcelain was first commercially produced (1710) in Meissen, Germany. The English have strengthened porcelain with bone ash since 1750.

Porcupine. Member of either of two Rodent families, characterized by having some of its hairs modified as bristles, spines, or quills. Loosely attached, the quills pull out easily, remaining imbedded in predators that come in contact with them. The North American, or Canadian, tree porcupine (*Erethizon dorsatum*) is found in wooded areas over most of North America, excluding the SE U.S. It forages in trees at night. Members of most species weight 50 to 60 lb (23 to 27 kg).

Porpoise. Small Whale of the family Phocaenidae, distinguished from the Dolphin by its smaller size (4-6 ft/120-180 cm) and rounded, beakless head. Black above and white below, porpoises are found in all oceans and prey on fish. In North America dolphins are often called porpoises.

Positive Column. A large luminous column near the anode in a discharge tube.

Positive Rays. Streams of positive ions, as produced in a discharge tube.

Positivism. In philosophy, a system of thought opposed to

Metaphysics and maintaining that the goal of knowledge is simply to describe the phenomena experienced. Its basic tenets are contained in the works of Francis Bacon, George Berkeley, and Hume. The term itself was coined by Comte, whose doctrines influenced the development of much of 19th and 20th cent. thinking, especially that of Logical Positivism.

Positron. The antiparticle of the electron: an elementary particle with a poistive charge equal in magnitude to the electron's charge and a mass equal to the mass of the electron. Positrons are produced during pair production and in one form of beta decay.

Positronium. A short-lived entity consisting of an electron and a positron revolving around a common centre of mass, analogous to a hydrogen atom. Positronium has a mean life of the order of 10^{-7} second, undergoing annihilation to photons.

Potash. Any of various compounds of potassium, including potassium carbonate, potassium hydroxide (*caustic potash*), and potassium aluminium sulphate (*potash alum*).

Potassium. Symbol: K. A soft silvery metallic element that is widely distributed in nature, large amounts being found in salt deposits such as carnallite and kainite. Potassium can be obtained by electrolysis. The element has few uses but it is essential to plant growth and its salts are extensively used in fertilizers. It is one of the alkali metals. A.N. 19; A.W. 39.102; m.pt. 63°C; b.pt. 770°C; r.d. 0.86; valency 1.

Potassium (K). Metallic element, discovered in 1807 by Sir Humphry Davy, who decomposed potash with an electric current. It is a soft, silver-white extremely reactive Alkali Metal. Potassium is the seventh most abundant element in the earth's crust and the sixth most abundant of the elements in solution in the oceans. It is an essential nutrient for plants and animals. Potassium compounds are used in fertilizers, soaps, explosives, glass, baking powder, tanning, and water purification.

Potassium-argon Dating. A method of dating geological samples by using the decay of potassiun-40, a radioisotope occurring in small amounts in stable potassium. It decays slowly with a half-life of 1.28 x 10^9 years to give argon-40. Measurement of the ratio of the two isotopes present gives an estimate of the sample's age. The method is useful for ages of up to about 10^{10} years.

Potassium Bicarbonate. Potassium hydrogen carbonate.

Potassium Bromide. A white crystalline solid, KBr, made by heating a mixture of iron bromide and potassium carbonate. It is used in photography. M.pt. 730°C; b.pt. 1435°C; r.d. 2.7.

Potassium Carbonate (potash). A white deliquescent alkaline powder, K_2CO_3 extracted from natural sources of potassium chloride by treatment of a solution with carbon dioxide. It is used in glass manufacture and the preparation of other potassium salts. M.pt. 891°C; r.d 2.4.

Potassium Chlorate. A colourless or white crystalline substance, $KClO_3$, made from potassium chloride and calcium chlorate. It is used in making explosives and matches. M.pt. 356°C; r.d. 2.3.

Potassium Chloride. A white crystalline solid, KCl, occurring naturally in carnallite and in certain brines. It is used in fertilizers and as a source of other potassium salts. M.pt. 790°C: sublimes at 1500°C; r.d. 1.9.

Potassium Chromate. A yellow crystalline solid, K_2CrO_4, made by roasting chromite, potassium carbonate, and limestone, then treating the product with aqueous potassium sulphate. It is used as an analytical reagent, mordant, and pigment. M.pt. 971°C; r.d. 2.7.

Potassium Cyanide. An extremely poisonous white deliquescent crystalline solid, KCN, with a faint odour of bitter almonds. It is made by heating potassium carbonate with carbon in a current of ammonia and is used in case-hardening steel and in the extraction of gold and silver. M.pt. 634°C; r.d. r.d. 1.5.

Potassium Dichromate. A poisonous red crystalline solid, $K_2Cr_2O_7$, made by adding sulphuric acid to aqueous potassium chromate. It is a strong oxidizing agent and is used in electroplating, dyeing, tanning, and the production of many chemicals. M.pt. 396°C; r.d. 2.7.

Potassium Hydrogen Carbonate (potassium bicarbonate). A colourless crystalline slightly alkaline powder, $KHCO_3$, made by passing carbon dioxide into aqueous potassium carbonate. It is used in soft drinks and fire extinguishers. R.d. 2.17.

Potassium Hydrogen Tartrate (cream of tartar). A colourless crystalline compound, $C_4H_5O_6K$, used as an ingredient of baking powders. It is deposited as argol in wine vats.

Potassium Hydroxide (caustic potash). A white deliquescent strongly alkaline solid, KOH, manufactured by electrolysis of concentrated potassium chloride solution and used in making soft soaps. M.pt. 360°C; r.d. 2.0.

Potassium Iodide. A white crystalline solid, KI, prepared by the reaction of hydrogen iodide and potassium hydrogen carbonate. It readily liberates iodine and is used in medicine, photography, and as a fuel additive. M.pt. 686°C; b.pt. 1330°C; r.d. 3.1.

Potassium Nitrate (saltpetre, nitre). a white crystalline solid, KNO_3, used as a source of oxygen in fireworks, a preservative for food, and a nitrogen fertilizer. M.pt. 33°C; r.d. 2.1.

Potassium permanganate. A purple crystalline solid, $KMnO_4$, with a metallic sheen. It is prepared by oxidation of potassium manganate and is used as an oxidizing agent and disinfectant. Decomposes at 240° C; r.d. 2.7

Potassium Sodium Tarate (Rochelle salt). A colourless efflorescent crystalline substance, $KNaC_4H_4O_6.4H_2O$, used in baking powders. Crystals of potassium sodium tartrate show the piezoelectric effect. R.d.1.77.

Potassium Sulphate. A colourless or white crystalline substance, K_2SO_4, used in fertilizers and the manufacture of glass. M.pt. 1072°C; r.d. 2.7.

Potato or **White Potato.** Perennial plant (*Solanum tuberosum*) of the Nightshade family, and its swollen underground stem (tuber), a widely used vegetable. Probably native to the Andes, the potato was introduced (via Europe) into North America c. 1600. With its high carbohydrate content, it is a primary food for Western peoples; it is also a source of Starch, flour, Alcohol, dextrin, and fodder. Potatoes are usually propagated by planting pieces of the tubers that bear two or three "eyes," the buds of the underground stems. The Sweet Potato is unrelated.

Potential. Symbol: *V*. The work done in bringing unit electric charge from infinity to a specified point in an electric field. The electric *potential difference* between two points is the work done in moving the charge from one point to the other. It is independent of the path taken. Electric potential and potential difference is measured in volts.

The potential can also be defined for magnetic and gravitational fields by the work necessary to bring unit pole or unit mass to the point.

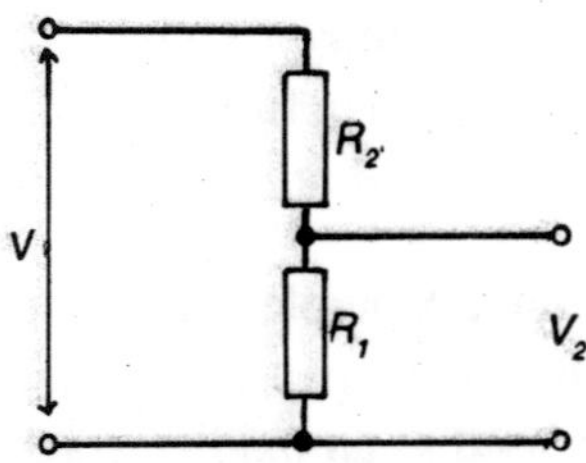

Fig. P-12. Potential divider.

Potential Divider (voltage divider). A resistor or chain of resistors with a voltage applied across it and a tapping connection so that a fraction of this voltage can be drawn. In the diagram, $V_2 = R_1V/(R_1 + R_2)$. The connection may be fixed to give a constant fraction of the voltage or may be variable as in a potentiometer.

Potential, Electric. Work per unit electric charge expended in moving a charged body from a reference point to any given point in an electric field. The potential at the reference point is considered to be zero, while the reference point itself is usually chosen to be at infinity. The change in potential associated with moving a charged body is independent of the actual path taken and depends only on the initial and final points. Potential is measured in Volts and is sometimes called voltage.

Potential Energy. Symbol: E_p. The energy that a system has by virtue of its position or state, determined by the work necessary to change the system from a reference position to its present state. Thus, a stretched spring has potential energy equal to the work done in stretching it from its equilibrium position, which is taken as the reference state. An object a height *h* above the ground has a potential energy of *mgh*, where *m* is its mass and *g* is the acceleration of free fall.

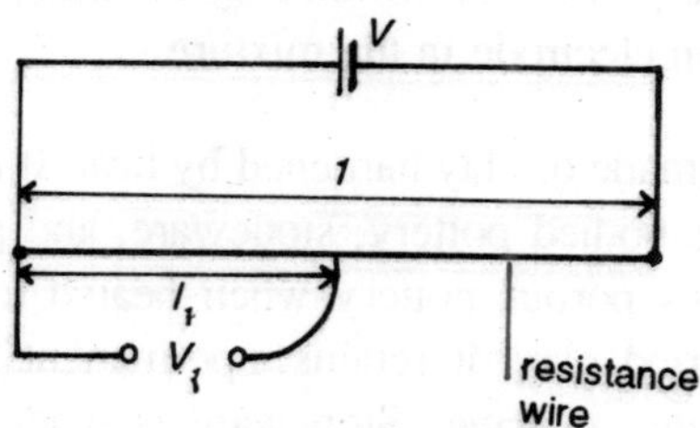

Fig. P-13. Potentiometer.

Potentiometer. 1. An apparatus for measuring potential difference by use of a potential divider. A simple form has a straight length of uniform resistance wire with a constant voltage connected across it. The potential difference to be measured is connected between one end of the resistance wire and a sliding contact, which is moved along the wire until no current flows as indicated by a galvanometer. The unknown voltage V_1 is given by $V_1 = l_1V/1$. Usually a standard cell with voltage V_2 is introduced to find a new balance, l_2: then $V_1/V_2 = l_1/l_2$. The device gives the true potential difference or e.m.f. because at balance no current is drawn.

2. A circuit component consisting of a resistance with a sliding contact, suitable for use in giving a controlled variable voltage.

Potentiometer or **Voltage Divider.** Manually adjustable variable electrical resistor that has a Resistance element attached to an Electric Circuit by three contacts, or terminals. The ends of the resistance element are attached to the two input voltage conductors of the circuit, and the third contact, attached to the output of the circuit, is usually a movable terminal that slides across the resistance element, dividing it into two resistors. Because the position of the movable terminal determines what percentage of the input voltage is applied to the circuit, a potentiometer can be used to vary the magnitude of the voltage, *e.g.*, in radio-volume and television-brightness controls.

Potentiometric. Denoting an experimental technique that depends on measurements of potential. A *potentiometric titration* is one in which the end point is determined by monitoring the electrode potential of an electrode in the mixture.

Pottery. Utensils made of clay hardened by heat. It is of three main types: porous-bodied pottery, stoneware, and porcelain. Raw clay becomes porous pottery when heated to about 500°C. Unlike sun-dried clay, it retains a permanent shape and does not disintegrate in water. Stoneware is made by raising the temperature, and Porcelain is baked at still higher heat. This makes the clay glassy (vitrified) and adds strength. Pottery is shaped while the clay is in its pliant form. Either a long rope of clay is coiled (hand-building) or the clay is shaped by the hands as it is spun on a potter's wheel (used in Egypt before 4000 B.C.). After drying to leatherlike hardness, the piece is fired in a Kiln. The type of finish, or glaze, determines the number of firings. Early unglazed pottery was coloured by firing alone. In most parts of the world pottery was the earliest craft. Prehistoric remains of pottery in Europe and the Americas have great importance in Archaeology. Notable ancient pottery includes the Greek vases of 800-300 B.C.; Chinese porcelain of the 6th-10th cent. A.D.; and the ceramic tiles used in Islamic architecture of the 15th cent. Majolica and stoneware were the major types of European pottery until the advent of porcelain in the 18th cent. Pottery made by the prehistoric peoples of the Americas reveals great artistic and technical skill, and the traditional designs are still being used today. North American colonists were making baked-clay bricks and tiles by 1612; by the late 19th cent. many pottery firms had been established. Today, despite mass production and the proliferation of utensils made of synthetic materials, the demand for hand-crafted pottery continues.

Pound. 1. Symbol: lb. An imperial unit of mass equal to 0.453 592 37 kilogram. This is the unit in avoirdupois weight: it was formerly defined as the mass of a block of platinum, the change to a definition based on the kilogram occurring in 1963.

2. Units of mass in troy and apothecaries weight.

3. Pound-weight.

Poundal. Symbol: pdl. An FPS unit of force equal to the force that would give a mass of one pound an acceleration of one foot per second per second. it is equivalent to 0.138 255 newton.

Pound-weight. Symbol: lbwt. A unit of force equal to the force that would give a body of mass one pound an acceleration equal to the local acceleration of free fall. The force in poundals is the force in pounds-weight multiplied by the local value of g in feet per second. The unit, which is sometimes loosely called the *pound*, varies slightly from place to place. The *pound-force* (symbol: lbf) is the force that would give a mass of one pound an acceleration equal to the standard acceleration of free fall (32.174 ft s^{-2}): 1 lbf = 32.174 pdl.

Power. 1. The rate of performing work, measured in units such as watts and horsepower.

2. A measure of the magnifying ability of an optical system, equal to the ratio of the size of the image produced to the size of the object.

3. The number of times a number or expression is multiplied by itself. For example a to the power 4 is $a \times a \times a \times a$, written a^4, where 4 is the exponent.

Power. In physics, the time rate of doing work or of producing or expending Energy. The unit of power in the Metric System is the watt, which equals 1 joule per second. It is also the amount of power that is delivered to a component of an electric circuit when a current of 1 ampere flows through the component and a voltage of 1 volt exists across it. The English Unit of Measurement is the horsepower, which equals 550 foot-pounds per second or 746 watts.

Power, Electric. Is the rate per unit of time at which electric Energy is consumed or produced. Electric Power is usually measured in watts or kilowatts (1,000 watts). The energy supplied by a current to an appliance enables it to work or to provide other forms of energy such as light or heat. The amount of electric energy an appliance uses is found by multiplying its power rating by the operating time. Units of electric energy are usually watt-seconds (joules), watt-hours, or kilowatt-hours (the choice for commercial applications). Generally, practical electric-power-generating systems convert mechanical energy into electric energy. Whereas some electric plants obtain mechanical energy from moving water (Water Power or hydroelectric power), the vast majority derive it from heat engines in which the working substance is steam generated by heat from combustion of fossil fuels or nuclear reactions. Although the conversion of mechanical energy to electric energy may approach 100% efficiency, the conversion of heat to mechanical energy is about 41% efficient for a fossil-fuel plant and about 30% for a nuclear plant. It is thought that a magnetohydrodynamic generator, which operates by using directly the kinetic energy og gases produced by combustion, would have an efficiency of about 50%. Although Fuel Cells develop electricity by direct conversion of hydrogen, hydrocarbons, alcohol, or other fuels, with an efficiency of 50 to 60%, their high cost has restricted their use to space programs. Solar Energy has been recognized as a feasible power source. It can be exploited through wind Turbines, Photovoltaic Cells, and heat engines, as well as through both conventional and low-head hydroelectric power plants. Research and development is bringing down the costs. An important problem in utilizing solar energy is related to the variable nature of sunlight and wind. To minimize energy losses from heating of conductors and to economize on the material needed for conductors, electricity is usually transmitted at the highest voltages possible. As modern Transformers are virtually loss free, the necessary steps upward or downward in voltage are easily accomplished. Electric utilities producing power are tied together by transmission lines into large systems called power grids. They are thus able to exchange power, so that a utility with low power demand can assist another with a high demand.

Pragmatism. Method of philosophy in which the truth of a proposition is measured by its correspondence with experimental results and by its practical outcome. Thus pragmatists hold that truth is modified as discoveries are made and that it is relative to time and place and purpose of inquiry. C.S. Peirce and William James were the originators of the system, which influenced John Dewey.

Prairie Dog. Short-tailed, ground-living, herbivorous Rodent (genus *Cynomys*) of the Squirrel family, found in the W U.S. and N Mexico and related to ground squirrels, Chipmunks, and marmots. Prairie dogs, named for their barking, cries, are 12 to 15 in. (30 to 36 cm) long with short, buff-coloured fur. Black-tailed prairie dogs live in colonies of connected burrows; these may extend many miles and include thousands of individuals.

Praseodymium. Symbol: Pr. A soft silvery metallic element belonging to the lanthanide series. A.N. 59; A.W. 140.907; m.pt. 931°C; b.pt. 3212°C; r.d. 6.7; valency 3 or 4.

Praseodymium (Pr). Metallic element, discovered in 1855 by C.A. von Welsbach. It is a soft, malleable, ductile, silver- yellow

Rare-Earth Metal in the Lanthanide Series. Its compounds are used in carbon electrodes for arc lighting and to colour enamel and glass.

Praying Mantis. Large, slender, slow-moving winged Insect, so called because of its habit of standing on its hind legs, with the forelegs raised as if in prayer. Mantises range in length from 1 to 5 in. (2.5 to 12.5 cm). Their typically green or brown Protective Coloration camouflages them among leaves and twigs. Voracious predators, they feed on insects and other invertebrates; the female often eats the male after mating.

Precession. 1. The motion of a body that is spinning on an axis (OA) in which one end of the axis (A) moves around a second axis passing through O. The precessional motion results from an applied torque. It is seen in tops and gyroscopes.

2. This phenomenon as it occurs in planets, particularly the earth. Because of precession the vernal and autumnal equinoxes are moving westward along the ecliptic; during the 26 000 year cycle over which the earth's axis describes its small circle as a result of precession, the Pole Star changes. At present the Pole Star is represented by Polaris, the brightest star in Ursa Minor. 5000 years ago the Pole Star was Alpha Draconis. In general, precessional effects are caused by the gravitational influence of the moon and sun, but the influence of the other planets also causes precession of the earth.

Precession of the Equinoxes. Westward motion of the Equinoxes along the Ecliptic. The precession, first noted (c. 120 B.C.) by Hipparchus and explained (1687) by Isaac Newton, is primarily due to the gravitational attraction of the moon and sun on the earth's equatorial bulge, which causes the earth's axis to describe a cone in somewhat the same fashion as a spinning top. As a result, the equinoxes, which lie at the intersections of the celestial equator and the ecliptic, move on the celestial sphere. Similarly, the celestial poles move in circles on the celestial sphere, so that at various times different stars occupy positions at or near one of these poles. After about 26,000 years the equinoxes and poles lie again at nearly the same points on the celestial sphere. The gravitational influences of the other planets cause a much smaller amount of precession.

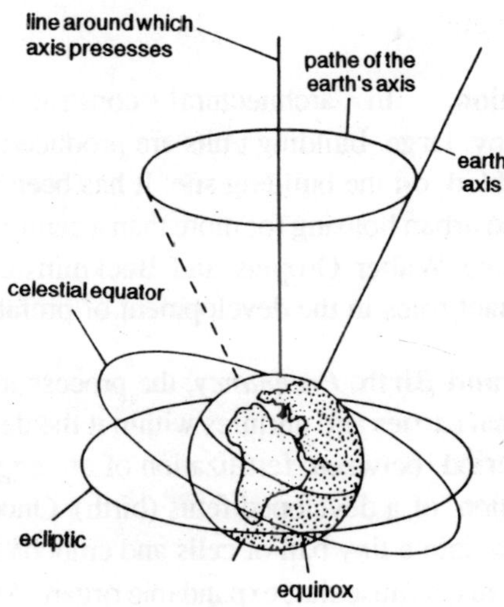

Fig. P-14. Precession of the equinoxes (the points at which the earth's celestial equator intersects its ecliptic is due to the slow rotation of the earth's axis around a perpendicular to the ecliptic

Precipitate. A visible suspension of solid particles produced in a solution by chemical reaction. For example, if sodium chloride and silver nitrate solutions are mixed together, the insoluble silver chloride appears as a white precipitate, which eventually settles to the bottom of the vessel.

Precipitate. A solid obtained by precipitation, *i.e.,* the separation of a substance from a Suspension, sol, or Solution. In a suspension, such as sand in water, the solid spontaneously precipitates (settles out) on standing. In a sol the particles are precipitated by coagulation. A solute may be precipitated by evaporation or by the addition of a compound that reacts with the with the solute to form an insoluble precipitate. In each case, the precipitate formed may settle out spontaneously or may be collected by filtration or centrifugation.

Precipitation. 1. The process of forming a chemical precipitate.

2. A process occurring in certain alloys that are supersaturated solid solutions, in which a separate phase is formed. The process is the basis of a method of hardening alloys (*precipitation hardening*).

Prefabrication. In architectural construction, a technique whereby large building units are produced in factories to be assembled on the building site. It has been applied in various ways to urban housing for more than a century. Major architects, including Walter Gropius and Buckminster Fuller, have had important roles in the development of prefabrication.

Pregnancy and Birth. *Pregnancy,* the process in which the female mammal carries and nurtures within it the developing young, is the period between fertilization of an egg (conception) and expulsion of a developed fetus (birth). Once fertilized, an egg divides into a tiny ball of cells and embeds itself in the wall of the uterus, a muscular, expandable organ. At this early stage of development, the growing organism is called the embryo. Some cells of the embryo develop into the placenta (a disk of tissue that brings nourishment from the mother and accepts waste substances from the fetus); the umbilical cord (the cord between embryo and placenta, carrying nutrients and waste); and the protective amniotic sac and fluid surrounding the embryo. During the course of human pregnancy, changes take place in the mother's body to meet the fetal demands on it: The uterus and breasts enlarge and blood composition and volume change. It takes about 280 days for the embryo (in later stages called the fetus) to develop the structures and organs of a fully formed human. *Birth* is the process by which the fetus is expelled from the uterus. Rhythmic contractions of the uterus, called labor, dilate the neck (cervix) of the uterus and expel the baby and later the placenta. *Natural childbirth* is birth with little or no use of painkilling drugs. In the Lamaze method, prospective parents are taught physical and mental techniques for reducing pain and discomfort and participating more fully in the birth process. Birth by the "Leboyer method" emphasizes extremely gentle handling of the newborn. Cesarean Section is delivery of an infant by removing it from the uterus surgically.

Presbyopia. A defect of the eye in which near objects cannot be focused clearly. It is caused by hardening of the lens so that the ciliary muscles cannot produce enough accommodation and the near point is further away than the normal 33 centimetres. Presbyopia can be corrected with convex spectacle lenses worn for close work. It is sometimes called *far sight,* which should not be confused with long sight (hypermetropia).

Pressure. Symbol: *p*. The force acting per unit surface area. The pressure at a point below the surface of a fluid acts in all directions and is given by the height of the fluid multiplied by its density. Pressure is measured in pascals. Other units are the bar, the atmosphere, and the mmHg.

Pressurized-water Reactor. A type of fission reactor in which water is used as a coolant, maintained at a high pressure to prevent it from boiling.

Priestley, Joseph. 1733-1804, English theologian and scientist. Trained for the Presbyterian ministry, he later adopted Unitarian views and wrote widely on theological and philosophical issues. His improved techniques for studying gases led to his discovery of sulfur dioxide, ammonia, and what he called "dephlogisticated air," the gas that Antoine Lavoisier named Oxygen. He was an opponent of orthodox doctrines, England's colonial policy, and the slave trade. Due to his sympathy with the aims of the French Revolution, his house was wrecked and his library and laboratory were destroyed. In 1794 he emigrated to the U.S.

Primary. 1. The body around which a satellite orbits. The earth is the moon's primary in the earth-moon system; the sun is the earth's primary in the solar system.

2. Transformer.

Primary Cell. Cell.

Primary Colours. A set of three colours than can be mixed in suitale proportions to produce any other colour. Red, green, and blue are in example: another set of primary colours is cyan (greenish-blue), magenta, and yellow.

If red, green and blue lights are mixed in equal proportions, white light is produced by an additive process. Similarly, three pigments of primary colours mixed in equal proportions combine by a subtractive process to give black.

Primate. Member of the mammalian order Primates, which includes human beings, Apes (*e.g.*, Gibbons, Orangutans, Chimpanzees, and Gorillas), Monkeys (*e.g.,* Baboon and Mandrill), and prosimians, or lower primates (*e.g.*, Tree Shrews, Lemurs). Nearly all inhabit warm climates and are arboreal, although a few are terrestrial. Unspecialized anatomically, primates are distinguished by social organization and evolutionary trends within the order tending toward increased dexterity and intelligence.

Prime Number. A number that has no factors other than itself and one.

Primrose. Name for the genus *Primula* of the family Primulaceae, low perennial herbs found on all continents. The family includes the primroses, cyclamens (genus *Cyclamen*), and pimpernels (genus *anagallis*). Species of all these genera are cultivated as rock-garden, border, and pot plants. The primrose is a common and favored wildflower of England, and several species are indigenous to North America. The Evening Primrose is not a true primrose. Cyclamens are chiefly native to the Alps; *C. indicum* is a common florists' pot plant in the U.S. The scarlet pimpernel (*A. arvensis*), native to Eurasia but naturalized in the U.S., has flowers that close on the approach of bad weather.

Principal Axis. Mirror, lens.

principal Focus. Mirror, lens.

Principal Quantum Number. Atom.

Printed Circuit. An electrical circuit in which the connections between components are made by thin channels of metal coated onto an insulating board.

Printed Circuit. Electric Circuit in which the metallic conducting paths connecting circuit components are affixed to a flat, insulating base board typically composed of plastic, glass, ceramic, or some other Dielectric. Conducting paths may be placed on the base by etching, die stamping, or spraying the pattern through a stencil or mask. The circuit components, such as Resistors and Capacitors, are not usually produced by these processes but rather are mounted afterwards on the base.

Printing. The reproduction of lettered or illustrated matter through the use of mechanical, photographic, or electrostatic devices. (For the invention of printing and the development of the earliest forms of type). In addition to the letters themselves, the elements involved in printing including the press and various methods for setting type and for making the reproducing printing surface, or plate. The press used by Johannes Gutenberg in 15th cent. Germany was a hand press, in which ink was rolled over the raised surfaces of hand-set letters held within a form and the form was then pressed against a sheet of paper. The hand press remained in use for all forms of printing until the early 19th cent. A steam-powered press with a flat type-bed was used by *The Times* of London beginning in 1814. In 1847 the American Richard Hoe developed a high-speed rotary press in which the printing surface was wrapped around a cylinder. Later presses used continuous rolls of Paper and incorporated folding, cutting, and paper-moving devices that vastly increased printing speed. The first mechanical type-setter, the Linotype machine, was invented by Ottmar Mergenthaler in 1884. Operated by a typewritter-keyboard, it assembles brass matrices into a line and casts the line as a single metal slug. Other machine-set type system were developed, notably the Monotype (first used in 1897), which casts individual characters from a punched tape produced by a keyboard operator. The recent development of photo-composed "cold type" (as distinguished from "hot," or cast-metal, type) has further increased typesetting speeds. electronic typewriters and computers project the type images onto a film that is then used to make a plate. Letterpress is printing from a raised, inked surface. Various inventions have improved the letterpress printing surface, especially the stereotype, which can produce a large metal plate of a newspaper page from a mold made of pulp; and the later electrotype, which uses an electrolytic process to create a curved metal plate. In *Intaglio,* the design to be produce is cut below the surface of the plate, and the incised lines are filled with ink that is then transferred to paper. Photogravure is an intaglio process in which the plate is produce photographically. *Offset,* or *planographic, printing,* derived from Lithography, uses a plate treated so that ink will adhere only to the areas that will print the design. The plate transfers its ink to a rubber cylinder, which in turn offsets it onto paper. Colour printing is achieved by photographically separating four basic colors (black, magenta, yellow, and cyan, a blue-green) from the original picture, making a plate for each colour, then using the plates to print the colours consectively over one another.

Prism. 1. A solid geometrical figure with two identical parallel faces, consisting of congruent polygons with their corresponding sides parallel, the other faces being formed by joining corresponding vertices of these polygons.

2. A piece of glass, quartz, rock salt, or other transparent material cut in the shape of a prism and used for dispersing or chainging the directtion of light or ultraviolet or infrared radiation. The commonest type are traingular prisms used for dispersing light. Prisms are also used for inverting images or for internal reflection.

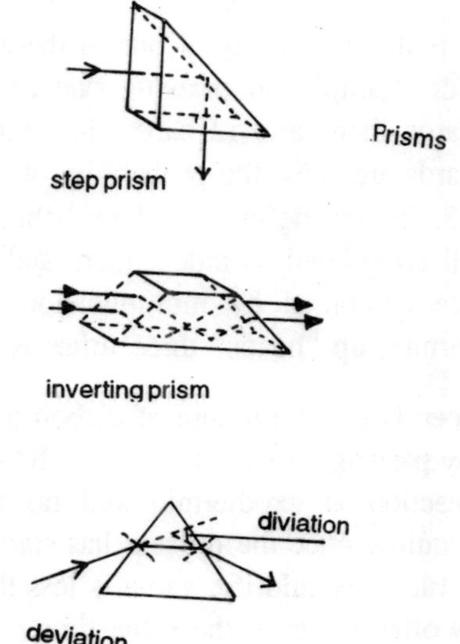

Fig. P-15. Prisms.

Prism. In optics, a piece of translucent glass or crystal with a triangular cross section, used to form a Spectrum of light by separating it according to colours. Prisms are also used in optical instruments for changing the direction of light beams by total internal reflection.

Prismatic. Denoting optical instrumentts that contain prisms. *Prismatic binoculars,* for example, are binoculars in which the path length is increased and the image erected by pairs of prisms.

Prismatic Colours. The colours produced when white light is split up by dispersion. The prismatic colours are the colours of the rainbow: violet, indigo, blue, green, yellow, orange, and red.

Probability. A numerical expression of the likelihood that an event will occur. Probabilities are usually written as fractions. If b is the total number of ways in which an event can occur and of these, a events can occur in a specified way, then $a/$

b is the probability of one of these specified events occurring. For example, in drawing one card from a pack of playing cards there are 52 cards that may be drawn. Of these, 4 cards are aces: the probability of drawing an ace is 4/52 = 1/ 13. In this definition of probabilitty the possible events are all considered as independent and qually likely. Probabillities are combined by multiplication: the probability of a coin turning up "heads" three times is 1/2 x 1/2 x 1/2 = 1/8.

Producer Gas. A mixture of carbon monoxide and nitrogen made by passing air over hot coke : $2C + O_2 + 4N_2$ $2CO + 4N_2$. The reaction is exothermic and no further external heating is required once the process has started. Producer gas is used as a fuel: its calorific value is less than that of water gas and it is often made at the same time.

Product. 1. The result of multiplying two or more numbers or quantities together.

2. A compound produced in a chemical reaction, as distinnguished from the reactants.

Progeria. Syndrome of unknown etiology causing premature aging in children; also known as Hutchinson-Gilford syndrome. Affected children appear normal up to the first year of life. Gradually, characteristic of the disorder become apparent—retarded growth and physical development, loss of body fat, absence of body hair, and baldness. Eventually the child develops dry, wrinkled skin, Arteriosclerosis, and other conditions associated with old age. There is no treatment, and death occurs by the second decades.

Progesterone. Female sex hormone that induces changes in the lining of the uterus essential for implantation of a fertilized egg. In a pregnant woman, progresterone prevents spontaneous abortion (miscarriage) and prepares the Mammary Glands for milk production.

Program. A set of instructions fed into a computer.

Prolate. *See* ellipsoid.

Promethium. Symbol : Pr. A radioactive element belonging to the lanthanide series. It does not occur naturally, being obtained from nuclear reactors. The most stable isotope, ^{145}Pr. has a half-life of 17.7 years. A.N. 61; valency 3.

Promethium (Pm). Artificially produced radioactive element, first identified definitely by J.A. Marinsky and colleagues in 1945 by ion-exchange chromatograpy. It is a Rare-Earth Metal in the Lanthanide Series. The promethium-147 isotope is used in the making of phosphorescent materials, in nuclear-powered batteries for spacecraft, and as a radioactive tracer.

Prominence. A towering eruption of gas in the upper chromosphere of the sun that throws matter out into space in vast streamers. Prominences, whhich are composed of incandescent hydrogen, helium. and calcium, climb high above the sunn's surface and appear to stand there motionless over long periods. They are associated with sunspotts and are subject to the 11-year cycle.

Proof. A measure of the strength of alcoholic drinks. *Proof spirit* contains 49.28% by weight of pure alcohol (52.1% by volume at 16°C). Alcoholic strengths can then be specified as being *under proof* or *over proof.* Each degree under or over proof corresponds to 0.5 per cent of alcohol by volume. Usually liquors are described as having *degrees of proof.* The number of degrees of proof is equal to the percentage of proof spirit in a mixture of proof spirit and water : 70° proof contains 70 parts of proof spirit and 30 parts of water. The percentage of alcohol by volume is converted into degrees proof by multiplying by 1.7535. Thus pure alcohol is 175.35 degrees proof. In America a similar system is used with the difference tthat proof spirit is defined as containing 50% of alcohol by volume.

Propanal. *See* propionaldehyde.

Propane. A colourless flammable gaseous alkane, C_3H_8 obtained from petroleum and used as a fuel. M.pt. -190°C; b.pt. -42°C; r.d. 1.56 (air = 1).

Propanol. One of two isomeric alcohols C_3H_7 OH, which are colourless flammable liquids used as solvents and flavourings. *Propan - 1-ol* (n-propyl alcohol) $CH_3CH_2CH_2OH$, is made by oxidizing natural-gas hydrocarbons. M.pt. -127°C; b.pt. 97°C; r.d. 0.8. *Propan-2-ol* (iso-propyl alcohol). $(CH_3)_2CHOH$, is made by treating propene with sulphuric acid followed by hydrolysis. It is the principal source of acetone and its derivatives. M.pt. –86°C; b.pt. 82°C; r.d. 0.8.

Propellant. 1. The fuel, including the oxidant, used in a rocket.

2. An explosive charge used to fire a bullet or shell.

3. A volatile inert gaseous substance, liquefied under pressure, used to expel the contents of an aerosol can.

Propene (propylene). A colourless slightly soluble gaseous alkene, $CH_3CH{:}CH_2$, made by cracking petroleum hydrocarbons and used in making polypropylene. M.pt. –185.2C;; b.pt. –48°C; r.d. 1.46 (air = 1).

Propionaldehyde (propanal). A colourless liquid aldehyde. C_2H_5CHO. M.pt. —81°C; b.pt. 48.8°C; r.d. 0.81.

Propionic Acid Propanoic Acid). A colourless liquid carboxylic acid, C_2H_5COOH. M.pt. –20.8°C; b.pt. 141°C; r.d. 0.99.

Propyl Group. The group, C_3H-. The term is often used for the *n-propyl group,* $H_3CCH_2CH_2$-. The other isomer is the iso-propyl group, $(CH_3)_2$**CH-**.

Protactinium. Symbol: Pa. An actinide element found in uranium ores. Its most stable isotope, ^{231}Pa, has a half-life of 33000 years. The element was formerly called *protoactinium.* A.N. 91: m.pt. about 1200°C; b.pt. about 4000°C; r.d. 15.4 (calc).

Proteins. Any of a large number of naturally occurring oraganic compounds found in all living matter. Proteins consist of chains of amino acids joined by peptide linkages. They have high molecuclar weights (up to 10 million) and complex structures. The particular nature of a protein depends on the amino acids present and the sequence in which they are bound. It also depends on the way the chains are linked and folded.

Protium. The isotope of hydrogen with mass number one.

Proton. An elementary particle with a positive charge equal in magnitude to the charge of the electron and a rest mass of 1.67262×10^{27}kg (about 1836 times the mass of the electron). Protons are stable particles. They form the nuclei of hydrogen atoms and together with neutrons, are present in all other atomic nuclei. Positive hydrogen ions are protons although in solution these are solvated by water molecules.

Proton Number. *See* atomic number.

Prussian Blue. Any of a number of blue pigments containinng ferrocyanide or ferricyanide ions.

Prussic Acid. *See* hydrogen cyanide.

Pseudoaromatic. Denoting a chemical compound that has molecules containing a ring of conjugated double bonds but is not an aromatic compound. *Cyclooctater traene*, C_8H_8, for instance, has an eight-membered ring of alternating double and single bonds yet behaves like an alkene. It does not have a planar ring of carbon atoms.

Pseudohalogen. Any of a small group of compounds that resemble the halogens in their reactions. Pseudohalogens have a formula of the type R_2, where R is a simple group of atoms, and they form R^- ions. Cyanogen, $(CN)_2$, and thiocyanogen, $(SCN)_2$, are the commonest examples.

Psi Particle. Any of three heavy short-lived elementary particles. The psi particles have masses of 3.095 GeV, 3.684 GeV, and 4.1 GeV and have zero charge. It is thought that they may be formed from charmed quarks.

Psychobiology. Study of anatomical and biochemical structures and processes and their effect on behaviour. It is closely related to physiological psychology. Area of investigation include hormonal and biochemical changes in nerves, glands, and muscles, and how these changes influence development, emotions, and learning.

Psychopharmacology. Science of the relationship between drugs and behaviour, usually applied to the study of four basic categories of drugs used in the treatment of psychiatric illness: antipsychotic drug, Antinaxiety Drugs, Antidepressants, and psychtomimetic.

Psychrometry. The measurement of atmospheric humidity.

Ptolemaic system. The system of celestial mechanics developed by the Alexandrian astronomer. Ptolemy (Claudius Ptolemaeus: c. A.D. 127-151), and based upon the concept that the earth lay at the centre of the universe while the sun, moon, planets, and stars orbited around it within a set of concentric spherical shells. The paths of the celestial bodies in these spheres were connsidered to be circular and the more complex motions of the planets were accounted for by an elaborate system of epicycles (small circles around which the planets moved and whose centres orbited around the earth). This theory, which was the crystallization of contemporary cosmological thinking, was later superseded by the revolutionary Copernican system.

Puddling. The process of heating cast iron with iron oxide to reduce the carbon content and produce wrought iron.

Puffin. Migratory diving Bird of the Auk family. Puffins have dumpy bodies, short legs set far black, small wings, and a large brilliantly coloured bill adapted for carrying several fish at once. Clumsy on land and in flight, they are expert swimmers. Puffins nest in colonies in burrows or rock cavities on northern islands.

Pulsar. A star that acts as a source of regularly fluctuating electromagnetic radiation, the period of the pulses usually being very rapid, occasionally even as high as 50 times a second. These are probably neutron stars, which, as they rotate, emit radiation in directions that are sttrictly controlled by the directions of their magnetic fields. The energy output is only detectable when the "beam" is turned towards the observer.

Pulsar. In astronomy, a celestial object that emits brief, sharp pulses of radio waves. The pulses recur at precise intervals; the time between pulses ranges from 1.56 millisec. to 3.75 sec. Pulsars are believed to be rotating Neutron Stars that emit radio waves in a narrow beam. Because of the rotation of the star, the beam will sweep across the line of sight, causing the observed pulses.

Pulsating Star. A star that is subject to variations in the light emitted by it owing to periodic and regular expansion and contraction of its surface atomosphere.

Pulse. A sudden increase in magnitude of a physical quantity shortly followed by a decrease.

Pulse. Alternate expansion and contraction of artery walls a Heart action varies blood volume within them. Artery walls are elastic and become distended by the increase in blood volume when the heart contracts. When the heart relaxes, this volume decreases and the walls contract, propelling the blood through the arteries. The reulst is a pressure wave, with a pulsation for each heartbeat.

Pulse-height Analyser. An electronic circuit for sorting voltage pulses according to their amplitude. The pulses obtained from certain types of counter have an amplitude proportional to the energy of the original particle, and the use of a pulse-height analyser can yield an energy spectrum of the radiation.

Purple of Cassius. A purple pigment made by mixing tin IV oxide and colloidal gold. It is used for colouring glass.

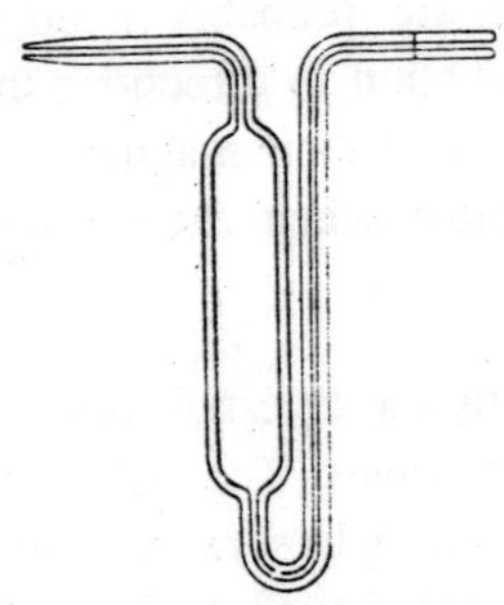

Fig. P-16. Pyknometer.

Pyknometer. A bulb with two attached capillary tubes used for measuring the density of liquids. It is filled by sucking the liquid into the bulb until it reaches the graduation on one tube, thus giving a known volume of liquid.

Pyramid. A solid geometrical figure whose base is a polyhedron and whose other faces are triangles with a common apex. If this apex lies above the centroid of the base, the solid is a *right pyramid.*

Pyrimidine. Type of organic compound based on a six-membered carbon-nitrogen ring. Virtually all animal and plant tissue contains the pyrimidines thymine, cytosine, and uracil as part of certain Coenzymes and in the Uncleic Acids.

Pyrene. A colourless hydrocarbon, $C_{16}H_{10}$, obtained from coal tar. M.pt. 156°C; b.pt. 404°C; r.d. 1.3.

Pyrex. ® A type of heart-resistant borosilicate glass widey used in laboratory appartus, cooking utensils, etc.

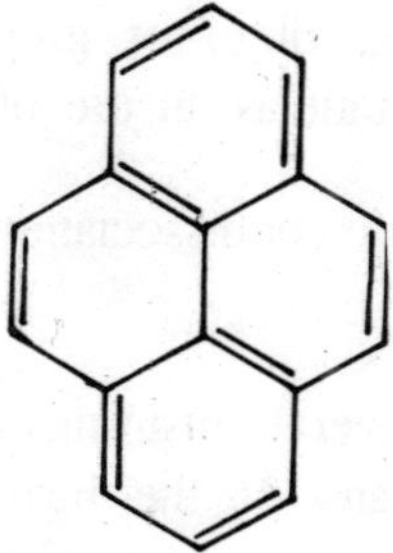

Fig. P-17. Pyrene.

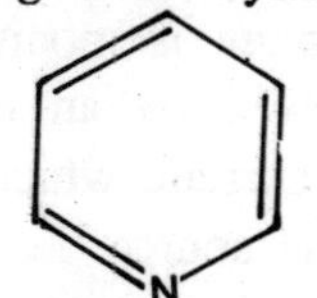

Fig. P-18. Pyridine.

Pyridine. A flammable colourless liquid, C_5H_5N, with a strong disagreeable odour. It is a heterocyclic aromatic base, extracted from coal tar and used as a solvent and raw material for the manufacture of other organic chemicals. M.pt -42°C; b.pt. 115°C; r.d. 0.9.

Pyrite or **Iron Pyrite.** The most common sulfide mineral (FeS_2), brass yellow in colour, found in crystals and massive, granular, and stalactite forms. In spite of its nickname, "fool's gold," it is often associated with gold. Pyrite, widely distributed in rocks of all ages, is used chiefly as a source of Sulfur for sulfuric acid.

Pyrites. Any of several naturally occurring metal sulphides, especially iron pyrites, FeS_2.

Pyrochlore. A brown or black naturally occurring oxide. $NaCaNb_2O_5F$, used as an ore or niobium.

Pyroelectricity. The production of opposite electric charges on opposite faces of certain crystals when they are heated. Tourmaline is an example of a pyroelectric substances.

Pyrolusite. A soft black, blue, or greyish naturally occurring oxide, MnO_2, important as an ore of manganese.

Pyrolysis. The decomposition or dissociation of a chemical compound as a result of heat.

Pyrometer. Any of several instruments for measuring high temperature by means of the heat radiation emitted. The *optical pyrometer* consists of a tube with an objective lens, which produces an images of the source at a point inside the tube. A tungsten filament is mounted in the focal plane of this lens and its viewed by an eyepiece. The filament is heated by an electric current, which is varied until it has the same brightness as the source, i.e. until it "disappear". The size of the current gives a measure of the temperature: the instrument is precalibrated. In the *total-radiation pyrometer*, all the radiation is focused onto a bolometer, thermopile, or similar detector.

Pyrometry. The science of measurement of high temperatures. The term is usually understood to mean measurement of high temperature by means of the radiation emitted.

Pyrophoric. Denoting materials that emit sparks when struck. Pyrophoric alloys, such as Misch metal, are used in lighter flints.

Pyroxene. Name for a group of widespread magnesium, iron, and calcium silicate minerals. They are commonly white, black, and brown, but other varieties occur. Found chiefly in igneous and metamorphic rocks, they are abundant in lunar rocks.

Pythagoras's theorem. The theorem that in a right-angled triangle the square of the length of the hypoteneuse is equal to the sun of the squares of the lengths of the other two sides: *i.e.* $BC^2 = AB^2 + AC^2$, with the right angle at A. [After Pythagoras of Samos, *c.* 580-500 B.C., Greek philospher and mathematician.]

Python. Nonvenomous constrictor Snake of the Boa family, found in tropical regions of the Old World and the South Pacific islands. It kills its prey by squeezing them in its coils. The reticulated, or royal, python (*Python reticulatus*), one of the largest snakes in the world, may reach a length of 30 ft (9 m) or more.

Quadrant. A segment of a circle equal in area to one quarter of the circle: *i.e.* having an angle of 90°.

Quadrant Electrometer. A type of electrometer in which a light metal vane on a torsion wire can move inside four metal quadrants. Pairs of opposite quadrants are connected together. A high steady potential (about 120 volts) is applied to the vane. A low potential difference applied across the pairs of quadrants causes a deflection which is directly proportional to the potential difference. The instrument draws no current. It is now hardly ever used.

Quadrature. The relative positions of two celestial objects, in practice the sun and the moon or a planet, having a difference in celestial longitude of 90°. *Eastern and western quadrature* coincide roughly with the first and last quarters of the moon respectively.

Quadrilateral. Any plane figure bounded by four straight lines.

Quadrivalent (tetravalent). Having a valency of four.

Quail. Name for small, extremely popular game Birds of the Pheasant family, including the New World quails, the Old World qualis and Partridges, and the true pheasants. In parts of the U.S. the Bobwhite is called quail. The California quails (*Lophortyx californicus*), grayish with black and white markings and a forward-curving head plume, can be found in city parks. Quails have great reproductive potential, laying 12 to 15 eggs per clutch. They eat harmful insects and seeds and travel in flocks called coveys.

Quality (timbre). A property of a sound arising from the fact that it is a combination of a fundamental frequency and a number of partials (or overtones) of higher frequency. Notes produced by different musical instruments may have the same loudness and frequency and yet be distinguishable to a listener because they differ in quality. A pure tone has a single frequency.

Quantized. Denoting a physical quantity that can only have certain discrete values in a particular system. The value of the quantity cannot vary continuously but must change in steps.

Quantum. A discrete amount of energy: the smallest amount of energy by which a system can change. The photon, for example, is the quantum of electromagnetic radiation, a phonon is a quantum of vibrational energy in a crystal etc.

Quantum Electrodynamics. The quantum mechanical theory of electromagnetic interactions between particles and between particles and electromagnetic radiation.

Quantum Electronics. The application of quantum mechanics to the study of the generation of microwave power in solid crystals.

Quantum Mechanics. A mathematical theory that developed from quantum theory and is used to explain the behaviour of atoms, molecules, and elementary particles. In early quantum theory, the basic idea that energy is quantized was introduced in an *ad hoc* manner to explain particular phenomena (black-body radiation, the photoelectric effect, atomic spectra, etc.). De Broglie's theory that a particle can also behave like a wave was used by Schrodinger in his formulation of wave mechanics, in which the wave-like nature of matter leads directly to the idea that only certain energy levels are possible, corresponding to the existence of standing waves. At about the same time an alternative formulation of quantum mechanics was developed based on the Heisenberg uncertainty principle.

Quantum Number. An integer or half integer giving the possible values of a quantized property of a system. For example, an electron in the Bohr atom can have angular momenta of *nh*/2Π, where *n* can be 1 ,2, 3 etc.: *n* is the quantum number which determines the possible values of the angular momentum.

Elementary particles also have quantum numbers to describe their properties. The charge of a particle can be zero or can be equal to the charge of the electron and of equal or opposite sign. These states can be described by quantum numbers 0, –1, or 1, characterizing the charge. Other properties, such as spin, baryon number, hypercharge, and parity, are similarly described by quantum numbers.

Quantum Theory. Modern physical theory that holds that energy and some other physical properties often exist in tiny, discrete amounts. The older theories of classical physics assumed that these properties could vary continuously. Quantum theory and the theory of Relativity together form the theoretical basis of modern physics. The first contribution to quantum theory was the explanation of Blackbody radiation in 1900 by Max Planck, who proposed that the energies of any harmonic oscillator are restricted to certain values, each of which is an integral multiple of a basic minimum value. The energy *E* of this basic quantum is directly proportional to the frequency *v* of the oscillator; thus $E = hv$, where Planck's constant *h* is equal to 6.63×10^{-34} J-sex. In 1905 Albert Einstein, in order to explain the Photo-Electric Effect, proposed that radiation itself is also quantized and consists of light quanta, or Photons, that behave like particles. Niels Bohr used the quantum theory in 1913 to explain both atomic structure and atomic spectra. The light or other radiation emitted and absorbed by atoms is found to have only certain frequencies (or wavelengths), which correspond to the absorption or emission lines seen in atomic spectra. These frequencies correspond to definite energies of the photons and result from the fact that the electrons of the atoms can have only certain allowed energy values. or levels. when an electron changes from one allowed level to another, a quantum of energy is emitted or absorbed whose frequency is directly proportional to the energy difference between the two energy levels E_1 and E_2; thus $E_2 - E^1 = hv$. Quantum mechanics, the application of the quantum theory to the motions of material particles, was developed during the 1920s. In 1924 Louis de Broglie proposed that not only does light exhibit particle-like properties but also particles may exhibit wavelike properties. The observation, by Clinton Davisson and Lester Germer in an 1927 experiment, of the diffraction of a beam of electrons analogous to the diffraction of a beam of light confirmed this hypothesis. A particularly important discovery of the quantum theory is the uncertainty principle, enunciated by Werner Heisenberg in 1927; it places an absolute, theoretical limit on the combined accuracy of certain pairs of simultaneous, related measurements.

Quantum Theory. A mathematical theory involving the idea that the energy of a system can change only in discrete amounts (quanta), not continuously. The concept was first introduced by Max Planck in explaining the form of the curves of intensity against wavelength for the radiation emitted from a black body. He assumed that a radiant energy could only be produced in discrete units. Other early applications of quantum theory were Einstein's explanation of the photoelectric effect and the theory of the Bohr atom.

Quark. Any of three hypothetical particles that are postulated, together with their antiparticles, as constituents of the known hadronic particles. Any particle can be explained as a combination of quarks and antiquarks. Despite many searches, no quark has as yet been discovered.

Quark. Any of a group of hypothetical entities, believed to the basic constituents of all hadrons. Quarks and leptons are thought to be the most elementary classes of particles. Quarks have fractional charges of 1/3 or 2/3 of the basic charge of the electron or proton. The baryons, a subgroup of the hadrons, are assumed to consist of three quarks Three antiquarks make up the antibaryons. Mesons, the other subgroup, are believed to consist of one quark and one antiquark. There is strong evidence for five kinds, or flavors, of quarks—up, down, strange, charm, and bottom—and a sixth, labeled top, is predicted. Each flavor of quark is believed to come in three varieties, differing in a property called colour.

Quarrying. Open, or surface, excavation of rock used for various purposes, including construction, ornamentation, and road building. The methods depends on the desired shape of the stone and its physical characteristics. Sometimes the rock is shattered by the use of explosive (*e.g.* for road-beds). For building stones, a process called broaching, or channeling, is used, whereby holes are drilled and wedges inserted and hammered until the stone splits off. This method was probably used by the ancient Egyptians and the Incas.

Quartz. One of the commonest rock-forming minerals (SiO_2), and one of the most important constituents of the earth's crust. It occurs in crystals, often distorted and commonly twinned. Varieties are classified as crystalline (*e.g.*, Amethyst) and cryptocrystalline (having crystals of microscope size, *e.g.*, Chalcedony and Chert). Clastic quartzes are Sand and Sandstone.

Quartz. A colourless or white natural crystalline form of silica, SiO_2. Quartz is one of the commonest minerals. It is used in high temperature laboratory apparatus and in optical devices for its transparency to ultraviolet radiation.

Quartz Clock. A type of clock in which the frequency of an oscillator circuit is stabilized by coupling it to a crystal of quartz using the piezoelectric effect.

Quartzite. Rock composed of firmly cemented Quartz grains, commonly resulting from the Metamorphism of sandstone and distinguished from the latter by fracturing across, rather than along the line of cementing material between, its constituent grains of sand.

Quasar (quasi-stellar object). Any of several hundred recently discovered extragalactic sources of very strong electromagnetic radiation. These objects are starlike in appearance and have enoromous red shifts. If these are interpreted as Doppler effects, the velocity of recession of quasars must be well over half the speed of light and their distances must be several thousand million light-years. Consequently their energy output appears to be in the region of three or four hundred times that of a galaxy, perhaps even more. How such energy outputs occur is so far unexplained.

Quasar or **Quasi-stellar Object**. One of a class of faint blue celestial objects, starlike in appearance, that are currently believed to be the most distant and most luminous objects in the universe. The spectral lines of quasars have enormous Red Shifts that seem to imply that they are receding from our galaxy with speeds as great as 80% of the speed of light. If Hubble's Law for the expansion of the universe is extrapolated to include quasars, they may be as far as 8 billion light-years away and consequently as luminous intrinsically as 100 galaxies combined.

Quaternary Ammonium Compound. Any compound with the general formula $R_4N^+X^-$, where X is an acid radical and R is either an organic group or a hydrogen atom.

Queen Anne's lace or **wild carrot**. Herb (*Daucus carota*) of the Carrot family, native to the Old World but naturalized in North America. Similar in appearance to the cultivated carrot (which is believed to be derived from this plant), it has feathery foliage but a woody root. The tiny white flowers bloom in a lacy, flat-topped cluster or umbel.

Quenching. 1. Fast cooling of a hot metal, used to harden steel, etc.

2. Sudden termination of the discharge in a Geiger counter.

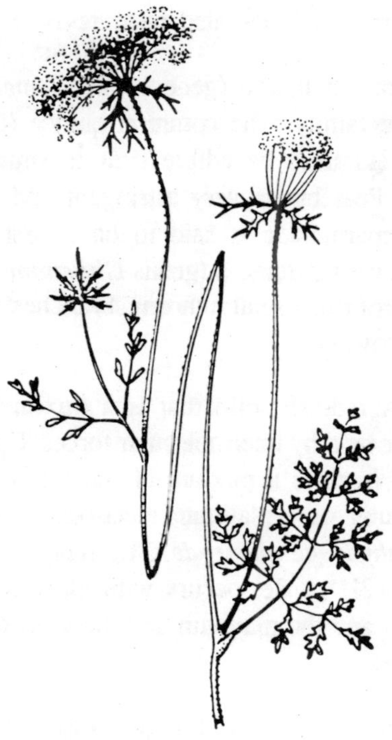

Fig. Q-1. Queen Anne's lace, Daucus carota.

3. A process in which the lifetime of excited atoms, ions, etc., is reduced by addition of a substance that deactivates them.

Quiet Sun. The sun when it has no sunspots, flares, or prominences.

Quilting. Form of needlework in which two layers of fabric separated by an interlining are sewn together, usually in a pattern of back or running stitches. A traditional means of adding warmth to fabric in Europe and Asia, it has been used in America since colonial times to make bed covers (quilts). Until c.1750 quilts were of the pieced, or patch-work, type; later, applique and trapunto (padding or cording) also became popular. Old quilts are prized by collectors. The art survives in rural Southern areas and has also been revived by textile artists.

Quince. Shrub or small tree (genera *Chaenomeles and Cydonia*) of the Rose family. The common quince (*Cydonia oblonga*) is a spineless tree. Its edible fruit is similar to the related Apple and Pear but is very astringent and is used mainly in preserves; marmalade is said to have first been made from quince. Flowering quinces (genus *Chaenomeles*) are cultivated for their profuse, usually thorny branches and scarlet, pink, or white flowers.

Quinhydrone. A reddish solid that is a mixture of hydroquinone and quinone held by intermolecular forces, C_6H_4 $(OH)_2.C_6H_4O_2$. In solution it gives a mixture of the two compounds. A half cell consisting of a platinum electrode in such a mixture is called *quinhydrone electrode*. The redox reaction $C_6H_4(OH)_2 = C_6H_4O_2 + 2H^+ + 2e$, occurs with electron transfer between the solution and the platinum and the electrode can be used to measure pH.

Quinine. A white bitter alkaloid, $C_{20}H_{24}N_2O_2.3H_2O$, extracted from cinchona bark and used as an antimalarial drug, often in the form of a soluble salt. M.pt. 57°C.

Fig. Q-2. Quinone.

Quinine. Alkaloid isolated from the bark of several species of *Cinchona* trees. Prior to the development of synthetic agents, quinine was the specific drug treatment of Malaria. It is used in a soft drink known as tonic, which is often mixed with alcoholic beverages. Excessive use may cause cinchonism, a condition characterized by ringing of the ears, dizziness, and

Quinone. **1. (benzoquinone)**. A yellow crystalline solid, $C_6H_4O_2$, prepared by the oxidation of aniline with chromic acid and used in manufacturing dyes and quinol. M.pt. 116°C; r.d. 1.3.

2. Any of a class of compounds in which a CH group in a benzene ring has been replaced by a CO group.

Quotient. The result of dividing one number by another; a ratio.

Q-value. The amount of energy released in a specified nuclear reaction.

ENCYCLOPAEDIC DICTIONARY OF SCIENCE

ENCYCLOPAEDIC DICTIONARY OF SCIENCE

Vol. 4

[R—Z]

Edited by
SATISH ANAND

ANMOL PUBLICATIONS PVT. LTD.
NEW DELHI-110 002 (INDIA)

ANMOL PUBLICATIONS PVT. LTD.
4374/4B, Ansari Road Daryaganj
New Delhi-110 002

Encyclopaedic Dictionary of Science

First Edition 1995
Reprint 1997
ISBN 81-7488-003-8(Set)

PRINTED IN INDIA

Published by J.L. Kumar for Anmol Publications Pvt. Ltd., New Delhi-110 002 and Printed at Mehra Offset Press, Delhi.

Preface

The Encyclopaedic Dictionary of Science is an effort to keep pace with continuing rapid developments and expanding vocabularies in various branches of science. This dictionary has been encyclopaedic in its contents and style of representation. Throughout the process of compilation and editing the volumes, the main effort was to make the dictionary serve as a ready reference for undergraduate and postgraduate students of science and for those appearing for competitive examinations, research scholars, teachers and authors. The entries have been written in a clear and explanatory style to provide both straight forward definitions and invaluable background information. This approach, combined with an extensive cross-reference system, enables the reader to place the each entry into a broader scientific context. At certain appropriate places, the diagrams have been included whenever the meaning of a word can be best understood by means of a diagram.

The presentation throughout has been aimed at sustaining the interest of readers while enriching their vocabulary and comprehension of technical terms and expressions of various branches of science.

This encyclopaedic dictionary will be of immense value to the students of science and to scientists and research scholars working on science projects.

In compiling a set of this kind it becomes necessary to draw upon the work of many authorities and seek the advice of colleagues to all of whom the editor is deeply indebted. All the comments from users on omissions or shortcomings will be most welcome.

The Editor

Vol. 4

R — Z

R

Rabbit. Mammal of the family Leporidae, which includes the Hare. Rabbits and hares have large front teeth, short tails, and hind legs and feet adapted for running and jumping. The term *rabbit* generally refers to small, running animals that give birth to blind, naked young; *hare* refers to larger hopping forms with longer ears and legs whose young are born furred and open-eyed. Wild rabbits are up to 16 in. (41 cm) long and have grayish-brown fur. The European common rabbit (*Oryctolagus cuniculus*), native to S Europe and Africa, is now found worldwide. It lives in elaborate, adjoining warrens and is chiefly nocturnal, feeding on vegetation. All domestic rabbits belong to this species.

Rabi, Isidor Isaac. (*Rob'e*), 1898-, American physicist; b. Austria. He taught physics at Columbia Univ. from 1929 and was chairman (1952-56) of the general advisory committee to the U.S. Atomic Energy Commission. Rabi is known for his work on magnetism, molecular beams, and quantum mechanics. He won the 1944 Nobel Prize in physics for his discovery and measurement of the radio-frequency spectra of atomic nuclei whose magnetic spin has been disturbed.

Rabies hydrophobia. Acute, often fatal, disease of mammals, caused by a virus transmitted from an animal to another animal or to humans through infected saliva, usually through a bite. After a variable incubation period, rabies produces fever, headache, nausea, and pain at the site of the bite, followed by convulsions, inability to drink fluids, apathy, and death. A Vaccination of antirabies vaccine is administered to the bite victim to prevent the disease from developing.

Raccoon. Nucturnal New World Mammal (genus *Procyon*) with a heavily furred body, pointed face, and handlike forepaws. It has mixed gray, brown, and black hair, a black face mask, and black rings on the tail. The common raccoon of North America (*Procyon lotor*) is found from S Canada to South America, except in parts of the Rocky Mts. and in deserts. Highly omnivorous, it is adaptable to civilization and will feed on garbage. Its fur is commercially valuable.

Race. Biological grouping or subspecies of human beings, distinguished by such physical traits as skin colour, hair type and colour, shape of body, head, and facial features, and blood traits. These characteristics are transmitted by heredity through Genes (the term *race* is inappropriate when applied to national, religious, or cultural groups), but they are highly variable—not every member of a race will exhibit all distinguishing traits. Races arose in response to Mutation, selection, geographic adaptation, and genetic drift. In the 19th and early 20th cent., Joseph Arthur Gobineau and Houston Stewart Chamberlain attributed cultural and psychological values to race, proposing theories of racial superiority. By limiting the criteria to certain physical traits, anthropologists at one time agreed on the existence of three relatively distinct groups of people, namely, Caucasoid, Mongoloid, and Negroid. Today, however, anthropologists stress the heterogeneity of world population, and many reject the concept of race outright.

Racemic. Denoting a substance composed of equal amounts of its optically active isomers. A racemic mixture is not optically active. Racemic compounds are denoted by *dl*-, as in *dl*-tartaric acid.

Racemic Acid. The racemic form of tartaric acid.

Rad. A unit of dose of ionizing radiation equal to the dose equivalent to an absorbed energy of 0.01 joule per kilogram of material.

Radar. A system for detecting distant objects by reflecting electromagnetic waves off them. A beam, consisting of regular pulses of microwaves, is transmitted from a movable aerial. Pulses hitting a distant object are reflected back to the aerial and the distance of the object is measured by the time taken between transmission of a pulse and reception of the reflected pulse. In a typical radar system, the aerial moves round and the detected signals are fed to a cathode-ray screen which gives a display of the surrounding area.

If the object is moving the difference between the frequency of the transmitted pulse and that of the reflected pulse (caused by the Doppler effect) can be used to distinguish motion of the object and to determine its velocity (*Doppler radar*). The term is an acronym for *radar detection and ranging.*

Radar. (Radio detection and ranging), system or technique for detecting the positon,motion, and nature of a remote object by means of radio waves reflected from its surface. Radar systems transmit pulses of electromagnetic waves by means of directional Antennas; some of the pulses are reflected by objects in the path of the beam. Reflections are received by the radar unit, processed electronically, and converted into images on a Cathode-Ray Tube. The distance of the object from the radar source is determined by measuring the time required for the radar signal to reach the target and return. The direction of the object with respect to the radar unit is determined from the direction in which the pulses were transmitted. In most units, the beam of pulses s continuously rotated at a constant speed, or it is scanned (swung back and forth) over a sector at a constant rate. The velocity of the object is sometimes determined by the Doppler Effect: if the object is approaching the radar unit, the frequency of the returned signal is greater than the frequency of the transmitted signal; if the object is receding, the returned frequency is less; and if the object is not moving relative to the radar unit, the frequency of the returned signal is the same as the frequency of the transmitted signal. Most radar units operate on microwave frequencies.

Radial Velocity. The speed with which a star moves toward or away from the sun. It is measured from the Red Shift or blue shift in the star's spectrum.

Radian. The angle subtended at the centre of a circle by an arc whose length is equal to the radius of the circle. A complete revolution is 2Π radians. Π radians is equivalent to 180 degrees.

Radiation. Particles or waves emitted fro a source, in particular electromagnetic radiation, sound waves, or streams of particles thrown out by radioactive materials.

Radiation. The emission or transmission of energy in the form of waves through space or through a material medium; the term also applies to the radiated energy itself. The term includes electromagnetic, acoustic, and particle radiation, and all forms of ionizing radiation. According to Quantum Theory, Electromagnetic Radiation may be viewed as made up of Photons. Acoustic radiation is propagated as sound waves. Examples of particle radiation are alpha and beta rays in Radioactivity, and Cosmic Rays.

Radiation Belts. *See* Van Allen belts.

Radiation Pressure. A pressure exerted on a surface by the impact of electromagnetic radiation.

Radiation Sickness. Illness caused by the effects of radiation on body tissues. It may be acute, delayed, or chronic and may occur after repeated (cumulative) exposure to small doses of radiation (as in a plant, a laboratory, or the environment); undue exposure to solar radiation; or exposure to a nuclear explosion. Symptoms may be mild and transitory, or severe, depending on the type of radiation, the dose, and the rate at which exposure is experienced. They include weakness, loss of appetite, vomiting, diarrhea, a tendency to bleed, increased susceptibility to infection, and—in severe cases—brain damage and death. Mild radiation sickness is a common side effect of radiation therapy for cancer. Exposure to radiation is of concern even in small doses because of possible long-term genetic effects.

Radiative. Involving the production of electromagnetic radiation. Thus a *radiative collision* is a collision resulting in the emission of a photon.

Radical. 1. *See* free radical, group.

2. A quantity that is a root of another quantity. Thus ✓ x is a radical: ✓ is a radical sign.

Radio. The communication of signals by means of radio- frequency electromagnetic waves. The term embraces all forms of communication, including television, but is sometimes used specifically for sound broadcasting. A radio transmitter generates a carrier wave of fixed frequency. The information to be transmitted is converted into an electrical signal (as by a microphone or television camera) which modulates this carrier wave. The modulated carrier wave is projected from an aerial and carried, either as a ground wave of sky wave, to the aerial of the receiver where the signal is demodulated.

Radio. Transmission or reception of Electromagnetic radiation in the radio frequency range from one place to another without wires. For the propagation and interception of radio waves, a transmitter and receiver are employed. A radio wave carries information-bearing signals; the information may be encoded directly on the wave by periodically interrupting its transmission or impressed on the carrier frequency by a process called Modulation, *e.g.*, amplitude modulation (AM) or frequency modulation (FM). In its most common form, radio transmits sounds (voice and music) and pictures (Television). The sounds (or images) are converted into electrical signals by a microphone (or camera tube), amplified, and used to modulate a carrier wave that has been generated by a transmitter. The modulated carrier is also amplified, then applied to an Antenna that converts the electrical signals to electromagnetic waves that radiate into space at the speed of light. Receiving antennas intercept part of this radiation, convert it back into electrical signals, and feed it to a receiver. Once the basic signals have been separated from the carrier wave, they are fed to a Loudspeaker or Cathode-Ray Tube, where they are converted into sound and visual images, respectively. Some celestial bodies and interstellar gases emit relatively strong radio waves that are observed with radio telescopes composed of very sensitive receivers and large directional antennas. Long-range radio signals enable communications between astronauts and ground-based controllers and carry information from Space Probes as they travel to and encounter distant planets. The invention of the Transistor and other microelectronic devices led to the development of portable transmitters and receivers. Military applications of radio include the proximity fuse and various types of Reconnaissance Satellites. Citizens band (CB) radios, operating at frequencies near 27 megahertz, are used in vehicles for communication while traveling.

Radio. Prefix indicating a radioactive isotope of an element.

Radioactive. Exhibiting radioactivity.

Radioactive Series. A series of radioisotopes such that each member of the series is produced by alpha or beta decay from the preceding members. Thus, in the *uranium series* the first member is uranium-238, which undergoes alpha decay to thorium-234. This undergoes beta decay to protactinium-234, which in turn undergoes beta decay to uranium-234, and so on. Three other series exist: the *thorium series*, the *actinium series*, and the *neptunium series*. All end with stable isotopes of lead.

Radioactivity. Spontaneous disintegration of the nuclei of certain isotopes with emission of beta rays (electrons), alpha rays (helium nuclei), or gamma rays. Electrons are produced by beta decay (in some cases positrons are emitted). They have kinetic energies up to the MeV range and are not very penetrating. Alpha particles are produced by alpha decay. They produce more ionization than beta rays and have energies up to about 10 MeV. Like beta rays, they are also easily absorbed by matter but are less penetrating. Gamma rays are photons of electromagnetic radiation. They are formed when beta or alpha decay leads to the production of a nucleus in an excited state, which decays to its ground state with emission of a gamma-ray photon. They are not efficient at ionizing matter and, in consequence, are highly penetrating, being capable of passing through several centimetres of lead.

In any sample of radioactive material the number of disintegrations per unit time, and thus the rate of decomposition, is proportional to the amount of material present. Thus the rate follows and equation of the form $-dN/d\pm = \lambda N$, where N is the number of atoms and λ is a constant for particular substance-its *decay constant*. The number of atoms present follows the equation $N = N_o \exp(-\lambda t)$, where/N_o is the initial number: the amount of radioactivity shows an exponential decay. The reciprocal of the decay constant is the *mean life* of the isotope (τ), which is equal to $T_{1/2}/0.693$, where $T_{1/2}$ is the *half-life* — the time for half the sample to decay. The activity of an isotope is measured in curies.

Natural radioactivity is exhibited by natural isotopes of heavy elements: these are arranged in radioactive series. Artificial radioisotopes may also be produced by bombardment with particles.

Radioactivity. Spontaneous disintegration or decay of the nucleus of an atom by emission of particles, usually accompanied by Electromagnetic radiation. Natural radioactivity is exhibited by several elements, including Radium and Uranium. The radiation produced is of three types: the *alpha particle*, which is a nucleus (two protons and two neutrons) of an ordinary helium atom; the *beta particle*, which is a high-speed electron or, is some cases, a positron (the electron's antiparticle); and Gamma Radiation, which is a type of electromagnetic radiation with very short wavelengths. The rate of disintegration of a radioactive substance is commonly designated by its half-life, which is the time required for one half of a given quantity of the substance to decay. Radioactivity may be induced in stable elements by bombardment with particles of high energy.

Radio Astronomy. Study of celestial bodies by means of the radio waves they emit and absorb naturally. These waves are received by specially constructed antennas called radio telescopes. The most common design consists of a parabolic "dish" of open metal latticework that focuses the radio waves into a concentrated signal that is filtered, amplified, and, finally, analyzed using a computer. The radio signals received from outer space are extremely weak, and long observing times are required to collect a useful amount of energy. There are several basic types of galactic radio emissions. The type first discovered, accidentally in 1931 by Karl Jansky, is spread over a wide band of radio frequencies. It is produced when free electrons are scattered by collisions with heavier ions in the ionized interstellar gases surrounding hot, bright Stars. A second type, called synchrotron radiation, is emitted by energetic electrons as they rapidly spiral within the strong magnetic fields existing in the vicinity of Supernova remnants. A third type, originating in Interstellar Matter, radiates at discrete frequencies characteristic of the quantum jumps made by electrons in the atoms and molecules, *e.g.*, atomic hydrogen and formaldehyde, in the interstellar medium, Pulsars, stellar radio sources that regularly radiate periodic bursts of energy, were discovered in 1968. Radio waves also come from outside our Milky Way galaxy. Some extragalactic sources are detected only by their radio emission, but others have been identified with optically observed Galaxies and Quasars. In addition to localized radio sources, there is uniform low-level radio noise from every direction in the sky. This cosmic background radiation is believed to be an indication that the universe began with an explosive "big bang" rather than having always existed in an unchanging steady state.

Radio Astronomy. The branch of astronomy involving the use of radio telescopes.

Radiocarbon Dating (carbon-14 dating). A method of dating archaeological specimens that are made of wood, cotton, or other once-living matter. The carbon dioxide in the air contains a small amount of carbon-14, a radioisotope produced by bombardment of nitrogen with cosmic rays. This is incorporated into living matter during the lifetime of the organism. Once the matter dies, the amount of carbon-14 present begins to decrease due to beta decay to give nitrogen-14. Thus, measurement of the amount of radiocarbon present, by using a counter, gives an estimate of the object's age. The half-life is 5730 years making the technique suitable for dating objects up to about 10,000 years old.

Radiochemistry. The branch of chemistry concerned with compounds containing radioisotopes, their use in separating isotopes, chemical reactions, application to tracer studies, etc.

Radiofrequency. A frequency between 3 kilohertz and 300 gigahertz.

Radiofrequency Radiation. Electromagnetic radiation with frequencies lying in the range 3 kilohertz to 300 gigahertz. Radiofrequency radiation is the portion of the electromagnetic spectrum with wavelengths longer than those of infrared radiation. It includes microwaves and is divided into frequency bands.

Radiogenic. Resulting from or produced by radioactive disintegration.

Radiography. The use of radiation, usually X-rays, to investigate internal structure. The X-rays are passed through the object investigated and form a shadow image (*radiograph*) on a fluorescent screen of photographic plate.

Radioimmunoassay (RIA). Highly sensitive technique used to measure the concentration of substances in the body, including hormones and drugs present in very low concentrations. The substance or antigen (a foreign substance in the body which causes the production of antibodies) to be measured is injected into an animal, causing it to produce antibodies. Serum containing the antibodies is withdrawn and treated with a radioactive antigen and later with a non-radioactive antigen. Measurements of the amount of radio-activity are the used to determine the amount of antigen present. The technique was developed by Solomon Berson and Rosalyn Yalow. Yalow was awarded the 1977 Nobel Prize in physiology or medicine for her work.

Radio interferometer. *See* radio telescope.

Radioisotope. An isotope that is radioactive.

Radiology. The science of X-rays, particularly their effects and applications.

Radiolucent. Partially transmitting radiation. The term is usually applied to materials that partially transmit X-rays or gamma rays, distinguishing them from *radiotransparent* materials, which allows almost all the radiation through, and *radio-opaque* material, which block most of the radiation.

Radioluminescence. Fluorescence caused by radioactive decay. The radiation—electrons, alpha particles, or gamma rays—causes excitation of the atoms and light emission results. Radioluminescence is the process responsible for the glow of self-luminous paints.

Radiolysis. The process of decomposing a substance by irradiation with high-energy particles or X-rays.

Radiometer. Instrument for the detection or measurement of Electromagnetic Radiation, particularly Infrared Radiation. Radiometers that function by increasing the temperature of the device are called thermal detectors; examples include the Bolometer and the Thermocouple.

Radionuclide. A nuclide that is radioactive.

Radio-opaque. *See* radiolucent.

Radiosonde. Group of instruments for simultaneous measurement and radio transmission of meteorological data. The instrument package is usually carried into the atmosphere by a Weather Balloon; it may, however, also be carried by a rocket (rocketsonde) or dropped by parachute (dropsonde).

Radio Source. A source of radio waves in space. Radio sources include super-novae, pulsars, quasars, and some stars, including the sun.

Radio Telescope. An instrument for detecting radio waves from a particular direction in space. One type has a large parabolic aerial that can be moved to point to different areas of the sky. The *radio interferometer* has two or more separate aerials connected to the same radio receiver. The position of the radio source is determined by the interference of radio waves reaching the different aerials.

Radio telescopes are used in studying the intensity and nature of radio waves from inside and outside the Galaxy. Many celestial objects have been discovered that are not visible with optical telescopes.

Radiotransparent. *See* radiolucent.

Radish. Herb (*Raphanus sativus*) of the Mustard family, with an edible, pungent root used as a relish. There are many varieties, with red, white, or black roots of different shapes and sizes, some quite large. Horseradish is not a radish but a related plant of the mustard family.

Radium. Symbol: Ra. A white radioactive metallic element occurring as a product of radioactive decay of heavier elements. Its most stable isotope, ^{226}Ra, has a half-life of 1622 years. The element is used as a radioactive source in research and medicine. A.N. 88; A.W. 226.05; m.pt. 700°C; b.pt. 1140°C; valency 2.

Radium (Ra). Radioactive metallic element, discovered in pitchblende in 1898 by Pierre and Marie Curie. It is a rare, lustrous, white Alkaline-Earth Metal that resembles barium in its chemical properties. Radium compounds are found in uranium ores. The Radioactivity of radium and its compounds is used in the treatment of cancer. Radium compounds are mixed with a phosphor in luminous paints. In its radioactive decay, radium emits alpha, beta, and gamma rays and produces heat.

Radius. The line or distance between the centre of a circle or sphere and the circumference of surface.

Radius of Gyration. Symbol: k. A distance associated with a rotating body, such that $I = mk^2$, where m is the mass of the body and I is its moment of inertia.

Radius Vector. *See* polar coordinates.

Radix. The base of a number system. For example, 10 is the radix of decimal notation.

Radon (Rn). Gaseous radioactive element, discovered by Ernest Rutherford in 1899. A colourless, chemically unreactive Inert Gas, it is the densest gas known. Highly radioactive (emitting alpha rays), it is used chiefly in the treatment of cancer by radiotherapy.

Raffinate. A liquid that has been purified by solvent extraction.

Raffinose (melitose). A white crystalline powder, $C_{18}H_{32}O_{16}.5H_2O$, with a sweet taste, extracted from sugar beet. It is a trisaccharide sugar used in medicine and in making other saccharides. M.pt. 118°C; r.d. 1.5.

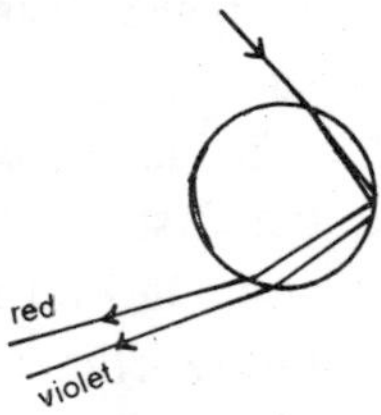

Fig. R-1. Dispersion by a raindrop.

Ragweed. Weedy herb (genus *Ambrosia*) of the Composite family, mostly native to America. They have greenish flowers and subdivided leaves. Their Pollen is one of the primary hay fever irritants, especially that of the common ragweed (A. *artemisiifolia*) and great ragweed (A. *trifida).*

Rail. Name for some marsh Birds of the family Rallidae, cosmopolitan in distribution, except in polar regions. Their extremely slender bodies are protectively coloured in drab browns and reds. There are two major types: the long-billed rails, which include the king and water rails; and those with short, conical bills, including the black rails. Rails, including the gallinule and coot, are probably the most widely distributed bird family.

Railroad or **Railway**. Form of transportation most commonly consisting of steel rails, called tracks, on which freight cars passenger cars, and other rolling stock are drawn by one or more powered Locomotives. Other types of railroads include single, self-propelled cars; cable-drawn units, used on steep grades; and monorails, whose cars travel along a single rail. Crude railways pulled by horses along wooden rails were used for mining purposes as early as the 16th cent, but the modern railroad began with the steam locomotives pioneered by the Englishmen Richard Trevithick and George Stephenson in the early 1800s. Steam-powered freight and passenger services were first provided by the Stockton and Darlington Railway in England in 1825. In the U.S., the Baltimore & Ohio RR began operation in 1828 with horse-drawn cars, but after the successful run (1830) of the *Tom Thumb*, a locomotive built by Peter Cooper, steam power was used. Many short-run railroads began to appear in the U.S. in the 1840s. Early improvements in railroading included sleeping cars, patented (1856) by G.M. Pullman; the first all-steel car (1859); and the use of steel rather than wrought iron rails (by 1863). U.S. railroad building reached a historic climax with the completion of the first transcontinental railway. The Union Pacific, building westward from Nebraska, and the Central Pacific, building eastward from California, met a Promontory Point, Utah, on May 10, 1869. Railroad tracks, which had varied in width for many years (preventing cars from passing from one line to another), were standardized in the mid-1880s at 4 ft $8^1/_2$ in. (1.44 m). After the Civil War, which gave tremendous impetus to railroads, the great battles of the railway financiers began, involving such men as Cornelius Vanderbilt, Jay Gould, James Fisk, J.J. Hill, and E.H. Harriman. Competition from other sources led to the decline of U.S. railroads beginning in the 1920s, and after World War II railroad companies claimed annual deficits in passenger service. Their plight was dramatized when the huge Penn Central filed for bankruptcy in 1970. In that year Congress acted to create Amtrak, a quasi-public agency that took over virtualy all intercity passenger service in the U.S., and in 1979 the federal government set up the Con Rail system to operate six bankrupt railroads in the northeast. Despite sharp cutbacks in service and track mileage, both systems suffered persisting financial problems. In most countries other than the U.S. railroads have long been either nationalized or heavily subsidized by government.

Rain. Liquid form of precipitation. It consists of drops of water falling from Clouds; if the drops are very small, they are collectively termed *drizzle*. Clouds contain huge amounts of tiny water droplets. Raindrops are formed as additional water vapor from the air condenses on the droplets. When the raindrops become too heavy to be supported by the air currents in the cloud, they slowly fall. Collisions with other raindrops continue to increase their size until they hit the ground.

Under warm, dry conditions, raindrops can evaporate completely before they land; such raindrops are called *virga*. Rainfall is one of the primary factors of Climate. Average annual rainfall can vary from less than 1 in. (2.5 cm) in an arid Desert to over 400 in. (1,000 cm) in some rainforests. Factors that control rainfall include the belts of converging-ascending air flow, air temperature, moisture-bearing winds, ocean currents, location and elevation of mountains, and the proximity of large bodies of water.

Rainbow. An arc of coloured bands seen in the sky when sunlight is refracted and dispersed by raindrops. The rainbow is composed of spectral colours with the violet band on the inside of the bow. It is produced by light that undergoes total internal reflection inside the drop and the light reaching the observer makes an angle of 40-42° with the direction of he observer's shadow. Sometimes a faint secondary bow is seen lying outside the primary bow: this is formed by two internal reflections and has its colours reversed.

Rainbow. Arc showing the colours of the spectrum, which appears when sunlight shines through water droplets. It often appears after a brief shower late in the afternoon. The sun, the observer's eye, and the center of the arc must be aligned—the rainbow appears in the part of the sky opposite the sun. It is caused by the refraction and reflection of light rays from the sun: a ray is refracted as it enters the raindrop, reflected from the drop's opposite side, and refracted again as it leaves the drop and passes to the observer.

Raman Effect. The inelastic scattering of light or ultraviolet radiation by molecules. The scattered radiation and the incident radiation differ in frequency by discrete amounts, corresponding to vibrational and rotational changes in energy level of the molecules. The spectrum of the scattered radiation (*Raman spectrum)* can thus be used to give information on molecular structure. [After Chandrasekhara Raman (1880-1970), Indian Physicist.]

Ramanujan, Srinivasa. (Rama Noojen), 1889-1920, Indian mathematician. A self-taught genius i pure mathematics, he made original contributions to function theory, power series, and number theory.

Ramsay, Sir William. 1852-1916, Scottish chemist. He synthesized (1876) pyridine from acetylene and prussic acid. Ramsay discovered helium and was codiscoverer (with Lord Rayleigh) of argon and (with Morris Travers) of krypton, neon, and xenon. Knighted in 1902, he won the 1904 Nobel Prize in chemistry for his work on gases.

Raney Nickel. A form of nickel made by dissolving the aluminium from an aluminium-nickel alloy with sodium hydroxide. The nickel produced has a spongy texture and is an extremely effective hydrogenation catalyst.

Rankine Scale. A scale of temperature based on the Fahrenheit scale. Temperatures are measured in degrees Rankine (°R) which are equivalent to Fahrenheit degrees. Absolute zero is 0°R (-459.67°F). °R = °F + 459.67. [After William John Rankine (1820-72). Scottish engineer.]

Raoult's Law. The principle that the fractional decrease in vapour pressure occurring when a solute is dissolved in a solvent is equal to the ration of the number of solute molecules to the total number of molecules: *i.e.* $(p - p_1)/P = N_1/(N + N_1)$, *where* P_1 is the vapour pressure of the solution, p that of the pure solvent, N_1 the number of solute molecules, and N the number of solvent molecules. [After Francois Raoult (1930-1901), French chemist.]

Rare-earth Metals. Group of chemical elements including those in the Lanthanide Series, usually Yttrium, sometimes Scandium and Thorium, and rarely Zirconium. Promethium, which is not found in nature, is not usually considered a rare-earth metal. The metals occur together in minerals as their oxides (Rare Earths) and are difficult to separate because of their chemical similarity. The cerium metals are a subgroup, consisting of the elements with atomic numbers between 57 and 63 and ytterbium.

Rare Earths. Oxides of the Rare-Earth metals. The name of an earth is formed from the name of its element by replacing -*um* with -*a*-. Once thought to be very scarce, they are widely distributed and fairly abundant in the earth's crust. Rare-earth minerals include bastnasite, cerite, euxenite, gadolinite, Monazite, and samarskite. Mixed rare earths are used in glassmaking, ceramic glazes, and glass-polishing abrasives, and as catalysts for petroleum refining. Individual purified rare earths are used in lasers and as color-television picture-tube phosphors.

Rare Gas. *See* inert gas.

Rat. Any of various stout-bodied Rodents, usually having a pointed muzzle and long, scaly tail. The name refers particularly to the two species of house rat, the brown, or Norway, rat (*Rattus norvegicus*) and the black, or Alexandrine, rat (*R. rattus*). The brown rat is the larger of the two, growing up to 10 in. (25 cm) long, excluding the tail, and some-times weighing more than a pound (.5 kg). Rats spread human diseases and destroy food supplies; efforts to exterminate them have been relatively unsuccessful. Many other rodents are also called rats, *e.g.*, the Muskrat.

Ratio. The quotient of two numbers, often used for comparison. For example, in a school with 500 children and 25 teachers, the teacher-pupil ration is $^{25}/_{500} = ^{1}/_{20}$.

Rational Number. A number that is an integer or can be expressed as a quotient of integers.

Rattan. Climbing Palm (genera *Calamus and Daemonorops*) of tropical Asia. Rattan leaves, unlike those of most palms, are not clustered in crown, and they have long barbed tips by which the plant climbs to treetops. Commercial rattan, a tough, flexible Cane of uniform diameter used for wickerwork, is obtained from the plant's long stem.

Rattlesnake. Poisonous New World Snake of the Pit Viper family, distinguished by a rattle at the end of the tail. The rattle, a series of dried, hollow segments of skin, makes a whirring sound when shaken, serving to warn attackers. Most species are classified in the genus *Crotalus*, including the largest and deadliest, the eastern diamondback rattle- snake, *C. adamanteus*, of the S and SE U.S. The Sidewinder is a North American desert species.

Raven. The largest member of the Crow family, found in arctic and temperate regions of the Northern Hemisphere. The raven (*Corvus corax*) is glossy, black scavenging bird about 26 in. (66 cm) long with a call resembling a guttural croak; it can be taught to mimic human speech.

Ray or **Wray, John**. 1627-1705, English naturalist. He was extremely influential in laying the foundations of systematic biology. Together with his pupil Francis Willughby, he planned a complete classification of the animal and vegetable kingdoms. His work—the botanical part of the project—includes the important *Historia plantarum* (3 vol., 1686-1704). Ray was the first to name and make the distinction between monocotyledons and dicotyledons and the first to define and explain the term *species* in the modern sense of the word.

Ray. Flat-bodied cartilaginous marine Fish (order Batoidea), related to the Shark. Most rays have broad, flat, winglike pectoral fins along the sides of the head, whiplike tails, eyes and spiracles on top of the head, and mouth and Gill slits underneath. Many rays, such as the skates, lie on the sea floor and feed on smaller animals, *e.g.*, snails. Fertilization is internal, and most bear live young. The largest rays are the top-swimming mantas, which may be up to 22 ft. (7 m) wide and weigh up to 3,000 lb (1,360 kg). Rays also include stingrays and sawfishes.

Ray. A line used to represent the path of light or other radiation.

Rayon. 1. Either of two types of textile fibre made from cellulose. *Viscose rayon* is produced by dissolving wood pulp in a mixture of sodium hydroxide and carbon disulphide to form *viscose*, a thick brown solution of cellulose xanthate. This is forced through fine holes into a bath containing acid, which decomposes the xanthate to regenerate cellulose fibres. *Acetate rayon* is made by treating wood pulp with a mixture of acetic anhydride with acetic and sulphuric acids to form cellulose acetate, which is dissolved in a solvent and forced through fine holes into the air. The solvent evaporates leaving the cellulose acetate fibre. Rayon was formerly called *artificial silk.*

2. Synthetic fiber made from Cellulose, or textiles woven from such fiber. Silklike rayon, the first synthetic Textile Fiber, was produced in 1884 by a French scientist, Hilaire de Chardonnet. In rayon manufacturing, cellulose, chiefly from wood pulp, is dissolved by chemicals and forced under pressure through minute holes in a metal spinneret, emerging as filaments. The filaments are doubled and twisted into silky yarns or cut into staple lengths and spun. Spun rayon can be treated to simulate wool, linen, or cotton.

Reactance. Symbol: X. The imaginary part of the impedance of an electrical circuit. The reactance of an inductance L is ωL and that of a capacitance C is $1/\omega C$, where ω is the angular frequency. Reactance is measured in ohms.

Reactant. A substance taking part in a chemical reaction, as distinguished from a product of the reaction.

Reaction. 1. *See* chemical reaction.

2. An equal and opposite force occurring when a force acts on a body.

Reactive Dye. *See* dye.

Reactor. 1. An apparatus for producing energy by nuclear fission or fusion.

2. An capacitance, inductor, or other element producing reactance in an electrical circuit.

Reagent. A substance used for chemical analysis or synthesis.

Real Image. *See* image.

Reaumur, Rene Antoine Ferchault de. (Ra'emyoor), 1683-1757, French physicist and naturalist. He invented an alcohol thermometer and the Reaumur temperature scale. He studied the composition of Chinese porcelain, which led him to develop an opaque glass, and the composition and manufacture of iron and steel, and published (1734-42) an exhaustive study of insects.

Reaumur Scale. A temperature scale in which the melting point of ice is 0° and the boiling point of water is 80°. Temperatures are measured in *degrees Reaumur* (°R). [After Rene Antoine Reaumur (1683-1757), French scientist.]

Reciprocal. One divided by a number or quantity: the reciprocal of 4 is 1/4.

Reciprocal Ohm. *See* siemens.

Reciprocal Proportions, Law of. The principle that if two elements combine individually with a third element then they combine with each other in the same relative proportions as they react with this third element. Oxygen and hydrogen, for example, both combine with carbon. The ratio C:O is 12:32 and the ratio C:H is 12:4, these being proportions by weight. In accordance with the law, hydrogen and oxygen combine together in the ratio 4:32 (or 2:16). The three compounds involved are CO_2, CH_4, and H_2O. The law involves the idea that elements have combining weights (or equivalent weights) and is sometimes called the *law of equivalent proportions*. It is one of three laws of chemical combination.

Recitative. Musical declamation for solo voice, used in Opera and Oratorio for dialogue and narration. Its development at the end of the 16th cent., enabling words to be clearly understood and natural speech rhythms to be followed, made possible the rise of opera. In the 17th cent., the rapid patter of *recitative secco*, punctuated by occasional occompanying chords, served only to advance the action; by the 18th cent., more strict measure and full orchestral accompaniment helped to highlight recitative passages of dramatic interest. In Wagner's *Sprechgesang*, the melody is completely molded to the text.

Reconnaissance Satellite. Artificial Satellite launched by a country to provide intelligence information on the military activities of foreign countries. There are four major types. *Early-warning satellites* detect enemy missile launchings. *Nuclear-explosion detection satellites* are designed to detect and identify nuclear explosions in space. *Photo-surveillance satellites* provide photographs of enemy military activities, e.g., the deployment of intercontinental ballistic missiles (ICBMs). There are two subtypes: close-look satellites provide high-resolution photographs that are returned to earth via a reentry capsule, whereas area-survey satellites provide lower-resolution photographs that are transmitted to earth via radio. Later satellites have combined these two functions. *Electronic-reconnaissance (ferret) satellites* pick up and record radio and radar transmissions while passing over a foreign country. The U.S. and the USSR have launched numerous reconnaissance satellites since 1960.

Record Player or **Phonograph**. Device for reproducing sound that has been recorded as a spiral, undulating groove on a disk. The disk, or phonograph record, is placed on the record player's motor-driven turntable, which rotates the record at a constant speed. A tone arm, containing a pickup at one end, is placed on the record and touches the groove with a stylus, or needle, that vibrates as the record revolves. The Transducer, also part of the pickup, converts these vibrations into corresponding electrical signals, which are increased in size by an Amplifier and then passed to a Loud-Speaker that converts them into sound. The first phonograph was built by Thomas Edison in 1877.

Rectangle. A parallelogram whose angles are all right angles.

Rectifier. 1. An electronic component or circuit for converting an alternating current into a unidirectional current. The commonest types are semiconductor diodes.

2. An apparatus for separating liquid mixtures by fractional distillation.

Rectifier. Component of an Electric Circuit that changes alternating current to direct current. Rectifiers operate on the principle that current passes through them freely in one direction, but only slightly or not at all in the opposite direction.

Red Giant. A type of giant star emitting red light.

Red Lead. *See* lead oxide.

Redox. Relating to simultaneous oxidation and reduction. A redox reaction is one in which one reactant is oxidized and the other is reduced.

Red Shift. The displacement of the spectral lines emitted by a moving body towards the red (longer wavelength) end of the visual spectrum. It is caused by Doppler effect and, when observed in the spectrum of distant stars and galaxies, it indicates that the body is receding from the earth. The greater the displacement, the higher is the velocity of recession. Displacement of lines to the red end of the spectrum can also result from the Einstein shift, caused by the high gravitational field of a star.

Red Shift. The systematic increase in the wavelength of all light received from a celestial object; it is observed in the shifting or individual lines in the Spectrum of the object toward the red, or longer-wavelength, end of the visible spectrum. Most observed red shifts are the result of the Doppler Effect; they are also produced by gravitation in accordance with the general theory of Relativity.

Reduced Equation. An equation of state of a gas in which the pressure (*p*), volume (*V*), and temperature (*T*) are replaced by the ratios p/p_c, V/V_c, and T/T_c, where p_c, V_c, *and* T_c are the critical pressure, volume, and temperature of the substance. The ratios are the reduced pressure, volume, and temperature.

Reducing Agent. A compound that causes reduction, either by abstracting oxygen, supplying hydrogen, or donating electrons. Many compounds act as reducing agents in particular circumstances but the term is also used for compounds that are particularly effective reducing agents. Examples are hydrogen, carbon, carbon monoxide, and ferrous salts.

Reducing Sugar. *See* sugar.

Reduction. A process in which oxygen is removed from or hydrogen is combined with a compound. Thus, the reaction Fe_2O_3 + 3C= 2Fe + 3CO is a reduction of iron oxide and $3H_2 + N_2 = 2NH_3$ is a reduction of nitrogen. The term has been extended to any reaction in which electrons are gained, as in the reduction of ferric ions to ferrous ions: $Fe^3 + e = Fe^{2+}$. *Compare* oxidation.

Reed. Name for several plants of the Grass family. The common American reed (*Phragmites communis*) is a tall perennial widely distributed in wet places. It has stout, creeping rootstalks and a large, plumelike panicle. The giant reed (*Arundo donax*), similar in appearance but native to the Mediterranean region, has long been used to make reed instruments, *e.g.*, the panpipe of Pan.

Refining. The process of purifying a substance.

Reflecting Telescope (reflector). A telescope in which the primary image is formed by a concave mirror. Reflecting telescopes are used in astronomy. Unlike refracting telescopes they do not suffer from chromatic aberration and spherical aberration is reduced by use of parabolic mirrors. Several types exist, including the Newtonian, Cassegrainian, Gregorian, and Schmidt telescopes.

Reflection. The process in which light or other radiation that is incident on a surface bounces off the surface back into the original medium rather than passing into the second medium. There are *two laws of reflection:*

(1) the incident ray, the reflected ray, and the normal to the surface at the point of incidence all lie in the same plane;

(2) the angle of incidence equals the angle of reflection. The laws only apply for specular reflection.

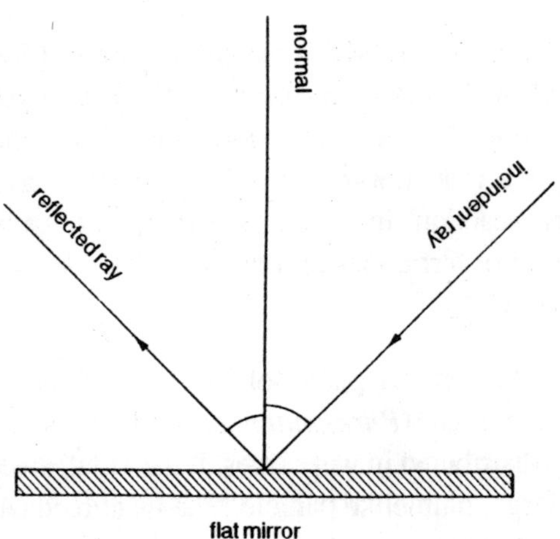

Fig. R-2. Reflection from a flat mirror.

Reflection. Return of a wave, such as light, from a surface it strikes into the medium through which it has traveled. The law of reflection states that the angle of reflection (the angle between the reflected ray and the normal, or line perpendicular, to the surface at the point of reflection) is equal to the angle of incidence (the angle between the incident ray and the normal).

Reflector. *See* reflecting telescope.

Reflex Angle. An angle lying between 180° and 360°.

Reflux. To boil a liquid in a vessel fitted with a condenser so that the vapour is continuously returned to the vessel.

Reforming. The process of converting petroleum hydrocarbons into hydrocarbons suitable for use in petrol by temperature, pressure, and use of catalysts.

Refracting Telescope (refractor). A telescope in which the image is formed by lenses rather than mirrors: examples are the Kepler telescope and the Galilean telescope. Refractors are used as terrestrial telescopes. They are less suitable for use in astronomy than reflecting telescopes because of their chromatic and spherical aberration and because of the difficulties involved in grinding large lenses.

Refraction. The change in direction of a ray of light or other electromagnetic radiation as it passes from one medium to another medium in which it has a different velocity. The two *laws of refraction* are:

(1) the incident ray, the refracted ray, and the normal at the point of incidence all lie in the same plane;

(2) the sine of the angle that the incident ray makes with the normal divided by the sine of the angle that the refracted ray makes with the normal is a constant. Thus $\sin i/\sin r = \mu$ (this is *Snell's law).*

Refraction. The deflection of a wave on passing obliquely from one transparent medium into a second medium in which its speed is different, as the passage of a light ray from air into glass. The index of refraction of a transparent medium is equal to the ratio of the speed of light in a vacuum to the speed of light in the medium. Snell's law states that the ratio of the sine of the angle *i* of incidence (angle between the incident ray and the normal, or line perpendicular to the boundary between the two mediums at the point of refraction) to the sine of the angle *r* of refraction (angle between the refracted ray and the normal) is equal to the ratio of the refracting medium's index of refraction ***n***, to the original medium's index of refraction *ni*.

Refractive Index. A measure of the extent to which a medium refracts light. When light passes from medium 1 to medium 2, the ratio sin *i/sin r* is a constant for the two media. It is called the ***relative refractive index*** and is given the symbol n_{12}. It is equal to the ratio of the velocities of light in the two media (*i.e.* c_1/c_2). It follows that $n_{12} = 1/n_{23}$and n_{13}.. The *absolute refractive index*, which has the symbol *n*, is the value for a medium when the first medium is free space. The absolute value for air is 1.000 29, and values of refractive index measured in air are thus close to absolute values. The refractive index depends on the wavelength of light used and values are commonly given for the yellow D lines in the sodium emission spectrum (λ = 589.3 nm).

Refractive index can be found by a variety of methods. An object immersed in a liquid or viewed through a rectangular block of solid has an apparent depth d_a related to its true depth d_t by $d_t\ d_a = n$. This is used in a simple method of measuring the refractive index of liquids or solids by using a travelling microscope to focus on the apparent position of an object. The refractive index of a solid in the form of a prism of known angle can be measured by using a spectrometer to determine its deviation. This method can also be applied to a liquid contained in a hollow prism. Another technique is that of determining the angle at which total internal reflection occurs. Small samples of solid can be investigated by immersing them in a liquid mixture and changing the composition of the mixture until the sample "disappears". The liquid and solid then have the same refractive index. Other techniques depend on interference measurements.

Refractometer. Any of various instruments or pieces of apparatus used to measure refractive index.

Refractor. *See* refracting telescope.

Refractory. Denoting a solid that has a high melting point and can withstand high temperatures.

Refrigerant. A substance that is used as the working fluid in a refrigerator. Refrigerants are usually volatile liquids or easily liquefied gases.

Refrigeration. Process for drawing heat from substances to lower their temperature, often for purposes of preservation. Mechanical refrigeration systems (first patented in 1834) are based on the principle that absorption of heat by a fluid (refrigerant) as it changes from a liquid to a gas lowers the temperature of the objects around it. In the compression system, employed in electric home refrigerators, a Compressor, controlled by a thermostat, exerts pressure on a refrigerant gas (usually Freon or Ammonia), forcing it to pass through a condenser, where it loses heat and liquefies. When the liquid is circulated through refrigeration coils, it vaporizes, drawing heat from the air surrounding the coils. The refrigerant gas then returns to the compressor, and the cycle is repeated. In the absorption system, widely used in commercial installations, ammonia is used to cool brine, which is then piped into the refrigerated space.

Regnault's Method. A method of measuring the density of gases by weighing a large bulb of known volume, first filled with the sample at a known temperature and pressure and then evacuated. [After Henri Victor Regnault (1810-78), French chemist.]

Relative Density (specific gravity). Abbreviation: r.d. The ratio of the density of a substance to the density of a reference substance under specified conditions. Relative densities of solids and liquids are measured with reference to water, the density of the solid or liquid being taken at a specified temperature and the density of the water being taken at 4°C, the temperature at which its density is a maximum. The temperature of the substance is usually understood to be 20° C. The relative density measured in this way is numerically equal to its density in grams per cubic centimetre. Any substance with a relative density less then 1 will float on water. The relative density of gases is often measured with reference to air, both gases being at standard temperature and pressure.

Relative Humidity. *See* humidity.

Relativistic. Involving the theory of relativity or effects predicted by this theory.

Relativistic Mass. The mass of a body as predicted by the theory of relativity. The relativistic mass of a particle moving at velocity v is $m_o(1 - v^2/c^2)^{-1/2}$, where m_o is the rest mass.

Relativistic Velocity. A velocity sufficiently close to that of light for effects predicted by the theory of relativity to be significant.

Relativity. A system of mechanics introduced by Einstein to replace Newtonian mechanics. The theory was developed in two parts.

The *special theory of relativity* deals with uniform motion. Up to the early part of the twentieth century the motion of bodies and the action of forces had been based on Newton's laws of motion, involving such ideas as the constancy of mass and the simple method of obtaining relative velocities by vector addition of the velocities of the observer and the object observed. In particular, it was assumed that light was propagated through a stationary medium, the ether, at a fixed velocity *c* relative to the ether. At the beginning of this century several results were obtained that appeared to contradict the principles of Newtonian mechanics. The best known is the Michelson-Morley experiment, which indicated that the velocity of light was the same when measured in the direction of the earth's rotation and perpendicular to this direction. The observer always appears to be at rest. Several attempts were made to explain this phenomenon and this experiment wa the factor that inspired Einstein's theory, published in 1905. He assumed that the velocity of light is a constant for all observers, no matter how they move, and that there is no standard of absolute rest. The constant velocity of light is an experimental fact derived from the Michelson-Morley experiment.

The assumption leads to several non-classical results for observations made on a system that is moving at a constant velocity relative to the observer. An object of length, l_o when at rest relative to an observer will appear to have a length $l_o(1 - v^2/c^2)^{1/2}$ when moving at velocity v. The apparent shortening is equivalent to a Lorentz-Fitzgerald contraction. A similar change occurs in the mass: an object having a mass m_o when at rest relative to an observer has a mass $m_o(1 - v^2/c^2)^{1/2}$, when moving at a velocity v. This increase in mass is only significant at high velocities: it is the effect producing the limiting velocity of particles in the cyclotron. The increase of mass with velocity led to the idea that mass and energy are inter-convertible: *i.e.* that mass can be "destroyed" with the production of an equivalent amount of energy and that mass can be produced with loss of energy. The two are related by Einstein's equation, $E = mc^2$. Another relativity effect is that of *time dilation*. If an observer A measures the passage of time on a clock and another observer B is moving at a velocity v relative to A, it appears to A that time as measured on B's clock is given by $t(1 - v^2/c^2)$. In other words, it appears to A that time is running more slowly for B. The effect is observed in the motion of some mesons in cosmic rays, which appear to have longer lifetimes because they are moving at relativistic velocities. One of the consequences of relativity theory is that there is no absolute value of time. Two observers moving relative to one another would assign different times to the occurrence of an event. In relativity theory, events are specified in a space-time continuum.

The *general theory of relativity* deals with accelerated relative motion and is particularly concerned with gravitational effects. An observer who is travelling in a circular path experiences an acceleration directed towards the centre of his path and he appears to be subjected to an outwards, centrifugal, force. This force is proportional to his mass in the same way that the weight of an object resulting from gravitational attraction is proportional to the object's mass. Indeed, if the observer were in a sealed vehicle and were not aware of his motion, he might ascribe the force to a gravitational attraction by a body outside the vehicle. In other words, the effects of gravity thought of as action at a distance due to a body's gravitational field, can alternatively be described by a particular motion in space. This idea is used in the general theory to describe gravitational attraction as a property of space-time. Thus, the apparent force experienced by a body moving close to a large mass is explained by assigning non-Euclidean geometrical properties to space-time, which is said to be "curved" by the presence of the mass. Tests of general relativity are usually based on astronomical measurements: an example is the Einstein shift.

Relativity. Physical theory, introduced by Albert Einstein, that discards the concept of absolute motion and instead treats only relative motion between two systems or frames of reference. Space and time are no longer viewed as separate, independent entities but rather as forming a four-dimensional constinuum called Space-Time. In 1905 Einstein enunciated the special relativity theory, in which the hypothesis that the laws of nature are the same in different moving systems also applies to the propagation of light, so that the measured speed of light is constant for all observes regardless of the motion of the observer or of the source of light. From these hypotheses Einstein reformulated the mathematical equations of physics. In most phenomena of ordinary experience the results from the special theory approximate those based on Newtonian dynamics, but the results deviate greatly for phenomena occurring at velocities approaching the speed of light. Among the assertions and consequences of the special theory are the propositions that the maximum velocity attainable in the universe is that of light; that mass increases with velocity; that mass and energy are equivalent; that objects appear to contract in the direction of motion; that the rate of a moving clock seems to decrease as its velocity increases; that events that appear simultaneous to an observer in one system may not appear simultaneous to an observer in another system. Einstein expanded the special theory of relativity into a general theory (completed in 1915) that is principally concerned with the large-scale effects of Gravitation. The general theory recognizes the equivalence of gravitational and inertial mass, and asserts that material bodies produce the curvature of the space-time continuum and that the path of a body is determined by this curvature. The theory predicts that a ray of light is deflected by a gravitational field; observations of starlight passing near the sun, first made by Arthur Eddington and colleagues during a 1919 eclipse of the sun, confirmed this. The theory also predicts a Red Shift of spectral lines of substances in a gravitational field, a result confirmed by observation of light from white dwarf stars. Finally, the theory also accounts for the entire observed perihelion motion of the planet Mercury, only part of which could be explained by Newtonian Celestial Mechanics.

Relay. An electrical device in which an electrical current in one circuit can open or close a separate circuit. In the simplest type the first circuit contains a coil which produces a magnetic field to open or close a mechanical switch in the other circuit. Other types have electronic switching using semiconductors or valves.

Relay. Electromechanical Switch in which the variation of current in one Electric Circuit controls the flow of electricity in another circuit. A relay consists of a movable contact connected to an Electromagnet by a spring. When the electromagnet is energized by the controlling current, it exerts a force on the contact that overcomes the pull of the spring and moves the contact so as to either complete or break a circuit. When the electromagnet is de-energized, the contact returns to its original position.

Remanence. The magnetization of a material remaining when the external field has been reduced to zero in a hysteresis cycle. Substances such as steel, with high remanences, make good permanent magnets. Soft iron and other materials with low remanence are used in electromagnets.

Reproduction. The ability of living systems to give rise to new systems similar to themselves. The term may refer to self-duplication of a single cell, of a group of cells and organs, or of a complete organism. Reproductive processes vary tremendously, but two fundamental types may be distinguished: asexual reproduction, in which a single organism separates into two or more parts, each genetically identical to the parent; and sexual reproduction, in which a pair of specialized reproductive (sex) cells fuse, creating an individual that combines two sets of genetic characteristics. Asexual reproduction is found in all plants and in some one-celled and invertebrate animals. In one-celled organisms it takes the form of Mitosis, the division of one individualy into two new, identical individual. Primitive filamentous organisms, such as Flatworms, reproduce by fragmentation, in which a piece of the parent breaks off and develops into a new individual. Many Protozoa and plants reproduce by means of spores. In budding, the means by which Yeasts and animals such as the Hydra reproduce, a small protuberance (bud) on the parent increases in size until a wall forms to separate the new individual. Sexual reproduction occurs in many one-celled organisms and in all multicellular plants and animals; it involves the Fertilization of one sex cell (gamete) by another, producing a new cell (zygote), which develops into a new organism. In higher plants and animals two clearly different kinds of sex cells (distinguished as Ovum and Sperm) fuse. Multicellular plants alternate reproducing sexually and asexually. Although asexual reproduction ensures that beneficial combinations of characteristics will be passed on unchanged, sexual reproduction permits the offspring to inherit endlessly varied combinations of characteristics (because of the fusion of two different parental nuclei), providing for new variations that may improve a species and further enhance its chances of survival.

Reproductive System. In animals, the organs concerned with production of offspring. In humans and other mammals the female reproductive system produces the female reproductive cells (eggs, or ova) and includes an organ (uterus) in which the fetus develops. The male reproductive system produces the male reproductive cells (sperm) and includes an organ (penis) that deposits the sperm within the female. In the female, the mature egg, or Ovum, passes from the ovary into the fallopian tube, where fertilization occurs if sperm are present. From the fallopian tube, the ovum passes into the uterus, or womb. If the egg has not been fertilized, the endometrium (lining of the uterus) degenerates and sloughs off, and Menstruation occurs. If the egg has been fertilized, it becomes embedded in the lining of the uterus about one week after fertilization. The lower end of the uterus, called the cervix, is connected to the vagina, a passage joining the uterus with the external genitals. The vagina receives sperm

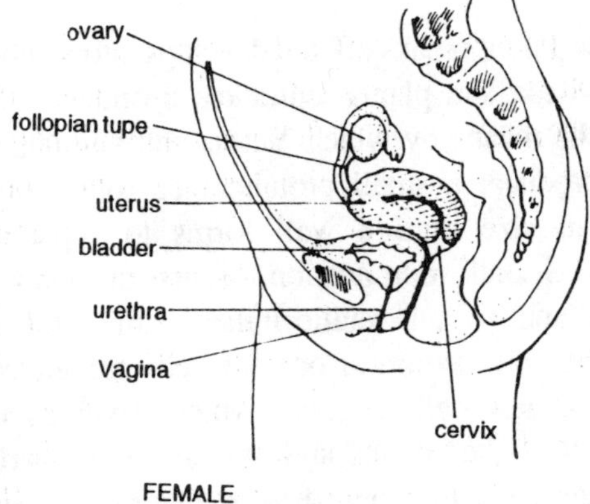

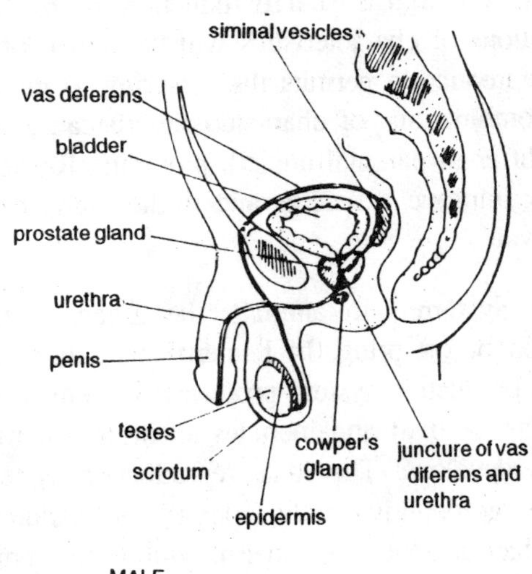

Fig. R-3. Reproductive system.

during sexual intercourse and is the passageway through which menstrual blood is eliminated and birth takes place. In the male reproductive system, Sperm are produced in the testes, two organs contained in the scrotum, an external sac in the groin. The testes also produce the hormone Testosterone and a portion of the seminal fluid, the liquid in which the sperm are carried. From the testes the sperm move into a passage (epididymis) and then into a long duct (vas deferens); fluids from the Prostate Gland and seminal vesicles also enter this duct. Just before ejaculation, contractions along the ducts mix the sperm with the seminal and prostatic fluids to form semen. During ejaculation semen is propelled into the urethra (the canal through the penis) and discharged.

Reptile. Dry-skinned, usually scaly, cold-blooded vertebrate of the class Reptilia. Reptiles are found in a variety of habitats in warm and temperate zones and range in size from 2 in. (5 cm) long lizards to 30 ft (9 m) long snakes. Typically they have low-slung bodies with long tails, supported by four short legs; snakes are limbless. They are mostly terrestrial, although a few are aquatic; all breathe air through lungs and have thick, waterproof skins. Unlike Amphibians, they do not possess Gills at any stage of their development. Nearly all lay porous, shelled eggs or bear their young on land. The living orders of reptiles are the Turtles; Alligators, caimans, Crocodiles, and gavials; Lizards and Snakes; and the Tuataras. Reptiles evolved from amphibians and were the dominant fauna in the Mesozoic era, often called the Age of Reptiles.

Resin. Any of a class of amorphous solids or semisolids. Natural resins occur as plant exudations (*e.g.*, of pines and first), and are also obtained from certain scale insects. They are typically yellow to brown in colour, tasteless, and translucent or transparent. Oleoresins contain Essential Oils and are often sticky or plastic; other resins are exceedingly hard, brittle, and resistant to most solvents. Resins are used in Varnish, Shellac, and lacquer and in medicine. Synthetic resins, *e.g.*, Bakelite, are widely used in making Plastics.

Resin. A solid or semisolid natural or synthetic polymeric substance.

Resistance. Symbol: *R*. A measure of the ability of a piece of material or an electrical component or circuit to resist the flow of current, equal to the potential difference across the circuit divided by the current it produces. In alternating-current circuits the resistance is the real part of the impedance. The quantity is measured in ohms.

Resistance. Property of an electric conductor by which it opposes flow of electricity and dissipates electrical energy away from the Electric Circuit, usually as heat. Resistance is basically the same for alternating and direct-current circuits. A high-frequency alternating current, however, tends to travel near the surface of a conductor. Because such a current uses less of the available cross section of the conductor, it meets with more resistance than direct current. The unit of resistance is the Ohm.

Resistance Thermometer. A type of thermometer that depends for its action on changes in resistance with temperature. Usually a coil of fine platinum wire is used incorporated in a Wheatstone-bridge circuit. The resistance of the coil increases with temperature.

Resistivity. Symbol: p. A measure of the ability of a material to resist the flow of an electric current, equal to the resistance per unit length of the material taken for unit cross-sectional area. Resistivity is given by the equation: $p = RA/l$, for a resistor of length l and constant cross-sectional area A. Its units are ohm-metres. Resistivity is sometimes incorrectly called *specific resistance.*

Resistor. A component used in an electrical circuit to introduce resistance. Resistors are made of a coil of resistance wire or of carbon.

Resistor. Two-terminal Electric Circuit component that generates heat by offering opposition to an electric current. The most common forms of resistors are made from fine wires of special alloys wound onto cylindrical forms or from a molded composition material containing carbon and other substances in varying counts. Resistors are rated for the maximum amount of power that they can safely handle.

Resolution. 1. The separation of a vector into components.

2. *See* resolving power.

Resolving Power (resolution). The ability of a microscope, telescope, or other optical instrument to produce distinguishable images of closely spaced objects. The resolving power depends on the wavelength of the light and on the amount of light collected by the instrument. The resolving power of a spectrometer is its ability to distinguish between closely spaced wavelengths, energies, etc.

Resonance. 1. (in mechanics). The condition occurring when a vibrating system is subjected to a periodic force that has the same frequency as the natural vibrational frequency of the system. At resonance, the amplitude of vibration is a maximum.

2. (in chemistry). The phenomenon in which the molecular structure of a chemical compound is not described by any single simple formula but can be represented by a combination of two or more structures. For instance, the polar molecule HCl has a character between that of a covalent molecule H-Cl and an ionic molecule H^+Cl^- and can be described as a *resonance hybrid* of the two. Benzene, in which the bonds have a character between single bonds and double bonds, can be written as a resonance hybrid of two Kekule structures. In chemistry, the separate resonance structures, which do not have any independent existence, are often written with a double-headed arrow (↔) between them.

3. (in electrical circuits). The condition occurring in an electrical circuit when its impedance is a minimum and depends only on the resistance of the circuit, not on its reactance. Resonance occurs at the frequency f for which $2\Pi fl = {}^{1}/_{2}\Pi_{f}C$, where L is the inductance and C the capacitance. At this frequency the reactance is zero and the circuit is said to be *tuned*.

4. (in particle physics). Any of a large number of elementary particles with very short lifetimes. The resonances are excited states of more stable particles.

5. (in nuclear physics and spectroscopy). The absorption of a photon with the correct energy to excite a nucleus, atom, etc., from one energy level to a higher level.

Resorcinol (1, 3-dihydroxybenzene). A white crystalline solid phenol, $C_6H_4(OH)_2$, used in the manufacture of dyes and celluloid. M.pt. 111°C; b.pt. 281°C; r.d. 1.3.

Respiration. Process by which an organism exchanges gases with its environment. The term commonly refers to the overall process by which oxygen is taken from the air and transported to the cells for the oxidation of organic molecules, while the products of Oxidation, carbon dioxide and water, are returned to the environment. In single-celled organisms, gas exchange occurs directly. The cells lose their high concentration of carbon dioxide to the environment by simple diffusion, while the environment provides its higher concentration of oxygen to the cells, also by diffusion. In complex animals, where internal cells are distant from the external environment, respiratory systems facilitate the passage of gases to and from internal tissues. In lower animals such as Flatworms, exchange occurs through a moist surface membrane. In Fish, blood vessels in the gills are exposed for direct exchange with the external (aquatic) environment. In plants, gas exchange occurs in the stomates, respiratory organs found mostly in leaves. In human beings and other vertebrates, gas exchange takes place in the Lungs. In breathing—the mechanical procedure for getting air to and from the lungs—muscles enlarge the chest cavity to force air in and reduce it to expel air. Actual gas exchange in the lungs occurs in cup-shaped air sacs called alveoli. Oraganisms that utilize respiration to obtain energy are aerobic, or oxygen-dependent. Organisms able to live in the absence of oxygen are called anaerobic; they obtain energy from fuel molecules solely by Fermentation or Glycolysis. In biochemistry, *respiration* refers to the series of biochemical oxidations in which organic molecules—such as carbohydrates, amino acids, and fatty acids—are converted to carbon dioxide and water. The chemical energy thus obtained is trapped and stored for later use by the cells in Adenosine-Triphosphate (ATP).

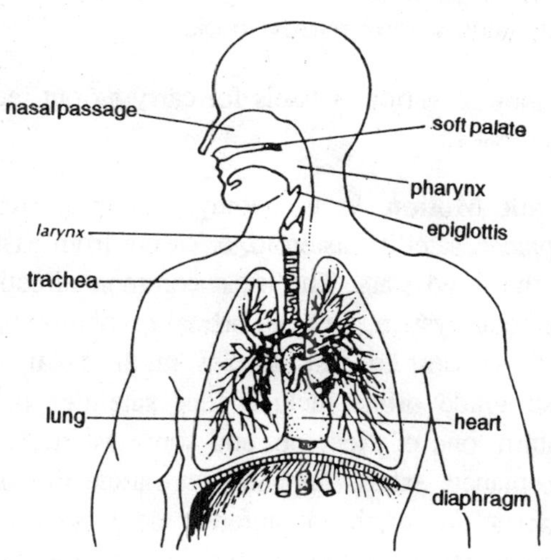

Fig. R-4. Respiratory system.

Restitution. Elasticity. The *coefficient of restitution* (symbol: *e*) for the impact of two spheres is the ratio of the relative velocity of the bodies before impact to that after impact, the velocities being taken along the line between the centres of the spheres at impact.

Rest Mass. The mass of a body when it is at rest relative to its observer, as distinguished from its relativistic mass.

Resultant. A vector produced by the combination of two or more other vectors.

Resurrection. Arising again from death to life. The resurrection of Jesus is the cornerstone of Christianity. It guaranteed his mission and promised the resurrection of all people. The

Christian belief is that on Judgment Day people's souls will be reunited with their risen (but glorified and immortal) bodies.

Retina. *See* eye.

Retort. 1. A piece of laboratory apparatus consisting of a glass bulb with a long narrow neck.

2. Any of various vessels for carrying out industrial chemical processes.

Retrograde Motion. In astronomy, real or apparent movement of a planet, satellite, asteroid, or comet from east to west relative to the fixed stars. The most common direction of motion in the solar system, for both orbital revolution and axial rotation, is from west to east. Bodies in the solar system with real restrograde orbits include four satellites of Jupiter, one of Saturn, one of Neptune, and some asteroids and comets. All the planets exhibit apparent retrograde motion when they are nearest the earth (at inferior conjunction for the inferior planets and at opposition for the superior planets) because of the relative speeds of the planets in their orbits about the sun.

Reverbatory Furnace. A type of furnace in which the fuel is burnt in one section and the heat is directed down onto the contents by a curved roof made of refractory material. Reverbatory furnaces are used for smelting or refining metals in such a way that the fuel is not mixed with the material used.

Reversible. 1. Denoting a chemical reaction that can proceed in either direction. An example is the hydrolysis of esters: $CH_3COOC_2H_5 + H_2O = CH_3COOH + C_2H_5OH$. The reverse reaction—esterification of an alcohol by an acid—also occurs. In general, such reactions reach a state of chemical equilibrium which can be displaced in one direction or the other by changing the conditions or by continuously removing one compound from the system as it is formed. Reactions that proceed completely in one direction are said to be *irreversible.*

2. Denoting a change in the conditions of a system in which the system is always at thermodynamic equilibrium during the change. Reversible processes are theoretical ideals: they would have to be performed infinitely slowly and are never realized in practice. All real change are *irreversible.*

Reynolds number. Symbol: *Re.* A number used in problems involving the flow of fluids, equal to vpl/η, where v is the fluid's velocity, p is the density, η the viscosity, and l is a chosen dimension of the system through which the fluid is moving. For example, l may be the radius of a pipe. At a certain value of the Reynolds number a change occurs from laminar flow to turbulent flow. [After Osborne Reynolds (1842-1812). British physicist and engineer.]

Rhea. South American flightless Bird (order Rheiformes), superficially resembling the Ostrich. Weighing from 44 to 55 lb (20 to 25 kg) and standing up to 60 in. (152 cm) tall, rheas lack the ostrich's plumelike tail feathers. A herbivore, the rhea inhabits the Pampas and Savannas, often feeding with cattle.

Rhenium. Symbol: Re. A silvery-white lustrous transition element obtained as a by-product in refining molybdenum. Alloys of rhenium with molybdenum are superconducting at very low temperatures. The element dissolves in concentrated nitric and sulphuric acids but not in hydrochloric acid. It reacts with oxygen at high temperatures. A.N. 75; A.W. 186.2; m.pt. 3180° C; b.pt. 5620° C; r.d. 21.0; valency 1-7.

Rhenium (Re). Metallic element, discovered in 1925 by Walter Nodack and colleagues. It is a very dense, high-melting, silver-white metal occurring in platinum and molybdenum ores, and in many minerals. It gives improved ductility and high-temperature strength to its alloys, which are used in electrical contacts, electronic filaments, thermocouples, and photographic flash lamps.

Rheumatic Fever. Serious inflammatory disease occurring as a complication of infection by the streptococcal bacterium. Appearing chiefly in children, the disease in characterized by fever, Arthritis, Chorea, and inflammation of the Heart that can produce permanent damage (rheumatic Heart Disease). The course of the disease and its effects on the heart vary greatly. Antibiotics (*e.g.,* Penicillin), Aspirin, Cortisone, and rest are the usual treatments.

Rheology. The study of the flow properties of fluids.

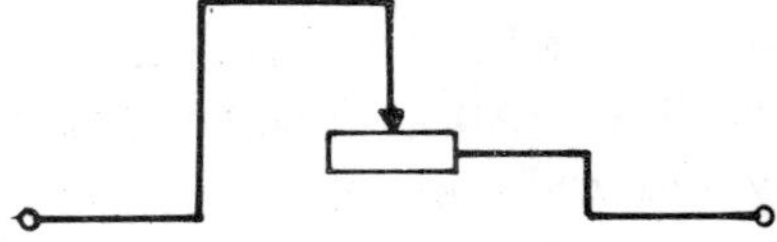

Fig. R-5. Rheostat.

Rheostat. A variable resistor, typically consisting of a coil of bare resistance wire with one end connected in the circuit, the other connection being made by a sliding contact at points along the coil.

Rh Factor. Protein substance present on the surface of the red blood cells of most (85% or more) people and capable of inducing an intense antigen-antibody reaction. When Rh-positive blood is given to an Rh-negative person or when fetal Rh-positive blood (inherited from the father) is mixed with maternal Rh-negative blood during pregnancy, the Rh-negative person develops antibodies to the foreign Rh factor. A serious or even fatal reaction may occur in subsequent mixing of two blood types, as in repeat transfusions or other pregnancies involving an Rh-positive fetus. In the latter case, the possibility of a fetal or newborn reaction (called erythroblastosis fetalis) can be prevented by the administration to the Rh-negative mother of a special immune globulin that suppresses antibody formation. In other cases, when the immune globulin is not given, the reaction must be treated by total blood exchange shortly before or after birth.

Rhinoceros. Massive, hoofed Mammal of Africa, India, and Se Asia, characterized by one or two horns on the snout made of congealed hair. A thick-skinned vegetarian, it has poor vision but excellent senses of smell and hearing. Solitary and unpredictable, the rhinoceros feeds at night and rests in the shade in the daytime. It has been hunted to near extinction for its horns, sold powdered as an aphrodisiac.

Rhizome or **Rootstock**. Fleshy, creeping, underground Stem by means of which some plants propagate themselves. Buds that form at the joints produce new shoots. If a rhizome is split, it does not die (unlike a Root) but becomes several plants instead of one. Common rhizome-producing plants are ginger and iris.

Rhodium. Symbol: Rh. A silvery-white soft ductile transition element occurring associated with platinum. It is used as a catalyst and constituent of some platinum alloys. The element is not attacked by acids and only slowly dissolves in aqua regia. It reacts with oxygen, chlorine, and other nonmetals at red heat. The main rhodium compounds are complexes of rhodium III and, to a lesser extent, rhodium IV. A.N. 45; A.W. 102.9; m.pt. 1966°C; b.pt. 3727°C; r.d. 12.4; valency 3 or 4.

Rhodium (Rh). metallic element, discovered in 1804 by William Wollaston. A silver-white, lustrous, brilliant, tarnish-resistant, and chemically-resistant metal, rhodium is used to plate jewellry and searchlight reflectors. Its major use in platinum and iridium alloys, to which it gives improved strength and the ability to withstand higher temperatures.

Rhodochrosite. A red or pink naturally occurring carbonate, $MnCO_3$, used as a source of manganese.

Rhododendron. Shrub (genus *Rhododendron*) of the Health family, found chiefly in mountainous areas of the arctic and north temperate zones. They typically have large, shiny, leathery evergreen leaves and clusters of large pink, white, or purplish flowers. North American species include the great laurel, or rose bay (*R. maximum*), West Virginia's state flower; and the Western rhododendron (*R. californicum*), Washington's state flower. Azaleas are in the same genus.

Rhombohedral. *See* crystal system.

Rhombus. A parallelogram whose sides are all equal. A square is a rhombus that is also a rectangle.

Rhythm. Basic element of music concerned with the duration of tones and the stresses or accents placed upon them. The formulation in the 12th cent. of basic rhythmic patterns (modes) led to the development of meter, the division of a composition into units of equal time value. The rhythmic characteristics are a major factor in the analysis of the style of a composer or period.

Rice. Cereal Grain (*Oryza sativa*) of the Grass family, probably native to the Ganges, Tigris, Yangtze, and Euphrates river deltas. Requiring warmth and abundant moisture, the plant is an annual, from 2 to 6 ft (61 to 183 cm) tall, with a round, jointed stem, long, pointed leaves, and seeds borne in a dense head on separate stalks. Thousands of rice strains are now known, both cultivated and wild. It is estimated that half the world's population subsists wholly or partially on rice. About 90% of the world's crop is grown in India, China, and Japan, and most of it is consumed domestically. Brown rice has greater food value than white, since the outer coatings, which are polished away to yield white rice, contain protein and minerals, while the white endosperm is chiefly carbohydrate. As a food, rice is low in fat and (compared to other grains) protein. The beverage Sake is brewed from rice.

Rickets. Bone disease caused by a deficiency of vitamin D, often resulting in knock-knee, bowlegs, and deformities of the chest and pelvis.

Rigidity Modulus. *See* modulus.

Ring. A closed circle of atoms in the structure of a molecule.

Ring Closure. *See* cyclization.

Ringworm or **Tinea.** Superficial eruption of the skin, caused by a fungus. Although any area of the skin may be affected, the most common site is the feet (the condition called athlete's foot). Ringworm infection causes dry, scaly patches or blisterlike elevations, which usually burn or itch. It is treated with local antifungal ointements.

River. Stream of water larger than a brook or creek. Runoff after precipitation flows downward by the shortest and steepest course. Runoffs of sufficient volume and velocity join to form a stream that, by the Erosion of underlying earth and rock, deepens its bed. It becomes perennial when it cuts deeply enough to be fed by groundwater or when it has an unlimited source (*e.g.,* the Saint Lawrence flowing from the Great Lakes). Sea level is the ultimate base level for a river, but the floor of a lake or basin into which a river flows may become a local and temporary base level. Rivers modify topography by both erosion and deposition. Young streams have steep-sided valleys, steep gradients, and irregularities in the bed. Mature rivers have valleys with wide floors and a more smoothly graded bed. Old rivers have courses graded to base level and run through broad, flat areas.

RMS. Root mean Square.

RNA or **Ribonucleic Acid.** Nucleic Acid, found mostly in the cytoplasm of Cells, that is important in the synthesis of proteins. The amount of RNA varies from cell to cell. RNA, like the structurally similar DNA, is a chain made up of subunits called nucleotides. In protein synthesis messenger RNA replicates the DNA code for a protein and moves to sites in the cell called ribosomes. There transfer RNA assembles Amino Acids to form the protein specified by the messenger RNA. Most forms of RNA (including messenger and transfer RNA) consist of a single nucleotide strand, but a few forms of viral RNA that function as carriers of genetic information (instead of DNA are double-stranded.

Robotics. The design and construction of machines (robots) that sense (by means of vision or some other type of sensor) aspects of their environment and that are capable physically (*e.g.,* by means of a mechanical arm) of acting upon it. The Czech dramatist Karel Capek coined the word *robot* (from the Czech *robota*, drudgery) in his 1921 satirical play R.U.R., and robots, or automatons, have long played a role in fantastic literature. Present-day robots, which bear little resemblance to the often humanoid figures of Science Fiction, are essentially computer-controlled machine tools that can be programmed to perform any of a number of functions, such as welding an automobile chassis or assembling machine parts. Robots can perform dangerous, uncomfortable, tiring, or monotonous tasks, and do them with greater speed and accuracy than can human beings. Robots play an increasingly significant role in the movement toward industrial automation, especially in Japan, the world's leader in the production and use of robots.

Rochelle Salt. *See* potassium sodium tartrate

Rock. Aggregate of solid matter composed of one or more of Minerals forming the earth's crust. Rocks are commonly divided into three major classes depending on their origin. *Igneous rocks* (*e.g.,* Basalt, Granite, Obsidian, Porphyry, Pumice) result from the cooling and solidification of molten matter from the earth's interior. If formed below the surface, such rock is said to be intrusive (as in a Batholith). If formed at the surface, it is extrusive. *Sedimentary rocks* (*e.g.,* Chalk,

Clay, Coal, Limestone, Sand, Sandstone., Shale) originate from the consolidation of sediments deposited chiefly through the action of Erosion on older rocks of all kinds. The characteristic feature of sedimentary rocks is their Stratification. *Metamorphic rocks* (*e.g.*, Gneiss, Marble, Quartzite, Schist, Slate) originate from the alteration of the texture and mineral constituents of existing rocks of any type under extreme heat and pressure within the earth.

Rock Crystal. A transparent form of quartz, SiO_2.

Rocket. Any vehicle propelled by ejection of the gases produced by combustion of self-contained propellants. Tremendous pressure is exerted on the walls of the combustion chamber, except where the gas exits at the rear; the resulting unbalanced force, or thrust, on the front interior wall of the chamber pushes the rocket forward. The most vital component of any rocket is the propellant, which accounts for 90 to 95% of the rocket's total weight. A propellant consists of two elements, a fuel and an oxidant; engines that are based on the action-reaction principle and that use air instead of carrying their own oxidant are properly called jets. Liquefied gases, *e.g.*, hydrogen as fuel and oxygen as oxidant, are more powerful propellants, whereas solid explosives, *e.g.*, Nitroglycerin as oxidant and nitrocellulose as fuel, are more reliable. The chemical energy of the propellants is released in the form of heat in the combustion chamber. The rocket's exit nozzle usually converges to a narrow throat, then diverges to obtain maximum energy from the exhaust gases moving through it. No currently practical single-stage rocket can reach orbital velocity (5 mi/sec, or 8 km/sec) or th earth's Escape Velocity (7 mi/sec, or 11 km/sec). Hence Space Exploration requires multistage rockets; two or more rockets are assembled in tandem and ignited in turn; as their fuel is used up, each of the lower stages detaches and falls back to earth. When extremely large thrust is required, several rockets may be clustered and operated simultaneously. Rocket navigation is usually based on inertial guidance; internal Gyroscopes are used to detect changes in the position and direction of the rocket.

History. The invention of the rocket is generally ascribed to the Chinese, who as early as A.D. 1000 stuffed Gunpowder into sections of bamboo tubing to make effective weapons. The astronautical use of rockets was cogently argued in the early 20th cent. by the Russian Konstantin E. Tsiolkovsky; the American Robert H. Goddard, who launched the first liquid-fuel rocket in 1926; and the German Hermann Oberth. During World War II, a German team under Wernher Von Braun developed the V-2 rocket, the first long-range guided missile. After the war, rocket research in the U.S. and the USSR intensified, leading to the development of the modern array of intercontinental ballistic Missiles and spacecraft-launching rockets. Important U.S. expendable rockets have included the Agena, Atlas, Delta, Saturn, and Titan carriers. Saturn V, the largest rocket ever assembled, during the Apollo program delivered a payload of 44 tons to the moon. The U.S. Space Shuttle, the first reusable space vehicle, achieves orbit through a combination of several orbiter liquid-propellant engines and two expendable solid-rocket boosters.

Rock Salt. A transparent naturally occurring crystalline form of sodium chloride, NaCl.

Rodent. Member of the largest mammalian order, Rodentia, characterized by front teeth adapted for gnawing and cheek teeth adapted for chewing. The approximately 1,8000 species of Rodentia are worldwide in distribution and are divided into three suborders. The Sciuromorpha, or squirrellike rodents, include Squirrels, Chipmunks, and Beavers. The Myomorpha, or mouselike rodents, include a variety of Mouse and Rat species, as well as species of Hamster, Lemming, and Gerbil. The Hystricomorpha, or porcupinelike rodents, include the porcupine, Guinea Pig, and Chinchilla.

Rontgen. Symbol: R. A unit of exposure dose equal to the dose that produces ions of one sign carrying a charge of 2.58 x 10^{-4} coulomb in air. [After Wilhelm Konrad Rontgen (1845-1923), German physicist.]

Rontgen Rays. *See* X-rays.

Root. 1. One of two or more identical factors of a given number. The *square root* of a number is a number that, multiplied by itself, gives the specified number. Thus the square root of 16, written 16 or $(16)^{1/2}$, is either +4 or -4. since 4 x 4 = 16 and -4 x -4 = 16. The *cube root* is the number that gives the specified number when multiplied by itself twice. In general, the nth root of x is the number that multiplied by itself $n-1$ times gives x.

2. A value that satisfies a given equation. For example, the equation $x^2 + x - 2 = 0$ is satisfied by $x = 1$ and $x = -2$: these are its roots. In general, a quadratic equation has two roots, a cubic equation three roots, etc.

3. (In botany). The descending axis of a plant (as contrasted with the Stem), usually growing underground but also growing in air and water. Roots serve to absorb and conduct water and dissolved minerals; to anchor the plant; and, often, to store food. Some plants have a main root (tap root) that is larger than the branching roots; others have many slender root branches. Roots grow primarily in length, with growth occurring at the tip, which is protected by a cap of cells (root cap) that break off and are replaced as the root probes through the soil. Root hairs, tiny cellular projections from the surface of the growing portion, absorb water and minerals from the soil. Root systems help prevent soil Erosion.

4. (In mathematics). The number x for which an equation $f(x) = 0$ holds true, where f is some Function. For example, $x = 3$ and $x = -4$ are the roots of the equation $x^2 + x - 12 = 0$. In the special case where $f(x) = x^n - a = 0$ for some number a, x is called the nth root of a , denoted by $\sqrt[n]{a}$ α $a^{1/n}$. For example, 2 is the third, or cube, root of 8 ($\sqrt[3]{8} = 2$), because it satisfies the equation $x^3 - 8 = 0$.

Rose. Common name for some members of the family Rosaceae, herbs, shrubs, and trees distributed over most of the earth. Roses are often thorny, and many typically have a fleshy fruit, such as a rose hip or an apple. The largest genera are *Rubus*, including the blackberry, raspberry, loganberry, and other Brambles; *Rosa,* the true roses; and *Prunus,* including the Almond, Apricot, Cherry, Nectarine, Peach, and Plum. Other members of the family, also of economic importance, are the Apple, Pear, Quince, and Strawberry. The true roses, the most popular ornamentals of the family, are esteemed as cultivated plants and as cut flowers for their often fragrant, showy blossoms. Attar of rose, a Perfume oil, is obtained from the damask rose, and the rose hips of *R. rugosa* are a source of Vitamin C. Favorite flowers in many lands since prehistoric times, roses have been used medicinally and eaten in preserves and salads. New York, North Dakota, lowa, and the District of Columbia have adopted different species of rose as their emblems.

Rosemary. Evergreen, shrubby perennial (*Rosmarinus officinalis*) of the Mint family, native to the Mediterranean region. It has small blue flowers and aromatic leaves; the later are used for seasoning. An extract of the flowers is used in perfumes and medicines.

Rose's Metal. An alloy of bismuth (about 50%), lead (about 25%), and tin. It has a low melting point (94°C) and is used for bending metal pipes, in constant temperature baths, and similar applications.

Rotary Converter. A device for converting an altenating current into a direct current, consisting essentially of an a.c. motor driving a d.c. generator.

Rotifer. Microscopic aquatic or semiterrestrial Roundworm. As a rule, only female rotifers are seen; in some species the male has never been observed. Eggs usually develop parthenogenetically, *i.e.*, without fertilization, to produce only females.

Rotor. The moving part of a generator or electric motor. It can be either the field winding or the armature.

Roundworm. Wormlike invertebrate of the phylum Aschelminthes, having a noncellular coat, or cuticle, and a fluid-filled cavity (pseudocoelom) separating the body wall from the gut. Roundworms are widely distributed in aquatic and terrestrial habitats. The phylum includes Rotifers, Nematodes, and horsehair worms.

Rubber. 1. A natural polymeric material obtained from the sap of the tree *Hevea braziliensis*. The sap (*latex*) is first coagulated by acetic acid and either dried in air (*crepe rubber*) or in wood smoke (*smoke rubber*). This material is moulded and vulcanized with addition of a filler, such as carbon black. Natural rubber is a polymeric material made up of isoprene units.

2. Any of various synthetic elastomers made by polymerizing or copolymerizing styrene, butadiene, neoprene, and similar compounds.

Rubber. Any solid substance usually elastic, that can be vulcanized to improve its elasticity and add strength; the term includes natural rubber, or caoutchouc, and a wide variety of synthetic rubbers, which have similar properties. Rubbers of composed chiefly of Carbon and Hydrogen, but some synthetics also have other elements, *e.g.*, chlorine, fluorine, nitrogen, or silicon. All are compounds of high molecular weight; each consists of a series of one kind of molecule (*e.g.*, isoprene in natural rubber) hooked together in a long chain to form a very flexible, larger molecule, the Polymer. Natural rubber is obtained as latex, a milky suspension of rubber globules found in a large variety of plants, chiefly tropical and subtropical. An important source is the Para Rubber tree. Latex can be shipped for processing either as a liquid or coagulated by acid and rolled into sheets. For most purposes rubber is ground, dissolved in a solvent, and compounded with other ingredients, *e.g.*, fillers, Pigments, and plasticizers. Known by pre-columbian Indians of South and Central America, rubber first attracted interest in Europe in the 18th cent. Vulcanization, a process invented (1839) by Charles Goodyear, revolutionized the rubber industry. It usually involves heating raw or compounded rubber with Sulfur, causing sulfur bridges to form between molecules. The product is nonsticky, elastic, and resistant to heat and cold. Natural rubber is used chiefly to make tires and inner tubes because it is cheaper than synthetic rubber and has greater resistance to tearing when hot. Natural rubber can be treated to make foam rubber and sponge rubber. The first synthetic rubber was made in Germany in World War I. Today synthetics, *e.g.*, Buna S, neoprene butyl, and nitrile, account for most of the world's rubber production. Made from Coal, Petroleum, Natural Gas, and Acetylene, Synthetic rubbers are resilient over a wider temperature range than natural rubber and are more resistant to aging, weathering, and attack by certain substance, notably, oil, solvents, oxygen, and ozone. Silicone rubbers are used in insulation. Polyurethanes are used in tires, in shoes, and as foams. Neoprene is used for making hose and tank linings. Butyl rubber is used in inner tubes and as insulation.

Rubber Plant. Name for several plants, *e.g.*, Para Rubber Tree, Castila tree, Ceara tree, and guayule, all of which are sources of the milky fluid, called latex used to make Rubber. The India-rubber tree (*Ficus elastica*), a common house-plant, is an Asiatic Fig.

Rubella or **German Measles.** Acute infectious viral disease of children and young adults, causing a rash and fever. It is mild and uncomplicated unless contracted during the first three months of pregnancy, when it can cause serious damage to the fetus. Vaccination against rubella is given to children and advised for young women who have not had the disease.

Rubidium (Rb). Metallic element, discovered spectroscopically by Robert Bunsen and Gustav Kirchhoff in 1861. A very soft, silver-white Alkali Metal, it is extremely reactive, and must be kept out of contact with air and water. It has few commercial uses.

Rubidium. Symbol: Rb. A soft silvery-white metallic element occurring in some minerals and brines. It is a made by electrolysis of the fused chloride. Rubidium is an alkali metal. A.N. 37; A.W. 85.47; m.pt. 38.4° C; b.pt. 688° C; r.d. 1.53; valency 1.

Rubidium-strontium Dating. A method of dating geological samples by using the decay of rubidium-87, a radioisotope occurring in some rocks. It decays with a half-life of 5×10^{11} years to give strontium-87. Measurement of the ratio of the two isotopes present gives an estimate of the sample's age. The method is useful for ages of over 10^9 years.

Ruby. A naturally occurring red form of aluminium oxide, Al_2O_3, containing small amounts of chromium, used in lasers.

Ruby. Gem, transparent red variety of Corundum, found chiefly in Burma, Thailand, and Sri Lanka. Star rubies (showing an internal star when cut with a rounded top) are rare. Synthetic rubies are produced by fusing pure aluminium oxide.

Rue. Common name for various members of the family Rutaceae, mostly woody shrubs or small trees, often evergreen and spiny, of temperate and tropical regions. They are typified by glands that produce an essential oil widely utilized for flavorings, perfume oils, and drugs; the foliage, fruits, and flowers are noticeably fragrant. Chief in importance are the Citrus Fruits. Several species of the family yield lumber used in cabinetmaking, *e.g.*, the Orange and the species called satinwood. More specifically, the name *rue* refers to the shrubby herbs of the genus *Ruta*, found from the Mediterranean to E Siberia. The common rue (*R. graveolans*) has greenish-yellow flowers, blue-green leaves, and a strong odor. Its leaves are sometimes used in flavorings, beverages, and cosmetics.

Rum. Fermented, distilled, and aged spirituous liquor made from the molasses and foam that rise to the top of boiled sugarcane juice. Naturally colourless, it becomes brown when caramel is added and the liquor is stored in casks. The best-known producer is Jamaica, but rum is also made in such places as Cuba, Brazil, Trinidad, Puerto Rico, and the U.S.

Rumford, Benjamin Thompson, Count. 1753-1814, American-British scientist and administrator; b. Woburn, Mass. After leaving America (1776), he served as the British under-secretary of the colonies (1780-81). In contrast to the prevalent belief that Heat was substance, he stated that it was produced by the motion of particles. Rumford made valuable experiments with gunpowder, introduced improved methods of heating and cooking, and founded the Royal Institution in England.

Ruminant. Any of group of even-toed mammals that chew their cud, *i.e.,* regurgitate and rechew already swallowed food. Ruminants have three or four chambered stomachs. In the first chamber (the rumen), food is mixed with fluid to form a soft mass (the cud or bolus), which, after regurgitation and rechewing, passes through the rumen into the other stomach chambers for digestion. Goats, sheep, cows, camels, and antelope are ruminants.

Rush, Benjamin. 1745?-1813, American physician; b. Byberry (now in Philadelphia), Pa. He was the first professor of chemistry in the colonies, at the College of Philadelphia. He signed the Declaration of Independence and was a member (1776-77) of the Continental Congress. He established (1876) in Philadelphia the first free dispensary in the U.S. and became (1792) professor of medicine at the Univ. of Pennsylvania. His bleeding and purging of patients aroused controversy. A pioneer in psychiatry and a prolific writer, he also founded the first American antislavery society and served (1797-1813) as treasurer of the U.S. mint in Philadelphia.

Rush. Any plant of the family Juncaceae; more loosely, a tall, grasslike, often hollow-stemmed plant of various families. The common, or bog, rush (*Juncus effusus*) is widely distributed in swamps and moist places of the Northern Hemisphere; the slender rush (*J. tenuis*) is found in drier surroundings. Some wood rushes (genus *Luzula*), which grow on dry ground, are relished by livestock. Rushes are used for basketwork, mats, and chair seats, and as wicks. The scouring rush is a Horsetail.

Rust. A hydrated form of iron III oxide, $Fe_2O_3H_2O$, formed by the action of moist air on iron or steel.

Ruthenium. Symbol: Ru. A hard white transition element found associated with platinum. It is used as a catalyst and a constituent of some platinum alloys. The metal is not attacked by acids, dissolves in fused alkalis, and combines with oxygen and halogens at high temperatures. It form large numbers of complexes in several oxidation states. A.N. 44; A.W. 101.07; m.pt. 2310°C; b.pt. 3900°C; r.d. 12.41; valency 1-8.

Ruthenium (Ru). Metallic element, discovered in 1827 by G.W. Osann in crude platinum ore. It is a hard, lustrous, silver-gray metal that is usually alloyed with other metals to provide hardness and corrosion resistance. Its compounds are used to color ceramics and glass.

Rutherford, Ernest Rutherford, Ist **Baron**. 1871-1937, English physicist; b. New Zealand. He taught at McGill Univ., Montreal (1898-1907) and the Univ. of Manchester (1907-19) and was director from 1919 of the Cavendish Laboratory, Cambridge. Rutherford discovered and named alpha and beta radiation and helped propose a theory of radioactive transformation of atoms for which he received the 1908 Nobel Prize in chemistry. On the basis of experiments carried out under his direction, he concluded (1911) that the Atom is a small, heavy nucleus surrounded by orbital electrons. Rutherford was the first to split atomic nuclei artificially.

Rutile. Mineral, one of three forms of titanium dioxide (TiO_2). It occurs in crystals, often in twins or rosettes, and is typically brownish red, although there are black varieties. Rutile is found in igneous and metamorphic rocks, chiefly in Switzerland, Norway, Brazil, and parts of the U.S.

Rutile. A reddish-brown to black mineral consisting of titanium dioxide, TiO_2, used as an ore of titanium.

Rydberg Constant. Symbol: R. The constant with a value $me^4/8h^3\varepsilon^2_0$ appearing in equations for the wavelengths of lines in the Rydberg series and similar spectral series (*m* and *e* are the mass and charge of the electron, *h* is the Planck constant, and ε_0 is the electric constant). For hydrogen, is value its $1.096\,77 \times 10^7\ m^{-1}$. {After Johannes Robert Rydberg (1854-1919), Swedish Physicist.]

Rydberg Series. A series of lines appearing in the absorption spectrum of a gas a visible and ultraviolet radiation. The lines get progressively closer together as the wavelength decreases until they merge into a continuum. The series is produced by excitation from the ground state to various excited states and the onset of the continuum corresponds to complete removal of the electron to form an ion. *Rydberg spectra* can thus be used to measure ionization potentials.

Rye. Cereal plant (*Secale cereale*) of the Grass family, important chiefly in central and N Europe. The Grain, or seed, is used for pumpernickel and the lighter-coloured rye bread (made from a mixture of rye and wheat flour), as a stock fee, and in the distillation of Whiskey and Gin. The plant, which grows well in areas where the soil is too poor and the climate too cool for wheat, is also grown for hay and pasturage and as green manure and a Cover Crop. The USSR leads in world production. Ergot, a fungous infection, is poisonous and can make rye unsafe for use.

S

Saccharide. A carbohydrate: a monosaccharide, disaccharide, polysaccharide, etc.

Saccharimeter. A form of polarimeter designed for measuring the strength of sugar solution. *Compare* saccharometer.

Saccharin. A white crystalline powder, $C_7H_5ONHSO_4$, with a taste five hundred times as sweet as cane sugar. It is used, often in the form of the sodium salt, as a sweetener. M.pt. 228°C.

Saccharometer. A hydrometer that is designed and calibrated for measuring the strengths of sugar solution. *Compare* saccharimeter.

Safflower. Eurasian thistlelike herb (*Carthamus tinctorius)* of the Composite family; also called false saffron. It has long been cultivated in S Asia and Egypt for food and medicine, and as a substitute for true Saffron dye; in the U.S. it is more important as the source of safflower oil, used in cooking.

Saffron. Fall-flowering plant (*Crocus sativus*) of the IRIS family, and the yellow dye obtained from it. Native to Asia Minor, it has long been cultivated for the orange-yellow stigmas of its pistils, which yield saffron powder, the source of the dye. The powder is also used in perfumes and medicines and for flavoring. One ounce of saffron powder requires stigmas from about 4,000 flowers.

Sage. Aromatic herb or shrub (genus *Salvia*) of the Mint family. The common sage of herb gardens (*S. officinalis*), native from S Europe to Asia Minor, is a strongly-scented shrubby perennial; its dried leaves are used as a seasoning and in a tea. Ornamental sages, popularly called salvia, include the scarlet sage (*S. solendens),* noted for its neat spikes of usually red flowers.

Sagebrush. Name for several species of the genus *Artemesia*, deciduous shrubs of the Composite family, especially abundant in arid regions of W North America. The common sagebrush (*A. tridentata*) is a silvery-gray, low shrub with a pungent odor of sage, although it is unrelated to true Sage. Common in the West, sagebrush is an important forage plant. It is the state flower of Nevada.

Sailfish. Marine game and food Fish of the family Istiophoridae, related to the Swordfish and Marlin. Its high, wide dorsal fin, or sail, is deep blue with black sports. The average length of a sailfish is 6 ft (180 cm). The Atlantic species (*Istiophorus americanus*), found as far north as Cape Cod in summer, average 60 lb (27 kg). The Pacific *I. orientalis* grows to 100 lb (45 kg).

Sakharov, Andrei Dmitriyevich. 1921-, Soviet nuclear physicist an human-rights advocate; first Soviet citizen to win the Nobel Peace Prize (1975). From 1948 to 1956 he helped to develop the USSR's hydrogen bomb. In the late 1960s he emerged as an outspoken critic of the arms race and of Soviet repression. His banishment to the city of Gorky in 1980 inspired worldwide protest.

Salamander. Amphibian (order Urodela) having a tail and small, weak limbs that can regenerate. Found in damp regions of the northern temperate zone, they are abundant in North America. Most are under 6 in. (15 cm) long, although the giant salamander of Japan may reach 5 ft (1.5 m). Usually nocturnal and feeding on small animals such as insects and worms, salamanders are mostly terrestrial as adults, although some are aquatic and a few arboreal. *Newts* are a large, widely distributed family of salamanders.

Sal Ammoniac. *See* ammonium chloride.

Salicylate. Any salt or ester of salicylic acid.

Salicylic Acid (1-hydroxybenzoic acid). A white crystalline powder, $C_6H_4(OH)(COOH)$, that occurs naturally in several plants. It is used to make aspirin and dyes and as a food preservative. M.pt. 157°C; b.pt. 211°C (20 mmHg); r.d. 1.4.

Saline. Denoting a solution that contains a salt or salts, especially one containing sodium chloride or another alkali metal halide.

Salmon. Marine Fish of the family Salmonidae that spawn in fresh water, including salmon, trout, and char. The herringlike Salmonidae are characterized by soft, rayless, adipose fins and live in cold, oxygen-rich waters. They are generally a uniform silver in colour. The Atlantic salmon (genus *Salmo*), once abundant, is threatened by pollution, damming, and overfishing. It feeds on Crustaceans at sea and small fish while spawning and reaches 15 lb (6.8 kg). The Pacific salmon (genus *Oncorhynchus*), comprising five species, is commercially the most important. The largest, the chinook, may reach 100 lb (45 kg); the blueback forms the bulk of canned salmon. Pacific salmon are famed for their grueling journeys of hundreds of miles to their headwater breeding grounds. The genus *Salvelinus* includes the European chars and the North American brook trout.

Salmonellosis. Infection caused by intestinal bacteria of the genus *Salmonella*. The most common form is gastroenteritis, an intestinal disease usually resulting from contaminated food. Symptoms of food poisoning include nausea, abdominal pains, diarrhea, and fever. Treatment includes rest and replacement of lost body fluids. The generalized form is often called paratyphoid fever because it resembles Typhoid Fever. Treatment includes the use of Antibiotics.

Salt. Chemical compound (other than water) formed by neutralization reactions between Acids and Bases; by direct combination of metal with nonmetal, *e.g.*, sodium chloride (common table salt); by reaction of a metal with a dilute acid; by reaction of a metal oxide with acid; by reaction of a nonmetallic oxide with a base; or by reaction of two salts with each other to form two new salts. Most salts are ionic compounds. The chemical formula indicates the proportion of atoms of the elements making up the salt. A salt is classified as acidic, basic, or normal if it has, respectively, hydrogen (H), hydroxyl (OH), or neither in its formula. A salt undergoes Dissociation when dissolved in a polar solvent, *e.g.*, water.

Salt. 1. Any of a class of compounds produced, together with water, by reaction between an acid and a base. Salts are derived from acids by replacement of one or more hydrogen atoms with metal atoms or ions or with other positive ions. Typically, they are crystalline water-soluble solids consisting of mixtures of positive and negative ions, as in sodium chloride, Na^+Cl^-, or ammonium nitrate, NH_4^+ NO_3. Some metal salts are not purely ionic. Aluminium chloride, for example, is a low-melting solid containing Al_2 Cl_6 molecules, in which the atoms are held together by covalent bonds. Salts such as $NaHCO_3$, in which acidic hydrogens are present, are *acid salts*. Similarly, salts such as $Al(OH)_2Cl$ are basic salts.

2. (common salt). *See* sodium chloride.

Salt bridge. A tube containing a solution of potassium chloride in a gel, used to obtain electrical contact between two half cells without mixing of the electrolytes.

Saltcake. *See* sodium sulphate.

Salt Lake City. City (1980 pop. 163,033), state capital and seat of Salt Lake co., north-central Utah, on the Jordan R., near the Great Salt Lake, at the foot of the Wasatch Range; inc. 1851. Food-processing, oil-refining, smelting, and electronic industries are important. Founded (1847) by Brigham Young as the capital of the Mormon community, it is still its economic hub. The gigantic Temple (1853-93) is at the city's heart. Other points of interest include the state capitol (1914), and the Brigham Young home (1877) and Memorial (1897).

Saltpeter or **Potassium Nitrate.** Chemical compound (KNO_3), occurring as colourless prismatic crystals or as a white powder. When heated, it decomposes to release oxygen. Salt-peter has been used in gunpowder manufacture since about the 12th cent; it is also used in explosives, fireworks, matches, and fertilizers, and as a food preservative.

Salvage Archaeology. Division of archaeology concerned with removal of artifacts threatened by construction of highways, dams, buildings, pipelines, and other projects. One of the greatest feats of salvage archaeology occurred in the mid-1960s, when an international effort (funded by more than 50 countries) succeeded in preserving the temples at ABU Simbel, Egypt, from flooding by the Aswan High Dam. The two massive structures were moved to higher ground.

Samarium. Symbol: Sm. A silvery metallic element belonging to the lanthanide series. A.N. 62; A.W. 150.4; m.pt. 1072°C; b.pt. 1778°C; r.d. 7.5; valency 3.

Samarium (Sm). Metallic element, discovered in 1879 by P.E. Lecoq de Boisbaudran. It is a lustrous, silver-white Rare-Earth Metal of the Lanthanide series. Uses include pyrophoric alloys in cigarette-lighter flints, magnets in computer memories, and nuclear-reactor control rods.

Sand. Rock material occurring in the form of loose, rounded or angular grains, varying in size from 0.06 to 2 mm in diameter. It is formed by the Weathering and decomposition of all types of Rock, its most abundant mineral constituent being Quartz. Sand has numerous uses, particularly in the manufacture of bricks, cement, glass, concrete, explosives and abrasives.

Sandalwood. Name for several fragrant tropical woods, especially that of *Santalum album*, a partially parasitic evergreen tree native to India. Oil distilled from the wood is much used as a perfume and in medicine. Red sandalwood, obtained from *Pterocarpus santalinus,* is used in dyes.

Sand Dollar. Marine invertebrate animal (an Echinoderm), related to the Starfish. The sand dollar has a rigid, flattened, circular shell consisting of numerous small skeletal plates. Small spines enable the animal to burrow under the surface of the sand. Tube feet on the underside of the body are used for locomotion and to convey food particles to the mouth.

Sandmeyer Reaction. A chemical reaction in which a diazonium group is replaced by a halogen atom or cyanide group by reaction with the corresponding copper I salt. For example. $C_6H_5N_2^+X^- + CuCl = C_6H_5Cl + N_2 + CuX$. In a variation of the method, the *Gatterman reaction,* the copper salt is replaced by a mixture of copper powder and a hydracid (HCI, HBr, etc.). [After Traugott Sandmeyer (1854-1922), German chemist.]

Sandpiper. Name for some shore Birds of the family Scolopacidae, including the Snipe and Curlew. Sandpipers are small wading birds with long, slender bills for probing in sand or mud for small invertebrates. Their plumage is usually brown or gray streaked above and buff with streaks or spots below. Most are found in flocks on seacoasts throughout the Northern Hemisphere.

Sandstone. Sedimentary Rock formed by the cementing together of Sand grains. The hardness varies depending on the cementing material. Sandstone is widely used in construction and industry. It can also be crushed and used like sand.

Sandwich Compound. A type of transition-metal complex in which a metal atom is "sandwiched" between two parallel hydrocarbon rings. The original example is ferrocene which may be regarded as an iron atom coordinated to two cyclopentadienyl ions, $C_5H_5^-$. The coordination is to the pi-electrons of the whole ring, not to the carbon atoms, and in fact the compound obeys the Huckel rule and is aromatic. Other metallocenes are known in addition to sandwich compounds formed with other ring compounds. Chromium, for example, forms a complex with benzene, Cr $(C_6H_6)_2$.

Sap. Plant fluid consisting of water and dissolved substances. The term is generally applied to all the fluid that moves through the xylem and phloem of the Stems of higher plants. Water with dissolved minerals ascends to the leaves via the xylem vessels; the solution with food produced by the leaves descends to other parts via the phloem.

Saponification. The hydrolysis of an ester using an alkali. The reaction yields an alcohol and the salt of the carboxylic acid. It fats are saponified, a soap is produced. The *saponification number* of a fat or oil is the number of milligrams of potassium hydroxide necessary to completely saponify one gram of fat.

Sapphire. A naturally occurring blue form of aluminium oxide, Al_2O_3, containing small amounts of cobalt. It is used as a gemstone.

Sapphire. Gem, transparent blue variety of corundum, found chiefly in Thailand, India, Sri Lanka, and Burma. Like rubies, some sapphires show an internal star when cut with a round top. Synthetic stones are made by fusing aluminium oxide, with titanium oxide added for colour.

Saprophyte. Any plant that relies on dead organic matter for its food; most do not have Chlorophyll and therefore cannot photosynthesize. Saprophytes include most Fungi and a few flowering plants, *e.g.*, some Orchids. They aid in the breakdown of dead plants and animals.

Sarsaparilla. Name for various plants and for an extract from their roots, used in medicine and beverages. True sarsaparilla is obtained from tropical American species of the genus *Smilax* of the Lily family; these have thick rootstalks and thin roots several feet long. Wild sarsaparilla (*Aralia nudicaulis*), related to Ginseng, is used as a substitute source for the extract.

Saros. *See* moon.

Satellite. A smaller celestial body that orbits around a larger one. The earth is a satellite of the sun, whereas the moon is a satellite of the earth. In the solar system, of the planets lying within the asteroid belt, only the earth and Mars are known to have satellites. The giant planets, by contrast, have numerous satellites, Jupiter having twelve, Saturn ten, Uranus five, and Neptune two.

Saturated. 1. Denoting a vapour that is in equilibrium with its liquid or solid. In a saturated vapour, atoms or molecules are condensing from the vapour at the same rate as they are evaporating. The amount of substance in the vapour. Phase depends on the temperature and increases with rising temperature.

2. Denoting a solution in which the solute is in equilibrium with undissolved substance. Atoms or molecules are leaving a saturated solution at the same rate as they are dissolving. The concentration of dissolved substance depends on the temperature. It increases with temperature for liquids and solids and decreases with temperature for gases.

3. Denoting a magnetic material in which the amount of magnetization is a maximum: *i.e.* it can increase no further no matter how strong the magetizing field. At saturation, all the domains are aligned in the direction of the external field.

4. Denoting a chemical compound that contains no double or triple bonds, so that no further addition can occur.

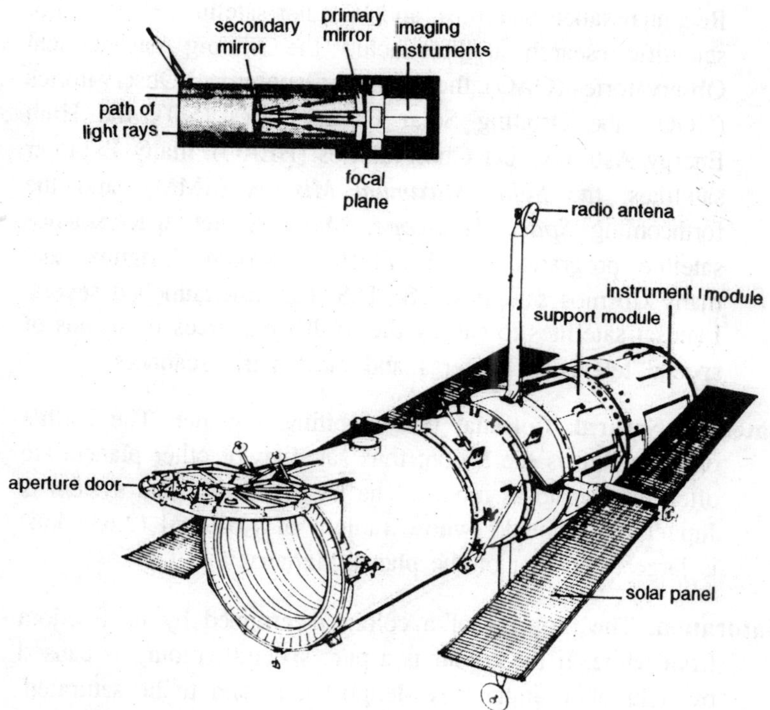

Fig. S-1. *Artificial satellite:* The Space Telescope, scheduled to be put in orbit by the U.S. Space Shuttle in 1985, will be the largest astronomical telescope (94 in. / 2.4 m in diameter) ever orbited.

Satellite, artificial. Object launched by a Rocket into orbit around the earth or, occasionally, another solar-system body. A satellite in circular orbit at an altitude of 22,300 mi (35,880 km) has a period of exactly 24 hr, the time it takes the earth to rotate once on its axis; such an orbit is called synchronous. If such an orbit also lies in the equatorial plane, it is called geostationary, because the satellite will remain stationary over one point on the earth's surface. The first artificial satellite, *Sputnik* 1, was launched by the USSR on Oct. 4, 1957. *Explorer 1*, the first American satellite, was launched on Jan. 31, 1958. The principal types of application satellites are Communications Satellites, Navigation Satellites,

Reconnaissance Satellites, and Weather satellites. Major U.S. scientific research satellites include the Orbiting Astronomical Observatories (OAO), the Orbiting Geophysical Observatories (OGO), the Orbiting Solar Observatories (SO), the High Energy Astronomical Observatories (HEAO), many Explorer satellites, the *Solar Maximum Mission* (SMM), and the forthcoming *Space Telescope*. Major Soviet space-science satellite programs include Electron, Proton, Prognoz, and many Cosmos satellites. The U.S. has also launched several Landsat satellites to survey the earth's resources by means of special television cameras and radiometric scanners.

Satellite, Natural. Celestial body orbiting a planet. The earth's only satellite is the Moon; thus satellites of other planets are often referred to as moons. The largest in the solar system is Jupiter's Ganymede, whose radium of 1,639 mi (2,638 km) is larger than that of the planet Mercury.

Saturation. The property of a colour determined by its freedom from white. If the colour is a pure spectral colour (as caused by light of a single wave-length) it is said to be saturated. Colours that have the same hue but differ in saturation are called *tints*.

Saturn. The sixth planet in succession outwards from the sun and the second largest planet in the solar system. Lying at a mean distance of 1427 million kilometres from the sun. Saturn completes its orbit in 29.46 years. Like Jupiter it is a giant planet, with a diameter of 119 300 km, a mass 95.14 times the earth's, and a mean relative density of 0.7. Like the other giant planets it has a dense atmosphere with latitudinal cloud belts consisting mainly of hydrogen, methane, and ammonia, and the planet's surface temperature has been estimated at about -160°C. Its rotation period is about 10 hours 14 minutes. Saturn has ten satellites, the largest being *Titan*, which is as large as Mercury. The most remarkable feature of Saturn is its unique system of rings, formed, it is believed, by solid particles that have failed to coalesce into a satellite because they are too close to the planet. Three rings have so far been definitely sighted, arranged concentrically and girdling and planet's equator. The outer diameter of the largest ring is about 272 000 km and the rings themselves are probably no more than 16 km thick, so that they almost disappear.

Saturn. In astronomy, 6th Planet from the sun, at a mean distance of 886.7 million mi (1.4270 billion km). It has an equatorial diameter of 74, 980 mi (120, 660 km) and an atmosphere of hydrogen, helium, and traces of methane and ammonia. Like Jupiter, Saturn has counterflowing easterly and westerly winds. Saturn's most remarkable feature is its ring system, composed of billions of water-ice particles orbiting around the planet. The main rings are at distance ranging from 4,140 to 49,880 mi (6,670 to 80,270 km) above the cloud tops, and two other tenuous rings orbit much more distantly. Saturn also has at least 17 natural satellites. The largest is *Titan* (diameter: 3,200 mi/5,150 km); discovered by Christiaan Huygens in 1655, it is the only natural satellite in the solar system with a substantial atmosphere. Saturn has six major icy satellites. The most prominent feature of heavily cratered *Mimas,* the innermost, is a large impact crater about one third the diameter of the satellite. Certain broad regions of *Enceladus* are uncratered, indicating geological activity that has somehow resurfaced the satellite within the last 100 million years. *Tethys* also has a very large impact crater, as well as an extensive series of valleys and troughs that stretches three quarters of the way around the satellite. Both *Dione* and *Rhea* have bright, heavily cratered leading hemispheres and darker trailing hemispheres with wispy streaks thought to be produced by deposits of ice inside various surface troughs or cracks. *Lapetus,* the outermost of the large icy satellites, has a dark leading hemisphere and a bright trailing hemisphere. The remaining ten satellites, some sharing orbits with other, are smallers in size. The two largest of these, dark-surfaced *Phoebe* and irregularly shaped *Hyperion*, orbit far from the planet. Saturn has been encountered by three Space Probes:

Pioneer 11 (1979), *Voyager 1* (1980), and *Voyager* 2 (1981). Photos and other data from *Voyager 2* indicate the possibility of at least six more small satellites.

Savanna or **Savannah**. Tropical or subtropical grassland lying on the margin of the trade-wind belts. Its climate is characterized by a rainy period in summer and a dry winter when the grass withers. The largest savannas are in Africa.

Sawtooth Waveform. A waveform that repeatedly increases steadily to a peak value and drops suddenly to zero.

Saxifrage. Name for some members of the Saxifragaceae, a diverse family of herb, shrub and small trees of cosmopolitan distribution. The true saxifrages (primarily the genus *Saxifraga*) comprise a group of low rock plants often cultivated as rock-garden and border plants, *e.g.,* the strawberry geranium (*S. sarmentosa*). The Eastern early saxifrage (*S. virginiensis)* and the Western umbrella plant (*S. peltata*) are American wildflowers. The arctic and alpine *S. oppositifolia* is one of the northernmost of flowering plants. The genus *Ribes,* a group of berry-bearing shrubs including the goose-berry and Currant, is of minor economic importance. The mock orange, or syringa (genus *Philadelphus*), has white, sometimes fragrant flowers similar to orange blossoms, and the hydrangea (genus *Hydrangea*), another member of the family, has flat-topped clusters of white, pink, or blue flowers.

Saxons. Germanic people living in S Jutland in the 2d cent. In the 3d and 4th cent. they raided the coasts of the North Sea, and by the 5th cent. they had settled in N Gaul. Raiding Saxons began (c. 450) to settle in Britain. By the 6th cent. they and the Angles were founding Anglo-Saxon kingdoms, with Wessex dominant. Continental or "Old Saxons" in 566 became tributaries of the Franks, and were finally conquered (804) by Charlemagne. The treaty of Verdun (843) included Saxon lands in the area that is modern Germany.

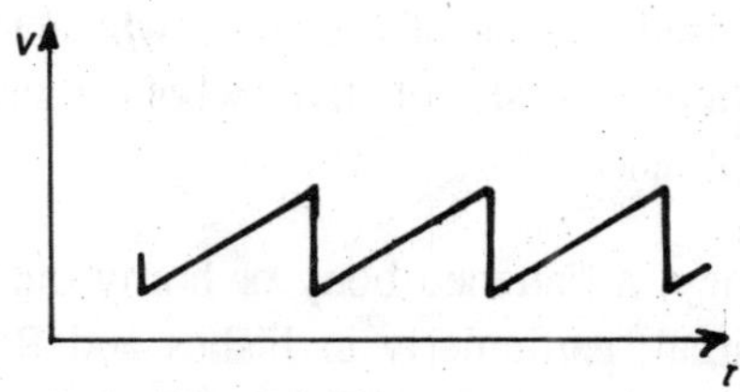

Fig. S-2. Sawtooth waveform.

Scalar. A physical quantity that is characterized only by its magnitude and does not depend on directional measurement. *Compare* vector.

Scalar Product (dot product). The product of two vectors yielding a scalar whose magnitude is the product of the magnitudes of the vectors and the cosine of the angle between them. The scalar product is written *a.b.* and read as "a dot b". Thus, if $a.b = c$, then c is a scalar of magnitude $|a|\ |b| \cos \theta$. A physical example of a scalar product is that a force F making an angle θ with a given direction. The work W done by the force moving a distance d along this direction is $F.d = Fd \cos \theta$. If the vectors are represented in Cartesian coordinates, their scalar product has the form: $a.b = a_x b_x + a_z b_z$. Two vectors can also be multiplied together to give a vector product.

Scale. In music, any series of Tones arranged in a step-by-step rising or falling of Pitch. The scale most used in Western musical composition until the end of the 19th cent. was the diatonic scale, a series of seven tones. (The addition of a final top note, with a frequency twice that of the lowest note, defines this sequence as an *octave.*) The intervals of the diatonic scale were defined by Pythagoras in the 6th cent. B.C. as five whole tones (t) and two semitones (s) in the order ttsttts. By the time of J.S. Bach, the chromatic scale of 12 equal semitones (as in the white and black keys of a keyboard scale) had become established, and the scales beginning on these notes remained the basis of Western Tonality until the innovations of such modern composers as Schoenberg.

Debussy used a scale of six equal whole tones in his works. The pentatonic scale of five whole tones is prevalent in Oriental music.

Scale. In zoology, a flattened bony or horny outgrowth of the skin of an animal, particularly in Fishes and Reptiles, that serves primarily as protection. Fish scales, composed mostly of bone, form directly in the skin membrane; the type and number of scales figure in the identification of a species. Most reptiles have horny scales, or scutes, but some (*e.g.,* the Crocodile) have both horny and bony scales. In snakes, scales aid in movement. Birds have scales on the feet and sometimes the legs. Some mammals (*e.g.,* the rat and the mouse) have tail scales.

Scaler. An electronic circuit or device for counting pulses, giving a single output pulse as a result of a certain number of input pulses. Scalers are used in counting particles of radiation.

Scandium. Symbol: Sc. A silvery metallic element occurring as thortveittie (Sc_2SiO_7). Scandium belongs to group III of the periodic table but is usually classified with the lanthanides. A.N. 21; A.W. 44.96; m.pt. 1539°C; b.pt. 2727°C; r.d. 3.02; valency 3.

Scandium (Sc). Metallic element, discovered in 1879 by L.F. Nilson. Soft, silver-white Rare-Earth Metal, it is in relatively greater abundance in the sun and certain stars than on earth. It is used in nickel alkaline storage batteries and as a radioactive tracer.

Scanning. The process of traversing an area, volume, range of values, etc. Examples are the repeated movement of a beam of electrons in a television camera or in a cathode-ray tube, the continuous movement of a radar aerial, and the continuous change in position of a spectrometer grating to cover the range of values of a spectrum.

Scanning Electron Microscope. *See* electron microscope.

Scarab Beetle or **Scrab.** Heavy-bodied, oval, often brightly coloured or iridescent Beetle. Some, called dung beetles, feed on animal dung and are important in recycling organic matter and disposing of disease-breeding wastes. Another group, which feeds on living plants and includes the Japanese and June beetles, are major crop and garden pests. Representations of scarabs, carved in stone or other materials, were popular in ancient Egypt and Rome.

Scarlet Fever. Acute, contagious respiratory infection caused by streptococci bacteria. Symptoms include a sore throat, fever, headache, white-coated tongue with red spots (strawberry tongue), and a skin rash. The disease usually occurs before age ten and is spread by direct contact. Antibiotics such as Penicillin are used to treat severe cases and reduce the possibility of complications.

Scattering. Deflection of incident particles (electrons, photons, etc.) by other particles in their path. Scattering is said to be elastic or inelastic depending on the type of interaction. The type of interaction also determines the direction in which scattering occurs. For instance, charged particles can be scattered by interaction with the electrostatic field of atomic nuclei. High energy neutrons, and other particles, may be scattered as a result of strong interactions.

The scattering of electromagnetic radiation can occur in several ways. If the particles are relatively large, reflection occurs in random directions. This happens with particles of dust, etc. If the particles are small compared with the wavelength of the radiation, diffraction occurs and the amount of deflection depends on the wavelength (proportional to $1/\lambda^4$). This phenomenon is responsible for the blue of the sky—air molecules scatter blue light more than red light. Compton scattering involves the interaction of photons with electrons.

Scheelite. A white, yellow, green, or brown naturally occurring tungstate of calcium, $CaWO_4$, used as an ore of tungsten.

Scheele, Karl Wilhelm. (Sha'le), 1742-86, Swedish chemist. A pharmacist, he prepared and studied Oxygen (c. 1773), but his published account (1777) appeared after Joseph Priestley's. He discovered Nitrogen independently of Daniel Rutherford and showed it to be a constituent of air. His treatise (1774) on Manganese aided in the discovery of that element and also of barium and chlorine. He isolated glycerin and many acids.

Schiff's Base. Any of a class of compounds with the general formula $ArN{:}CR_2$, where Ar is an aryl group and R is an aryl or alkyl group or a hydrogen atom. Schiff's bases are crystalline substances obtained by reacting an aromatic amine ($ArNH_2$) with an aldehyde or ketone (R_2CO). [After Hugo Schiff (1834-1915), German chemist.]

Schiff's Reagent. A solution of the dye magenta that has been decolorized by sulphur dioxide. The colour is restored by aliphatic aldehydes, which oxidize the reduced form of the dye back to its original coloured form. Aromatic aldehydes and aliphatic ketones restore the colour more slowly: aromatic ketones do not react.

Schist. Metamorphic Rock having a foliated (plated) structure in which the component flaky minerals are visible to the naked eye. Schists' mineral crystals are larger than those of Slates and smaller than those of Gneisses (the other two foliated metamorphic rocks).

Schizophrenia. Severe mental disorder, or psychosis, characterized by faulty thought processes, bizarre actions, unrealistic behaviour dominated by private fantasies, and incapacity to maintain normal interpersonal relationships. It occurs most commonly, but not always, in late Adolescence or early adulthood. There may be a hereditary predisposition that makes some individuals likely to develop the disorder when exposed to particular environmental stresses. In the past a high percentage of patients required long-term hospitalization, but recent advances in the use of Tranquilizers have made the outlook for remission of schizophrenia much more hopeful than it was formerly.

Schizophyta. Division of the plant kingdom consisting of the Bacteria and blue-green Algae.

Schleiden, Matthias Jakob. 1804-81, German botanist. With Theodor Schwann, he is credited with establishing the foundations of the cell theory. An 1838 paper, though mistaken in some aspects, recognized the significance of the nucleus in cell propagation.

Schlieren Photography. The use of highspeed photography to demonstrate and study turbulent flow in fluids. Variations in the density of the fluid cause transient streaks, which are visible because of local variations in the refractive index. [From German, *Schliere*: a streak.]

Schmidt Telescope. A type of reflecting telescope in which a curved photographic plate is used at the centre of curvature of the primary concave mirror, thus reducing coma, astigmatism, and spherical aberration of the image. [After Bernhard Schmidt (1879-1935), German Astronomer.]

Schnauzer. Sturdy, wirehaired dog. Developed in S Germany, it is divided into three breeds. The standard—shoulder height, 17-20 in. (43.1-50.8 cm); weight, 27-37 lb (12.3-16.8 kg)—is a Working Dog that dates to the 15th cent., when it was used as a ratter, farm dog, and guardian. The giant—shoulder height, $21^1/_2$ - $25^1/_2$ in. (54.6-64.8 cm); weight,, 65-78 lb (29.5-35.4 kg)—and the miniature, a Terrier—shoulder height, 12-14 in. (30.5-35.6 cm); weight, 13-15 lb (5.9-6.8 kg)—were both developed at the end of the 19th cent. All three breeds may be pepper and salt, silver, or black in colour.

Schottky Sefect. *See* vacancy.

Science Fiction. Literary genre to which a background of science of pseudoscience is integral. Although fantastic,it contains elements within the realm of future possibility. Science fiction began with the late-19th cent. work of Jules Verne and H.G. Wells. The appearance of the magazines *Amazing Stories* (1926) and *Astounding Science Fiction* (1937) encouraged good writing in the field, which was further spurred by post-World War II technological developments. Contemporary writers of science fiction include Robert Heinlein, Isaac Asimov, A.E. van Voght, Alfred Bester, Arthur C. Clarke, Frederik Pohl, Stanislaw Lem, and Ursula K. LeGuin. The gnere's effectiveness as an instrument for social criticism is seen in Aldous Huxley's *Brave New World* (1932), Ray Bradbury's *Fahrenheit 451* (1953), and Kurt Vonnegut's *Cat's Cradle* (1963).

Scientific Notation. Means of expressing very large or very small numbers in a compact form, to simplify computation. In this notation any number is expressed as a number between 1 and 10 multiplied by the appropriate power of 10. For example, 32,000,000 in scientific notation is 3.2×10^7, and 0.00526 is 5.26×10^{-3}.

Scintillation. The production of small flashes of light in certain substances (*scintillators*), caused by the impact of high-energy particles. The light is emitted by fluorescence.

Scintillation Counter. A type of counter in which particles of radiation are detected by the flashes of light that they produce in a scintillator. This is mounted on the front of a photomultiplier, which produces an output pulse for every flash of light. The pulses are counted by a scaler. The heights of the photomultiplier pulses depend on the energies of the incident particles and a pulse-height analyser can be used to study these energies. The device is then a *scintillation spectrometer*.

Scintillator. *See* scintillation.

Sclerometer. Any instrument for measuring the hardness of materials. An example is the apparatus used in the Brinell test. Another type of sclerometer, the *scleroscope*, determines hardness by dropping a standard ball from a prescribed height and measuring the height to which it rebounds.

Scorpion. Invertebrate animal (order Scorpionida) with a pair of powerful, pincerlike claws and a hollow, poisonous stinger at the tip of the tail; an Arachnid. Most are 1-3 in. (2.5-7.6 cm) long, but some measure as much as 6 in. They seize and crush prey with their large claws, immobilizing it by stinging if necessary. With the exception of the fatal stings of *Androctonus australis* of the Sahara and several Mexican species, scorpion stings, although painful, are not usually dangerous to humans.

Scotopic Vision. Vision by the eye when the rods of the retina are the main receptors of light. This is the dominant process at very low luminance levels. The rods cannot distinguish different colours. *Compare* photopic vision.

Scottish Terrier. Short-legged Terrier; shoulder height, c.10 in. (25 cm); weight, 18-22 lb (8.2-10 kg). Its wiry coat may be gray, brindle, grizzle, black, sandy, or wheaten. Perfected in Scotland in the mid-19th cent., the scottie was used to hunt small game, particularly badgers.

Screen Grid. A grid placed between the anode and control grid of a thermionic valve and held at constant potential in order to reduce the capacitance between the anode and the control grid.

Screening. Shielding of a piece of apparatus in order to protect it from electric or magnetic fields. Electrical screening is effected by surrounding the apparatus with a conducting container. Magnetic screening uses a container of high permeability, such as Mumetal.

Scruple. *See* Appendix.

Sea Anemone. Predominantly solitary marine polyp, usually attached to submerged objects. Many of these are animals are beautifully coloured and look like flowers when their tentacle-encircled feeding end is open. They are mostly 1-4 in. (2.5-10 cm) long; a few are 3 ft (90 cm) in diameter. Most sea anemones are predators, immobilizing their prey with stringers located in the tentacles.

Seaborg, Glenn Theodore. 1912-, American chemist. Professor and later chancellor at the Univ. of California, Berkeley, he worked at the Univ. of Chicago during World War II on the development of the atomic bomb ad later was chairman (1961-71) of the U.S. Atomic Energy Commission. He shared with Edwin M. McMillan the 1951 Nobel Prize in chemistry for work on transuranium elements. Seaborg is codiscoverer of the elements Plutonium, Americium, Curium, Berkelium, Californium, Einsteinium, Fermium, Mendelevium, and Nobelium.

Sea Cucumber. Flexible, elongated invertebrate animal with a cucumber-shaped, leathery body; an Echinoderm. Most are under 1 ft (30.5 cm). Many sea cucumbers eject most of their internal organs when sufficiently irritated, later regenerating a new set.

Seahorse. Small, bony-plated fish of the family Syngnathidae, usually found in warm waters. Its elongated head and snout, flexed at right angles to its body, suggest those of a horse. Members of different species range from 2 to 8 in. (5 to 20 cm). Weak swimmers, they anchor themselves by curling their thin, prehensile tails around seaweed. While mating, the female seahorse injects eggs into a pouch on the underside of the male, where they are fertilized and then develop until the young are expelled. The related pipefish look more fishlike.

Seal. Fin-footed Mammal (pinniped) of the family Phocidae, usually marine, with front and hind feet modified as flippers. Seals have streamlined bodies with a thick, subcutaneous layer of fat, and most inhabit cold or temperate regions. All species leave the water at least once a year to breed; some species migrate. Seals live on fish and shellfish; many dive deep to feed, navigating by means of echolocation. True seals lack external ears. They are polygamous and gregarious, and most live in one of three geographical regions: northern, antarctic, and the warm waters of the Mediterranean, Caribbean, and Hawaiian seas. Seals are extensively hunted for food, fur, hides, and oil.

Sea Lion. Fin-footed marine Mammal (pinniped) of the eared seal family. The sea lion has external ears (unlike the true seal), a long, flexible neck, supple forelimbs, and hind flippers. Males may reach 8 ft (2.4 m) in length. Sea lions live close to shore in the oceans of the Southern Hemisphere and in the N Pacific, feeding on fish and Squid. To breed they gather in colonies, where the males assemble harems. Most species are protected.

Seaplane. Airplane designed to take off from and alight on water. The two most common types are the floatplane, whose fuselage is supported by struts attached to two or more pontoon floats, and the flying boat, whose boat-hull fuselage is constructed with buoyancy and strength necessary to alight and float on water. First built and flown in 1911 by Glenn Curtiss, the seaplane developed rapidly in the 1920s and 30s and for a time was the largest and fastest aircraft in the world.

Search Coil. A flat coil of wire used to measure magnetic flux. The coil is removed from the position in the magnetic field to position outside the field or is turned through 180° in the field and the total induced current is measured by a ballistic galvanometer.

Searle's Method. A technique for measuring the thermal conductivity (λ) of good conductors. The specimen, in the form of a bar, is lagged and mounted horizontally with a steam bath connected to one end. Thermometers are placed to measure the temperatures (θ_1 and θ_2) at two positions a known distance (*d*) apart along the bar. Cooling water is passed through a copper spiral at the other end of the bar. The inlet and outlet temperatures of the water (θ_3 and θ_4) are also measured. In the steady state, the thermal conductivity is calculated from the equation:

$$\lambda A(\theta_1 - \theta_2)/d = m\ (\theta_4 - \theta_3),$$

where *A* is the cross-sectional area of the bar and *m* the mass of water flowing per unit time.

Sea Slug. Usually brightly coloured marine Gastropod mollusk that lacks a shell as an adult. Most sea slugs, or nudibranchs, creep along the sea bottom or cling to vegetation below the tide line; a few swim in the open ocean. Most are under 1 in (2.5 cm) long.

Sea Spider. Long-legged, spiderlike marine invertebrate animal with at least four pairs of walking legs; an Arthropod. Most sea spiders are tiny and live near the shore, crawling around on the surface of animal colonies or seaweeds.

Sea Urchin. Sphere-shaped marine invertebrate animal (an Echinoderm) with a rigid body wall and long, sharp, movable, sometimes poisonous spines used for protection and locomotion. Sea urchins prefer shallow waters and rocky bottoms. The roe of some species is considered a delicacy.

Seaweed. Multicellular marine Algae. Most have a stemlike basal disk (holdfast) and a leaflike frond of varying length and shape. The simplest seaweeds are blue-green and green algae and occur in shallow waters as threadlike filaments, irregular sheets, or branching fronds. Brown algae, which are the largest and most numerous of the seaweeds, grow at depths of 50-75 ft (15-23 m), and include the large kelps—sources of iodine and potassium salts, potash, fertilizer, and medicines. Red algae, some of which are fernlike, are found at the greatest depths (100-200 ft/30-61m); commercial AGAR is obtained from a red alga. Some seaweeds provide food for marine animals; all provide oxygen through Photo-Synthesis. Seaweed, especially species of red algae, is an important part of the human diet in some regions.

Secant. 1. *See* trigonometric function.

2. A line cutting off an arc of a curve.

Second. 1. Symbol: s. The SI unit of time, equal to the interval of time occupied by 9, 192, 631, 700 periods of vibration of the radiation produced by the transition between two hyperfine levels in the ground state of the caesium-133 atom. This standard of the second is obtained using a caesium clock. Formerly, the unit was defined by astronomical measurement.

2. One sixtieth of a minute of angle. The symbol is placed to the right of the number.

Second. (sec or s), fundamental unit of time in all systems of measurement. In practical terms, the second is 1/60 of a minute and 1/3,600 of an hour. Since 1967 it has been calculated by atomic standards to be 9,192,631,770 periods of vibration of the radiation emitted at a specific wavelength by a caesium-133 atom.

Secondary. *See* transformer.

Secondary Cell. *See* accumulator.

Secondary Emission. Emission of electrons (*secondary electrons*) from a solid as a result of the impact of other electrons. If the incident primary electron has sufficient energy it can often eject several secondary electrons, an effect utilized in the electron multiplier.

Secondary Standard. *See* unit.

Sector. A part of a circle bounded by two radii and the arc contained between them. A sector of a sphere is a sector of a circle rotated about a diameter.

Sedge. Common name for the Cyperaceae, a family of grass-like and rushlike herbs found worldwide but especially in subarctic and temperate marshes. More specifically, the name *sedge* is used for the genus *Carex* of this family. Sedges differ from true Grasses in having solid, usually triangular stems. Most are perennial, reproducing by Rhizomes. Bulrushes are sedges of the genus *Scirpus;* some species are grown as ornamentals. Other members of the family include Papyrus and the Oriental water chestnut (*Eleocharis tuberosa*), valued by the Chinese for its edible Tubers. Some sedges are woven into mats and chair seats; a few provide coarse hay.

Sedimentation. The settling of particles suspended in a liquid. The rate at which this occurs, under gravity or centrifugal force, yields information on the average particle size.

Seebeck Effect. The phenomenon in which an electromotive force is produced in a circuit containing two conductors when both junctions are at different temperatures. [After Thomas Johann Seebeck (1770-1831), German physicist, inventor of the thermocouple.]

Seed. A small crystal used to induce crystallization from a saturated or supersaturated solution or from a super-cooled melt.

Seed. Ripened ovule of the pistil of a Flower, consisting of the plant embryo, stored food material (endosperm), and a protective coat. True seeds vary in size from dustlike (as in some orchids) to very large (the coconut seed). Seeds undergo a period of dormancy before Germination. Many seeds are frequently confused with the Fruit enclosing them, as in the Grains and Nuts. The seed-bearing plants (Angiosperms and

Gymnosperms) are the highest plants in the evolutionary scale; lower plants (Mosses and Ferns) propagate by means of spores.

Segment. 1. A part of a circle bounded by a chord and the arc it cuts off. A spherical segment is a part of a sphere contained between two parallel planes, one of which may be tangential to the sphere.

2. A part of a line or curve between two points on it.

Seismograph. An instrument, effectively a large pendulum, for detecting and recording earthquakes, underground nuclear explosions, etc.

Seismology. The scientific study of earthquakes.

Seismology. Study of Earthquakes and related phenomena using instruments called seismographs. In general, a recording device is connected to a heavy mass that, when earth tremors occur, remains still due to inertia. The relative motion between the Earth and the instrument is magnified and recorded on a rotating drum by a stylus. Through the use of three such instruments the location and severity of earth quakes can be detected. Seismology is also used to locate oil (by analyzing waves from detonated explosions), to detect underground nuclear tests, and to determine the configuration and depth of the ocean floor.

Selenate. Any salt or ester of selenic acid.

Selenic Acid. A strong oxidizing acid, H_2SeO_4. M.pt. 57-58°C; b.pt. 172°C; r.d. 2.95.

Selenide. A compound of selenium with a more electropositive element or with an organic group.

Selenium. Symbol: Se. A nonmetallic element existing in several allotropic forms, including a grey crystalline form, a red crystalline form, and a red or black amorphous form. It is obtained from the anode sludge produced in electrolytic copper refining and is used in semiconductors and particularly in photoelectric applications, such as photocells and xerography. Its chemistry is similar to that of sulphur: it is one of the chalconide elements. A.N. 34; A.W. 78.96; m.pt. 217°C (grey); b.pt. 695°C (grey); r.d. 4.79 (grey); valency 2, 4, or 6.

Selenium (Se). Nonmetallic element, discovered by Jons Jakob Berzelius in 1817. Selenium is used in pigments, photographic exposure meters, electronics, and xerography.

Selenium Cell. A type of photocell consisting of a metal support carrying a layer of selenium covered by a thin transparent layer of gold. Light falling on the cell produces a potential difference between the gold and the selenium by the photovoltaic effect.

Selenium Rectifier. A type of rectifier consisting of alternate layers of iron and selenium.

Selenium Sulphide. A brown powder, SeS_2. Decomposes at 100°C.

Selenology. This scientific study of the moon.

Semiconductor. A solid with an electrical conductivity that is intermediate between those of insulators and metals and that increases with increasing temperature. In contrast, the electrical conductivity of metals and other good conductors falls as the temperature rises. Semiconductors are usually metalloid elements or their compounds. Examples are germanium, silicon, lead telluride, and similar materials. They form covalent crystals, in which the atoms are held together by covalent bonds. At low temperatures their electrons cannot move through the solid because they are held in position by the bonds. As the temperature is increased some electrons gain enough energy to escape from the covalent bonds and the "free" electrons can move through the crystal under the influence of an applied electric field, thus producing an electric current.

In addition, the removal of electrons from the bonds between the atoms leaves a number of *holes: i.e.* missing electrons. These also contribute to the conductivity: an electron can migrate from one bond to occupy the hole in an adjacent bond, thus leaving another hole behind it. This is filled by another electron from the next bond, and so on. The net effect is of a positively charged hole moving in the opposite direction to the current. An alternative description of semiconductivity is given by the band theory of solids. A semiconductor of this type is called an *intrinsic semiconductor*.

Extrinsic semiconductors are materials that owe their semiconductivity to the presence of small amounts of impurities. Often these are introduced in controlled amounts—a process known as doping. *Donor* impurities are atoms that lose electrons and form positive ions, thus releasing electrons that can carry the current. The remaining electrons are strongly bound to the positive ions of the donor so hole conductivity does not occur. Thus the main part of the current is carried by electrons (known as the *majority carriers*). The remainder of the current is transported by *minority carriers*. A semiconductor of this type, in which most of the current is carried by negative electrons, is called an *n-type* semiconductor. An example is a silicon crystal doped with phosphorus atoms: the phosphorus has five outer electrons and can form four bonds to silicon atoms and release the extra electron.

Acceptor impurities are atoms that require extra electrons to form bonds. Boron, for example, has three outer electrons and in accepting an extra electron can form four bonds to neighbouring atoms. The ionization of the acceptor produces positive holes, which carry the current. A semiconductor in which the majority carriers are positive holes is called a *p-type* semiconductor.

Semiconductors are extensively used in a variety of electronic devices, including thermistors, transistors, phototransistors, and integrated circuits.

Semiconductor. Solid material whose electrical conductivity at room temperature lies between that of a conductor and that of an insulator. At high temperatures its conductivity approaches that of a metal, and at low temperatures it acts as an insulator. In a semiconductor there is a limited movement of electrons, depending upon the crystal structure of the material used. Incorporation of certain impurities in a semiconductor enhances its conductive properties. The impurities either add free electrons or create holes (electron deficiencies) in the crystal structures of the host substances by attracting electrons. Thus there are two semiconductor types: the *n-type* (negative), in which the current carriers (electrons) are negative, and the *p-type* (positive), in which the positively charged holes move and carry the current. Compounds such as indium antimonide, gallium arsenide, and aluminum phosphide are semiconductors. Semiconductors are used in electronic devices such as Computers, Rectifiers, Transistors, and Photoelectric cells.

Semiconductor Junction. A junction between two different layers of semiconductor. A *p-n* junction is a junction between a layer of p-type and a layer of n-type material.

Semipermeable Membrane. A barrier that permits the passage of some substances but is impermeable to others.

Semipolar Bond. *See* coordinate bond.

Sense. Faculty by which external or internal stimuli are conveyed to the brain. The special sense (sight, hearing, smell, and taste) convey external stimuli via receptors in the Eye, Ear, Nose, and taste buds. Most somatic Sensations, such as touch, heat, cold, and pain, originate with receptors in the skin; a few, such as hunger, result from internal stimuli.

Septicemia. Invasion of the bloodstream by bacteria (usually streptococci or staphylococci), a grave condition commonly known as blood poisoning. Primary causes include local infections that the body's defenses are unable to contain and progressing tissue infections. Symptoms are fever, chills, prostration, and skin eruption. The condition is treated with massive doses of Antibiotics.

Sequence. A set of numbers or mathematical expressions in a fixed order, each term being determined by its position in the sequence. For example: 2, 6, 12, 20, 30.... is a sequence for which the nth term is $(n^2 + n)$. Sequences can be convergent of divergent. They are sometimes called *progressions.*

Sequence. In mathematics, ordered set of mathematical quantities, called *terms.* A sequence can be finite, like 1, 2, 3..., 50, which has 50 terms, or infinite, like 1, 2, 3, ..., which has no final term. An infinite sequence may or may not have a Limit. Frequently there is a rule for determining the terms in the sequence, as in the Fibonacci sequence and in various types of Progressions.

Sequestration. The process of forming chemical complexes with metal ions to make them ineffecitive. Sequestration does not necessarily remove the ions but converts them into an inactive form. For example, sequestering agents, which are usually chelating agents, can be used to counteract the effect of lime in soil by forming complexes with the calcium ions. Other applications include water softening and use in analytical chemistry.

Sequoia. Name for the redwood (*Sequoia sempervirens*) and for the big tree, or giant sequoia (*Sequoiadendron giganteum*), both huge, evergreen Conifers of the Bald Cypress family, and for extinct related species. The redwood, which grows to 385 ft (117 m) in height, is probably the world's tallest tree; it is found along the central Pacific coast of the U.S. Its trunk is 10 to 25 ft (3 to 7.6 m) in diameter. The giant sequoia, which grows to 325 ft (99 m) in height, with a trunk of 10 to 30 ft (3 to 9.1 m), is found in California on the western slopes of the Sierra Nevadas; some are believed to be 3,000 to 4,000 years old. The reddish, decay-resistant heartwood of both species is valued for outdoor construction. Sequoias are now protected in parks.

Series. In mathematics, the indicated sum of Sequence. A series may be finite or infinite, depending on the number of terms in the sequence. As one takes sums of progressively more terms in an infinite series, these partial sums form a new sequence of values that may or may not approach a certain value, called the limit of the series. If they do, the series is said to converge to that limit; if not, the series diverges

Series-wound. *See* electric motor.

Serotonin. Biochemical first recognized as a powerful vasoconstrictor occurring in blood serum, and subsequently found in wasp and scorpion venom and in various fruits and nuts. The role of the compound in humans remains obscure; its structural similarity to certain mind-altering drugs, such as LSD, has prompted speculation that serotonin may have a role in certain mental disorders.

Serpentine. Widely distributed hydrous magnesium silicate mineral ($3MgO. 2SiO_2. 2H_2O$), formed by the alteration of other minerals or rocks containing magnesium. Usually green, it may also be reddish, yellowish, black, or nearly white. It is sometimes used as a gem; massive varieties are used like marble for decoration, although they are too easily damaged by exposure to be used for exteriors. Fibrous serpentine is chrysotile, a commercial Asbestos.

Servomechanism. A mechanism in which mechanical motion requiring a large amount of power is controlled by a motion requiring a smaller amount of power. The system may involve feedback.

Sesame. Herb (*Sesamum indicum*) of the family Pedaliaceae cultivated for its seeds since ancient times, found chiefly in the Old World tropics. Sesame seeds, or bennes, are black or white and yield an oil that resists turning rancid. The oil is used extensively in India for cooking, soap manufacture, food, and medicine, and as an adulterant for Olive oil. The seeds are used in baking and confectionery.

Sesqui-. Prefix indicating a ratio of 2:3. A *sesquioxide,* for example, is one with the formula M_2O_3.

Sesquiterpene. *See* terpene.

Set. In mathematics, collection of entities, called the elements or members of the set, that may be material objects or conceptual entities. Braces, { }, are commonly used to enclose the listed elements of a set, *e.g.,* if A is the set of even numbers between 1 and 9, then $A = \{2, 4, 6, 8\}$. The elements may also be described within braces, *e.g.*, the set B of real numbers that are solutions to the equation $x^2 = 9$ can be written as $B = \{x: x^2 = 9\}$, which is read "the set of all x such that $x^2 = 9$." In fact, $B = \{3, -3\}$. There are three basic set operation: intersection, union, and complementation. The intersection of two sets, denoted by the symbol $\cap$, is the set containing the elements common to both. The union of two sets, denoted by $\cup$, is the set of all elements belonging to at least one of the original sets. Thus, if $C = \{1, 2, 3, 4\}$ and $D = \{3,4,5\}$, then $C \cap D = \{3,4\}$ and $C \cup D = \{1,2,3,4,5\}$. In any discussion the set of all elements under consideration is the universal set U; the complement of A, written A', is the set of all elements of the universal set that are not in A; if $U = \{1,2,3,4,5\}$ and $A = \{1,2,3\}$, then $A' = \{4,5\}$. A set with no elements, *e.g.,* the set of all foreign-born U.S. presidents, is called the null, or empty, set and is symbolized by ϕ Membership in a set is indicated by the symbol ε; thus, $x \varepsilon A$ means that x is a member of the set A. If the set B contains at least all the elements of the set A, then A is a subset of B, written $A\ C\ B$. Set theory is involved in many areas of mathematics and has important application in such other fields as computer technology and atomic and nuclear physics.

Sewing Machine. Device that stitches cloth and other materials. The machine's invention is attributed primarily to Elias Howe (1846) and Isaac M. Singer (1850). In the typical home sewing machine, an eye-pointed needle, raised and lowered at great speed, pierces cloth lying on a steel plate, casting a loop of thread on the underside of the seam. A second thread, fed from a shuttle under the plate, passes through the loop and is interlocked with the upper thread as it is drawn up by the rising needle. Modern electric-powered machines, some controlled electronically, are capable of producing many types of stitches, and specialized machines have been devised for sewing particular items.

Sextant. Instrument (invented 1731) for measuring he altitude of a celestial body. The image of the body is reflected from the index mirror to the mirror half of the horizon glass and then into the telescope. If the movable index (or image) arm (to which the index mirror is attached) is then adjusted so that the horizon is seen through the transparent half of the horizon glass, with the reflected image of the body lined up with it, the body's attitude can be read from the index-arm position on the graduated arc.

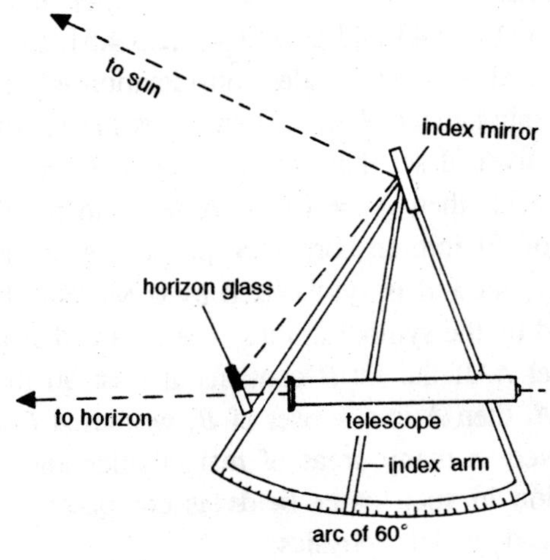

Fig. S-3. Sextant.

Sex Therapy. Treatment of sexual disorders and dysfunction, including Impotence, orgasmic dysfunction, premature ejaculation, and lack of sexual responsiveness. Pioneered by Masters and Johnson in the early 1970s, sex therapy is based on the premise that sexual behaviour is learned and that problems can be alleviated through sex education, sensitization exercises, and improved communication between sexual partners. Treatment may include individual, couple and/or group therapy.

Shad. Fish of the Herring family, found along the Atlantic cost of North America and Successfully introduced on the Pacific coast. One of the largest of the herring (6 lb/2.7 kg average), the shad has delicious but bony flesh. Its roe is valued as a delicacy.

Shade. *See* colour.

Shadow. A dark region formed when an opaque body blocks part of the light falling onto a surface. If the light does not come from a point source, two regions are distinguished: the *umbra* in which all the light is obstructed and the *penumbra*, surrounding it, in which only part of the source's light is obstructed.

Shadow Bands. Parallel bands of shadow moving rapidly across the earth sometimes seen just before and after totality in a total eclipse. They are caused by a refraction effect in the earth's atmosphere.

Shale. Sedimentary Rock formed by the consolidation of mud or Clay, having the property of splitting into thin layers parallel to its bedding planes. Shales comprise an estimated 55% of all sedimentary rocks and often contain large numbers of fossils. Oil shales are widely distributed in the W U.S. and may be a future source of petroleum

Shamrock. Plant with leaves composed of three leaflets. According to legend, it was used by St. Patrick to explain the Trinity. The identity of the true shamrock has long been debated, but the plants most often designated as shamrocks are the white Clover, *Trifolium repens*; the small hop clover, *T. procumbens*; and the wood sorrel, *Oxalis acetosella.* The shamrock is the emblem of Ireland.

Sharecropping. Farm tenancy system once common in parts of the U.S. that arose from the cotton plantation system after the Civil War. Landlords provided land, seed, and credit; croppers-initially former slaves—contributed labor and received a share of all crop's value, minus their debt to the landlord. The system's abuses included emphasis on single cash crops, high interest charges, and cropper irresponsibility. Farm mechanization and reduced cotton acreage have virtually ended sharecropping.

Shark. Predatory cartilaginous Fish (order Selachii), found in all seas but most abundant in warm waters. About 250 species exist, ranging from the 2 ft (60 cm) pygmy shark to the 50 ft (15 m) whale shark. Sharks have pointed snouts and crescent-shaped mouths with several rows of sharp, triangular teeth. The most feared is the white shark, or maneater (*Carcharodon carcharias*), reaching 20ft (6 m) in length, which feeds on large fish and other animals and is known to attack swimmers and boats without provocation. Unlike other sharks, the whale shark and basking shark are harmless plankton feeders. Shark meat is nutritious, and shark oils are used in industry. Tanned sharkskin is a durable Leather.

Shear. Deformation of a body in which planes within the solid move parallel to one another. A rectangular block, for example, would be deformed so that one pair of opposite faces became parallelograms.

Sheep. Wild and domesticated ruminant Mammals (genus *Ovis*). Wild sheep, which include the Bighorn, are agile rock climbers with large, spiraling horns. Present-day domesticated sheep are thought to be derived chiefly from the mouflon of Sardinia and Corsica and the urial of Asia. An adult male is called a ram; a female, a ewe; an infant, a lamb. Sheep are bred for their wool, meat (mutton or lamb, according to age), skins, and milk (from which Cheese is made). Important breeds include the Cotswold, Hampshire, Leicester, Lincoln, Merino, Rambouillet, and Suffolk.

Shell. A region around an atomic nucleus in which all the electrons have the same principal quantum number, n. In a simple model of the atom, the electrons can only move in a certain number of circular orbits of fixed radius, each orbit having a fixed energy. Thus the electron shells may be thought of as a series of concentric spheres: in more advanced theories they are regions in space characterized by the quantum number and their energy. Shells with principal quantum number 1, 2, 3, ... are designated by letters K, L, M, etc., so the K shell ($n = 1$) is the one nearest the nucleus. Each shell contains a maximum of $2n^2$ electrons.

Shellac. Solution of lac, a Resin exuded by a scale insect, in alcohol or acetone. The colour ranges from light yellow to orange; the darker shellacs are the less pure. When bleached it is known as white shellac. Applied to surfaces, *e.g.* wood, shellac forms a hard coating when the solvent evaporates. Shellac is used as a spirit Varnish, as a protective covering for drawings and plaster casts, for stiffening felt hats, and in electrical insulation.

Shell Model. A theoretical model of the nucleus in which it is assumed that the nucleons move in a central field (in the same way that electrons in the atoms move in the field of the nucleus). The application of quantum mechanics leads to the idea that the nucleons move in shells, analogous to the electron shells in the atom. The filled shells of particular stability correspond to nuclei that have magic numbers.

Sheradizing. A technique for coating steel with a layer of zinc, used to protect the steel from corrosion. The surface is covered with zinc powder and heated to just below the melting point of zinc, at which temperature a zinc-iron alloy forms at the interface and this is covered by a layer of pure zinc.

Sherry. Naturally dry, fortified wine containing 15% to 23% alcohol; originally made from grapes of the Jerez de la Frontera region in Spain. The term now includes wines of S Spain, the U.S., Latin America, and South Africa. After fermentation, sherry is fortified with Brandy and aged. Blending and, in some cases, sweetening produce a wide variety of sherries.

SHF. *See* super-high frequency.

Shih Tzu. Active Toy Dog; shoulder height, 8-11 in. (20.3-27.9 cm); weight, 9-18 lb (4.1-8.2 kg). Its long, soft double coat may be any colour. Probably descended from the Lhasa Apso, it originated in Tibet and was sent as a gift by Dalai Lamas to Chinese emperors as early as the 16th cent.

Shingles or **Herpes Zoster.** Infection of a ganglion, or nerve center, with severe pain and blisterlike eruption in the area of the nerve distribution, due to the same virus that causes Chicken Pox. Most common in people past age 50, it involves the area of the upper abdomen and lower chest, but may appear along other nerve pathways.

Ship. Large craft in which persons and goods can be conveyed on water. The term *boat* properly applies only to smaller craft, but some vessels may be called by either name. Ancient ships were propelled by oars, sails, or both; the trireme used by the Greeks and the Romans was the most famous warship of ancient times. In the Middle Ages, Viking ships, propelled by both oars and sails, carried Leif Ericsson to America. The introduction of the mariner's Compass and the transoceanic voyages of the Portuguese and of Columbus and other explorers of the New World gave impetus to the building and navigation of ships. With differences in the number and positions of masts, and sails either square-rigged or fore-and-aft, a number of different types of ships appeared. Building wooden ships became an important industry, especially in Britain and the U.S. Later, the Steamship replaced the sailing ship and steel replaced wood, making possible the construction of much larger ships. The steam engine was followed by the steam Turbine and, early in the 20th cent, by the Diesel Engine. In the 1950s nuclear marine engines were introduced. Today, some freight ships are equipped with cargo-handling machines that rival the power of any on the docks, and the latest generation of mammoth oil-carriers (supertankers) includes the largest ships that ever put to sea. Although the airplane has let to the virtual demise of the great ocean liner, luxurious cruise ships continue to be built. The pivotal vessels of modern warfare are the Aircraft Carriers and the Submarine, but any sizable Navy still includes destroyers, cruisers, and frigates.

Shipworm or **Teredo.** Wormlike marine Bivalve mollusk that bores in wood. It is a greatly elongated Clam, with two ridged shells that function as boring tools enclosing only the front of the body. Shipworms burrow in submerged wood, feeding on wood particles and minute organisms; they do enormous damage to piers and ships.

Shock. Condition in which the cardiovascular system in unable to provide adequate Blood circulation to boy tissues. It may be due to inadequate pumping by the Heart (heart failure), hemorrhage, extensive Burns, or low Blood Pressure. Symptoms include weakness, pallor, cold and moist skin, thirst, nausea, and in severe cases, unconsciousness. Because shock can be fatal, emergency treatment, such as Blood Transfusions and administration of fluids and oxygen, should be given while the underlying cause is being diagnosed.

Shock Therapy. Treatment of severe mental disorders with electricity (called electroconvulsive therapy, or ECT) or chemical agents (such as insulin), usually producing convulsions or coma. Although there is no general agreement as to overall value of shock therapy, electric shock has been used successfully with certain Depression-related disorders, and insulin shock has produced a very limited number of remission in Schizophrenia. Shock treatment has largely been replaced by the use of Tranquilizers.

Shock Wave. Wave formed of a zone of extremely high pressure within a fluid, especially one such as the atmosphere, that propagates through the fluid at supersonic speed, *i.e.,* faster than the speed of Sound. Shock waves are caused by the sudden, violent disturbance of a fluid, such as that created by a powerful explosion or by the supersonic flow of a fluid over a solid object.

Shock Wave. A very narrow region in which the flow of a fluid changes from subsonic to supersonic flow. Shock waves are caused when an object moves through a fluid at supersonic velocities. They are propagated as sound waves.

Sooting. Firing with rifle, shotgun, pistol, or revolver at fixed or moving targets. In the Sport of small-bore rifle shooting, the targets range in distance from 50 ft to 200 yd (15.24-182.88 m); in pistol and revolver, from 50 ft to 50 yd (15.24-45.72 m); and in long-range rifle, from 200 to 1,000 yd (182.88-914.4 m). Competitors shoot from four positions with the rifle—prone, sitting, kneeling, and standing. In skeet or trapshooting marksmen fire shotguns at small disks ("clay pigeons") hurled into the air by a mechanical device (the trap). Major U.S. tournaments are sponsored by the National Rifle Association (formed 1871).

Shooting Star. *See* meteor.

Short Circuit. A direct electrical connection of low resistance between two points in a circuit causing the current to take a reduced path and by-pass part of the circuit.

Short Sight. *See* myopia.

Short Takeoff and Landing Aircraft. (STOL), Heavier than air-craft capable of taking off or landing with only a short length or runway (1,000 ft/305 m), but incapable of taking off vertically. STOL aircraft have large wings that are equipped with slotted flaps, drooped leading edges, and auxiliary spoilers that augment lift, increase stability, and improve the effect of control surfaces.

Showboat. Steamship equipped with a stage and carrying a troupe of actors that brought entertainment to 19th cent. U.S. river communities. Showboats flourished along the Mississippi and Ohio rivers until and Civil War.

Shower. A large number of particles originating from a single collision of a primary particle.

Shrew. Small, insectivorous Mammal of the family Soricidae, of Eurasia and the Americas. Related to Moles, they include the smallest mammals (under 2 in./5.1 cm). Light-boned and fragile, with mouselike bodies, shrews are terrestrial and nocturnal and can produce a protective musky odor. They have the highest known metabolic rate of any animal and must eat incessantly to survive.

Shrimp. Small Crustacean with 10 jointed legs and a nearly cylindrical, translucent body. Unlike the closely related Lobsters and Crabs, which are crawlers, shrimp are primarily swimmers. Shrimp are widely distributed in temperate and tropical salt and fresh waters. Some grow to 9 in. (23 cm) long, but most are smaller. One of the most popular crustacean foods, shrimp are fished throughout the world.

Shrub. Any woody, Perennial, bushy plant that branches into several stems or trunks at the base and is smaller than a Tree. Tree species may grow as shrubs under unfavorable conditions. In regions of extreme climatic conditions, *e.g.*, the Arctic, where trees do not thrive, shrubs provide valuable food and wood. Common shrubs include the Lilac, mock orange, viburnum, Forsythia, and Azalea.

Shunt. A resistor with a low value connected in parallel with an ammeter or similar measuring instrument so that a known fraction of the current is allowed to flow through the instrument. In this way, the instrument's operating range can be changed.

Shunt-wound. *See* electric motor.

Siamese Twins. Congenitally united organisms that are complete or nearly complete individuals, developing from a single fertilized ovum that has divided imperfectly. Siamese twins are attached at the abdomen, chest, back, or top of the head, depending on where division of the ovum failed, and can often be separated surgically after birth.

Siberian Husky. Muscular Working Dog; shoulder height, 20-23$^1/_2$ in. (50.8-59.7 cm); weight, 35-60 lb (15.9-27.2 kg). Its weather-resistant, dense coat is usually black, white, tan, gray, or combinations of these colours. Its origins date back thousands of years in N Siberia, where is was raised by the Chukchi people to pull sleds. The term *husky* is often used to designate any mixed-breed arctic sled dog.

Side Chain. A group of two or more atoms attached at some point along a chain of atoms in a molecule. In isobutane, for example, the methyl group attached to the central carbon atom is regarded as a side chain attached to the main straight chain of three carbon atoms. Isobutane thus has a branched chain.

Side Reaction. A chemical reaction that takes place at the same time as the main chemical reaction and yields small amounts of other products.

Sidereal. Denoting measurements taken with respect to the star.

Sidereal Time. Time measured with reference to the stars. The *sidereal day* (23 hours 56.06 minutes) is the time between successive transits of a chosen star across the observer's meridian. The *sidereal month* is the time between two successive conjunction of the moon with a star. The *sidereal year* is the time for the sun to make one revolution of the celestial sphere, moving relative to the fixed stars. It is measured by the time between successive conjunction with a chosen star. 1 sidereal year = 365.256 36 mean solar days.

Sidereal Time. Time measured relative to the fixed stars. The *sidereal day* is the period during which the earth completes one rotation on its axis, so that some chosen star reappears on the observer's celestial meridian; it is 4 min shorter than the solar day because the earth moves in its orbit about the sun.

Sidewinder. Desert Rattlesnake (*Crotalus cerastes*) of the SW U.S. This 2 ft (60 cm) pale yellow and pink snake is named for its method of locomotion: It throws out successive loops at oblique angles so that it appears to move sideways. This form of movement allows it to traverse sand, which has little traction.

Siegbahn Unit. *See* X-unit. [After Karl Siegbahn (b. 1886), Swedish chemist.]

Siemens (mho, reciprocal ohm). Symbol: S. A unit of conductance equal to the conductance between two points of a conductor when a potential difference of one volt between these points causes a current of one ampere to flow. [After Sir William Siemens (1823-83), British engineer born in Germany.]

Siemens, Sir William. 1823-83, English electrical engineer; b. Germany. He went to Britain to introduce his and his brother Ernst's electroplating device and became a British citizen in 1859. He was head of the English branch of the Siemens firm, which made electrical apparatus. Siemens invented (1851) a water meter and developed, with his brother Frederick, a regenerative furnace that was the prototype for the open-hearth steelmaking process.

Siemens-Martin Process. *See* open-hearth process.

Sigma Orbital (a-orbital). A type of molecular orbital that is completely symmetrical about a line between the two atoms. Sigma orbital are responsible for the single bonds (*sigma bonds*) between atoms in covalent compounds.

Sigma Particle. Symbol: Σ. An elementary particle having either zero charge and a mass 2333 times that of the electron, a positive charge and a mass 2328 times that of the electron, or a negative charge and a mass 2343 times that of the electron. The three sigma particles are classified as hyperons.

Sigma Pile. A source of neutrons together with a moderator but no fissile materials. Such piles are used to study the properties of the moderator.

Signal. A changing electric current, waveform, or other agency that carries information.

Silane. Any of a class of hydrides of silicon with the general formula $Si_n H_{2n+2}$. They are named according to the number of silicon atoms: SiH_4 (silane), Si_2H_6 (disilane), Si_3H_8 (trisilane), etc. The silanes are formally similar to the alkanes but less stable due to the weakness of the Si-Si bonds: only the first six members of the series are known. A mixture of silanes is produced by the action of acid on magnesium silicide, Mg_2Si.

Silica. Silicon dioxide, SiO_2. It is a hard crystalline colourless material that occurs in several natural crystalline forms, including quartz, flint, and sand. *Vitreous silica* is a glassy transparent material made by fusing silica. It has a low coefficient of thermal expansion and is used in heat-resistant apparatus.

Silica or **Silicon Dioxide.** Chemical compound (SiO_2). It is widely and abundantly distributed throughout the earth, both in the pure state (colourless to white) and in silicates (*e.g.*, Quartz, Opal, Sand, Sandstone, Clay, Granite); in skeletal parts of various plants and animals, such as certain protozoa and Diatoms; and in the stems and other tissue of higher plants. Because silica has a low thermal coefficient of expansion, it is used in objects subjected to wide ranges of heat and cold. Unlike ordinary glass, it does not absorb infrared and ultraviolet light.

Silica Gel. A form of silica produced by coagulation of a sodium silicate sol. This gives a gelatinous solid, which can be dehydrated to give a hard granular material. It is used as an absorbent, drying agent, and catalyst support. Cobalt salts absorbed in the gel give it a blue colour when dry and a pink colour when moist.

Silicate. Any of a large number of oxy compounds of silicon, many of which occur naturally in rocks and minerals. The simplest type are the *orthosilicates* with formula M_4SiO_4, where M is a monovalent metal. These contain the ion SiO_4^{4-}. An example of an orthosilicate is zircon, $ZrSiO_4$, although such compounds are probably not completely ionic.

More complex silicates contain SiO_4 tetrahedra linked together by sharing corners. Two such units are present in the *pyrosilicate* ion, Si_2O^{6-}, which is found in the scandium ore thortveitite, Sc_2SiO-. The ions $Si_3O_9^{6-}$ and $Si_6O_{18}^{12-}$ both have cyclic structures in which the SiO_4 units are linked into a ring. The latter ion is found in beryl, $Be_3Al_2Si_6O_{18}$. The *pyroxenes* are a class of silicates containing infinite chains of SiO_4 units. They contain ions of the type $(SiO_3^{2-})_n$ together with sufficient positive ions to ensure neutrality. The *amphiboles* also have infinite chains with two strands held together by cross linkages. They contain ions of the type $(Si_4O_{11}^{6-})_n$. The various types of asbestos are amphiboles. The SiO_4 tetrahedra can also be joined together in infinite sheets, leading to the presence of $(Si_2O_5^{2-})_n$ ions. Such sheets are held together by the positives ions. This type of layer structure is found in micas. Finally the SiO_4 units characteristic of silicates can be linked together to form complex three-dimensional structures. In such lattices, some of the silicon atoms may be replaced by aluminium atoms, giving the *aluminosilicates*. The feldspars and the zeolites are examples of such compounds.

The compounds known as *metasilicates*, M_2SiO_3, do not contain SiO_3^{2-} ions. Instead they are either pyroxene compounds, having infinite chains, or compounds containing cyclic anions.

Silicic Acid. Any of a number of oxy acids of silicon. A mixture of silicic acids is produced by the action of acids on sodium silicates. It is best described as a mixture of hydrated forms of silica ($SiO_2.nH_2O$), from which no distinct acids can be isolated. The compounds *pyrosilicic acid*, H_6SiO_7, and *orthosilicic acid*, $H_4Si\ O_4$, are known in the form of their salts and esters. *Metasilicic acid*, H_2SiO_3, does not exist in this form: it is formally the parent acid of the polymeric metasilicates.

Silicide. Any compound of silicon with a more electropositive element.

Silicon. Symbol: Si. A nonmetallic element occurring widely as silicates in many minerals and as the oxide in quartz and sand; the second most abundant element in the earth's-crust (25.7% by weight). The element can be obtained as a brown amorphous powder or a brown crystalline electrically conducting material. It is made by reducing sand (SiO_2) with carbon in an electric furnace. High-purity silicon is obtained by the Van Arkel-de Boer process. The element is used in semiconductors and the production of some silicon compounds.

Chemically, it is fairly unreactive: the amorphous form will ignite in oxygen at high temperatures; both forms react with halogens and are soluble in alkali and hydrofluoric acid, although not in other acids. Unlike carbon, silicon does not form a large number of compounds with hydrogen although it does have many oxygen compounds, such as the silicates and silicones. A.N. 14; A.W. 28.086; m.pt. 1410°C; b.pt. 2355°C; r.d. 2.33 (crystalline); valency 4.

Silicon (Si). Nonmetallic element, discovered by Jons Jakob Berzelius in 1824. Silicon has brown amorphous and dark crystalline allotropic forms. The second most abundant element (28% by weight) of the earth's crust, it occurs in compound from as Silica, Silicon Carbide, and silicates. Silicon adds strength to alloys, and is used in transistors and other semiconductor devices. It is found a many plants and animals.

Silicon Carbide. Chemical compound (SiC) occurring as hard, dark iridescent crystals that are heat resistant, and insoluble in water and other common solvents. It is used as an Abrasive, in refractory materials, and i special parts for nuclear reactors. Silicon carbide fibers impart increased strength and stiffness to plastics and light metals.

Silicon carbide (carborundum). A bluish-black crystalline solid, SiC, that is one of the hardest substances known. It is made by heating coke with sand in an electric furnace using salt as a flux. Silicon carbide is used as an abrasive, a refractory material, and a semiconductor. Sublimes at 2210°C; r.d. 3.20.

Silicone. Polymer in which atoms of Silicon nd Oxygen alternate in a chain; bound to the silicon atoms are various organic groups, *e.g.*, the methyl group (CH_3). Silicones may be liquids, Rubbers, Resins, or greases. Water repellent, chemically inert, and stable at extreme temperatures, silicones are used as protective coatings and electrical insulators.

Silicone. Any of a class of synthetic polymeric compounds of the general formula $[R_2SiO]_x$, where R is an organic group. They contain rings or chains of alternating silicon and oxygen atoms with organic group attached to silicon. Many different types exist: they are used as lubricating oils, water repellants, electrical insulators, and constituents of polishes and lacquers. *Silicone rubbers* are synthetic polymers of dimethylsiloxane. $(CH_3)_2SiO$. They can be used at both high and low temperatures.

Silicon Tetrachloride. A colourless fuming liquid, $SiCl_4$, obtained by passing chlorine over hot silicon carbide. It is used to make silicones and as a source of pure silicon. M.pt. –70°C; b.pt. 58°C; r.d. 1.50.

Silk. Fiber produced by the Silkworm in making its cocoon, or textiles woven from such fiber. Legend has it that sericulture (the raising of silkworms) began in China in 2640 B.C. Raw silk was exported, but the export of silkworm eggs was punishable by death. Silkworm eggs and seeds of the mulberry tree, on which the worm feeds, were supposedly smuggled to Constantinople in A.D. c.550; thereafter Byzantium was famed for its silk textiles. In the 8th cent. the Moors brought sericulture and silk weaving to Spain and sicily, where exquisite silk fabrics were being woven by the 12th cent. Italy developed great silk-weaving centers-Lucca, Florence, and Venice- in the 13th and 14th cent. The French city of Lyons became a weaving center in the 15th cent. In 1685 Huguenots fleeing France after the revocation of the Edict of Nantes brought the art to England, where it became centered at Spitalfields, in London. In the American colonies, attempts to establish sericulture ultimately failed. The Asian *Bombyx mori*, which feeds on mulberry leaves, produces the finest silk and is thus the most widely raised silkworm. Wild silk is made by the tussah worm of India and China, which feeds on oak leaves. Silk manufacture begins with the reeling (unwinding) of the silk from the cocoons. In throwing, the raw silk is twisted and doubled to achieve various strengths and thicknesses. The silk is boiled in soap to remove the natural *gum*, then bleached or dyed. It is woven on delicate specialized looms to produce a wide variety of fabrics, *e.g.*, taffeta, faille, velvet, crepe. Modern silk production is highly mechanized, but the finest fabrics are still handwoven.

Silk-screen Printing. Multiple Printing technique, also called serigraphy, involving the use of stencils. Paint is applied to a fabric screen, penetrating areas not blocked by a stencil. Several stencils are used to produce a multicoloured print.

As a commercial medium, silk-screen printing has been used by such modern artists and Andy Warhol.

Silkworm. Name for the larva of various species of Moths, indigenous to Asia and Africa but now domesticated and raised for silk production throughout most of the temperate zone. After hatching, the larvae feed voraciously on Mulberry leaves. When they mature, they attach themselves to a twig and spin a thick, strong cocoon from a single continuous filament of pale yellow silk about a half-mile long. After about a week, the cocoon is unwound, and the silk processed. Only enough cocoons to ensure adequate reproduction are allowed to hatch. The most widely raised silkworm is the larva of *Bombyx mori.*

Siloxane. Any of a class of silicon compounds containing Si-O-Si bonds with organic alkyl or aryl groups bound to the silicon atoms. A simple example is $R_3SiOSiR_3$, where R is the organic group. The silicones are polymeric siloxanes.

Silt. Mostly Quartz mineral particles that are between sand and Clay in size and are formed by Weathering and decomposition of preexisting rock. Hardened silt becomes a sedimentary rock called siltstone.

Silver. Symbol: Ag. A white lustrous ductile and malleable transition element found native and as argentite (Ag_2S) and horn silver (AgCl). It is usually obtained during the refining of lead, copper, and nickel ores and a certain amount is recovered from the anode sludge produced in the electrolytic refining of copper. The metal is used for jewellery, coinage, decorative objects, and producing silver salts, especially those used in photography. It has the highest thermal and electrical conductivity of any element and is used in printed circuits. Silver is generally less reactive than copper, although it tarnishes in air because of attack by sulphur compounds. The most common compounds are those with silver in its +1 oxidation state: it forms silver I (*argentous*) ionic salts as well as numerous complexes. Silver II (*argentic*) compounds are nearly all complexes, although AgF_2 does contain Ag^{2+} ions. There are also a few silver III complexes. A.N. 47; A.W. 107.87; m.pt. 961.93°C; b.pt. 2212°C; r.d. 10.5; valency 1 or 2.

Silver (Ag). Metallic element, one of the first metals used by humans. Pure silver is nearly white, lustrous, soft, very ductile, malleable, and an excellent conductor of heat and electricity. It is used in mirrors, coins, utensils, antiseptics, jewellery, and for electrolytic plating of tableware. Silver nitrate is the most importance compound. Silver reacts with hydrogen sulfide in air to form silver sulfide (tarnish). Photographic emulsions contain silver halides,because of their sensitivity to light. Silver alloys are used in dentistry and electrical contacts. Sterling silver contains 92.5% silver and 7.5% copper. Silver is obtained from the ores argentite, cerarygrite, pyrargyrite, stephanite, and proustite.

Silver Bromide. A pale yellow crystalline solid, AgBr, that darkens on exposure to light. It is used as a light-sensitive material in photographic film. M.pt. 432°C; r.d. 6.5.

Silver Chloride. A white powder, AgCl, that darkens on exposure to light. It is used in silver plating and photography. M.pt. 445°C; b.pt. 1550°C; r.d. 5.6.

Silverfish. Primitive wingless Insect about 1/2 in. (1.27 cm) long, named for the tiny silvery scales on its body. Common indoors in cool, damp places, silverfish eat starch from bookbindings, wallpaper, and clothing.

Silver Glance. *See* argentite.

Silver Iodide. A pale yellow crystalline compound, AgI. M.pt. 556°C; b.pt. 1506°C; r.d. 6.01.

Silver Nitrate. A colourless poisonous crystalline solid, $AgNO_3$, made by dissolving silver in nitric acid. It is used in photography, in silvering mirrors, and as an analytical reagent. M.pt. 212°C; r.d. 4.3.

Silver Oxide. Either of two oxides of silver. *Silver I oxide*, Ag_2O, is a dark brown slightly soluble powder used as an oxidizing agent in chemical synthesis. Decomposes when heated; r.d. 7.14. It reacts with ozone to give *silver II oxide*, AgO.

Silver Point. *See* temperature scale.

Silversides. Small shore Fish (family Antherinidae) named for the silvery stripe on either side of its body, abundant in warmer waters of the Atlantic and Pacific. Related to mullets, silversides, also known as whitebait, feed on small Crustaceans and insects. Included in the family are the top and jack smelts and the California grunion (5-8 in./12.5-20 cm), which rides in on high tides to lay its eggs in the sand and is often found beached.

Silver Sulphide. A black insoluble solid, Ag_2S, precipitated from solution of silver salts by hydrogen sulphide. M.pt. 825°C; r.d. 7.32.

Silverwork. Term encompassing ecclesiastical an domestic utensils, jewellry, buttons, weapons, horse trappings, boxes, and other articles. It involves embellishments such as chasing, repousse, filigree, and inlaying. The art was highly developed in ancient times, as shown by Asian, Egyptian, Phoenician, Roman, and Byzantine examples. Much early Italian and French silverwork was melted down for reuse and thus lost. German, Swiss, Italian, Spanish, and English silverwork reached heights of ornamentation between the 15th and 18th cent. Much N European ecclesiastical silverwork was destroyed during the Reformation, and little early English silver survived the Wars of the Roses. Sheffield plate, an innovation of the mid-18th cent., led to important silverplate industries in England and, later, the U.S. In colonial America silversmithing was an important trade, attracting several hundred silverworkers from Europe. Colonial silver was simple in design and is prized by collectors. Silverwork is a traditional art of the indigenous peoples of the Americas and is still practiced by them today.

Simple Fraction. *See* fraction

Simple Harmonic Motion. Motion of a point moving along a path so that its acceleration is directed towards a fixed point on the path and is directly proportional to the displacement from this fixed point. The motion is thus described by the equation $d^2x/dt^2 = -k^2x$, x being the displacement from the point with due regard to sign and k being a constant. The solution has the form $x = A \sin (kt + b)$, where A and b are constants.

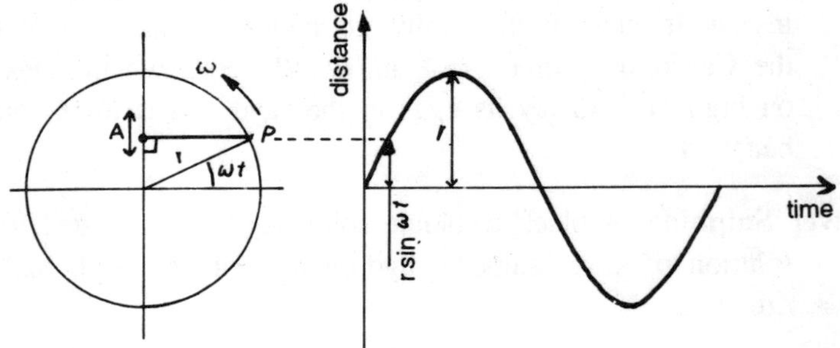

Fig. S-4. Simple harmonic motion of point A.

A representation of simple harmonic motion is shown in the diagram. The point P moves around the circle with constant angular velocity ω. The foot of a perpendicular from P onto a diameter moves with simple harmonic motion. The equation describing this motion is $x = r \sin (wt + a)$, where r, the radius of the circle, is the maximum displacement and α, the *epoch*, is the angular displacement at $t = 0$. The motion is a sinusoidal oscillation about O. The period of this oscillation is given by T = 2 Π/ ω.

It can be shown that simple harmonic motion occurs when a body or system moves under the influence of a restoring force F that is proportional to the displacement from its equilibrium position. Common examples of this are the motion of a pendulum (for small displacements) and the motion of a weight oscillating on the end of a vertical spring.

Simple Microscope. *See* microscope.

Simple Pendulum. *See* pendulum.

Simultaneous Equations. A set of two or more equations that together impose conditions on the same set of unknowns and whose solution is a value or set of values that satisfies all the equations. If solutions exist they are the points of intersection of the graphs of the individual equations.

Sine. *See* trigonometric function.

Sine Wave. A waveform that has an equation in which one variable is proportional to the sine of another. In simple harmonic motion, for example, the displacement is proportional to the sine of the time. The term is used with reference to the shape of the waveform: a waveform $y = A \cos x$ could also be described as a sine wave.

Singer, Isaac Merritt. 1811-75, American inventor; b. Rensselater co., N.Y. He patented a practical Sewing Machine (1851) that could do continuous stitching, and made 20 subsequent improvements (1851-65). Although he lost an infringement suit to Elias Howe, his company was so well established that it took the lead in a subsequent combination of manufactures and pooling of patents.

Single Bond. *See* covalent bond.

Single Crystal. A crystal that has a completely regular structure throughout.

Sintered Glass. Porous glass made by sintering. It is used for filtration.

Sintering. A process in which solid particles heated at a temperature below the melting point coalesce together to form a single mass. Glass and certain metals and ceramics can be sintered.

Sintering. Process of forming objects by heating a metal powder. When the metal powder is chemically or mechanically produced, compacted into the desired shape, and heated, the powder particles join together to form a solid object.

Sinus. Cavity or hollow space in the body, usually filled with air or blood. In humans, the paranasal sinuses, mucouslined cavities in the bones of the face, are connected by passageways to the nose and probably help to warm and moisten inhaled air. In invertebrates, spaces through which blood returns to the heart are also called sinuses.

Sinusoidal. Indicating a periodic quantity, motion, etc., that is described by a sine wave.

Siphon. A bent tube with one arm longer than the other, used for transferring liquid to a lower level over an intermediate barrier. The tube must be full of liquid for the siphoning action to begin. The siphon operates because of the excess weight of liquid in the longer arm.

Sirenian or **Sea Cow.** Large, aquatic Mammal of the order Sirenia. Living sirenians are the dugong and the manatee, both found in warm, sheltered waters, where they feed exclusively on sea vegetation. Their heavy, fishlike bodies end in a horizontally flattened fin; their skin is gray and virtually hairless. Sluggish and shy, they spend their entire lives in the water. Sirenians may reach 12 ft (3.6 m) in length and 600 lb (270 kg) in weight. The Florida manatee (*Trichechus manatus*) is protected.

Sisal Hemp. Important cordage fiber obtained from the leaves of several species of agave, of the Amaryllis family, and from related genera. The fiber, used especially for twine and considered second in strength to Manila Hemp, is obtained chiefly from the true sisal (*Agave sisalana*) and henequen (*A. fourcrlydes*).

SI Units (Systeme International d'Unites). A system of units used, by international agreement, for all scientific purposes. It is based on the metre-kilogram-second (MKS) system and consists of seven base units and two supplementary units. Measurements of all other physical quantities are made in derived units, which consist of combinations of two or more base units. Fifteen of these derived units have special names. Base units and derived units with special names have agreed symbols, which are used without a full stop.

Decimal multiples of both base and derived units are expressed using a set of standard prefixes with standard abbreviations. Where possible a prefix representing 10 raised to a power that is a multiple of three should be used (*e.g.* mm is preferred to cm).

Skeleton. Stiff supportive framework of the body. The two basic types are the exoskeleton and the endoskeleton. The shell of the clam is an exoskeleton that provides formidable protection but is bulky and severely restricts movement. The firm, flexible insect skeleton combines protective armor and a famework for attachment of muscles used in rapid movement. Such exoskeletons must be shed periodically to allow for growth (all exoskeletons are nonliving) and thus tend to limit size. The endoskeleton, a framework of living material enclosed within the body of vertebrates, permits larger size and freedom of movement. The general arrangement of skeletal parts into skull, spinal column, and ribs is the same in all vertebrates. The human skeleton consists of 206 bones, held together by flexible tissue consisting of cartilage and ligaments. In addition to its supportive function, the skeleton provides sites for the attachment of Muscles and protects vital organs, such as the Brain, Spinal Cord, Heart, and Lungs.

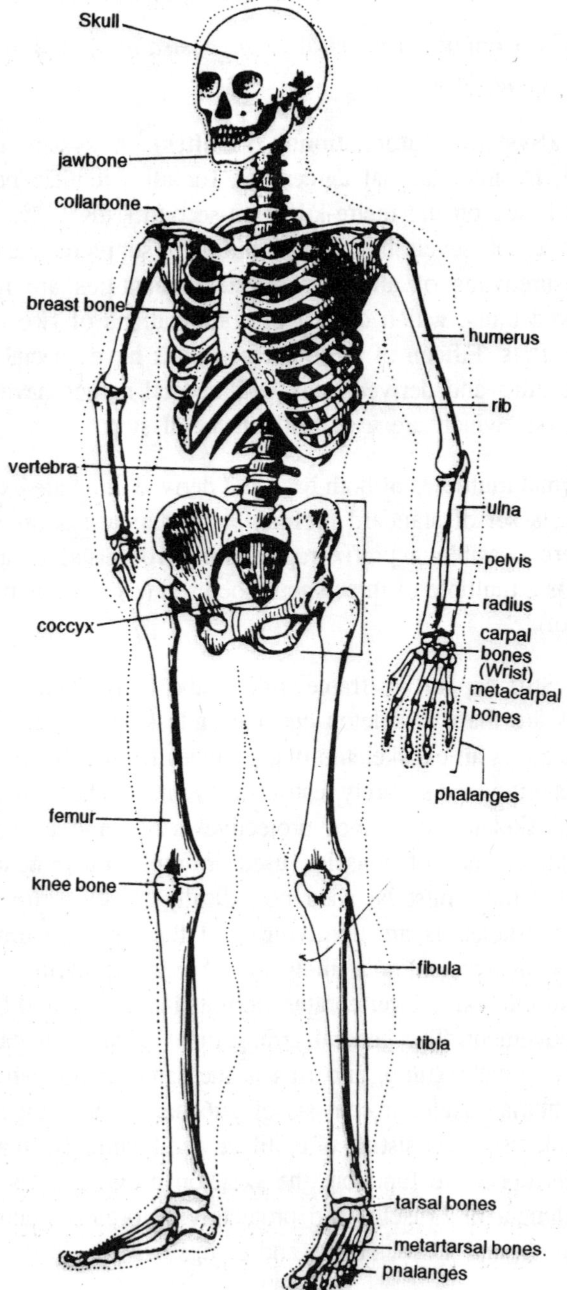

Fig. S-5. Human skeleton.

Skin. Flexible tissue enclosing the body of a vertebrate. In mammals, the skin is a complex organ of numerous structures serving vital protective and metabolic functions. It consists of two main cell layers: a thin outer layer (epidermis) and a thicker inner layer (dermis).The epidermis contains melanin, the pigment that gives the skin colour. Evolutionary adaptations of epidermis include horns, hooves, Hair, Feathers, and Scales. The dermis consists of Connective Tissue containing Blood vessels, lymph channels, nerve endings, sweat glands, sebaceous glands, fat cells, hair follicles, and Muscles. The nerve endings, called receptors, perform an important sensory function, responding to various stimuli, including light touch, pressure, pain, heat, and cold. The skin provides a barrier against invasion from outside organisms and protect underlying tissues and organs from abrasion and other injury. Its pigment shields the body from dangerous ultraviolet rays in sunlight. Skin prevents excessive loss of bodily moisture and in humans also performs functions that help maintain normal body temperature.

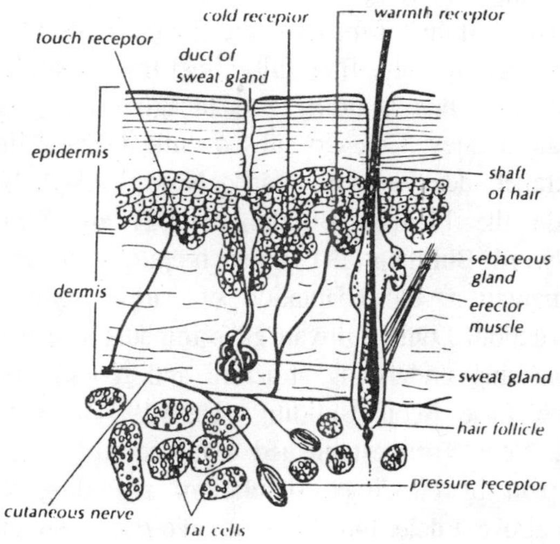

Fig. S-6. Cross section of skin.

Skunk. Several related New World Mammals of the Weasel family, characterized by their striking black-and-white striped fur and the strong, offensive odor they spray for defense. When severely provoked, the skunk squirts a mist from glands under the tail that causes choking and tearing of the eyes. Skunks are generally avoided by predators. The feed on rodents, insects, eggs, carrion, and vegetation, and live in one or more families in rock piles and abandoned burrows. Hunted for their fur, skunks range in size from 9 to 18 in. (23 to 46 cm), excluding the tail.

Skyscraper. Modern building of great height, constructed on a steel skeleton. The form originated in the U.S., and many late 19th cent. technological developments contributed to its evolution. In 1887 the first elevator was installed, and with the eventual perfecting of high-speed electric elevators skyscrapers were free to attain any height. The early tall masonry buildings required very thick walls in the lower stories, which limited floor space. The use of cast iron to permit thinner walls was followed by cage construction, in which iron frames supported the floors, while the walls were self-supporting. The first fully steel-frame building was the Home Insurance Building in Chicago (1883), designed by William Jenney. Chicago subsequently became the center of skyscraper development. Early New York City examples include the Flatiron Building (1902) and Cass Gilbert's Woolworth Building (1913), which epitomized, with is Gothic ornamentation, the adaptation of earlier styles to modern construction. Louis Sullivan gave impetus to a new aesthetic, emphasizing underlying structure and fenestration, *e.g.,* the Carson Pirie Scott Building (1899-1904), in Chicago. In 1916, New York established legal control over the height and plan of buildings; regulations regarding setting back walls above a determined height gave rise to the characteristic stepped profile. Skyscraper placement and design are a major concern of city planning. In 1982 the tallest skyscraper was the Sears Tower, Chicago, 110 stories, 1454 ft (473 m).

Major New York City skyscrapers are the twin towers of the World Trade Center, 110 stories, 1350 ft (442 m); the Empire State Building, 102 stories, 1250 ft (412 m); and the Chrysler Building, 77 stories, 1046 ft (349 m).

Sky Wave (ionospheric wave). A radio wave transmitted between points by reflection from the ionosphere rather than by direct transmission. *Compare* ground wave.

Slag. A nonmetallic material produced on the surface of the liquid metal in smelting and refining metals. It is usually a mixture of oxides, silicates, sulphides, phosphates, etc., produced by combination of a flux with the gangue of the ore or with impurities in the metal refined.

Slaking. The process of adding water, as in the production of slaked lime.

Slate. Fine-grained, characteristically gray-blue Rock formed by the Metamorphism of Shale. It splits into perfectly cleaved, broad, thin layers; this property is known as slaty cleavage. Better grades of slate are used for roofing.

Sleep. Resting state in which an individual becomes relatively quiescent and unaware of the environment. During sleep, most physiological functions, such as body temperature, blood pressure, and breathing rate, decrease. It is also a time of repair and growth, when some tissues proliferate more rapidly. Sleep occurs in cycles: so-called S sleep, characterized by large, slow brain waves, fills about three fourths of each short cycle. During the second stage, D sleep, parts of the nervous system are very active and rapid eye movements (REM) occur; Dreams apparently take place during REM sleep. Dream or sleep deprivation results in changes in human personality and in perceptual and intellectual processes.

Slide Rule. A mathematical instrument for simple calculations, consisting of fixed scales with a moving scale sliding between them. The scales are logarithmic: i.e. the distances along the scale are proportional to the logarithms of the index numbers. Numbers can be multiplied or divided by adding or subtracting distances using the moving scale.

Slime Mold or **Slime Fungus.** Organism usually classified with the Fungi, but showing equal affinity to the Protozoa. Slime molds have complex life cycles with an animallike motile phase, in which feeding and growth occur, and a plantlike immotile reproductive phase. The motile phase, commonly found under rooting logs and damp leaves, consists of either solitary amoebalike cells or a brightly colored multinucleate mass of protoplasm called a plasmodium, which creeps about and feeds by amoeboid movement. In the reproductive phase, slime molds are transformed into one or more reproductive structures, each consisting of a stalk topped by a spore-producing capsule. When the spores germinate they release amoebalike cells; in the plasmodium-forming plants, the cells grow and the nucleus subdivides to form a plasmodium.

Slip. Motion of one plane of atoms in a crystal relative to another in a direction tangential to the planes. This sliding motion occurs in the plastic deformation of solids.

Sloth. Tailless, arboreal Mammal found in tropical forests of Central and South America. The three-toed sloth (genus *Bradypus*), with three-toed front feet and five-toed hind feet, has a dense, gray-brown furry coat and is about the size of a house cat. Clinging to branches with powerful, curved claws, sloths eat, sleep, and travel upside down. Although sluggish, they can strike swiftly if attacked.

Slow Neutron. A neutron with an energy of only a few electron volts.

Slug. A unit of mass equal to the mass that would acquire an acceleration of one foot per second per second from a force of one pound-force. It is equivalent to 32.174 pounds (14.5939 kilograms).

Slug. Terrestrial Gastropod mollusk, a form of Snail, with a rudimentary shell and a lung for breathing air. Feeding at night, slugs devour both the roots and aerial portions of plants; some species are serious garden pests.

Slurry. A thin suspension of a solid in a liquid.

Smallpox. Acute highly contagious, sometimes fatal, disease causing a high fever and successive stages of severe skin eruptions. Caused by a virus that may be airborne or spread by direct contact, smallpox has occurred in epidemics throughout history. Edward Jenner, at the end of the 18th cent., demonstrated that cowpox virus was an effective vaccine against the disease. By the end of the 1970s, Vaccination programs, such as those conducted by the World Health Organization, had eliminated the disease world wide.

Smelt. Small, slender Fish of the family Osmeridae, allied to the grayling of the Salmon family. Most species are marine, but some spawn in fresh water and some are landlocked in lakes. The American smelt, or icefish *(Osmerus mordax)*, averages 10 in. (25 cm) in length and is valued as a food fish.

Smelting. 1. Any process of melting fusion especially to extract a metal from its Ore. Process vary depending on the ore and metal involved, but they are typified by the use of the Blast Furnace and the reverberatory furnace.

2. The metallurgical process of extracting metals from their ores by heating them in a furnace with a flux and usually a reducing agent, such as carbon. A melt is produced composed of molten metal with a layer of molten slag floating on top.

Smetic. *See* liquid crystal.

Smog. Dense, visible air Pollution, commonly of two types. The graw smog of older industrial cities like London and New York comes from the massive combustion of coal and fuel oil in or near the city, releasing ashes, soot, and sulfur compounds into the air. The brown smog characteristic of Los Angeles and Denver is caused by automobile emissions. Smog, which usually results in reduced visibility and can irritate the eyes and respiratory system, may be dangerous to people with respiratory ailments. It can also damage metal, rubber and other materials.

Smoke. A suspension of small solid particles in a gas.

Smooth. Denoting contact between two bodies in which there is considered to be no frictional force.

Smut. Name for plant diseases caused by Fungi of the order Ustilaginales. Smut produces sootlike masses of spores on the host, lowers its vitality, and often causes deformities. Serious threats to grain crops, smuts do not alternate hosts as do the Rusts. Severe annual crop losses are caused by the corn smut, oat smut, and loose smut of wheat. Bunt, the most serious smut, attacks young wheat seedlings and destroys the Grain.

Snail. Gastropod mollusk with a spirally coiled shell; there are thousands of species on land and in water. In aquatic species, respiration is carried on by gills; terrestrial forms often have lungs. Snails secrete a slimy substance over which they move by contractions of their muscular foot. Many species are eaten.

Snake. Limbless Reptile of the order squamata, which also includes the Lizards. The snake's extremely long, narrow body has many vertebrae, and paired internal organs are arranged linearly rather than side by side all snakes are deaf. some snakes—constrictors—crush their prey by wrapping their bodies around it and squeezing; others—venomous snakes—inject a toxic substance into their victims. The approximately 2,700 snake species, of which about four fifths are nonvenomous, are distributed throughout most temperate and tropical zones of the world. About two thirs of all species, most of them nonvenomous, belong to the family Colubridae, including garter and grass snakes. Most poisonous New World snakes belong to the Pit Viper family, while venomous Old World snakes are the true Vipers. The family Boidea includes the largest snakes, *e.g.*, Boas and Pythons, and the family Elapidae (snakes with inflatable neck hoods) includes the Cobra.

Snapdragon. Cultivated garden and greenhouse plant *(Antirrhinum majus)* of the figwort family, native to the Mediterranean area. Its showy blossoms, resembling a dragon's snout, display a wide range of colours and varieties.

Snapper. Carnivorous, spiny-finned Fish of the family Lutianidae, chiefly of tropical coastal waters. Snappers are active and voracious, with large mouths and sharp teeth. Best known is the red snapper, an important food fish. Found from the Gulf of Mexico N to Long Island, it grows up to 3 ft (90 cm) long and weighs up to 35 lb (16 kg).

Snell's Law. The principle that when refraction occurs the sine of the angle of incidence divided by the sine of the angle of refraction is a constant . [After Willebrord Snell (1591-1629), Dutch astronomer.]

Snipe. Shore Bird of the Sandpiper family, native to the Old and New World. The common, or Wilson's, snipe *(Capella gallinago)* is a game bird of marshes and meadows. The mud snipe, or woodcock *(Scolopax rusticola)*, is a nocturnal woodland bird.

Snow. Precipitation formed by the sublimation of water vapor into solid crystals at temperatures below freezing. A snowflake, like a raindrop, forms around a dust particle. Snowflakes form symmetrical (hexagonal) crystals, sometimes matted together if they fall through air warmer than that of the cloud in which they originated. Apparently, no two snow crystals are alike; they differ from each other in size, lacy structure, and surface markings. Snowfall has been produced artificially by introducing dry-ice pellets into clouds that contain unfrozen water droplets at temperatures below the freezing point. Melted, 10 in. (25 cm) of snow is approximately equal to 1 in. (2.5 cm) of rainfall.

Soap. Any of a group of organic compounds that are metallic salts of fatty acids. A soap of tallow and wood ashes was used as early as the 1st cent. A.D. by Germanic tribes. In the American colonies it was made from waste fats and lye, which is a strong alkali leached from wood ashes. The resulting chemical reaction, called saponification, remains the basis of soap manufacture today. Fats and Oils are heated with an alkali, *e.g.*, sodium hydroxide (which gives hard soaps) or potassium hydroxide (which gives hard soaps). Sodium or potassium may be replaced in the alkali by other metals, *e.g.*, aluminum, calcium, or magnesium, to make soaps used in industry as paint driers, ointments, and lubricating greases, and in waterproofing. After the alkali and fats have reacted, salt is added to form a curd of the soap. Glycerol (glycerin), a valuable by-product used as a solvent and sweetener, can then be removed by Distillation. Varying the composition or method of processing affects the lathering, cleansing, and water-softening properties. Soap can be formed as bars, chips flakes, beads, or powders and may contain perfumes, dyes, germicides, or so-called builders, which assist in rough cleaning. Like modern soapless detergents (usually sulfonated alcohols), soaps cleanse by lowering the surface tension of water, by emulsifying grease, and by absorbing dirt into the foam. Soap is less effective than detergent in hard water because the salts that make the water hard react with the soap to form insoluble curds (*e.g.*, the "ring" left in bathtubs).

Soap. A salt of a fatty acid. Normal household soap *(hard soap)* is mixture of sodium stearate, oleate, and palmitate. It is made by hydrolysing fats with caustic soda, thus converting the glycerides of stearic, oleic, and palmitic acids into the sodium slats and glycerol. *Soft soap* is a more liquid substance, produced by using potassium hydroxide instead of sodium hydroxide. Soaps owe their cleaning action to the structure of their anions, which consist of a long hydrocarbon chain attached to a carboxyl group. The hydrocarbon chain has an affinity for oil and grease and the charged carboxyl group has an affinity for water. Particles of grease or oil are thus emulsified in water. Insoluble salts of other metals with these acids are also called soaps. They are used as fillers and waterproofing agents.

Soapstone or **Steatite.** Metamorphic Rock in which the characteristic mineral is Talc. It is gray to green, has a soapy feel, and resists acids and heat. Soft and easily carved, it is a popular sculpture medium. Soapstone is also used to make sinks and laundry tubs. The chief deposits are in the U.S., Canada, and Norway.

Sociobiology. Application of the theory of Evolution to the study of animal and human social behaviour. Sociobiologists hold that behaviour patterns are genetically determined and are governed by the process of natural selection. The theories have been used successfully to explain animal altruism and reproductive and foraging behaviour, but they are controversial when applied to human behaviour in such areas as aggressiveness, sex differences, mate selection, and parenting behaviour. Edward O. Wilson's *Sociobiology* (1975) was instrumental in defining the field.

Socrates. 469-399 B.C., Greek philosopher of Athens, generally regarded as one of the wisest men of all time. It is not known who his teachers were, but he seems to have been acquainted with the doctrines of Parmenides, Heraclitus, and Anaxagoras. Socrates himself left no writings, and most of our knowledge of him and his teachings comes from the dialogues of his most famous pupil, Plato, and from the memoirs of Xenophon. socrates is described as having neglected his own affairs, instead spending his time discussing virtue, justice, and piety wherever his fellow citizens congregated, seeking wisdom about right conduct so that he might guide the moral and intellectual improvement of Athens. Using a method now known as the Socratic dialogue, or dialectic, he drew forth knowledge from his students by pursuing a series of questions and examining the implications of their answers. Socrates equated virtue with the knowledge one's true self, holding that no one knowingly does wrong. He looked upon the soul as the seat of both waking consciousness and moral character, and held the universe to be purposively mind-ordered. His criticism of the sophists and of Athenian political and religious institutions made him many enemies, and his position was burlesqued by Aristophanes. In 399 B.C. Socrates was tried for corrupting the moral of Athenian youth and for religious heresies; it is now believed that his arrest stemmed in particular from his influence on Alcibiades and Critias, who had betrayed Athens. He was convicted and, resisting all efforts to save his life, willingly drank the cup of poison hemlock given him. The trial and death of Socrates are described by Plato in the *Apology, Crito,* and *Phaedo.*

Soda. Any of various compounds of sodium, such as washing soda, caustic soda, and baking soda.

Soda Ash. *See* sodium carbonate.

Soda Lime. An off-white granular solid made by slaking calcium oxide with caustic-soda solution. It is mixture of the hydroxides of sodium and calcium and is used as a drying agent and absorbent for carbon dioxide.

Sodamide (sodium amide). A white crystalline solid, $NaNH_2$, with an odour of ammonia, made by passing ammonia over hot sodium. It is used in making sodium cyanide. M.pt. 210°C; b.pt. 400°C.

Soda Nitre. *See* sodium nitrate.

Soddy, Frederick. 1877-1956, English chemist. He worked under Lord Rutherford and Sir William Ramsay and was (1919-36) professor of chemistry at Oxford Univ. With others he discovered a relationship between radioactive elements and the parent compound, which led to his theory of Isotopes. For this work he won the 1921 Nobel Prize in chemistry. An advocate of technocracy and of the social credit movement, soddy wrote several books setting forth his views.

Sodium (Na). Metallic element, discovered in 1807 by Sir Humphry Davy; its compounds have been known since antiquity. A silver-white, very reactive Alkali Metal, it must be stored out of contact with air and water. The metal is used in arc-lamp lighting, as a heat-transfer liquid in nuclear reactors, and in manufacture of tetraethyl lead. Widely used compounds include Sodium Chloride (common salt), Sodium Bicarbonate (baking soda), Sodium Carbonate (soda ash), hydroxide (lye), nitrate, phosphates, and Borax. Soap is made with sodium hydroxide. Sodium compounds are widely distributed in rocks, soil, oceans, salt lakes, mineral waters, and salt deposits, and are found in the tissues of plants and animals. Sodium is an essential element of the diet.

Sodium Symbol: Na. A soft silvery-white metallic element; the sixth most abundant element in the earth's crust, occurring principally as sodium chloride in sea water and rock salt. It is manufactured by electrolysis of a fused mixture of sodium chloride and calcium chloride. Sodium is used as a reducing agent in the manufacture of some chemicals and as a cooling agent in nuclear reactors. A.N.11; A.W. 22.9898; m.pt. 97.5°C; b.pt. 892°C; r.d. 0.97; valency1.

Sodium Alginate. *See* alginic acid.

Sodium Amide. *See* sodamide.

Sodium Bicarbonate (sodium hydrogen carbonate). A white soluble powder, $NaHCO_3$, obtained by passing carbon dioxide into a saturated solution of sodium carbonate. It is used in effervescent drinks, baking powders, fire extinguishers, and as a medical antacid. Decomposes at 270°C; r.d. 2.16.

Sodium Bicarbonate, or **Sodium Hydrogen Carbonate.** Chemical compound ($NaHCO_3$), a white crystalline or granular powder, commonly known as bicarbonate of soda or baking soda. It is soluble in water and very slightly soluble in alcohol. Because it evolves carbon dioxide gas when heated above 50°C (122° F), It is used in baking powder. It is sometimes used medically to correct excess stomach acidity.

Sodium Carbonate. A white soluble salt obtained by the Solvay process. The commercial form *(soda ash)* is a white anhydrous powder, Na_2CO_3, used in the manufacture of glass, soaps, paper, and other chemicals. The decahydrate *(washing soda)* is a white crystalline solid, $Na_2CO_3.10H_2O$, used as a domestic cleanser and water softener. M.pt. 851°C (anhydrous); r.d. 2.5 (anhydrous), 1.4 (decahydrate).

Sodium Carbonate. Chemical compound (Na_2CO_3) soluble in water and very slightly soluble in alcohol. Pure sodium carbonate is a white, odourless powder that absorbs moisture from the air and forms a strongly alkaline water solution. One of the most basic industrial chemicals, it is usually produced by the Slovay process. The chief uses of sodium carbonate are in glassmaking and the production of chemicals.

Sodium Chlorate. A colourless crystalline solid, $NaClO_3$, obtained by electrolysing a concentrated acidic solution of sodium chloride. It is used as a powerful oxidant and bleach in manufacturing paper, herbicides, and explosives. M.pt. 248°C; r.d. 2.5.

Sodium Chloride (salt, common salt). A colourless crystalline solid occurring naturally in rock salt and in sea water. It is used to make many chemicals including ceramics, glasses, soaps, and fertilizers, and in seasoning and preserving foods. M.pt. 804°C; b.pt. 1413°C; r.d. 2.2.

Sodium Chloride (NaCl). Common salt. It is a chemical compound containing equal numbers of positively charged sodium and negatively charged chlorine Ions. The colourless-to-white crystals have no odour but a characteristic taste. When dissolved in water, the ions move about freely and conduct electricity. Salt is essential in the diet of humans and animals, and is a part of blood, sweet, and tears. Salt is widely used for the seasoning, curing, and preserving of foods. Its major use is in the production of Chlorine Sodium, and sodium hydroxide. Salt makes up nearly 80% of the dissolved material in seawater and is also widely distributed in solid deposits. Manufacture and use of salt is one of the oldest chemical industries.

Sodium Cyanide. A white poisonous deliquescent crystalline solid, NaCN, made by reaction of hydrogen cyanide with caustic soda. It has a variety of uses, including the case hardening of steel, the extraction of gold, and the manufacture of other chemicals. M.pt. 546°C; b.pt. 1500°C.

Sodium Dichromate (sodium bichromate). A red-orange deliquescent crystalline solid, Na_2Cr_2O-.$2H_2O$, manufactured by alkaline roasting of chromite ore, followed by leaching. It is used as an oxidizing agent and a source of chromium compound. M.pt. 357°C (anhydrous); r.d. 2.5. (dihydrate).

Sodium Fluoride. A poisonous colourless crystalline solid, NaF, used in the fluoridation of water supplies and as an insecticide. M.pt. 988°C; b.pt. 1695°C; r.d. 2.6.

Sodium Glutamate. *See* sodium hydrogen glutamate.

Sodium Hydrogen Carbonate. *See* sodium bicarbonate.

Sodium Hydrogen Glutamate (sodium glutamate, monosodium glultamate). A white crystalline powder, HOOC-$(CH_2)_2CH(NH_2)COONa$, with a meaty taste, extracted from sugar-beat waste by alkaline hydrolysis or synthesized from acrylonitrile. It is widely used in flavouring meat.

Sodium Hydrogen Sulphate (sodium bisulphate). An acidic white powder, $NaHSO_4$, obtained as a by-product in the manufacture of hydrochloric and nitric acids. It is used as a flux and an industrial cleaner. M.pt. 315°C; r.d. 2.7.

Sodium Hydrogen Sulphite (sodium bisulphite). A white crystalline powder, $NaHSO_3$, with an odour of sulphur dioxide. It is prepared by saturating sodium carbonate olution with sulphur dioxide, and is used principally as a reducing agent, antiseptic, and bleach. R.d. 1.5.

Sodium Hydroxide (caustic soda). A white deliquescent solid, NaOH, that is strongly alkaline in solution and is very corrosive to organic tissue. It is manufactured by the electrolysis of sodium chloride solution and is used in making rayon, paper, detergents, and many other chemicals. M.pt. 318°C; b.pt. 1390°C; r.d. 2.1.

Sodium Hhypochlorite. An unstable white crystalline solid, NaOCl, usually stored as a pale green aqueous solution. It is manufactured by electrolysis of cold dilute sodium chloride solution and is an oxidizing agent used in bleaching paper and textiles and as an antiseptic and fungicide. M.pt. 18°C.

Sodium Monoxide. *See* sodium oxide.

Sodium Nitrate (soda nitre). A colourless crystalline solid, $NaNO_3$, occurring naturally. It is used as an oxidizing agent and fertilizer and in making glass and pyrotechnics. The commercial salt is often called *Chile saltpetre or caliche.* M.pt. 308°C; b.pt. 380C (decomposes) r.d; 2.3.

Sodium Nitrite. A white or yellowish crystalline solid or powder, $NaNO_2$, made by heating sodium nitrate. It is used for making azo dyes. M.pt. 271°C; r.d. 2.2.

Sodium Oxide. Any of three oxides of sodium. *Sodium monoxide,* Na_2O, is a white powder produced by direct combination of the element with a deficiency of oxygen. It reacts with water to give sodium hydroxide. Sublimes at 1300°C; r.d. 2.27. In excess oxygen *sodium peroxide,* Na_2O_2, is formed. This is a yellowish white powder used as an oxidizing and bleaching agent and as a source of hydrogen peroxide. M.pt. 460°C; r.d. 2.8. A superoxide, NaO_2, also exists.

Sodium Peroxide. *See* sodium oxide.

Sodium Silicate. A mixture of silicates obtained by fusing sodium carbonate and sand (SiO_2) in an electric furnace. The product contains $x\text{Na}_2\text{O}.y\text{SiO}_2$. with the ratio x:y varying from 2:1 to about 1:4.The compounds form clear viscous colloidal solutions in water *(water glass)* and are used in making silica gel and detergents and in sizing textiles.

Sodium Sulphate. A white crystalline salt, Na_2SO_4, extracted from natural deposits and used in textile dyeing and detergents. The decahydrate, $Na_2DO_4.10\text{-}H_2O$, is known as *Glauber's salt.* M.pt. 888°C; r.d. 2.7.

Sodium thiosulphate. A colourless crystalline solid, $Na_2S_2O_35H_2O$, made by reacting sulphur with sodium sulphite or as a by-product in the manufacture of sodium sulphide. It is a reducing agent, used as a fixer in photography *(hypo)* and as an antichlor. M.pt. 48°C; r.d. 1.7.

Sodium-vapour Lamp. A type of fluorescent lamp in which an electrical discharge is passed through a low pressure of sodium vapour, causing it to emit a characteristic orange-yellow light.

Soft. 1. Denoting electromagnetic or particle radiation with relatively low energy and, as a consequence, low penetrating or ionizing power. Soft X-rays are X-rays with long wavelength and low photon energy.

2. Denoting a vacuum in which there is a relatively high pressure of gas, such that ionization of gas plays an important role in any electrical phenomena occurring in the vacuum.

Soft Iron. Iron that contains very little carbon. Soft iron has a low remanence, losing most of its magnetization when the external magnetic field is removed.

Soft Soap. *See* soap.

Soft Solder. *See* solder.

Software. The programs used in a computer, as distinguished from the equipment itself—the *hardware.*

Soft Water. *See* hardness.

Soil. Surface layer of earth containing organic matter and capable of supporting vegetation. A few inches to several feet thick, soil consists of fine rock material, Humus, air, and water. The arrangement of soil particles—soil structure—together with organic matter (including microorganisms and living roots)—determines the soil's capacity to retain gases, water, and plant nutrients. Soil varies with the type of vegetation, climate, and parent rock material. Soil fertility—determined in part by texture, chemical composition, water supply, and temperature—can be maintained or improved by Fertilizers or by cultivation practices.

Sol. A colloidal suspension of small particles of one substance in another. A *hydrosol* is a dispersion in water, an *aerosol* is a dispersion in air. A sol is distinguished from a gel by the fact that the dispersed particles are independent. There are two main types.

Lyophobic (or *emulsoid)* sols are dilute suspensions that are easily precipitated. An example is a colloidal suspension of small metal particles produced by striking an arc under water. Lyophobic colloids have little affinity for the dispersion medium and are irreversible. Smokes and mists are other examples.

Lyophilic (or *suspensoid)* sols are reversible systems in which there is affinity between the dispersed particles and the dispersion medium. An example is a solution of gelatin. This type of sol may be more concentrated and can be coagulated to form a gel.

When the dispersion medium is water the terms *hydrophobic* and *hydrophilic* are used.

Solar Cell. Any electrical device for converting solar energy directly into electrical energy. Solar cells are used to produce power in spacecraft. Essentially, they depend on the photovoltaic effect developed across a semiconductor junction. Another type of solar cell consists of an array of thermocouples heated by the sun.

Solar Constant. The energy per unit area per unit time received from the sun at a point that is the earth's mean distance from the sun away. The solar constant gives the energy flux that would be received by the earth in the absence of its atmosphere. It has the value 1400 joules per square metre per second.

Solar energy. Any form of Energy radiated by the Sun, including light, radio waves, and X rays. Solar energy is needed by green plants for the process of Photosynthesis, which is the ulitmate source of all food. The energy in fossil fuels (*e.g.*, Coal and Petroleum) and other organic fuels (*e.g.*, Wood) is derived from solar energy. Difficulties with these fuels have led to the invention of devices that directly convert solar energy into usable forms of energy, such as electricity. Solar batteries, which operate on the principle that light falling on photosensitive substances causes a flow of electricity, play an important part in astronautics but are presently too expensive to be in common use on the earth. Thermolectric generators convert the heat generated by solar energy directly into electricity. Heat from the sun is used in air-drying a variety of materials and in producing salt by the evaporation of sea water. Experimental solar heating system can supply heat and hot water for domestic use; heat collected in special plates on the roof of a house in stored in rocks or water held in a large container. Such systems, however, usually require a conventional heater to supplement them. Solar stoves, which focus the sun's heat directly, are employed in regions where there is much perennial sunlight.

Solar Flare. An explosive and violent surge or outburst of radiation apparently caused by electrical storms in the chromosphere of the sun. Flares are very bright and may last for up to half an hour, spreading horizontally throughout the chromosphere in a direction parallel to the surface of the photosphere. They emit large amounts of short-wave radiation and send out cosmic rays that can affect the terrestrial ionosphere causing magnetic storms. It is possible that they are associated with sunspots and faculae.

Solar System. The Sun, and the family of Planets, natural Satellites, Asteroids, Meteors, and Comets that are its captives. The principal members of the sun's retinue are the nine major planets; in order of increasing distance from the sun, they are Mercury, Venus, Earth, Mars, Jupiter, Saturn, Uranus, Neptune, and Pluto. All the planets orbit the sun in approximately the same plan (that of the Ecliptic) and move in the same direction (from west to east). Current theories suggest that the solar system was formed from a Nebula consisting of a dense nucleus, or protosun, surrounded by a thin shell of a gaseous matter extending to the present edges of the solar system. Because of gravitational instabilities, the nebula eventually broke up into whirlpools of gas, called protoplanets, within the rotating mass. In time the protoplanets condensed and accreted to form the planets.

Solar Time. Time defined by the position of the sun. The observer's *local solar time* is 0 hr (noon) when the center of the sun is on the observer's meridian. The *solar day* is the time it takes for the sun to return to the same meridian in the sky. The length of the solar day varies throughout the year because the earth moves with varying speed in its orbit and because the equatorial plane is inclined to the orbital plane. It is thus more convenient to define time in terms of the *mean solar time,* or average of local solar time; hence every mean solar day is of equal length. The *equation of time* is the difference between the local solar time and the mean solar time at a given location. *Civil time* is mean solar time plus 12 hr; the civil day begins at midnight, whereas the mean solar day begins at noon. *Greenwich mean time* (GMT) is the local civil time at the former site of the Royal Observatory in Greenwich, England, which is located on the Prime Meridian (0° longitude). *Standard time* is the civil time within one of the 24 times zones into which the earth's surface is divided. Within a zone all locations keep the same time, namely, the mean solar time of the central meridian (except when Daylight Saving Time is in effect). Zone times generally differ by a whole number of hours from GMT.

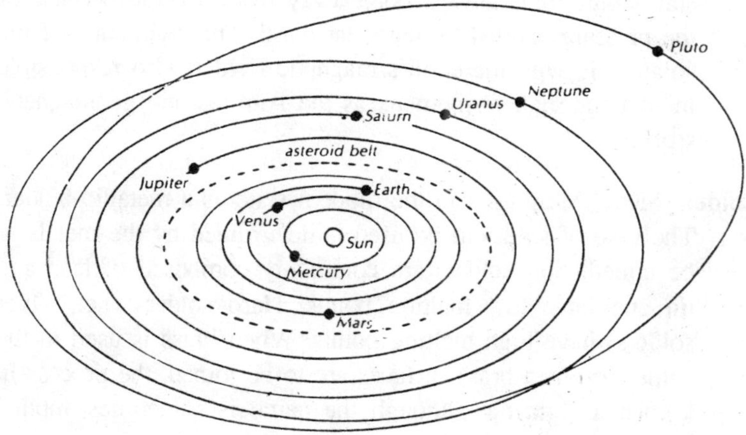

Fig. S-7. Solar system.

Solar Time. Time measured by reference to the sun. The *solar day* is the time between successive transits of the sun across the meridian. Averaging this time out over a year gives the *mean solar day.* The *solar month* is the time taken for the moon to complete one revolution of the earth, returning to the same longitude. The *solar year* is the time between successive arrivals of the sun at the first point of Aries. 1 solar year =

365.242 19 mean solar days; 1 solar month = 27.321 58 mean solar days.

Solar Wind. Stream of electrons and protons emitted by the sun. The solar wind is responsible for the formation of the Van Allen belts and the aurora. It also distorts the symmetry of the earth's magnetic field, causing it to extend further into space on the side of the earth that is distant from the sun. The intensity of the solar wind is greatest during the occurrence of solar flares and sunspots.

Solar Wind. Stream of ionized hydrogen and helium that radiates outwards from the sun, carrying away about 1 million tons of gas per sec. Near the earth the solar wind normally has a velocity of 450 mi/sec (700 km/sec). The wind is believed to extend to between 100 and 200 Astronomical Units from the sun. Comet tails always point away from the sun because of the pressure exerted by the solar wind. The interaction of the solar wind with the earth's magnetic field is also responsible in part for such phenomena as the Auroras and geomagnetic storms.

Solder. Metal Alloy used in the modern state as a metallic binder. The type of solder to be used is determined by the metals to be united. Soft solders are commonly composed of lead and tin and have low melting points. Hard solders (i.e., silver solders) have high melting points. When brass is used in the solder or when brass surfaces are to be joined, the process in known as brazing, although the name is sometimes applied also to other hard soldering.

Solder. Either of two alloys used for joining metals together. *Soft solder* is an alloy of tin (about 50%) and lead with a melting point in the range 200-300°C. The soft solder used in making electrical joints has a core of resin flux. *Hard solder* is an alloy of brass with a melting point above about 800°C. It is applied with a gas torch and a zinc chloride or similar flux—a process known as *brazing*.

Solenoid. A long spiral coil of wire, usually cylindrical, through which an electric current can be passed to produce a magnetic field. Solenoids are used in electromagnets and other devices.

Solenoid. Device made of a long wire wound many times into a tightly packed cylindrical coil. If current is sent through a solenoid made of insulated wire and having a length much greater than its diameter, a uniform magnetic field will be created inside the solenoid. The magnetic field can be intensified by inserting a ferromagnetic core into the solenoid.

Solid. 1. A state of matter characterized by incompressibility and resistance to shear stress. Unlike a gas, a solid has a fairly constant shape: unlike a liquid it will not flow to take the shape of its container.

The intermolecular forces involved are large enough to keep the atoms or molecules in fixed positions, about which they vibrate. Solids are classified into amorphous solids, in which there is no regular structure, and crystalline solids, in which the atoms have a regular arrangement.

2. A three-dimensional geometric figure, such as a cube or sphere.

Solid Angle. ω Symbol: or Ω The region subtended by an area at a point, measured by the ratio of the area that the solid angle intercepts on the surface of a sphere to the square of the surface's radius: $\omega = A/r_2$. Solid angles are measured in steradians. The total solid angle around a point is 4Π steradians.

Solid Solution. A homogeneous solid mixture of two or more substances, in which atoms, ions, or molecules of one substance occupy some of the lattice sites normally occupied by the other. The composition of solid solutions can vary within certain limits.

Solid-state. Denoting an electronic device that contains transistors or other semiconductor devices, rather than thermionic values.

Solid-state Physics. The experimental and theoretical study of the properties of the solid state, in particular the study of energy levels and the electrical and magnetic properties of metal and semiconductors.

Solid-state Physics. The study of properties exhibited by atoms because of their association and regular, periodic arrangement in Crystals. Besides mechanical and thermal properties, electric conductivity is one of the most important properties of solids. Metals are highly conductive and offer little resistance to electric currents. Most solid nonmetals are insulators; they offer virtually infinite resistance to electric currents. Semiconductors, which possess electrical conductivity that is neither very high nor very low, are used in Transistors.

Solstice. One of two points at which the sun appears to be at its furthest points north or south. The ecliptic is then at its maximum distances from the celestial equator. In the northern hemisphere the *summer solstice* occurs on about June 21 and the *winter solstice* occurs on about December 22. These correspond to the longest and shortest days respectively. The terms are used both for the points of maximum displacement and the times at which these occur. *Compare* equinox.

Solubility. The extent to which one substance will dissolve in another. Solubilities are expressed, under specified conditions of temperature and sometimes pressure, as the concentration of the saturated solution.

Solubility Product. The equilibrium constant of dissolved ions in equilibrium with undissolved solid. The concentration of the undissolved solid is taken to be unity, so the solubility product is simply a product of ionic concentrations at saturation. Silver chloride, for instance, has a solubility product given by $K_s = [Ag^+]\,[Cl^-]$. If the ionic product is exceeded, as by mixing silver nitrate with sodium chloride, silver chloride is precipitated until the ionic product falls to K_s. In general, the solubility product of a salt A^xB^y is $[A[^x[B]^y$.

Soluble. Denoting a substance that dissolves. Unless the solvent is specified, it is understood to be water.

Solute. *See* solution.

Solution. A homogeneous mixture of two or more substances. The atoms or molecules of the components are completely intermingled and the solution thus consists of only one phase. When a solid or gas is dissolved in a liquid, the liquid is the *solvent* and the dissolved substance is the *solute.* When one liquid substance is dissolved in another, the one in excess is the solvent.

Solvation. Association of molecules of a solvent with an ion in solution. In water, for example, a positive ion will be surrounded by a number of water molecules. In the case of transition metal ions, the closet water molecules are held by coordinate bonds. Any positive ion, however, will attract water molecules because of electrostatic forces between the ion and the polar water molecules. The process is known as hydration.

Solvay Process (ammonia soda process). A process for the manufacture of sodium carbonate from sodium chloride and calcium carbonate. The calcium carbonate is heated to give calcium oxide and carbon dioxide. This is bubbled into solution of sodium chloride that has been saturated with ammonia, thus precipitating sodium hydrogen carbonate and leaving ammonium chloride in solution. The sodium hydrogen carbonate is heated to yield sodium carbonate while the ammonium chloride is heated with calcium oxide, from the calcium carbonate, to regenerate the ammonia. [After Ernest Solvay (1833-1922), Belgian chemist.]

Solvent. *See* solution.

Solvent Extraction. Extraction of one component from a mixture by use of a solvent that dissolves the required substance without dissolving other components. The method is often used for extracting natural products from the tissues of plants. Mixtures may also be separated by partition between two different mutually immiscible solvents. The mixture is dissolved in one solvent, this is shaken with the other, and the two are then left to separate into layers. With a suitable choice of solvents, one layer contains the required substance.

Solvolysis. Reaction of a compound with its solvent. If the solvent is water the reaction is hydrolysis.

Sonar (asdic). A device for detecting objects under water by transmitting pulses of high-frequency sound and detecting the reflected pulses. The time difference between transmitting a pulse and collecting the reflected pulse gives a measure of the depth of the object. The system is similar to radar: the term is a shortened form of *sound navigation ranging*.

Sonar. Device for the location of submerged objects and for submarine detection and communication at sea. Capable of rotation, it can scan a surrounding area. The device projects subsurface sound waves and, as a Hydrophone, listens for returning echoes, determining the range and bearing of submerged targets. Returning signals may be audibly sounded through a loudspeaker and/or visually displayed by a Cathode-Ray Tube. Simpler sonar devices are used as depth finders and to locate schools of fish.

Sonic Boom. The loud bang produced by the shock wave from an aircraft travelling at supersonic speed.

Sorbite. A constituent produced in steels by tempering martensite above about 500°C. It consists of cementite in a matrix of ferrite (alpha iron).

Sorbitol. A white crystalline alcohol, $CH_2OH(CHOH)_4CH_2OH$, with a sweet taste. The D- compound is present in some fruits and can be made by reducing glucose. It is used as a sweetening agent and a raw material for the manufacture of vitamin C and polyurethane resins. M.pt. 110°C; b.pt. 295°C (3.5 mmHg); r.d. 1.5.

Sorghum. Tall coarse annual *(Sorghum vulgare)* of the Grass family, similar in appearance to Corn and probably indigenous to Africa. Valued for their drought resistance, its varieties include the sweet sorghums, which yield molasses from the cane juice; the broomcorns, which yield a fiber used for brooms; the grass sorghums, used for pasture and hay; and the grain sorghums, used primarily for stock and poultry feed and, in the Old World, for human food.

Sorption. Absorption or adsorption by a solid.

Sorption Pump A type of vacuum pump consisting of a vessel containing an absorbent material cooled in liquid nitrogen. Gas is absorbed on the material, usually activated charcoal or a molecular sieve, and removal from the system. Sorption pumps saturate in time and have to be regenerated by heating.

Sound. Mechanical longitudinal vibration of air, water, or other material, carrying energy through the medium. Sound is transmitted in the form of longitudinal waves: alternate compressions and rarefactions of the medium travelling at a characteristic velocity. It is produced by a vibrating object which creates the disturbance and cannot be transmitted through a vacuum. The term is often used for those waves that can be detected by the human ear: *i.e.* waves with frequencies between about 20 and 20,000 hertz. However, any analogous wave motion in any medium can also be designated as sound. Sound is characterized by the frequency of the wave motion (judged by its pitch), the intensity of the sound (judged by its loudness), and the quality of the sound. The velocity in gases is independent of the pressure: in air it is 332 metres per second at 0°C.

Sound. Pressure Waves that propagate through air or other media. Sounds are generally audible to the human ear if their frequency lies between 20 and 20.000 vibrations per second. Sound waves with frequencies below the audible range are called subsonic, and those with frequencies higher than the audible range are called ultrasonic. When a body, such as violin string, vibrates, or moves back and forth, its movement in one direction pushes the molecules of the air before it, crowding them together. When it moves back again past its original position and on to the other side, it leaves behind it a nearly empty space. The body thus causes alternately in a given space a crowding together of the air molecules (a condensation) and a thinning out of the molecules (a rarefaction). The condensation and rarefaction make up a second wave; such a wave is called longitudinal, or compressional because the vibratory motion is forward and backward along the direction that the wave is following. Because such a wave consists of a disturbance of particles of a material medium, sound waves cannot travel through a vacuum. The velocity of sound in air at 32°F(0°C) is 1,089 ft/sec (331.9 m/sec) but at 68°F (20°C) it is increased to about 1,130 ft/sec (344.4 m/sec). Sound travels more slowly in gases than in liquids, and more slowly in liquids than in solids. The pitch of a sound depends upon the frequency of vibraticn; the higher the frequency, the higher the pitch. Loudness, or intensity of sound, is measured in units called Decibels.

Soybean, Soya bean or **Soy pea.** Plant *(Glycine max)* of the Pulse family, natives to tropical and warm temperate regions of the Orient, where it has been a principal crop for at least 5,000 years. There are over 2,500 varieties in cultivation, producing high-Protein beans of many sizes, shapes, and colours. Soybeans are used in many forms, *e.g.*, oil, soybean meal, soy, sauce, soy milk, and bean curd, and as a Coffee substitute. Soybean oil is also valuable for its use in the manufacture of other products, *e.g.*, glycerin, soaps, and plastics. The green crop is used for forage and hay.

Soxhlet Apparatus. A laboratory apparatus for extracting soluble substances, consisting of a type of reflux condenser with a special container, usually a thimble of porous paper, into which the material to be treated is placed. Solvent, heated in a flask, circulates continuously over the material and carries away its soluble constituents.

Space Charge. A region of net electric charge caused by a collection of electrons or ions in a semiconductor or in a thermionic valve or other electron tube.

Space Exploration. The investigation of physical conditions in space and on stars, planets, and natural satellites through the use of artificial Satellites, Space Probes, and manned spacecraft. Although studies from earth using optical and radio Telescopes had accumulated much data on the nature of celestial bodies, it was not until after World War II that the development of powerful Rockets made direct space exploration a technological possibility. Manned spaceflight progressed from the simple to the complex, starting with suborbital and orbital flights by a single astronaut or Cosmonaut (Mercury and Vostok); subsequent highlight include the launching of several crew members in a single capsule (beginning with Gemini and Voskhod), rendezvous and docking of two spacecraft (beginning with Gemini and performed internationally in the *Apollo-Soyuz Text Program*), lunar orbit and landing (Apollo), the launching of space stations (Salyut and Skylab), and the launching of a reusable space vehicle, the space shuttle.

Space Law. Principle of law governing the exploration and use of outer space. The 1967 Outer space Treaty, signed by most nations, states that International Law applies to outer space and that while all states may freely explore and use outer space, territorial claims in space are prohibited. Other treaties dealing with rescue and return of astronauts, liability for damage caused by space objects, and registration of space objects became effective, respectively, in 1968, 1972, and 1976. A treaty on the UN in 1979, has been signed by several nations; the U.S., as of 1982, had not signed it.

Space Medicine. Study of the medical and biological effects of space travel on living organisms. The principal aim is to discover how well and for how long humans can withstand conditions encountered in space and to study their ability to readapt to the earth's environment after travel in space. Medically significant aspects of space travel include weightlessness, inertial forces experienced during liftoff, radiation exposure, absence of day-night cycle, and heat produced within the spacecraft. Participants in Space Exploration initially suffered from symptoms such as nausea, sensory disorientation, and poor muscular coordination, but astronauts and animals living a space are now able to adept to long periods of space travel without significant disability. Shielding and protective clothing prevent exposure to radiation from space and from nuclear reactors on board.

Space Probe. Unmanned space vehicle, usually carrying sophisticated instrumentation, designed to explore various aspects of the Solar System. Unlike an artificial Satellite, which is placed in more or less permanent orbit around the earth, a space probe is launched with enough energy to escape the gravitational field of the earth and navigate between planets. Radio contact between the control station on earth and the space probe provides a channel for transmitting data recorded by on-board instrument back to earth. A probe may be directed to orbit a planet, to soft-land instrument packages on a planetary surface, or to fly by one or more planets and/or natural satellites, approaching within a few thousand miles.

Space Shuttle. Reusable U.S. space vehicle. It consists of a winged orbiter, two solid-rocket boosters, and an external tank. Lift-off thrust is derived from the orbiter's three main liquid-propellant engines and the boosters. After 2 min the latter use up their fuel, are separated from the spacecraft, and—after deployment of parachutes—are recovered following splashdown. After about 8 min of flight, the orbiter main engines shut down; the external tank is then jettisoned and burns up as it reenters the atmosphere. The orbiter meanwhile enters orbit after a short burn of its two small Orbiting Maneuvering System (OMS) engines. To returns to earth, the orbiter turns around, fires it OMS engines to reduce speed, and, after descending through the atmosphere, lands like an airplane. Following four orbital test flights (1981-82) of the space shuttle *Columbia,* operational flights began in Nov., 1982. Missions of the space shuttle will include the orbiting within the shuttle of the Spacelab scientific workshop and the insertion into orbit of the space Telescope, the Galileo Space Probe, and a wide variety of communications, weather, scientific, and defense-related satellites.

Space Station or **Space Platform.** Artificial earth satellite, usually manned, that is placed in a fixed orbit and can serve as a base for astronomical observations, zero-gravity materials processing, satellite repair, or (possibly) weapons, or as a staging area for constructing large communications satellites to be placed in geosynchronous orbit. Early examples of space stations are the American *Skylab* and series of Soviet Salyut spacecraft. The American physicist Gerard O'Neill has proposed the construction of space colonies—evary large space stations built from lunar or asteroidal material and inhabited by several thousand people.

Space-time. Central concept in the theory of Relativity that replaces the earlier concepts of space and time as separate absolute entities. In space-time, events in the universe are described in terms of a four-dimensional continuum in which each observer located an event by three spacelike coordinates and a timelike coordinate. The choice of the last is not unique; hence, time is not absolute but is relative to the observer.

Space-time. The reference system of three dimension of space and one of time used in the theory of relativity to describe events. This system is sometimes called the *space-time continuum.*

Spallation. A type of nuclear reaction in which an atomic nucleus, hit by a high-energy photon or particle, emits several other particles or fragments.

Spark. A discharge of electricity occurring in air or other gases at normal pressures in the form of luminous wandering streamers between two points of opposite high electric potentials. A spark can only occur when the potential difference exceeds a certain value, the *sparking potential.* Usually the heating of the air by the spark causes a crackling noise.

Spark chamber. An apparatus for detecting and studying the behaviour of high-energy ionizing particles. The chamber is filled with a gas such as neon and contains a stack of many parallel metal plates separated by narrow gaps. Alternate plates are given a high potential so that a particle passing through triggers a succession of short sparks along its track. The trace can be photographed from the side of the chamber.

Spark Gap. A gap between two electrodes designed so that a spark occurs when their potential difference exceeds a certain value.

Spark Photography. A photographic technique in which the camera is set up in the dark with the shutter open and the object is illuminated by a spark of very brief but known duration, thus allowing photography of a rapidly moving object.

Sparrow. Small, perching New World Bird (genus *Passer*) of the Finch family. Field and hedge birds inconspicuously coloured in dull grays and browns, sparrows have stout, conical beaks adapted to seed eating, and are valuable to farmers in destroying weed seeds.

Special Relativity. The special theory of relativity.

Specific. Per unit mass. The use of *specific* in the name of a physical quantity indicates that it is the property of unit mass, and therefore a property of a substance or material rather than of an object or system. *Specific volume,* for example, is volume per unit mass. In the names of some physical quantities *specific* does not have this meaning, in particular in specific gravity and specific resistance.

Specific Activity. Symbol : *a*. The activity per unit mass of a radiosotope or radioactive material.

Specific Charge. The electric charge per unit mass. The term is usually used with reference to the charge-to-mass ratios of elementary particles.

Specific Heat Capacity. Symbol: *c*. The heat capacity of unit mass of a substance. It is the amount of heat required to raise the temperature of unit mass by unit temperature and is usually measured in joules per kilogram per kelvin. The heat required to raise the temperature of a gas depends on the way its pressure and volume change during the absorption of heat. Two *principal specific heat capacities* are used: one measured at constant pressure, c_p, and other measured at constant volume, c_v. The one at constant pressure is larger because some of the heat supplied is used in performing work. The ratio of these quantities, c_p/c_v, is given the symbol γ. Its value depends on the number of degrees of freedom of the gas molecules, being 1.66 for monatomic gases, 1.4 for diatomic gases, and approximately 1 for polyatomic gases.

Specific Resistance. *See* resistivity.

Specific Rotation. *See* optical activity.

Spectral Class. A classification of the stars by their Spectrum and Luminosity. The stars were originally divided into seven main classes designated by the letters O,B,A,F,G,K, and M; since 1924 four other classes (R,N,S, and W) have been added. Each of the letter classes has subdivisions indicated by the numerals 0 through 9. A Roman numeral—ranging from I (supergiant) to V (normal dwarf or main sequence)—is added to the spectral class to specify the luminosity, or intrinsic intensity, of a star.

Spectral Class: Characteristics

Type	*Colour*	*Temperature*	*Strong Lines*
O	blue-white	35,000°C	ionized helium
B	blue-white	21,000°C	helium
A	white	10,000°C	hydrogen
F	creamy	7000°C	ionized calcium
G	yellow	6000°C	calcium
K	orange	4500°C	titanium oxide
M	red	3000°C	titanium oxide

Spectral Series. A series of lines in an absorption or emission spectrum, in which the lines all arise from transitions from or to the same energy level. In the absorption spectrum of hydrogen, for example, an electron in the first shell can absorb photons with frequencies that cause transitions to the second, third, fourth shells, etc., thus causing a series of dark lines. This is a Rydberg spectrum. Similarly, if hydrogen is excited in some way it contains a mixture of hydrogen atoms with electrons in the second, third, fourth or higher orbits. These can decay with the emission of photons so that the electron falls back into the innermost shell giving a series of bright emission lines. This is the Lyman series. Emission series can also occur when electrons move to the second, third, or fourth shells from higher levels, thus forming the Balmer, Paschen, and Brackett series. Lines in such a spectral series have an equation of the type $1/\lambda = R(1/n^2 - 1/m^2)$, where R is the *Rydberg constant, n a constant integer and m takes values of $n+1$, $n+2$, etc.

Spectral Type (spectral class). A classification of a star according to the characteristics of its spectrum. The *Harvard classification* comprises several types of star. Type O, for example, have a blue colour and their spectrum is mainly absorption lines of helium ions. Type K have a red colour and their spectrum contains lines from many metals and some simple molecules.

Spectrograph. A spectrometer or spectroscope designed to give a photographic record (the *spectrogram)* of a spectrum.

Spectrographic Analysis. Analysis of substances by the spectra they produce. A routine method of analysis involves photographing the emission spectrum produced by vaporizing a small amount of sample in an electric arc or a flame. The lines in the spectrum are characteristic of the elements and the intensities of the lines can be used to determine the amounts present. More detailed analysis of spectra can often show the nature of the molecules.

Spectroheliograph. An instrument for photographing the sun using one of the wavelengths of light that it emits. It is essentially a monochromator for selecting light of one particular wavelength and focusing it onto a photographic plate.

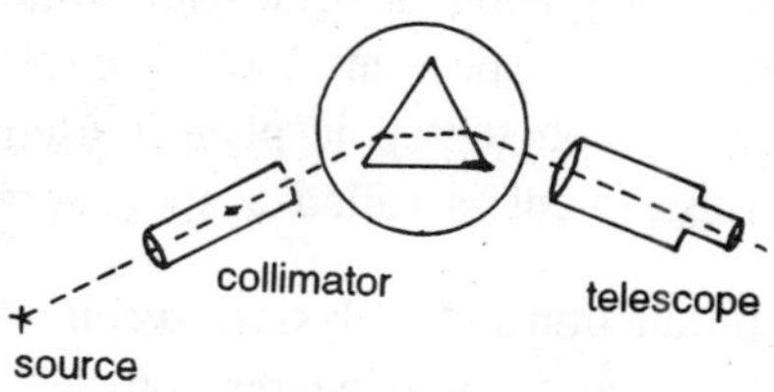

Fig. S-8. Prism spectrometer.

Spectrometer. Any of various instruments for producing and examining spectra. In a simple prism spectrometer, the light is first passed through a collimator onto the prism, which rests on a flat rotatable table. Refracted light is observed through the telescope, which can be moved round the prism. The instrument can be used for observations of simple emission and absorption spectra. It is also often used for measuring refractive indices and prism angles. Many other types of spectrometer exist for investigating absorption and emission spectra over almost the whole range of the electromagnetic spectrum. Spectrometers are often called *spectroscopes* or, when they are designed for ultraviolet or visible radiation, *spectrophotometers*. The term is also used for other instruments for analysing beams of electrons, alpha particles, ions, atoms, etc., in terms of their distribution of energy or mass.

Spectrophotometer. An instrument for producing a visible or ultraviolet spectrum and measuring the intensities of each wavelength.

Spectroscope. Any of various instruments for examining and measuring spectra.

Spectroscopic Binary. *See* binary star.

Spectroscope. Optical instrument for producing spectral lines and measuring their wavelengths and intensities, used in spectral analysis. In the simple prism spectroscope, a collimator, with a slit at the outer end and a lens at the inner end, transforms the light entering the slit into a beam of parallel rays. A prism or diffraction grating disperses the light coming from the collimator, and the spectrum formed is observed with a small telescope. If a photographic plate is used to record the spectrum, the instrument is called a spectrograph.

Spectroscopy. The production and analysis of spectra. Many different spectroscopic techniques exist depending on the type of radiation used. They give information on energy levels of atoms, molecules, ions, crystals, etc., and can be used to determine the structure and composition of molecules. Spectroscopy is also a valuable analytical technique. In astronomy it is used to investigate the composition and behaviour of stars, comets, and planets.

Spectrum. A particular distribution of some property over the components of a system. Thus, a beam of particles has a spectrum of kinetic energies which can be displayed as a graph of number of particles (as ordinate) against particle energy. A complex sound wave has a spectrum of frequencies which can be represented by a graph of intensity against frequency. An electromagnetic radiation has a spectrum of intensity against wavelength (or frequency). The spectrum of the light or other electromagnetic radiation emitted or absorbed by a substance is characteristic of the substance. Such spectra are classified as line, band, or continuous spectra according to their appearance. They are also designated by the type of electromagnetic radiation.

Spectrum. Arrangement or display of Light or other forms of Electromagnetic Radiation separated according to wavelength, frequency, energy, or some other property. Dispersion, the separation of visible light into a spectrum, may be accomplished by means of a Prism or Diffraction grating. Each different wavelength or frequency of visible light corresponds to a different Colour, so that the spectrum appears as a band of colours ranging from violet at the short-wavelength (high-frequency) end of the spectrum through indigo, blue, green, yellow, and orange, to red at the long wavelength (low-frequency) end of the spectrum. A continuous spectrum containing all colour is produced by all incandescent solids and liquids and by gases under high pressure. A low-pressure gas made incandescent by heat or by an electric discharge emits a spectrum of bright emission lines. A dark-line absorption spectrum is produced by white light passing through a cool gas and consists of a continuous spectrum with superimposed dark lines; each line corresponds to a frequency where a bright line would appear if the gas were incandescent. The absorption lines correspond to transitions of electrons from a lower energy level to a higher energy level when a Photon is absorbed by the atom, and the emission lines correspond to transitions from a higher to a lower energy level in the atom, accompanied by the emission of a photon. The frequency of each emission or absorption line is proportional to the difference in energy between the two energy levels involved (see Qualtum Theory). Both absorption and line spectra are useful in chemical analysis, because they reveal the presence of particular elements.

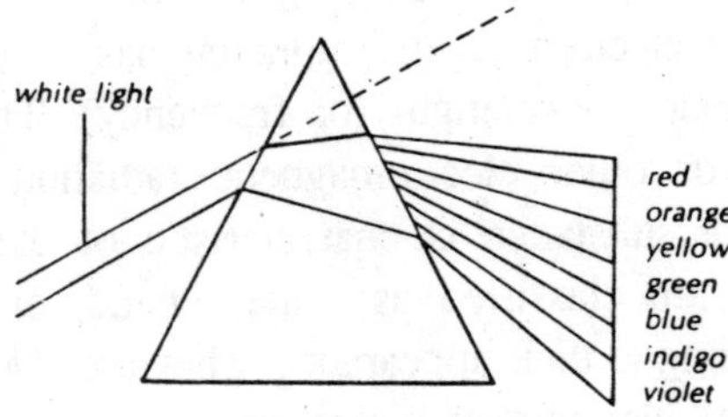

Fig. S-9. *Spectrum:* Dispersion of white light by a triangular prism.

Specular Reflection. Reflection in which the angle of reflection equals the angle of incidence. It is distinguished from diffuse reflection and occurs when the surface is flat.

Speculum. A mirror, especially one made of polished metal.

Speculum Metal. An alloy of copper and tin in the ratio 2:1, used for making metal mirrors.

Speed. 1. The ratio of distance travelled to time taken. Speed has the same units as velocity: unlike velocity it is not a vector quantity.

2. The extent to which a photographic film is sensitive to light.

3. The rate of operation of a pump, measured in litres per second, etc.

Speltar. Commercial zinc containing about 3% lead and other impurities.

Sperry, Elmer Ambrose. 1860-1930, American inventor; b. Cortland, N.Y. Best known for his work on the gyroscope, he also invented a gyrocompass (1910), a high-intensity searchlight, a street-lighting system, and many electrical devices. He founded the American Inst. of Electrical Engineers.

Sphagnum or **Peat Moss.** Economically valuable Moss (genus *sphagnum)* typically growing as a floating mat in freshwater bogs. Sphagnum are the principal constituent of Peat. They are highly absorbent and are commercially important as packing material and absorbent dressings.

Sphalerite. Zinc, sulfide mineral (ZnS), occurring worldwide, sometimes in crystals but more often in massive form, in a variety of colours. Often found in association with Galena, it is the most important source of Zinc.

Sphere. A surface or solid generated by rotating a circle about its diameter. The volume is (4.3)Πr^3 and the surface area is $4\Pi r^2$, where r is the radius.

Sphere. In geometry, a solid whose surface consists of points all at the same distance r (the radius) from a certain fixed point (the centre).The term *sphere* refers both to the surface and to the space it encloses. The area of the surface of a sphere is given by the formula $S = 4\Pi r^2$ and the volume of a sphere is given by $V = 4/3\ \Pi r^3$.

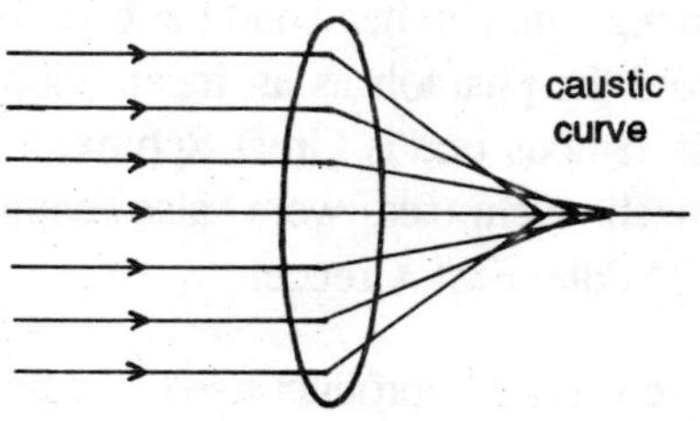

Fig. S-10. Spherical aberration.

Spherical Aberration. An aberration of lenses and mirrors with spherical surfaces, caused by the fact that rays striking the lens or mirror at points close to its edge are focused nearer than those close to the axis. Thus a point object is not focused as a point image but as a small disc, the *circle of least confusion.* Spherical aberration is a geometrical effects—it arises because of the shape of the surface. In mirrors, it can be reduced by using a parabolic shape for distant objects or an ellipsoid for nearer objects. In lenses, it is reduced by making the lens in such a way that the deviation of the ray is shared equally at the two faces of the lens.

Spherical Coordinates. *See* polar coordinates.

Spherical Triangle. A triangle drawn on a spherical surface with sides formed by the arcs of great circle. *i.e.* circles whose centres are the centre of the sphere. The branch of mathematics concerned with the properties of spherical triangles is *spherical trigonometry*.

Spheroid. *See* ellipsoid.

Spherometer. An instrument for measuring the curvature of a lens, mirror, or other spherical surface. It has three legs and a central screw, which is adjusted to touch the surface. The radius of the surface is calculated from the reading on the screw.

Sphinx. Mythical beast of ancient Egypt, usually represented in art as having a human head and the body of a lion. It frequently symbolized the pharaoh as an incarnation of the sun god RA. The most famous one is Great Sphinx, a colossal stone figure at Al Jizah. Sphinxes were also depicted throughout the ancient Middle East Greece.

Spice. Aromatic vegetable product used as a flavoring or condiment. The term was formerly also applied to pungent or aromatic foods, incense and perfume ingredients, and embalming agents. Spices include stimulating condiments (e.g., pepper and mustard), aromatic spices(e.g., cloves, nutmeg, and mace), and sweet herbs (*e.g.*, thyme and mint). They are used whole, as powders, and as tinctures.

Spider. Mostly terrestrial Arachnid, with a two-part body, four pairs of legs, and four pairs of eyes. Spinnerets (specialized organs under the abdomen) produce silk thread for binding prey or making webs, cocoons, and lines for floating. Spiders live chiefly on insects and other arthropods; some large species prey on small snakes, mammals, and birds. All spiders paralyze their prey with venom produced in poison glands under the head; several species, such as the Black Widow, have bites that are painful or even dangerous to humans.

Spin. Symbol: *s*, *J*, or *I*, A property of elementary particles corresponding to intrinsic angular momentum. The particle acts as if it were spinning on its axis and, according to quantum mechanics, only certain values of this angular momentum are possible. The angular momentum of a particle with spin can be represented by a vector along its axis. such a particle also acts as a small magnet with a magnetic moment. When an external magnetic field is applied precession occurs: the angular-momentum vector moves around the direction of the applied field. It can be demonstrated that only certain orientations are possible, namely those in which the components of the angular momentum along the field direction are $M_J h/2\pi$, where h is the Planck constant and M_j is called the *spin quantum number*. A particle with a spin *J* has M_J values of *J*, *J*–1, ...-(*J*–1), –*J*. Thus the electron, with spin of 1/2, has spin quantum numbers of 1/2 and –1/2.

Spinach. Annual plant *(Spinacia oleracea)* of the goosefoot family, probably persian in origin. The leaves are high in Vitamins and iron, and numerous varieties are cultivated.

Spinal Cord. Length of nerve-fiber bundles carrying information (electrical and chemical signals) through the Nervous System. The spinal cord carries sensory inpulses from the trunk and limbs to the brain; it returns commands from the brain to the muscles and glands. Anatomically, the spinal cord runs nearly the length of the trunk and merges with the brainstem. The spinal cord is housed within the *spinal column*, a bony column that also forms the main structural support of the Skeleton. The spinal column consists of segments (vertebrae) linked by flexible joints and held together by gelatinous disks of cartilage and by ligaments. Each vertebra has a roughly cylindrical body, winglike projections, and a bony arch. The arches, positioned next to one another, create the tunnellike space that houses the spinal cord.

Spinel. Any of a class minerals having the general formula $MO.M'_2O_3$, where M is a divalent metal and M' a trivalent metal. The ferrates are types of spinel.

Spinning. The drawing out, twisting, and winding of Fibers into a continuous thread or yarn. From antiquity until the Industrial Revolution, spinning was a household industry. The earliest tools were the distaff, a hand-held stick on which the cotton, flax, or wool fiber was wrapped; and the spindle, a shorter stick, held in the other hand, notched at one end and weighted at the other. The twirling of the spindle twisted fiber into thread. In Europe from the 14th to 16th cent. the distaff and spindle were replaced by the spinning wheel, a spindle set in a frame and turned by a belt passing over a wheel. The great wheel, also called the wool or walking wheel, was turned by hand; the more elaborate flax, or Saxony, wheel was operated by a foot treadle. In 18th cent. England, improvements in the Loom, increasing the demand for yarn, stimulated such inventions as James Hargreaves's spinning jenny (c.1765), which spun 8 to 11 threads at once; Richard Arkwright's spinning frame (1769), which made more tightly twisted, stronger threads; and Samuel Crompton's mule spinning frame (1779), which combined the best features of the two earlier machines. Using water power and, later, steam, spinning because a factory enterprise.

Spiral Galaxy. *See* galaxy.

Spirits of Salt. *See* hydrochloric acid.

Sponge. Aquatic invertebrate animal of the phylum Porifera. All but one family are marine. Colonies of adult sponges, often brilliantly coloured, live attached to rocks, corals, or shells, exhibiting so little movement that 18th cent. naturalists considered them plants. The sponge's body is like a sac. Water is drawn into a central cavity through many tiny holes in the body wall and expelled through a large opening at the top. Hard materials embedded in the body wall form a skeleton. The dried skeletons of colonial sponges have been used to hold liquid since ancient times. Natural sponges are light gray or brown when dried and irregular in shape.

Spontaneous Combustion. Phenomenon in which a substance unexpectedly bursts into flame without apparent cause. Spontaneous combustion occurs when a substance undergoes a slow oxidation that releases heat in such a way that it cannot escape the substance; the temperature of the substance consequently rises until ignition takes place.

Spore. Term applied both to a resistant or resting stage occurring among unicellular organisms such as bacteria and to an asexual reproudctive cell of multicellular plants that gives rise to a new organism without Fertilization. A spore is typically a mass of protoplasm containing a nucleus and surrounded by a cell wall that may be tough and waterproof, permitting the cell to survive unfavourable circumstances.

Sporting Dog. Class of dogs bred for pointing, flushing, and retrieving game. They hunt by air scent (as opposed to most Hounds, which are ground scenters), and their quarry is mainly game birds. Pointers stand rigidly, with nose and body pointing at their quarry, directing hunters to its location. Setters, originally trained to crouch in front of game, were later taught to point. Retrievers find and return killed game to the hunter.

Sports Medicine. Branch of medicine concerned with prevention, treatment, and study of injuries received during participation in sports. "Tennis elbow"; shoulder, knee, back, and leg injuries; stiffness and pain in joints; and tendinitis are some of the conditions involved. Treatment includes mechanical supports, specific exercise programs, Physical Therapy, and—

in severe cases—surgery. Sports medicine was initially practiced primarily by physicians associated with professional sports teams, but with increased interest in amateur sports and physical fitness programs in the 1970s and 80s, it grew rapidly.

Spring. In geology, natural flow of water from the ground or from rocks, representing an outlet for the water that has accumulated in permeable rock strata underground. Mineral springs have a high mineral content.

Spring Balance. A simple instrument for measuring weight by the amount by which it stretches a coiled spring.

Spruce. Evergreen tree or shrub (genus *Picea)* of the Pine family, widely distributed in the Northern Hemisphere. The needles are angular in cross section, not flattened as in the related Hemlocks and Firs. Spruces are a major source of pulpwood for Paper manufacture; the light, straight-grained wood is also used in construction. Common North American spruces include the red spruce *(P. rubens),* white spruce *(P. glauca),* and black spruces *(P. mariana)* of the East; the Engelmann spruce *(P. engelmanii)* of the Rockies; and the Sitka spruce*(P. sitchensis)* of the pacific forest belt. The siberian spruce *(P. obovata)* grows in the huge coniferous forests of the USSR.

Spurge. Common name for the family Euphorbiaceae, herbs, shrubs, and trees of greatly varied structure and almost cosmopolitan distribution, although most species are tropical. The spurges are of great economic importance; the sap of most species is a milky latex, and that of the Para Rubber Tree is the source of much of the world's natural Rubber. The genus *Manihot* includes Cassava, the source of tapioca and the most important tropical root crop after the Sweet Potato. The cactuslike euphorbias (genus *Euphorbia)* are among the most common Old World desert Succulents and comprise most of the species commonly called spurge. Many are cultivated for their often colourful foliage and the showy bracts enclosing their "naked flowers" (*i.e.*, Flowers lacking petals and sometimes sepals). The poinsettia, native to Central America and sometimes classed in a separate genus *(Poinsettia)*, is a popular Christmas plant with large rosettes of usually bright-red bracts.

Fig. S-11. Spurge: Snow-on-the-mountain, Euphorbia marginata, a plant of the spurge family.

Sputtering. A process in which atoms are ejected from a solid surface by the impact of high-energy ions. The effect can occur in gas discharges as result of ions hitting the cathode. It is a technique of depositing thin films of the sputtered material on a nearby surface.

Sputter-ion Pump. *See* ion pump.

Square. 1. A rectangle with all its sides equal.

2. The result of multiplying a number by itself: $x^2 = x \times x$.

Square root. *See* root.

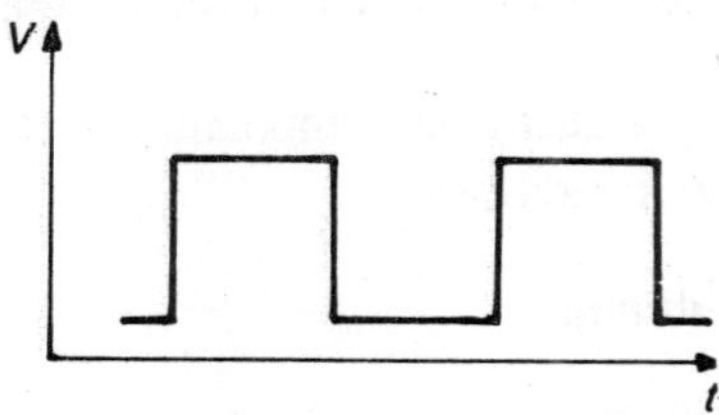

Fig. S-12. square wave.

Square Wave. A waveform that alternates between two steady values with (almost) instantaneous increases and decreases between these values and equal half periods.

Squid. Carnivorous marine mollusk with 10 sucker-bearing arms; a Cephalopod. Among the most highly developed invertebrates, the squid has eyes similar to those of humans, relatively sophisticated nervous and circulatory systems, and no external shell. Squids prey on fish, which they seize in their tentacles; when in danger, they emit a cloud of ink from a special sac. Squids range in size from 2 in. (5 cm) to the 50 ft (15.2 m) giant squid, the largest of all invertebrates. Squid is a favorite food in the Mediterranean region and the Orient.

Squill. Low, usually spring-blooming bulbous herb (genus *Scilla)* of the Lily family. The flowers, commonly deep blue but also white, rose, or purplish, are borne along a leafless stem; the leaves are usually narrow. Species of *Scilla* are used in rock gardens and borders.

Squirrel. Small or medium-sized Rodent of the family Sciuridae, found worldwide expect in Australia, Madagascar, and polar regions. Typical tree squirrels (genus *Sciurus),* including the Eurasian red squirrel *(S. vulgaris)* and the North American gray squirrels, are day-active animals with slender bodies, thick fur, and bushy tails. In addition to tree squirrels the family includes the Chipmunk, Woodchuck, Prairie Dog, and Flying Squirrel.

Stabilizer. A substance added to another substance to prevent or retard its deterioration or chemical decomposition.

Stable. 1. Denoting a chemical compound that does not decompose under normal conditions.

2. *See* equilibrium.

Stained Glass. Windows made of coloured glass. An art form of great antiquity in the Far East, used by Muslim designers in their intricate windows, it became one of the most beautiful achievments of medieval art. Christian churches had coloured glass windows as early as the 5th cent., but the art of stained glass reached its height in the Middle Ages, particularly 1150-1250. As the massive Romanesque wall was eliminated, the use of glass expanded. Integrated with the lofty verticals of Gothic architecture, large widows provided greater illumination that was regarded as symbolic of divine light. Early glaziers followed a Cartoon to cut the glass and fired the painted pieces in a kiln. Metallic oxides fused with the glass in the melting pot to produce the jewel-like colours of small pieces whose irregular surfaces created scintillating refractions of light. The pieces were fitted into channeled lead strips, the leads were soldered together, and the glass was installed in an iron framework. Outstanding examples of 12th cent. stained glass can be found in the windows of such churches as Saint-Denis, in Paris, and Canterbury, in England. Among the finest 13th cent. works are the windows at Charters and the Sainte-Chapelle in Pairs. With improved glassmaking many medieval qualities vanished, and by the 16th cent. a lesser art was produced with larger, smoother pieces and sophisticated painting techniques. In the 19th cent., Romanticism and the Gothic Revival caused renewed interest in stained glass. Important contributions to the art were made by William Morris, in England, and John La Farge and Louis Comfort Tiffany, in the U.S. In modern art the medium has been used Rouault, Matisse, and Chagall.

Stainless Steel. Any of a class of steels that resist corrosion and are not attacked by weak acids. Stainless steels have a variety of compositions but all contain chromium (10-25%), carbon (0.1-0.7%), and sometimes small amounts of nickel and niobium. They are used for cutlery, household goods, turbine blades, industrial chemical equipment, and many other applications.

Stalactite and Stalagmite. Mineral forms found in Caves; sometimes collectively called dripstone. A stalactite is an icicle-shaped mass of Calcite that hangs from the roof of a cave, formed by the precipitation of calcite from ground water. A stalagmite is a cone of calcite rising from the floor of a cave, formed by the same process. Stalactites and stalagmites often meet to form solid pillars. The many colours often seen in these formations are caused by impurities.

Stalloy. An alloy of iron and silicon used in transformer cores and similar applications requiring low hysteresis loss.

Standard Cell. Any cell with a constant reproducible e.m.f. that can be used as a standard. The Weston cell is usually used for this purpose.

Standard Deviation. The square root of the average of the squares of the deviations of a set of values.

Standard Solution. A solution of known concentration for use in a titration.

Standard Temperature and Pressure (S.T.P.). A temperature of 273.15 kelvins (0°) and 101 325 pascals (formerly 760 mmHg). used as standard conditions for properties of gases. The conditions are also called *normal temperature and pressure (N.T.P.).*

Star. Any of the vast numbers of huge celestial objects that radiate light, heat, and other forms of electromagnetic radiation and are grouped into galaxies. The stars, of which the sun is a typical example, are separated, even within the galaxies, by enormous distances and appear from earth only as points of light. In general, they fall into two classified groups, or *populations.* Stars of population I are young and are generally found in the outer arms of spiral galaxies; those of population II are much older and occur in the inner central region of galaxies and in globular clusters.

Stars range from *hot* (blue-white), with surface temperatures of up to 100,000°C, to *cool* (red), with surface temperatures of less than 2000°C. They can also be classified by their spectral type. Within our Galaxy, stellar diameters show huge variation. At extreme ends of the scale lie the pulsars and the supergiants. Brightness varies considerably among the stars. *Canopus,* one of the brightest stars in the heavens, is 80,000 times as luminous as the sun, and lies about 650 light-years away; on the other hand, the red dwarf, *Wolf 395,* is only 1/30,000 as bright as the sun, though it is one of our nearest neighbours, only 7.7 light-years away.

Because of its position towards the outer rim of the Galaxy, the sun has relatively few near neighbours. Only eight stars lie within ten light-years of us Proxima Centauri is the closest at a distance of only 4.3 light-years. Sirius, the bright as the sun, is only 8.7 light-years away.

Star. Hot incandescent sphere of gas (usually more than 90% hydrogen) that is held together by its own gravitation and emits light and other forms of electromagnetic radiation whose ultimate source is nuclear energy. The universe contains billions of galaxies, and each Galaxy contains billions of stars, which are frequency bunched together in star Clusters of as many as 100,000. The stars visible to the unaided eye are all in our six classes according to their apparent Magnitude. Stars differ widely in mass size, temperature, age, and Luinosity. About 90% of all stars have masses between one tenth and 50 times that of the sun. The most luminous stars (excluding supernovas) are about a million times more powerful than the sun, while the least luminous are only a hundredth as powerful. Variable stars fluctuate in luminosity. Red giants, the largest stars, are hundreds of times greater in size than the sun. At the opposite extreme, White dwarfs are no larger than the earth, and Neutron Stars are only a few kilometers in radius. The central region, or core, has a temperature of millions of degrees. At this temperature nuclear energy is released by the fusion of hydrogen to form helium. By the time nuclear energy reaches the surface of the star, it has been largely converted into visible light with a spectrum characteristic of a very hot body. The body of Stellar Evolution states that a star must change as it consumes its hydrogen in the nuclear reactions that power it. When all its nuclear fuel is exhausted, the star dies, possibly in a Supernova explosion.

Starch. White, odorless, tasteless Carbohydrate powder. It plays a vital role in the Biochemistry of both plants and animals. Made in green plants by Photosynthesis, it is one of the main forms in which plants store food. Animals obtain starch from plants and store it as Glycogen. Both plants and animals convert starch to Glucose when energy is needed. Commercially, starch is made chiefly from Corn and Potatoes. Corn syrup and corn sugar made from corn starch are widely used to sweeten food products. Starch is also used to stiffen laundered fabrics and to size paper and textiles.

Starch. A carbohydrate consisting of chains of glucose units, $(C_6H_{10}O_5)_x$. It has two forms *(amylose* and *amylopectin)* and is present in potatoes, rice, and cereals. In the body starch is hydrolysed to glucose.

The pure substance is a white powder, insoluble in cold water. It forms a bright blue compound with iodine.

Starfish or **Sea Star.** Star-shaped carnivorous marine invertebrate animal (an Echinoderm), found worldwide in shallow waters. The spiny body has five or more tapering arms radiating from a central areas. Usually dull yellow or orange but occasionally brightly coloured, starfish vary in size from under 1/2 in. (1.3 cm) to over 3 ft (90 cm).

Starling. Any of a group of originally Old World Birds of the family Sturnidae, now distributed worldwide. The common starling *(Sturnus vulgaris)* is found throughout North America. Insect eaters, starlings have iridescent black plumage and a long bill. They mimic bird songs and other sounds, and are considered pests since they collect in large, noisy flocks and drive away smaller, more desirable birds.

Stat-. Prefix placed in front of the name of a practical electrical unit to name the corresponding electrostatic unit. Thus the e.s.u. of current is the statampere.

Statcoulomb. The electrostatic unit of charge, equal to 3.3356 x 10^{-10} coulomb.

State of Matter. One of the three physical states—solid, liquid, or gas—in which matter may exist. The states of a particular substance change at definite transition temperatures. Plasmas, liquid crystals, glasses, colloids, and superfluids have all, at different times, been described as extra states of matter.

States of Matter. Forms of matter differing in several properties because of differences in the motions of and the forces between the molecules (or atoms or ions) of which they are composed. There are three common states of matter: solid, liquid, and gas. The molecules of a solid are limited to vibrations about a fixed position, giving a solid both a definite volume and a definite shape. When heat is applied to a solid, its molecule begin to vibrate more rapidly until, at a temperature called the Melting Point, they break out of their fixed positions and the solid becomes a liquid. Because the molecules of a liquid are free to move throughout the liquid but are held from escaping by intermolecular forces, a liquid has a definite volume but no definite shape. As more heat is added to the liquid, some molecules near the surface gain enough energy to evaporate, or break away completely from the liquid, and change to a gaseous state. Finally, at a temperature called the Boiling Point, molecules throughout the liquid become energetic enough to escape, forming bubbles of vapor that rise to the surface; the liquid thus changes completely to a gas. Because its molecules are free to move in every possible way, a gas has neither a definite shape nor a definite volume but expands to fili any container in which it is placed. The reverse processes of melting and boiling are, respectively, freezing and condensation.

Static Electricity. Electricity involving stationary electric charge.

Statics. The branch of mechanics concerned with the properties of bodies in equilibrium: *i.e.* systems of forces in which no motion occurs.

Statics. Branch of Mechanics concerned with the maintenance of equilibrium in bodies by the interaction of Forces upon them. In a state of equilibrium the resultant of all outside forces acting on a body is zero, thus keeping the body at rest.

Stationary Orbit. *See* synchronous orbit.

Stationary Point. A point on a curve at which the curve is neither increasing nor decreasing. At such a point the derivative is zero.

Stationary State. One of the possible quantized states of a system as described by quantum mechanics.

Stationary Wave. *See* wave.

Statistical Mechanics. The branch of physical science concerned with calculating the properties of matter from the statistical behaviour of large numbers of atoms and molecules.

Statistics. The branch of mathematics concerned with collecting numerical data and making inferences from the data on the basis of probability.

Statistics. Branch of applied mathematics dealing with the collection and classification of data by numerical characteristics and the use of these data to make inferences and predictions in uncertain situations. Generally, measurements taken from a small group, the sample, are used to infer the behaviour of a larger group, the population, as in television ratings and election predictions. The theory of Probability is necessary to determine how well the sample represents the population. The most widely used tools are the (arithmetic) mean and the median. The mean of a set of numbers is their sum divided by the number of elements in the set, *e.g.*, for the five numbers 7, 7, 8, 10, and 11, the mean is (7 + 7 + 8 + 10 + 11)/5 =43/5=8.6. The median of a set is the number that divides the set in half, so that as many numbers are larger than the median as are smaller. The median of 7, 7, 8, 10 and 11 is 8. (if the set has an even number of elements, the median is the number halfway between the middle pair). Another important statistical measure is the standard deviation, which indicates how closely the data are clustered about the mean. Statistics is used in scientific and social research, insurance, and many other fields.

Stator. The stationary part of a generator or electric motor. It can be either the field winding or the armature.

Steady-state Theory. The theory that the universe has always existed and that its expansion is compensated for by the continuous creation of matter in such a way that the average density remains constant. To maintain such a state, matter has to be produced at a rate of about 10^{-43} kilogram per cubic metre per second. The theory would be supported by observations of radio waves from far in space showing that the density is constant at all points: such evidence has not yet been obtained. Many observations tend to support the alternative big-band theory.

Steam. Water in the gaseous state at or above 100°C. Strictly, steam is an invisible gas but the term is also used for the white clouds which contain small droplets of liquid water.

Steam Distillation. A technique for separating immiscible liquids by bubbling steam through the hot mixture. The amount of a component present in the flow of steam is proportional to the product of its vapour pressure and its molecular weight.

Thus, relatively involatile compounds can be distilled by this method if they have high molecular weights.

Steam Engine. Machine for converting heat energy into mechanical energy, using steam as the conversion medium. When water is boiled into steam its volume increase about 1,600 times, producing a force that can be used to move a piston back and forth in a cylinder. The piston is attached to a crankshaft that converts the piston's back-and-forth motion into rotary motion for driving machinery. From the Greek inventor Hero of Alexandria to the Englishman Thomas Newcomen, many persons contributed to the work of harnessing steam. However, James Watt's steam engine (patented 1769) offered the first practical solution by providing a separate chamber for condensing the steam and by using steam pressure to move the piston in both directions. These and other improvements by Watt prepared the steam engine for a major role in manufacturing and transportation during the Industrial Revolution. Today steam engines have been replaced in most applications by more economical and efficient devices, *e.g.*, the steam Turbine, the electric Motor, the Internal-combustion Engine, and Diesel Engine.

Steam Point. The temperature at which the liquid and vapour of water are in equilibrium at standard pressure: *i.e.* the boiling point of water (100°C).

Steamship. Watercraft propelled by a Steam Engine or a steam Turbine. A number of experimental steam-powered vessels were built in the late 18th cent. and in 1807 Robert Fulton's *Clermont* made the 150-mi (240-km) trip from New York City to Albany in 32 hr. The *Savannah,* a full-rigged sailing ship fitted with engines and paddlewheels, made the first Atlantic Ocean crossing by a steam-propelled vessel in 1819. Two British Ships made the first crossing under steam power alone in 1838. By the late 1850s the screw propeller was replacing paddlewheels, and the steamship began to supplant the sailing ship. Great liners were plying the Atlantic on regular schedules by the end of the century. Modern leviathans of the sea such as the *Queen Mary* (1934), the *Queen Elizabeth* (1938), and the *United States* (1951) were turbine-powered steamships, as are most present-day passenger vessels, cargo ships, and warships.

Stearate. Any salt or egster of stearic acid.

Stearic acid (octadecanoic acid). A crystalline water-insoluble fatty acid, $CH_3(CH_2)_{16}COOH$, present, in the form of its glycerides, in many natural fats and oils. M.pt. 69.6°C; 376°C (decomposes); r.d. 0.94.

Stearin. *See* tristearin.

Steel. Alloy of Iron, Carbon, and small proportions of other elements. Steelmaking involves the removal of iron's impurities and the addition of desirable alloying elements. Steel was first made by cementation, a process of heating bars of iron with charcoal so that the surface of the iron acquired a high carbon content. The bars were then fused together, yielding a metal harder and stronger than the individual bars but lacking uniformity in these properties. The crucible method, consisting of melting iron together with other substances in a crucible, is one of the costlier steelmaking processes, employed only for making special steels (*e.g.*, the famous blades of Damascus). The Bessemer Process, the open-hearth process, and the basic oxygen process are more widely used. Steel is often classified by its carbon content: a high-carbon steel is hard and brittle; low-or medium-carbon steel can be welded and tooled. Alloy steels, now the most widely used, contain one or more elements that give them special properties. Aluminum steel is smooth and has a high tensile strength. Chromium steel is used in automobile and airplane parts because of its hardness, strength, and elasticity. Nickel steel is the most widely used of the alloys; it is nonmagnetic and has the tensile properties of high-carbon steel without the britleness. Stainless steel has a high tensile strength and resists abrasion and corrosion because of its high chromium content; it is used in kitchen utensils and plumbing fixtures.

Steel. Any of various forms of iron containing carbon (0.05-1.5%) and usually quantities of other elements. Two main types are distinguished: *carbon steels* contain mainly iron and carbon with only small amounts of other elements; *alloy steels* have larger amounts of other metals. If the carbon content is below 0.05% the material is iron. If it exceeds 1.5% it is usually described as cast iron. Steel is made by reducing the carbon content of pig iron in an open-hearth furnace or a Bessemer converter. Alloy steels are usually made from a ferroalloy in an electric-arc furnace.

Many different types of steel exist depending on the amounts of carbon and other elements present and on the heat-treatment given to the steel. Carbon steels are classified as *mild steel* (0.1-0.2% carbon), *medium-carbon steel* (0.3-0.4% carbon), and *high carbon steel* (above 0.5% carbon). Steels with low carbon content tend to be tougher and more ductile whereas a higher content of carbon produces a harder material. Alloy steels have a variety of properties depending on the elements present.

Steel has a complex variety of different phases. If molten iron containing carbon is solidified the material is a solid solution of carbon in gamma iron *(austenite)*. As the temperature is reduced, *cementite* (Fe_3C) is formed. Below a certain temperature the gamma iron changes to alpha iron *(ferrite)* and a mixture of ferrite and cementite is produced (called *pearlite)*. If the austenite is cooled quickly by quenching in water, the gamma iron changes to ferrite but the cementite does not precipitate. This yields a metastable structure (called *martensite)* consisting of a (supersaturated) solid solution of carbon in ferrite. Slower quenching produces a different structure *(troostite)* in which very fine grains of cementite and ferrite occur.

Stefan's Law. The principle that the energy radiated per unit area per second from a black body is proportional to the fourth powder of the thermodynamic temperature: i.e. $M = \sigma T^4$. The constant σ is *Stefan's constant.* It has the value 5.6697×10^{-8} $W m^{-2} K^{-4}$. [After Joseph Stefan (1835-93), Austrian physicist.]

Steinmetz, Charles Proteus. 1865-1923, American electrical engineer; b. Germany; came to U.S., 1889. He joined the General Electric Co. in 1892. His discovery of the law of hysteresis made it possible to reduce the loss of efficiency in electrical apparatus resulting from alternating magnetism. His method for calculating alternating current revolutionized electrical engineering.

Stellar Evolution. Life history of a Star. The initial phase of stellar evolution is contraction of the protostar from the interstellar gas. In this stage, which typically lasts millions of years, half the gravitational potential energy released by the collapsing protostar is radiated away and half goes into increasing the temperature of the forming star. Eventually the temperature becomes high enough for the fusion of hydrogen to form helium. The star then enters its longest period in stellar evolution. Because most stars are in this stage and fall along a diagonal line in the Hertzsprung-Russell Diagram, they are called main-sequence stars. As the star's helium content builds up, the core contracts and releases gravitational energy, which heats up the core and increases the rate of hydrogen consumption. The increased reaction rates cause the stellar envelope to expand and cool, and the star becomes a red giant. Eventually, the contracting stellar core will reach temperatures in excess of 100,000,000°K. At this point, helium burning sets in, and the stars shrinking in size. In the further course of evolution, the star may become unstable, possibly ejecting some of its mass and becoming an exploding nova or supernova or a pulsating Variable Star. The end phase of a star depends on its mass. A low-mass star may become a White Dware; an intermediate-mass star may become a Neutron

Star; and a high-mass star may undego complete Gravitational Collapse and become a black hole.

St. Elmo's Fire. *See* corona discharge.

Stem. Supporting structure of a plant, serving also to conduct and store food materials, stems of herbaceous and woody plants differ: herbaceous plants usually have pliant, green stems with relatively more pith and an almost inactive Cambium; woody stems are covered by Bark and increase in height and diameter because of an active cambium. Tendrils, thorns, and runners (stolons) are specialized aerial stems; Bulbs, Corms, Rhizomes, and Tubers are specialized underground stems. The stems of dicotyledons and Gymnosperms consist of upward-conducing xylem on the inside and downward-conducting phloem arranged on either side of the cambium. In monocotyledons, which generally lack cambium, bundles of xylem and phloem are scattered throughout the stem.

Stephenson, George. 1781-1848, English engineer and locomotive builder. He created a traveling engine to haul coal (1814) and the first Locomotive using the steam blast (1815). His locomotive *Rocket* won an 1829 contest and was used on the Liverpool-Manchester Railway. He became engineer for several of the new railroads. His son *Robert Stephenson,* 1803-59, and a nephew, *George Robert Stephenson*, 1819-1905, were railroad engineers and bridge designers.

Steradian. Symbol: sr. The SI unit of geometric solid angle, equal to the solid angle that cuts off an area on a sphere equal to the square of the sphere's radius, the vertex of the solid angle being at the centre of the sphere.

Stere. A unit of volume equal to one cubic metre.

Stereochemistry. The arrangement in space of the groups in a molecule and the effect this has on the compound's properties and chemical behaviour.

Stereochemistry. Study of the three-dimensional configuration of the atoms that make up a molecule and of the ways in which this arrangement affects the physical and chemical properties of the molecule. Central to stereochemistry is the concept of isomerism. Isomers are sets of chemical compounds having identical atomic composition but different structural properties. Stereochemistry is particularly important in Biochemistry and molecular biology.

Stereoisomerism. A type of isomerism in which two or more compounds have the same functional groups but different arrangements of these groups in space. Such compounds are said to be *stereoisomers*. Two types of stereoisomerism exist: geometrical isomerism and optical isomerism.

Stereophonic Sound. Sound recorded simultaneously by two or more Microphones placed in different positions relative to the sound source. The recorded sound is played back through Loudspeakers placed more or less as the recording microphones were placed. The voices or instruments composing the sound thus seem to be spread out as they would be naturally in the recording hall. In quadrophonic reproduction, which utilizes four microphones and four loudspeakers, the effect is further enhanced.

Stereoregular. Having a regular steric structure. The term is used for materials that are tactic polymers. *Stereoregular rubbers,* for example, are synthetic rubbers with a structural order resembling that found in natural rubbers.

Stereospecific. Involving particular steric arrangements of the atoms of molecules. *Stereospecific catalysis* is the use of catalysts to produce tactic polymers.

Sterility. Inability to reproduce; also called infertility. In the male, malfunctioning of the sex glands, or tests, usually results in the production of defective sperm or a decreased number of sperm, causing sterility. In the female, malfunctioning of the sex glands, or ovaries, disturbs ovulation (production of the egg cell). Structural deformity (*e.g.*, blockage of a tube), metabolic and infectious diseases, and psychological factors may also cause sterility. Voluntary sterilization is a form of birth Control.

Stern-Gerlach Experiment. An experiment in which a molecular beam of silver atoms was passed through a nonuniform magnetic field to demonstrate that the beam splits into two separate components. The experiment shows that the magnetic moments of the atoms can have only certain distinct orientations to the direction of the field.

Steroids. Class of organic compounds having a particular molecular structure based on four joined hydrocarbon rings. Steroids differ from one another only in the additional atoms attached to the central structure. One class of Hormones, consisting of steroid compounds, includes the sex hormones Testosterone, Estrogen, and Progesterone, as will as Cortisone, several forms of vitamin D, Cholesterol, and the Bile acids. Steroids are found in plants and invertebrates as well as in higher animals.

Stibine (antimony hydride). A colourless gas, SbH_3. It decomposes to antimony and hydrogen when heated. M.pt.–88°C; b.pt.–17°C.

Stibnite. A grey mineral consisting of antimony sulphide, Sb_2S_3. It is the principal ore of antimony.

Stibnite. Antimony sulfide mineral (Sb_2S_3), the most important ore of Antimony. It is found in many parts of the world, often in association with arsenic, calcite, gold, quartz, and silver. Silvery grey in colour, stibnite was used in ancient times by women to darken eyebrows and eyelashes. It is now used in alloys, in explosives, in vulcanizing rubber, and as an emetic.

Still. Any apparatus used for distillation.

Still Life. A pictorial representation of inanimate objects. Although detailed depictions of still-life subjects appeared early in Western art, *e.g.*, Hellenistic Frescoes and Mosaics, they were treated as subordinate until the Renaissance. Still life became an independent genre in Italy, but it was as a part of the religious works of Northern Europe that the study of still life was most complete. Such 15th cent. artists as the van Eycks, van der Weyden, and Campin observed and exactly recorded still-life objects. In the north, still life developed as a brilliantly painted separate genre in the works of such 17th cent. artists as Jan Bruegel, Rubens, Snyders, and Rembrandt. In France, still life was not used significantly until its extraordinary handling by Chardin in the 18th cent. French 19th cent. masters, including such diverse figures as Courbet and Cezanne, elevated still life as subject matter. In the U.S., the 19th cent. painters Harnett and John F. Peto used still life to display dazzling *trompe l'oeil* techniques, and in the 20th cent. It has been the most characteristic subject matter of many American and European artists, *e.g.,* the cubist works of Picasso, Braque, Gris, and Stuart Davis. Artists of many schools of abstract painting continued to develop the still life, concentrating on colour, form, and composition. Pop Art painters also used still-life elements, choosing such objects of popular culture as soup cans and comic strips. In the Far East, still-life subjects were depicted as early as the 11th cent. and were often given symbolic meaning.

Stimulated Emission. The process in which a photon colliding with an excited atom causes emission of a second photon with the same energy as the first. It is the phenomenon on which depend the action of lasers and masers.

Stochastic, Random. A *stochastic process* is one that involves random behaviour, so that it can be described in terms of probabilities.

Stoichiometric. Involving chemical combination in exact simple ratios. A *stoichiometric compound* is a normal compound:

i.e. one in which the elements have combined in small whole numbers. A *stoichiometric mixture* is one that would combine to give other substances with no excess of any reactant.

Stoichiometry. The branch of chemistry concerned with the proportions in which compounds react together and the use of these in finding atomic weights, chemical equivalents, etc. The term is often taken to mean the proportions themselves.

Stokes (stoke). Symbol: St. A CGS unit of kinematic viscosity equal to the kinematic viscosity of a fluid with a dynamic viscosity of one poise and a density of one gram per cubic centimetre. [After Sir George Gabriel Stokes (1819-1903), British mathematician and physicist.]

Stokes' Law. The principle that a small spherical particle falling under gravity in a viscous fluid reaches a terminal velocity given by $v = 2gr^2(\rho_1 - \rho_2)/9\varepsilon$, where g is the acceleration of free fall, r the radius of the sphere, ρ_1 and ρ_1 the densities of the sphere and the fluid respectively, and ε the viscosity of the fluid. The equation can be used in experimental method of determining the viscosities of liquids and gases.

Stomach. Saclike organ of the Digestive System located between the esophagus and the Intestines. The human stomach is a muscular, elastic, pear-shaped bag lying crosswise in the abdominal cavity, beneath the diaphragm. Food enters the stomach from the esophagus, through a ring of muscles known as the cardiac sphincter, and is converted into a semiliquid state by muscular action and by the digestive enzymes in the gastric juice secreted by the glands. The pyloric sphincter, which separates the stomach from the small intestine, remains closed until the food has been appropriately modified and is ready to be emptied into the duodenum, the first section of the small intestine.

Stone Age. Period beginning with the earliest human development, c.2 million years ago. It is divided into three periods. The

Paleolithic period, or *Old Stone Age,* was the longest phase of human history, roughly coextensive with the Pleistocene Geologic Era. Its most outstanding feature was the development of *Homo sapiens.* Paleolithic people were generally nomadic hunters and gatherers who sheltered in caves, used fire, and fashioned stone-tools. Their cultures are identified by distinctive stone-tool industries: Pre-Chellean, Abbevillian (or Chellean),and Acheulian in the Lower Paleolithic; Mousterian in the Middle Paleolithic, associated with Neanderthal Man; and Aurignacian, Solutrean, and Magdalenian in the Upper Paleolithic. By the Upper Paleolithic there is evidence of communal hunting, man-made shelters, and belief systems centering on Magic and the supernatural. Rock Carvings and Paintings reached their peak in the Magdalenian culture of Cro-Magnon Man. The *Mesolithic period,* or *Middle Stone Age,* began at the end of the last glacial era, over 10,000 years ago. Cultures included gradual domestication of plants and animals, formation of settled communities, use of the bow, and development of delicate stone microliths and pottery. Notable Mesolithic cultures were the early Azilian and Tardenoisian, over most of Europe; the middle Mesolithic Maglemosian, in the Baltic and N England; the late Ertebolle, or kitchen-midden culture; and the Natufiar, in the Middle East. The time periods and cultural content of the *Neolithic period,* or *New Stone Age,* vary with geographic location. The earliest known Neolithic culture developed from the Natufian in SW Asia between 8000 and 6000 B.C. People lived in settled villages, cultivated grains and domesticated animals, developed pottery and weaving, and evolved into the urban civilizations of the Bronze Age. In SE Asia a distinct type of Neolithic culture cultivated rice before 2000 B.C. New World peoples independently domesticated plants and animals, and by 1500 B.C. Neolithic cultures existed in Mexico and South America that led to the Aztec and Inca civilizations.

Stoneware. Hard Pottery made of siliceous paste fired at high temperature to make it glassy (vitrified). Stoneware is heavier and more opaque than Porcelain, and differs from Terra-Cotta in being nonporous. Fired stoneware is usually grayish, but colours may vary widely with the clay used. Produced in China ancient times, it was the forerunner of Chinese porcelain. Stoneware was made in Germany in the 12th cent. and elsewhere in Europe by the 14th. The English potter Josiah Wedgwood developed two important types. Today stoneware remains one of the most common types of pottery.

Stop. An aperture used to limit the width of a beam of light in an optical system.

Store. The part of a digital computer in which the information is held. Computer storage systems are classified into *direct-access* devices, in which any part of the stored file can be retrieved directly, and *sequential-access* devices, in which the file has to be scanned in sequence until the information is found. Direct-access devices have a much lower access time. In most digital computers more than one type of store is used. The *main store* (or memory) is used for holding the information that is required directly by the central processing unit. It consists of a large number of cores, either ferrite rings or semiconductor devices. The *backing store* is used for holding large amounts of data. The information is stored by magnetized region on a magnetic tape, drum, or disk.

Stork. Mute, long-legged wading Bird of the family Ciconidae. Found in most of the warmer parts of the world, it has long, broad, powerful wings. The American wood stork, a white bird c. 4 ft (122cm) tall with a greenish-black tail, is found in temperate and tropical regions.

Storm. Disturbance of the ordinary atmospheric conditions marked by strong winds and (usually) precipitation. Types of storms include the extratropical Cyclone, the tropical cyclone (Hurricane), the Tornado, and the Thunderstorm. The term is also applied to blizzards, dust storms, and sand-storms in which high wind is the dominant meteorological element.

S.T.D. *See* standard temperature and pressure.

Straight Chain. A chain of atoms, usually carbon atoms, in a molecule with no side chains attached. Butane, for example, is a straight-chain hydrocarbon, whereas isobutane has a branched chain.

Strain Gauge. A device for measuring strain. The term is commonly used for small devices attached to the surface of a body for sensing and measuring deformation at different points. The simplest type has a fine resistance wire attached to a piece of paper or thin plastic. Small changes in the length of the wire are followed by monitoring its resistance. Other types of strain gauge exploit changes in inductance or capacitance or make use of the piezoelectric effect.

Strain Hardening (work hardening). Hardening of a metal by subjecting it to strain, as by stretching or rolling. It results from the production of dislocations in the crystal structure.

Strangeness. Symbol: S. A property of certain elementary particles, originally assigned to account for the fact that they decay more slowly than would be expected. Kaons, for example, would be expected to decay in about 10^{-23} second whereas their actual lifetime is about 10^{-8}-10^{-10} second. To account for this, it is postulated that they have an additional property—strangeness—that is described by a quantum number (called the *strangeness number*- or simply the strangeness).

Normal particles, such as the proton, have S = 0: strange particles have integral values of S. The strangeness number is conserved in strong and electromagnetic interactions but not in weak interactions. A particle's strangeness number is its hypercharge minus its baryon number.

Stratification. Layered structure formed by the deposition of sedimentary Rocks. Changes between layers result from fluctuations in the intensity and persistence of the depositional agent (*e.g.*, currents, wind, or waves) or from changes of the source of the sediment. Intially, most sediments are deposited with essentially horizontal stratification, but the layers may later be tilted or folded by internal earth forces.

Stratosphere. The region of the earth's atmosphere extending from a height of about 10 kilometres to about 80 kilometres. It lies between the troposphere and the mesosphere and has an increasing temperature. The temperature rises (up to 10°C) near the top of the stratosphere—the maximum occurring at the *stratopause.*

Strawberry. Low, herbaceous perennial (genus *Fragaria*) of the Rose family, native to temperate regions. It is grown for its edible red fruits, which are used fresh, frozen, in preserves and confectionery, and for flavoring. The common strawberry (*F. chiloensis*) is believed native to Chile and W North America; it has probably hybridized to some extent with the wild strawberry (*F. virginiana*) of E North America. Most species propagate by runners, or stolons, slender horizontal stems.

Streamline. A line in a fluid such that, at any point in the fluid, the tangent to the line is the direction of the fluid's velocity at that point. Fluid flow in which continuous streamlines can be drawn through the fluid is called *streamline flow.*

Strength of Materials. The capacity of materials to withstand stress (the internal force exerted by one part of an elastic body upon an adjoining part) and strain (the deformation or change in dimensions occasioned by stress). When a body is subjected to a pull, it is said to be under tension, or tensional stress; when it is compressed, it is under compression, or compressive stress. Shear, or shearing stress, results when a force tends to make part of body slide past the other part. Torsion, or torsional stress, occurs when external forces tend to twist a body around an axis. The elastic limit is the maximum stress that a material can sustain and still return to its original form. The radio of tensile stress to strain for a given material is called its Young's modulus. Hooke's law states that, within the elastic limit, strain is proportional to stress.

Stress. The effect producing or tending to produce deformation of a body (strain), equal to the applied force per unit area. There are several different types of stress.

Tensile stress (symbol: σ) is a stress tending to stretch a material in a particular direction. It is the force in that direction divided by the cross-sectional area. *Hydrostatic* or *bulk stress* is the stress tending to compress the whole body: it is simply the pressure on the body. In these examples to the force are acting at right angles to the surface—they are examples of *normal stress.*

When the force tend to produce a twisting motion, the effect is a *shear stress* (symbol: T) which is given by the tangential force per unit area. Although stress has the same units as pressure it is always given in newtons per square metre—not pascals.

Striations. *See* glow discharge.

Stringed Instrument. Musical instrument whose tone is produced by vibrating strings. Those instruments played with a bow are principally of the Viol and Violin families. Those whose strings are plucked, either by finger or with a pick (plectrum), include the lyre, any of various ancient instruments with arms projecting from the sound box and having from 3 to 12 strings; the Harp; and a number of fretted instruments, in which narrow strips of wood or metal, called frets, mark the places on the keyboard where the player's fingertips should be applied to stop the strings and produce various notes. Fretted instruments include the balalaika, a Russian instrument having a triangular body, a long fretted neck, and usually three strings; the banjo, often used in Country And Western Music, having four to nine strings and a round body resembling a tambourine; the lute, popular in the Middle Ages and Renaissance, and the mandolin, both having a pear-shaped body and rounded back. Related to the lute is the guitar, which has six strings, a flat back, and a body curved inward to form a waist; it appeared in Spain as early as the 12th cent. and is used in both classical and folk music. Much smaller and more limited than the guitar, but developed from it, is the ukulele, which has four strings. The dulcimer, psaltery, and zither are stringed instruments constructed of a variable number of strings stretched over a flat sound box. The dulcimer is struck with small mallets; the psaltery and zither are plucked. The psaltery flourished in Europe from the 12th cent. until the late Middle Ages. The dulcimer, which originated in Middle East, was adopted in Europe in the Middle Ages. The zither is derived from the psaltery and dulcimer. Stringed instruments operated by a keyboard include the Piano and its predecessor, the clavichord, whose strings are struck by mallets or hammers. Keyboard instruments whose strings are plucked by means of quills or jacks include the spinet and virginal, small, legless instruments similar to the clavichord, and the harpsichord.

Strip Mining. Process of extracting coal (or certain metallic ores) in which the surface material is removed to expose a coal seam or bed.The coal is then usually removed in a separate operation. The environment can be protected by respreading soil and by seeding or planting grass or trees on the fertilized, restored surface. Sometimes the terms *open-pit*, *open-cast*, or *surface mining* are used in the same sense as *strip mining*.

Stroboscope. An instrument for viewing objects that have fast cyclic or vibrating motion by illuminating or viewing them intermittently at the same frequency as that of their motion, thus making them appear stationary. For example, if a vibrating bar makes n vibrations per second and is illuminated by a lamp giving n short bursts of light per second, then every time the bar is seen it is in the same position. If the light frequency differs slightly from that of the bar, the bar appears to move slowly. A similar method of producing a stroboscopic effect is by viewing the object through a hole in a rotating disc.

Stroboscopes can be used for inspecting moving parts in operation and for measuring frequencies of vibration and speeds of rotation. Stroboscopic effects are also produced in other ways: perhaps the best known is the behaviour of wheels in films. The film camera takes a rapid series of still images. If the wheel makes one complete rotation between each frame, it will appear to be stationary.

Stroke or **Cerebrovascular Accident** (CVA). Destruction of brain tissue due to impaired blood supply caused by intracerebral hemorrhage, Thrombosis (clotting), or embolism (obstruction caused by clotted blood or other foreign matter circulating in the bloodstream). It is a leading cause of death worldwide. Stroke is most common in the elderly, but may occur at any age; predisposing conditions include Arteriosclerosis, Diabetes, and Hypertension. Symptoms develop suddenly and can range from almost unnoticed clumsiness or headache to severe paralysis, speech and mental disturbances, and coma. Treatment depends on the cause of the stroke and may include Anticoagulants, surgery, and Physical Therapy.

Strong Acid or **Base.** *See* acid, base.

Strong Interaction. A type of interaction between elementary particles occurring at short range (about 10^{-15} m) and having a magnitude about 100 times greater than that of the electromagnetic interaction. Strong interactions occur between hadrons and account for the forces that hold nucleons together in the atomic nucleus. They are thought to be due to the exchange of virtual mesons between the interacting hadrons.

Strontia. *See* strontium oxide.

Strontianite. A mineral, white, grey, yellow, or green in colour, consisting of strontium carbonate, $SrCO_3$. It is a source of strontium salts.

Strontium. Symbol: Sr. A soft pale yellow metallic element occurring principally as strontianite ($SrCO_3$) and celestine ($SrSO_4$). It is made by electrolysis of fused strontium chloride and used in some alloys and as a getter. Strontium is one of the alkaline-earth metals. A.N. 38; A.W. 87.62; m.pt. 800°C; b.pt. 1300°C; r.d. 2.54; valency 2.

Strontium (Sr). Metallic element, first recognized as distinct from barium by A. Crawford in 1790. A soft, silver-yellow Alkaline-Earth Metal, it is stored away from air and water. Strontium-90 from nuclear fallout is absorbed in plants and animals, and may induce bone cancer and leukemia.

Strontium-90. The radioactive isotope of strontium with mass number 90. It is a particularly hazardous constituent of radioactive fallout, being metabolized like calcium and included in bone: its half-life is 28 years.

Strontium Oxide (strontia). A greyish-white powder, SrO, prepared by decomposing strontium carbonate or hydroxide. It is a basic oxide, used in the manufacture of strontium salts, pyrotechnics, and soaps, and as a desiccant. M.pt. 2430°C; b.pt. 3000°C; r.d. 4.7.

Structural Formula. *See* formula.

Structural Isomerism. A type of isomerism in which the isomers have different structural formulas. The compounds may be quite unrelated, as in the case of ethanol, C_2H_5OH, and dimethyl ether, CH_3OCH_3, which both have the formula C_2H_6O. Often structural isomers are similar, as in butane, $CH_3CH_2CH_2CH_3$, and isobutane, $CH_3CH\text{-}(CH_3)CH_3$. Sometimes such compounds differ only slightly in physical properties. Another example is the existence of isomers with functional groups in different positions, as in propan-1-ol, $CH_3CH_2CH_2OH$, and propan-2-

ol, $CH_3CH(OH)CH_3$. Derivatives of benzene with two substituted groups can have three isomers depending on the relative positions of the groups on the ring: *ortho,* in which the groups are adjacent, *meta,* in which they have one unsubstituted carbon atom between them, and *para* in which the groups are at opposite sides of the benzene ring.

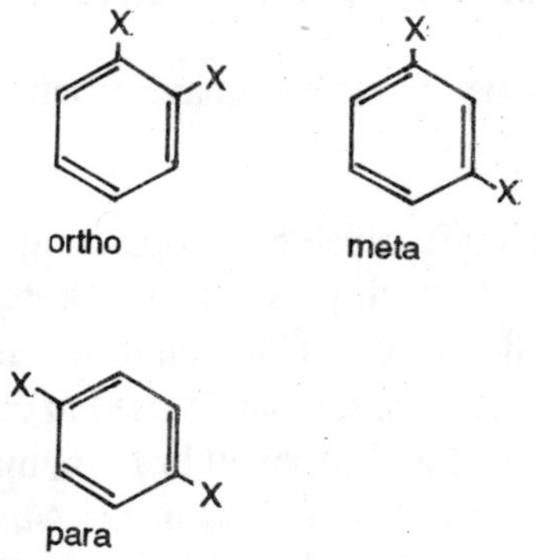

X X X ortho X meta X X para

Fig. S-13 Structural isomers.

Strychnine. A colourless crystalline highly poisonous alkaloid, $C_{21}H_{22}N_2O_2$, found in certain plants. M.pt. 270-80°C; r.d. 1.36.

Strychnine. Bitter Alkaloid drug derived from the seeds of a tree (*Strychnos nux-vomica*) native to Sri Lanka, Australia, and India. A potent stimulant, the drug has been used as a rat poison for five centuries. Strychnine poisoning is characterized by violent convulsions and is fatal unless promptly treated with barbiturate sedatives and Artificial Respiration.

Sturgeon. Primitive marine and freshwater Fish of N Eurasia and North America. It has reduced scalation, a mostly cartilaginous skeleton, upturned tail fins, and a toothless mouth set far back under its jaw. It sucks up its food—Crayfish, Snails, larvae, and small fish—from the water bottom. The largest is the Russian sturgeon, or beluga (*Acipenser huso*), reaching a length of 13 ft (396 cm) and a weight of 1 ton (908 kg). Smoked sturgeon and sturgeon caviar are valued as food.

Styrene. A colourless oily liquid with an aromatic odour, $C_6H_5CH{:}CH_2$, prepared by dehydrogenation of ethylbenzene. It is polymerized to make polystyrene. M.pt. –31°C; b.pt. 145°C; r.d. 0.9.

Subatomic. Smaller than an atom. *Subatomic particles* are particles such as protons and electrons that are found in atoms.

Subcritical. Involving a nuclear chain reaction that is not self-sustaining.

Submarine. Naval craft capable of operating underwater for an extended period of time. Cornelis Drebbel built(c.1620) a leather-covered rowboat that could remain under water as long as 15 hr. The first submarine used in combat was invented by the American David Bushnell in 1776; many of its principles were adopted by Robert Fulton in his *Nautilus,* a submarine successfully operated (1800-1801) on the Seine R., in France. Although the Confederates used several submersible craft in the Civil War, it was the work of John Holland and Simon Lake that advanced considerably the development of the modern submarine in the U.S. A Holland submarine became the first for the U.S. navy in 1900; Lake's *Argonaut* was (1898) the first submarine to navigate extensively in the open sea. E-boats, the first U.S. dieselengine submarines, appeared in 1912 and were the first to cross the Atlantic Ocean. Both sides used submarines extensively in World War I, especially Germany, whose 200-ton U-boats inflicted heavy damage on Allied shipping. Larger and improved submarines played a major role in World War II, in which the Allies and neutrals lost some 4,770 ships to these raiders. The advent of nuclear energy brought about major changes in submarine propulsion and striking power. The first nuclear submarine, the U.S.S. *Nautilus,* was completed in 1954. Nuclear submarines can remain submerged for almost unlimited periods of time and can fire long-range missiles from a submerged position. Such a submarine's primary mission is to strike at enemy land targets. Some nuclear attack submarines, for deployment against shipping, have also been built.

Sublimate. A solid deposited by sublimation.

Sublimation. The passage of certain substances from the solid state into the gaseous state and then back into the solid state, without any intermediate liquid state being formed. The term is often used to described the direct change from solid to vapour without reference to subsequent change back to solid.

Sub-shell. A subdivision of an electron shell, for which all the electrons have the same azimuthal quantum number. The sub-shells of a particular shell are designated by the letters *s*, *p*, *d*, or *f*, according to the value of their quantum number.

Subsonic. Relating to a speed less than Mach 1.

Substantivity. *See* dye.

Substituent. An atom or group that replaces another atom or group in a substitution reaction. The term is often used for an atom or group replacing hydrogen in a derivative of a compound.

Substitution. A chemical displacement reaction, especially one in which a hydrogen atom is displaced. An example is the chlorination of benzene, $C_6H_6 + Cl_2 = C_6H_5Cl + HCl$, in which the hydrogen atom is replaced by a chlorine atom. *Compare* addition.

Substrate. A substance that is acted upon in some way. For example, the substrate is the compound acted on by a catalyst, the material to which a dye is attached, or the solid on which a compound is adsorbed.

Subtend. To determine an angle at a given point; said of curves. Thus an arc AB subtends the angle AGB at O.

Subtractive Process. The process in which a colour is produced by combining coloured pigments, dyes, filters, or other absorbing materials. The final colour is produced by absorption of different wavelengths of light. Most colours can be produced by a suitable mixture of primary colours. *Compare* additive process.

Succinate. Any salt or ester of succinic acid.

Succinic Acid (butanedioic acid). A colourless crystalline weak carboxylic acid, $HOOC(CH_2)_2COOH$, made by fermentation of ammonium tartrate and used as a sequestrant and raw material for the manufacture of dyes, lacquers, and perfumes. M.pt. 185°C; b.pt. 235°C; r.d. 1.5.

Succulent. Any fleshy plant, typically with reduced leaves and an outer surface covered with a waxy substance (cutin) that reduces evaporation from the inner, water-storing tissue. Many are indigenous to dry regions. Species of Cactus, Aloe and Yucca are succulents.

Sucrose (sugar). A white sweet soluble crystalline disaccharide. $C_{12}H_{22}O_{11}$, obtained from sugar beet or sugar cane and used as a sweetener. M.pt. 185-6°C; r.d. 1.6.

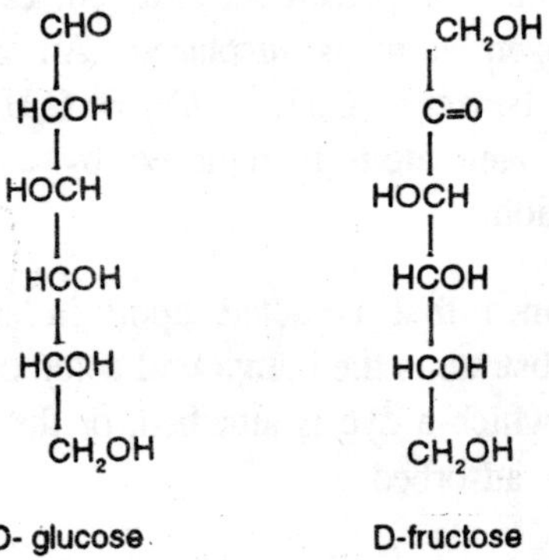

Fig. S-14. Glucose and fructose.

Sugar. 1. *See* sucrose.

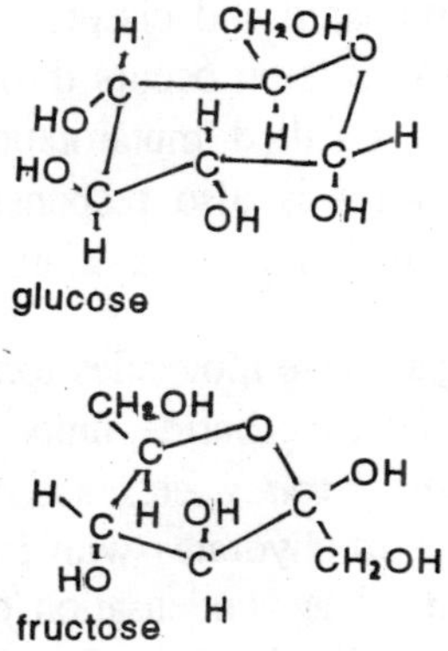

glucose

fructose

Fig. S-15. Ring forms of glucose and fructose.

2. Any of a class of simple carbohydrates. The *monosaccharides* or *simple sugars* have the general formula $(CH_2O)_n$ and their molecules contain hydroxyl groups and either an aldehyde group or a ketone group. Sugars containing an aldehyde group are *aldoses:* those containing a ketone group are *ketoses.* The simple sugars are also classified according to the number of carbon atoms in the molecule: *trioses* contain three carbon atoms, *tetroses* contain four, *pentoses* five, etc. The commonest types are the *hexose* sugars, with formula $C_6H_{12}O_6$. The different compounds with this formula differ in the relative arrangements of groups attached to the carbon atoms. Fig. 1 shows the formulas of glucose (an *aldohexose)* and fructose (a *ketohexose).*

In fact the monosaccharides do not normally exist in these straight-chain forms but in a ring form. The hydroxyl (alcohol) group of one end of the molecule can condense with the carbonyl group of the other to give a hemiacetal. Compounds such as glucose, which have six-membered rings, are called *pyranoses:* those that have five membered rings are *furanoses.*

In the solid state the sugar has a ring form and in solution this is equilibrium with a small amount of the straight-chain form. The sugars all show optical isomerism and if a pure optical isomer is prepared and dissolved in water the optical activity changes. Conversion occurs through the straightchain form—the process is called mutarotation. The presence of the straight-chain form is also responsible for the positive reaction of sugars with Fehling's solution.

More complex sugars have molecules formed by condensation of two or more monosaccharide units and are known as *disaccharides, trisaccharides,* etc., according to the numbers of units involved. Carbohydrates with large number of units are *polysaccharides.* The condensation of two simple sugars can be visualized as elimination of H_2O between two -OH groups, giving an -O- bond between two rings. This is known as a *glycosidic bond.* Sucrose, for instance, has a glucose ring bound to a fructose ring.

Sucrose is linked in such a way that it cannot form a noncyclic molecule containing an aldehyde or ketone group and thus does not reduce Fehling's solution. Such compounds are said to be *nonreducing sugars* to distinguish them from *reducing sugars.*

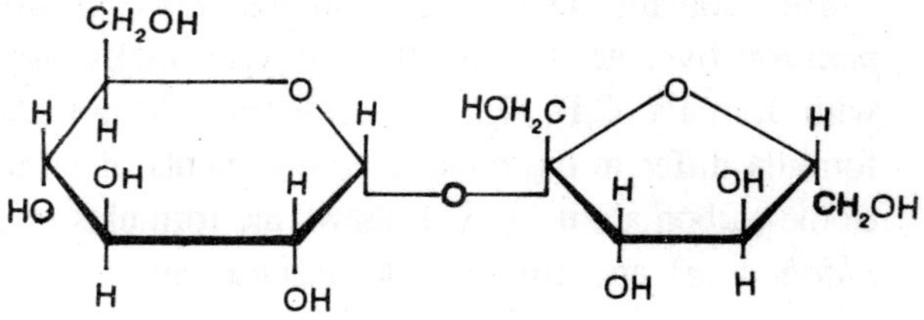

Fig. S-16. Sucrose.

Sucrose (empirical formula: $C_{12}H_{22}O_{11}$). Common table sugar, a white, crystalline solid with a sweet taste. Common names, which indicate the natural source, include cane sugar, beet sugar. and maple sugar. A disaccharide, sucrose can be hydrolized to yield invert sugar, a mixture of unequal amounts of Fructose and Glucose. Sucrose is obtained from the "juice" of sugarcane or sugar beets and from the sap of the sugar maple. It is evaporated to give first a brownish liquid, called molasses; further evaporation yields a brownish sugar. The colour, due to impurities, is removed by charcoal used in refining process.

Sudden Infant Death Syndrome (SIDS) or **Crib Death.** Sudden, unexpected, and unexplained death of an apparently well infant under one year (usually between 2 weeks and 8 months). SIDS accounts for 10% of infant deaths and (after accidents) is the second-highest cause of death in infancy. The risk is higher in males, in low-birth-weight infants, in lower socioeconomic levels, and during cold months. Current theories suggest that the infant may have immature lungs or problems with brain-stem control of breathing; SIDS victims are thought to have brief episodes of apnea (breathing stoppage) before the fatal one. An alarm system that detects breathing abnormalities is sometimes used with infants suspected of being prone to SIDS.

Sugar. Compound of Carbon, Hydrogen, and Oxygen, belonging to a class of substances called Carbohydrates. Sugars fall into three groups. Monosaccharides are the simple sugars, *e.g.*, Fructose and Glucose. Disaccharides, made up of two monosaccharides units, include Lactose, Maltose, and Sucrose. The less familiar trisaccharides, made up of three monosaccharide units, include raffinose, found in sugar Beets.

Sugarcane. Tall tropical perennial (genus *Saccharum*) of the Grass family, native to Asia. Sugarcane somewhat resembles Corn and Sorghum, with a large terminal panicle and a noded stalk. It and the sugar Beet are the major sources of Sugar. The cane is harvested by cutting down the stalks and pressing them to extract the juice, which is concentrated by evaporation. Refined sugar is produced by precipitating out nonsugar components; it is almost pure Sucrose. Cuba and India together produce over one third of the world's canesugar. Sugarcane by products include molasses, Rum, Alcohol, fuel, and livestock feed.

Sugar of Lead. Lead acetate.

Suicide. The deliberate taking of one's own life. It may be dictated by social convention, as with Hara-Kiri, in Japan, or by custom, as in those primitive societies in which nonproductive elderly individuals were expected to end their own lives for the welfare of the group. Long condemned by Judaism, Christianity, and Islam, suicide remains a crime in some countries. Britain abolished punishment for attempted suicide in 1961; in the U.S. attempted suicide and the act of helping someone to commit suicide remain illegal in some states. Suicide and attempted suicide are now more often considered the result of psychological factors, such as severe Depression, Guilt, and Aggression, or of chronic illness. Suicidal behaviour is also viewed as a form of communication, a cry for help. Statistics reveal patterns that defy easy explanation. Suicide is more common among men than women, and although severe depression exists among the very young and very old, suicide among these groups is rare; it has, however, increased among adolescents in Western countries in recent years. Emile Durkheim, Sigmund Freud, and Karl Menninger have studied and proposed theories of suicide.

Sulpha Drug. Any of a class of synthetic chemical substances derived from sulfanilamide and used to treat bacterial infections. These drugs inhibit the action of para-aminobenzoic acid, a substance bacteria need in order to reproduce. Sulfa drugs are used primarily in the treatment of urinary tract infections and ulcerative colitis; Antibiotics have largely replaced them in the treatment of other bacterial infections.

Sulphate. Any salt or ester of sulphuric acid.

Sulphation. The production of an insoluble layer of lead sulphate on the electrodes of a lead-plate accumulator when it is left discharged.

Sulphide. Any compound of sulphur with a more electropositive group or element. Inorganic sulphides are salts of hydrogen sulphide and contain the ion S^{2-}. *Polysulphides* can also be produced containing ions of the type S_x^{2-}, in which there are chains of several sulphur atoms.

Sulphite. Any salt or ester of sulphurous acid.

Sulphonamide. Any of a class of compounds that are amides of sulphonic acid: *i.e.* they contain the group $-SO_2NH_2$ or are derived from such a compound by replacement of one of the hydrogen atoms. An example is *sulphanilamide*, $H_2NC_6H_4SO_2NH_2$. The sulphonamides are an important class of drugs *(sulpha drugs)* active against bacteria.

Sulphonic Acid. Any of a class of organic compounds of the general formula $Ar.SO_2.OH$, where Ar is an aryl group. The simplest example is benzenesualphonic acid, $C_6H_5SO_3H$. The sulphonic acids tend to be strong water-soluble acids. They are made from chlorosulphonic acid: $ArH + 2ClSO_3H = ArSO_3Cl + HCl + H_2SO_4$. The acid chloride then reacts with water to give the sulphonic acid.

Sulphonyl Group. *See* sulphuryl group.

Sulphur. Symbol: S. A yellow nonmetallic element occurring in many mineral and also as the native element in large underground deposits in the U.S. It is obtained by the Frasch process. Sulphur exists in several allotropic forms. At room temperature the stable form has a rhombic structure and aboves 95.6°C this slowly changes into a monoclinic form. Both these solids contain S_8 rings. Another form of sulphur, containing s_6 rings and having a hexagonal structure, is obtained by adding sodium thiosulphate to hydrochloric acid and crystallizing the sulphur from toluene. Plastic sulphur is made by pouring molten sulphur at about 160°C into cold water. It is an amorphous solid containing S_8 molecules with atoms in helical chains. *Flowers of sulphur* is a fine powder produced by condensing sulphur vapour. Liquid sulphur also has several forms. Just above the melting point it is a mobile yellow liquid containing S_8 rings. Above 160°C it becomes brown and increases in viscosity as the temperature is increased. This is probably due to the formation of chains of sulphur atoms and the reaction of these chains with each other and with rings to form polymers. Above 200°C the viscosity again falls. Sulphur vapour probably contains an equililbrium mixture of S_8, S_6, S_4 and S_2 molecules. The element is used for making sulphuric acid, the most important compound, and paper and for vulcanizing rubber. It is also an agricultural fungicide. Chemically, it is one of the chalconide elements. A.N. 16; A.W. 38.06; m.pt. 112.8°C (rhombic), 119.0°C (monoclinic); b.pt. 444.674°C; r.d. 2.07 (rhombic), 1.96 (monoclinic); valency 2, 4, or 6.

Sulfur (S) or **Sulphur.** Nonmetallic element, known to antiquity as the biblical brimstone and recognized as an element by Antoine Lavoisier in 1777. Solid sulfur is yellow, brittle, odourless, tasteless, and insoluble in water. Sulfur is widely distributed in minerals and ores, some volcanic regions, and large underground deposits, and often occurs with coal, natural gas, and petroleum. It is found in most proteins and protoplasm of plants and animals. Sulfur is used in Gunpowder, matches, Rubber vulcanization, insecticides, and the treatment of certain skin diseases. Sulfuric Acid is its most important compound; others are used as disinfectants, refrigerants, organic solvents, and Sulfa Drugs.

Sulphur Chloride. Either or two chlorides of sulphur, both of which are readily hydrolysed, liquids used as chlorinating agents. *Disulphur dichloride,* S_2Cl_2,is yellowish in colour and is prepared by passing chlorine into molten sulphur. M.pt. –80°C; b.pt. 138°C; r.d. 1.7. *Sulphur dichloride,* SCl_2, is reddishbrown. It is made by passing chlorine into disulphur dichloride at a low temperature. M.pt. –78°C; r.d. 1.6.

Sulphur Dichloride. *See* sulphur chloride.

Sulphuretted Hydrogen. *See* hydrogen sulphide.

Sulphuric Acid (vitriol, oil of vitriol). A colourless hygroscopic oily liquid, H_2SO_4, made by the contact process. It is a strong acid in aqueous solution and also an oxidizing and dehydrating agent. Sulphur trioxide can be dissolved in sulphuric acid to form *fuming sulphuric acid,* $H_2S_2O_7$, which is also known as *oleum* and *pyrosulphuric acid.* Sulphuric acid is one of the most important commercial chemicals, being used in the manufacture of fertilizers, paints, rayon, explosives, and many other products as well as in petroleum refining and leadplate accumulators. M.pt. 10°C; b.pt. 315-338°C; r.d. 1.8.

Sulphurous Acid. A weak acid, H_2SO_3, made by dissolving sulphur dioxide in water. It is a reducing agent and is used in bleaching.

Sulfuric Acid. Chemical compound (H_2SO_4) colourless, odourless, extremely corrosive, oily liquid. It is sometimes called oil of vitriol. Concentrated sulfuric acid is a weak acid and a poor Electrolyte because relatively little is dissociated into ions at room temperature. When cold it does not react readily with such common metals as iron or copper. When hot it is an oxidizing agent. Hot concentrated sulfuric acid reacts with most metals and with several nonmetals, *e.g.,* sulfur and carbon. When concentrated sulfuric acid is mixed with water, large amounts of heat are released. Sulfuric acid is a strong acid and a good electrolyte when diluted. A very important industrial chemical, sulfuric acid is produced by the oxidation and dissolution in water of sulfur dioxide (SO_2).

Sulphur Oxide. Either of two acidic oxides of sulphur. *Sulphur dioxide,* SO_2, is the colourless pungent gas produced by burning sulphur. It is made by roasting pyrites in air and used to make sulphuric acid and as a preservative for foods, a bleaching agent, and a fumigant. It is also a reducing agent. In solution it yields a mixture of sulphuric and sulphurous acids. M.pt. —76°C; b.pt. —10°C; r.d. 1.4 (liquid at 0°C).

Sulphur trioxide is a white solid, SO_3, made from sulphur dioxide by the contact process. It is the acid anhydride of sulphuric acid. Three crystalline forms exist with different melting points.

Sulphur Point. *See* temperature scale.

Sulphur Trioxide. *See* sulphur oxide.

Sulphuryl Chloride. A colourless liquid, SO_2Cl_2, with a pungent odour, make by reacting sulphur dioxide with chlorine in the presence of activated carbon as a catalyst. It is used for chlorination and dehydration. M.pt. —54°C; b.pt. 69°C; r.d. 1.7.

Sulphuryl Group (sulphonyl group). The divalent group $=SO_2$.

Sumac or **Sumach.** Common name for the family Anacardiaceae, trees and shrubs native chiefly to the tropics but ranging into north temperate regions and typified by resinous, often acrid sap, a source of tannin. The sap of some plants in the family, *e.g.*, Poison Ivy, also contains an Essential Oil that is a toxic skin irritant. Pistachio, Mango, and Cashew are members of the family that provide food. The true sumacs belong to the genus *Rhus*. Most *Rhus* species contain tannin, and some are cultivated for it. Several North American species have brilliant fall foliage; the fruit of the common staghorn sumac (*R.typhira*) of the E U.S. is used to make wine and for medicinal purposes.

Sun. The star around which orbit the earth, the planets, and all other bodies in the solar system. The sun is a yellow dwarf, halfway in energy output between the hottest and brightest bluish-white stars and the coolest and dimmest red ones. In space it occupies a position about one third of the way from the perimeter to the central core of the Galaxy and revolves with the Galaxy at a velocity of 216 kilometres per second, completing one revolution in 200 million years. Lying at a mean distance of 149 million km from earth, its light and heat take eight minutes to reach us.

As with other evolving stars, its energy is provided by nuclear fusion reactions in its centre. Here, at a temperature of 15,000,000°C, hydrogen is turned first into deuterium and then helium. The surface is the *photosphere*. Its temperature is about 6000°C. Above it extends the solar atmosphere, the *chromosphere,* an incandescent gaseous envelope stretching outwards for some 15,000 km. Here prominences and solar flares occur. Above this the *corona,* split into the inner or K corona and the outer or F corona, extends into space. The corona, visible during a total eclipse as a halo, contains tenuous highly ionized gases at a temperature of about 1,000,000°C.

Sun. Intensely hot, self-luminous body of gases (mainly hydrogen and helium) at the center of the Solar System. The sun is a medium-size main-sequence Star. Its mean distance from the earth is defined as one Astronomical Unit. The sun is c.865,400 mi (1,392,000 km) in diameter; its volume is about 1,300,000 times, and its mass 332,000 times, that of the earth. At its center, the sun has a density over 100 times that of water, a pressure of over 1 billion atmospheres, and a temperature of about 15,000,000°K. This temperature is high enough for the occurrence of nuclear reactions, which are assumed to be the source of the sun's energy. Hans Bethe proposed a cycle of nuclear reactions known as the carbon cycle, in which carbon acts much as a catalyst, while hydrogen is transformed by a series of reactions into helium and large amounts of high-energy gamma radiation are released. The so-called proton-proton process is now thought to be a more important energy source: the collision of two protons ends with the production of helium atoms and the release throughout of gamma radiation. The bright surface of the sun is called the photosphere; its temperature is about 6000°K. During an Eclipse of the sun, the chromosphere (a layer of rarified gases above the photosphere) and the corona (a luminous envelope of extremely fine particles surrounding the sun, outside the chromosphere) are observed.

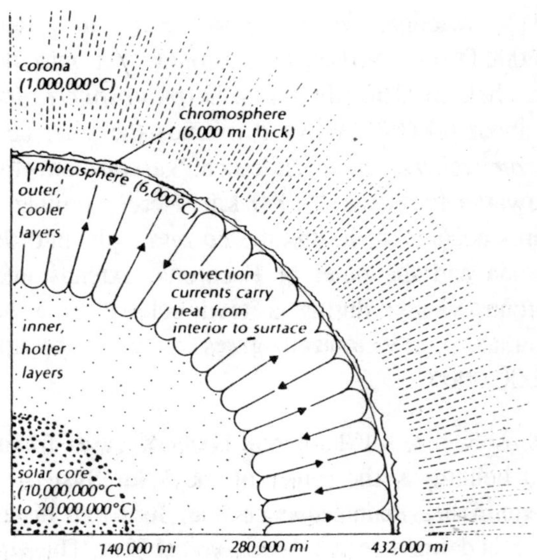

Fig. S-17. Structure of the sun.

Sun Belt or **Sunbelt.** Popular name for a region of the southern U.S.—a term that gained wide acceptance in the energy-conscious 1970s. The Sun Belt is generally considered to focus on Texas and California and to extend north and east as far as North Carolina. It thus embraces a diverse area that—because of rapid economic growth, high federal spending levels, and climatic advantages—has experienced rapid gains in population and political importance as a result of interregional migration from the so-called Frost Belt states of the N and NE U.S.

Sundial. Instrument that indicates the time of day by the shadow, cast on surface marked to show hours or fractions of hours, of an object on which the sun's rays fall. The shadow-casting objects is called a gnomon. Corrections must be made for the difference (which varies daily) between solar (or apparent) time and clock (or mean) time and for the difference in longitude between the position of a sundial and the standard-time meridian of a given locality.

Sunflower. Any annual or perennial herb (genus *Helianthus*) of the Composite family, native to the New World. The flower heads are commonly bright yellow and may reach 1 ft (30 cm) in diameter. Different parts of the common sunflower (*H. annuus*) are used in many ways: the seeds as a poultry food, bread grain, and a source of oil; the flowers for the production of nector; and the leaves for fodder. It is the state flower of Kansas. Other species are used for blood and as garden flowers.

Sunspot. A region of the sun's surface that is much cooler and therefore darker than the surrounding area, having a temperature of about 4000°C as opposed to 6000°C for the rest of the photosphere. Sunspots range from a few hundred kilometres to several times the earth's diameter and last from a few hours to a few months. A sunspot typically has a dark central core (the *umbra*) surrounded by a lighter region (the *penumbra*). The number of sunspots follows a pattern of fluctuation (the *sunspot cycle*) passing from maximum to maximum in about 11 years. High numbers of sunspots are associated with solar flares and prominences, giving out large numbers of charged particles that affect the earth's ionosphere.

Superconductivity. A phenomenon occurring in certain metals and alloys at temperatures close to absolute zero, in which the electrical resistance of the solid vanishes below a certain temperature (the *transition temperature*). A current induced in a ring of superconductor continues to flow when the magnetic field inducing it is removed — an effect used in high-current superconducting magnets.

Supercooling. The cooling of a system to a temperature below that at which a phase change should occur without any change taking place. For example, a pure liquid can sometimes be cooled slowly to a few degrees below its freezing point without freezing. Similarly, a solution of a solid in a liquid might be cooled to a temperature at which the solvent contains more solute than a saturated solution at that temperature. The solution is then *supersaturated.* Supercooling of liquids and solutions results in the production of a metastable state, in which the system is not at equilibrium. It occurs when no particles are present on which crystals may form. Crystallization may be induced by introducing a single crystal (a *seed)* on which the crystals can nucleate. Another method is to scratch the inner wall of the vessel—thus producing roughness on which the crystals can form. Supercooling can also be observed in vapours when they are free from dust or other small particles on which drops of liquid are produced.The vapour then exists at a temperature below its normal dew point. A *supersaturated vapour* results.

Superconductivity. The absence of electrical Resistance in certain substances when they are cooled to temperatures near absolute zero. The phenomenon, discovered by Kamerlingh Onnes in 1911, is displayed by some metals, including zinc, magnesium, and lead; in some alloys; and in certain compounds, such as tungsten carbide and lead sulfide. Current in a superconducting circuit will, continue to flow after the source of current has been shut off. Powerful Electromagnets, which, once energized, retain their magnetic field, have been developed, using, coils of superconducting metal.

Supercritical. Involving a nuclear chain reaction that is self-sustaining.

Superfluid. A fluid that flows without friction and has extremely high thermal conductivity, a phenomenon occurring in liquid helium at very low temperatures. Liquid helium makes the transition from the normal state to the superfluid state below 2.186 K.

Superfludity. The capability of liquid helium cooled below a temperature of 2.19°K (the lambda point) to flow freely, even upward, with no measurable friction and viscosity. Superfluid helium flows easily through capillary tubes that resist the flow of ordinary fluids, and a Dewar Flask filled with superfluid helium from a larger container will empty itself back into the original container because the liquid helium flows spontaneously in an invisible film over the surface of the flask.

Supergiant. Any of class of stars with higher absolute magnitude than ordinary giants, forming a small separate group above the giant branch on the Hertzsprung-Russell diagram.

Superheating. The process of heating a liquid to a temperature above its normal boiling point without occurring. Superheating can happen when there are no small particles of matter in the liquid or roughness on the vessel walls on which bubbles can from.

Superheterodyne. Denoting the combination of radiofrequency signals by the heterodyne principle to give a resultant signal with a frequency above the audiofrequency range. This effect is used in superheterodyne radio reception. The term is short for *supersonic heterodyne.*

Super-high Frequency (SHF). A frequency in the range 3 gigahertz to 30 gigahertz.

Superior Planet. Any of the planets Mars, Jupiter, Saturn, Uranus, Neptune, and Pluto, whose orbits lie beyond that of the earth. *Compare* inferior planet.

Supernova. A star that suffers an explosion, becoming up to 10^8 times brighter in the process and forming a large cloud of expanding debris (the supernova remnant). The Crab nebula is an example of the remnants of a supernova. The explosion is probably caused by a star exhausting its nuclear fuel and contracting under gravitation to an unstable condition. The outer layers are thrown off.

Supernova. Exploding star that suddenly increases it energy output as much as a billionfold and then slowly fades to less than its original brightness. At peak intensity, it can outshine the entire galaxy in which it occurs. Supernovas represent a catastrophic stage of Stellar Evolution; a sudden implosion of the core of certain massive stars produces a rapidly rotating collapsed stellar remnant, the explosive ejection of the stellar envelope at great velocity, and the release of enormious quantities of energy. Over 120 extended galactic radio sources have been indentified as supernova remnants. Of these, only four have been positively associated with explosions that were optically observed in recorded history; they occured in 1006, 1054 (the remnant of which is now visible as the Crab Nebula), 1572, and 1604.

Superoxide. An inorganic compound containing the O_2- ion. Potassium, rubidium, and a few other electropositive metals form superoxides by direct combination.

Superphosphate. A mixture consisting mainly of calcium hydrogen phosphate [$Ca(H_2PO_4)_2$] and calcium sulphate ($CaSO_4$), obtained by treating calcium phosphate with concentrated sulphuric acid. It is used as a fertilizer.

Supersaturated. Denoting a solution that contains a higher concentration of solute than a saturated solution at the same temperature. Similarly, the term describes a vapour that contains a higher pressure of vapour than a saturated vapour at the same temperature. Supersaturation is a consequence of supercooling.

Supersonic. Relating to a speed greater than Mach 1.

Supplementary Angle. An angle that together with a specified angle makes 180°. Thus, 30° is the supplementary angle of 150°.

Supplementary Unit. *See* SI units.

Suppressor Grid. The grid between the screen grid and the anode in a pentode valve.

Surd. An irrational root of a number. $\sqrt{2}$ and $\sqrt{3}$ are examples of surds; $\sqrt{4}$ (=2) is not.

Surface-active Agent. See surfactant.

Surface Tension. Symbol: γ. The property of liquids that makes them act as if they had an elastic skin at their surface. The effect is caused by forces between the molecules—a molecule in the interior is attracted by other molecules on all sides whereas one at the surface experiences a force tending to pull it into the liquid The phenomenon is responsible for the rise of certain liquids in vertical capillary tubes, the absorption of liquids by cloth, paper, and other porous substances, the ability of drops of liquid to stand on a surface without flowing, and the spherical shapes of soap bubbles, raindrops, etc.

Surface tension can be thought of quantitatively in two ways. It is the force acting tangential to the surface on one side of a line of unit length in the surface: *i.e.* its units are newtons per metre. Alternatively it can be thought of as the energy of the surface and defined by the work required to produce unit increase in surface area: its units are then joules per square metre. The two definitions are equivalent. The value of surface tension can be determined by capillary rise, Jaegar's method, or the Langmuir balance.

Surface Tension. The cohesion forces at the surface of a liquid. The molecules within a liquid are attracted equally from all sides, but those near the surface experience unequal attractions and thus are drawn toward the center of the liquid mass by this net force. A result of surface tension is the tendency of a liquid to reduce its exposed surface to the smallest possible area.

Surfactant (surface-active agent). A substance used to increase the spreading or wetting properties of liquid. Surfactants are often detergents, which act by lowering the surface tension. They are used in many processes including froth-flotation of ores, the dispersion of insoluble powders in liquids, and the dyeing of textiles (to increase the rate of penetration).

Surfing. Sport of gliding toward the shore on a breaking wave, done on a board from 4 to 12 ft (122-366 cm) long. The larger surfboards have a stabilizing fin in the rear. The surface paddles toward the beach until an incoming wave catches the board, then stands up and glides along or just under the crest of the wave. Developed in Hawaii, surfing spread to California in the 1920s and, by the 1960s, had become popular with youth in the U.S., Australia, and other countries.

Surgery. Branch of medicine concerned with the diagnosis and treatment of injuries and pathological conditions requiring manual or instrumental operative procedures. Surgery has been performed since prehistoric times. (bloodletting, opening of abscesses) and was practiced with great skill and cleanliness by the ancient Greeks and Romans. During the Middle Ages in Europe it fell into the hands of unskilled barber-surgeons, and postoperative infection and Gangrene were common. Surgery became more professional in the 18th cent. and entered its modern phase in the 19th cent. with the introduction of antiseptic techniques, sterilization, and Anesthesia. Twentieth-cent. advances include Blood Transfusion techniques, new diagnostic tools (X Ray, Cat, Scan, Ultrasound), Antibiotics and other Chemotherapies, microsurgery, and organ Transplantation.

Surveying. Accurate measurement of points and lines of direction on the earth's surface for the purpose of preparing maps or locating boundary lines. *Hydrographic surveying* records such features as bottom contours, buoys, channels, and shoals in bodies of water and along coastlines. *Land surveying* includes both *geodetic surveying,* used for large areas and taking into account the curvature of the earth's surface; and *plane surveying,* which deals with areas sufficiently small that the earth's curvature is negligible and can be disregarded.

Susceptance. Symbol: B. the imaginary part of the admittance of an electric circuit, equivalent to $—X/(R^2 + X^2)$, where R is the resistance and X the reactance.

Susceptibility. 1. (magnetic susceptibility). Symbol: Xm. A measure of the extent to which a substance is magnetized by an applied magnetic field, equal to the magnetization (M) per unit magnetic field strength (H): *i.e.* $M = XmH$. The susceptibility is related to the relative permeability (μ_r) by $Xm = 1 — \mu_r$. Diamagnetic materials have small negative susceptibilities, paramagnetic substances have small positive values, and ferromagnetic materials have large positive values. The susceptibility is a dimensionless ratio.

2. (electric susceptibility). Symbol: X_e. A measure of the extent to which a substance is polarized by an applied electric field, equal to the polarization (P) per unit electric field strength (E): $P = X_e E_o E$ where E_o is the electric constant. The susceptibility is related to the relative permittivity (E_r) by $X_e = 1—E_r$. The electric constant is used to make the susceptibility a dimensionless ratio.

Suspension. A mixture containing small particles of solid or liquid dispersed in a liquid or gas.

Suspension. In chemistry, mixture of two substances, one of which is finely divided and dispersed in the other. Particles in a suspension are larger then those in Colloids or solutions, and they precipitate if the suspension is allowed to stand undisturbed. Common suspensions include sand in water, dust in air, and droplets of oil in air.

Swallow. Common name for small perching Birds of the family Hirundinidae, of almost worldwide distribution. Swallows have long, narrow wings, forked tails, and weak feet. They are extremely graceful in flight, able to make abrupt changes in speed and direction. Their plumage is blue or black with a metallic sheen. American species include the common America barn swallow *(Hirundo rustica)* and the purple martin *(Progne subis)*.

Swamp. Shallow body of water in a low-lying, poorly drained depression, usually containing abundant plant growth. A notable example of U.S. Swamp is the Everglades in S Florida. Because the bottom of a swamp is at or below the water table, swamps serve to channel runoff into the groundwater supply, thus helping to stabilize the water table. During periods of very heavy rains, a swamp can act as a natural flood-control device. The increased use of drained swampland for urban development results in greater runoff and probability of flooding as well as the destruction of wildlife habitats.

Swan. Large aquatic Bird, related to the Duck and Goose, with a long, gracefully curved neck. The orange-billed white trumpeter swan *(Cygnus buccinator)*, seen in parks, breeds in the wild in parts of Euorpe, Asia, and the U.S. The trumpeter swan of North America *(Olor buccinator)*, which has a trumpetlike call during the breeding season, was once nearly extinct but is now protected.

Sweet Pea. Annual climbing plant *(Lathyrus odoratus)* of the Pulse family, native to S Europe but now widely cultivated for its fragrant flowers. There are three main types: dwarf, summer flowering (garden sweet peas), and winter flowering (florists'sweet peas). The flowers may be various colours; the vines climb by tendrils and require support.

Sweet Potato. Trailing perennial plant *(Ipomoea batatas)* of the Morning Glory family, native to the New World tropics. Raised mainly for human consumption, sweet potatoes are the most important tropical root crop and are grown in many varieties. Rich in Vitamin A, they yield starch, flour, glucose, and alcohol. The sweet potato, which is unrelated to the Potato, is often confused with the Yam, which belongs to another family.

Swift. Name for small, swallowlike Birds of the family Apodidae, found worldwide, chiefly in the tropics. Swifts are the most rapid flying animals known. In the U.S. the common Eastern species is the chimney swift *(Chaetura pelagica).* Western species include the black, Vaux's, and white-throated swifts. Nests of Oriental swifts, made entirely of a salivary secretion, are used in bird's-nest soup.

Swine. Cloven-hoofed Mammals of the family Suidae, native to the Old World and typified by long, mobile snouts, thick, bristly hides, and small tails. Domesticated swine, commonly called hogs or pigs (the latter more correctly reserved for young swine), are probably descended chiefly from the wild swine, or wild boar *(Sus scrofa),* of Eurasia and N Africa. Hogs are bred for their fat (lard), flesh (prepared as ham, bacon, and pork), and other products *(e.g.,* Leather for gloves and footballs, and bristles for brushes). Commercially raised swine are grouped as meat-type *(e.g.,* Hereford and Berkshire), lard-type *(e.g.,* Poland China, Duroc, and Spotted), or bacon-type *(e.g.,* Yorkshire and American Landrace). Hogs are highly susceptible to many diseases transmissible to humans, among them Trichinosis.

Switch. Electrical device having two states—on, or closed, and off, or open—and, ideally, having the property that when closed it offers a zero Impedance to a current and when open it offers infinite impedance to current. For many operations, as in digital computers, the operation of mechanical switches, which move contacts together and apart, is too slow. When faster switching is required, Transistors or vacuum tubes are used, operated in such a way that they conduct either heavily or very little.

Swordfish. Large food and game Fish *(Xiphias gladius)* of warmer

Atlantic and pacific waters, related to the Sailfish and Marlin. Its sharp, broad, elongated upper jaw is used to flail and pierce the smaller fish it eats. Swordfish, which breed as far north as Nova Scotia, may reach 15ft (457 cm) in length and 1,000 lb (454 kg) in weight. They are commercially valuable.

Sydenham, Thomas. 1624-89, English physician. A founder of modern clinical medicine and epidemiology, he conceptualized the causes and treatments of epidemics and provided classic descriptions of gout, smallpox, malaria, scarlet fever, hysteria, and chorea. He advocated observation instead of theorizing to determine the nature of disease and introduced the use of cinchona bark (containing quinine) to treat malaria and of laudanum to treat other diseases.

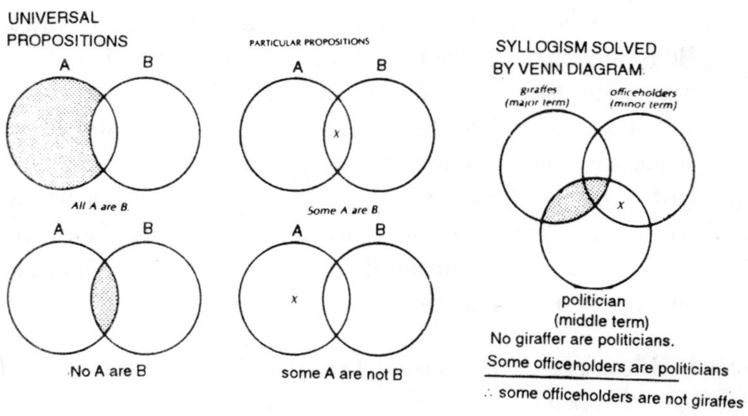

Fig. S-18. *Syllogism:* Categorical syllogisms can be analyzed by a Venn diagram. Three overlapping circles are drawn to represent the classes denoted by the three terms. Universal propositions (all A are B, no A are B) are indicated by shading the sections of the circles representing the excluded classes. Particular propositions (some A are B, some A are not B) are indicated by placing some mark, usually as "X," in the section of the circle representing the class whose members are specified. The conclusion may then be read directly from the diagram.

Syllogism. In logic, a made of argument that forms the core of the body of Western logical thought, consisting of a sequence of three propositions such that the first two imply the conclusion. Aristotle's formulations of syllogistic logic held sway in the Western world for over 2,000 years. The categorical syllogism comprises three categorical propositions, statements of the form *all A are B, no A are B, some A are B,* or *some A are not B.* A categorical syllogism contains precisely three terms: the major term, which is the predicate of the conclusion; the minor term, which is the subject of the conclusion and the middle term, which appears in both premises but not in the conclusion.

Symbiosis. Habitual cohabitation of organisms of different species. The term usually applies to a dependent relationship that is beneficial to both members (also called mutualism). Symbiosis includes parasitism, a relationship in which the Parasite depends on and may injure its host; Commensalism, an independent and mutually beneficial relationship; and helotism, master-slave relationship found among social animals. Symbiosis may occur between two kinds of plants *(e.g.,* Lichen-forming alga and fungus), two kinds of animal *(e.g.,* herbivores and cellulose-digesting gut microorganisms), or a plant and an animal.

Symmetry. A property of geometric figures, mathematical expressions, patterns, etc., whereby certain operations leave the structure unchanged. A square, for example, has symmetry about its centre and about certain lines drawn across it. Each point on one side of a diagonal has a corresponding point on the other side, as if it had been reflected in a mirror placed along this line. Another way of describing the symmetry of shapes is by rotation. If the whole square were rotated for 180° about a diagonal it would be unchanged. This is said to be a two-fold axis of symmetry—because two such rotations give a complete turn. The square also has a four-fold axis through its centre perpendicular to its plane. The symmetry of shapes can be described by such mirror planes and symmetry axes—known as *symmetry elements*. Patterns also have similar symmetry elements which involve translation operations. The properties of symmetry operations can be treated by group theory and are important in many branches of science. In particular in the study of the shapes and energy levels of molecules, the structure of crystals, and the properties of elementary particles.

Synchrocyclotron. A type of cyclotron in which the frequency of the accelerating electric field is slowly decreased in such a way that the particles stay in phase with the field as their velocity increases. The decrease in frequency compensates for the relativistic increase in mass. Synchrocyclotrons are able to produce particles with energies in the range 400-500 MeV. Higher energies can be obtained in the synchrotron.

Synchronous Motor. *See* electric motor.

Synchronous Orbit. The orbit of an artificial satellite that makes one revolution every 24 hours. If the orbit is in the same plane as the earth's equator the satellite remains stationary with respect to point on the earth. Communications satellites placed in such orbits *(stationary orbits)* are used for reflecting radio waves.

Synchrotron. A type of particle accelator in which the particles move in circular paths and are accelerated by alternating magnetic and electric fields. The *electron synchrotron* is similar to the betatron, having a torus-shaped vacuum chamber and a pulsed magnetic field. In addition a radiofrequency electric field is applied across gaps in a metal cavity inside the chamber. The electric field has a fixed frequency, synchronous with the orbiting motion of the electrons. Electrons are travelling at close to the velocity of light at energies in the MeV range and the relativistic increase in mass can be compensated for by varying the magnetic field. In the *proton sychrotron* this cannot be done and the electric-field frequency must also be varied, just as in the synchrocyclotron. Proton synchrotrons work at very high energies—up to 500 GeV.

Synchrotron Radiation. Electromagnetic radiation produced when electrons move at high velocities in circular paths in a magnetic field. The radiation is emitted in the direction of motion of the particle and is plane-polarized per-pendicular to the magnetic-field direction. Its frequency depends on the velocity and can lie anywhere in the electromagnetic spectrum from radio waves to X-rays. Synchrotron radiation is produced in synchrotrons and is also thought to be the cause of many astronomical radiation sources.

Synodic Period. Length of time it takes a solar-system body to return to an identical alignment *(e.g.,* conjunction or opposition; with another body as seen from the earth. Because the earth moves in its orbit around the sun, the synodic period differs from the *sidereal period,* the length of time a body takes to complete a orbit relative to a background star.

Syndiotactic. *See* polymer.

Syneresis. The process in which a liquid separates from a gel.

Synodic. Concerning two successive conjunctions or oppositions of the moon or a planet as viewed from the centre of the earth. *See also* lunar time.

Synthesis. The preparation of chemical compounds, usually from simpler compounds.

Synthetic. Denoting a substance or material that has been made artificially rather than naturally: *i.e.* a substance made by chemical reaction in a laboratory or in an industrial process.

Synthetic Elements. Radioactive chemical elements discovered not in nature but as artificially produced isotopes. They are Technetium Promethium; Astatine, Francium, and the Transuranium Elements. Some have since been found to exist in small amounts in nature as short-lived members of natural radioactive decay series.

Synthetic Textile Fibers. Artificial Fibers produced industrially by either synthesizing Polymers or altering natural fibers. Polyesters, *e.g.,* Dacron, produced by the polymerization of the product of an alcohol and organic-acid reaction, are strong and wrinkle-resistant. Nylon, a synthetic thermoplastic material introduced in 1938, is strong, elastic, resistant to abrasion and chemicals, and low in moisture absorbency. Orlon, the trade name for a polyacrylonitrile fiber made from natural gas, oxygen, and nitrogen, combines bulk, light weight, and resistance to acids and sun. Vinyl fibers, *e.g.,* Saran, are also widely used.

Syphilis. The most serious of the Venereal Diseases, caused by the spirochete *Treponema pallidum.* It is most commonly transmitted by sexual contact, although transmission can occur through infected blood or an open wound, or from mother to fetus. Primary syphilis is characterized by a chancre, a superficial skin Ulcer, at the site of infection; secondary, by generalized eruption of the skin and mucous membranes and inflammatiion of eyes, bones, and central nervous system; tertiary, by chronic skin lesions, damage to the heart and aorta, and central nervous system degeneration. Syphilis is treated with Penicillin, usually successfully unless extensive nervous system damage has occurred.

Syzygy. Alignment of three celestial bodies along a straight or nearly straight line. Viewed from one of these bodies, the other two will be either in conjunction (aligned in the same direction) or in opposition (aligned on opposite sides of the sky). An inferior planet, whose orbit lies inside the earth's, can, in reference to the sun as seen from the earth, be either in inferior conjunction (lying directly between the earth and the sun) or in superior conjunction (on the opposite side of the sun from the earth); unlike a superior planet, whose orbit lies outside the earth's, and unlike the moon, it can never be in opposition to the sun as seen from the earth.

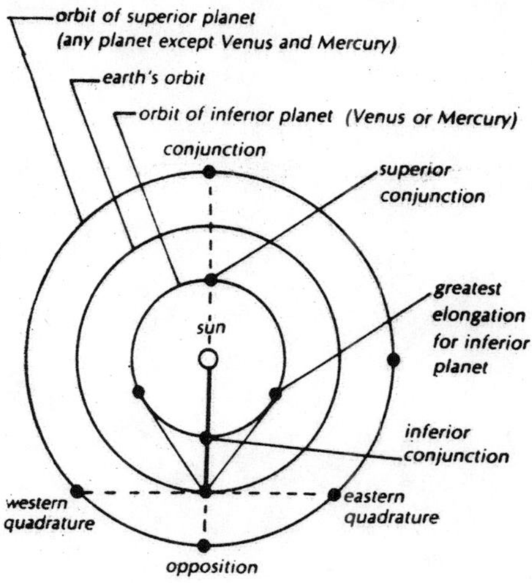

Fig. S-19. *Syzygy:* Alignment of celestial bodies necessary to produce syzygy (opposition, inferior conjunction, and superior conjunction) and quadrature conditions.

Szilard, Leo (Si'lard). 1898-1964, American nuclear physicist and biophysicist; b. Hungary. Working with Enrico Fermi at the Univ. of Chicago, he developed the first self-sustained nuclear reactor based on uranium fission. One of the first to realize that nuclear chain reactions could be used in bombs, and instrumental in urging the U.S. government to create the first atomic bomb, he later actively opposed nuclear warfare.

T

Tachometer. Any instrument for measuring speeds of rotation.

Tachyon. A hypothetical elementary particle that has the property of travelling faster than the speed of light. According to the special theory of relativity, this would be possible only if either the rest mass or the energy were imaginary. No tachyons have been detected.

Tactic. See polymer.

Tadpole. Larval, aquatic stage of Amphibians, also called polliwog. Hatching from the egg, the tadpole is gill-breathing and ledless and propels itself by mean of a tail. During Metamorphosis it develops lungs, legs, and other adult organs and, in the Frog and Toad, loses the tail.

Talc. A mineral consisting of hydrated magnesium silicate, $Mg_6Si_8O_{20}(OH)_4$. It is used as a lubricating agent.

Talc. Hydrous magnesium silicate mineral [$Mg_3Si_4O_{10}(OH)_2$], translucent to opaque, occurring in a range of colours and having a greasy, soapy feel. It is found in thin layers and in granular and fine-grained masses. Soapstone is a granular form of talc. Important sources include Austria, Canada, India, the U.S., and the USSR. Talc is used in making paper (as a filler), paints, powders, soap, lubricants, linoleum, electrical insulation, and pottery.

Tall Oil. An oily resinous substance containing fatty acids, obtained in the processing of wood pulp and used in making soaps and emulsions.

Tandem Generator. An electrostatic generator consisting of two Van de Graaff generators in series, thus producing twice the voltage. Tandem generators are used in accelerators.

Tangent. 1. *See* trigonometric function.

2. A line or plane touching a curve or surface at one point. The gradient of a tangent to a curve at a point is the derivative at that point.

Tangent Galvanometer. A type of galvanometer consisting of a vertical flat coil with a magnetic needle pivoted to swing horizontally and mounted at the centre of the coil. The plane of the coil is aligned in the direction of the earth's field and the current to be measured is passed through the coil, thus causing a deflection of the needle.

Tank, Military. Armored vehicle that has caterpillar traction and is armed with machine guns, cannon, rockets, or flame throwers. It was developed by the British and first used (Sept., 1916) in World War I. In World War II tanks and tank tactics were greatly improved. The German army, using large numbers of tanks, overran Poland in less than a month. In mass tank battles on the plains of Europe and N Africa the tide often swung toward the side with the best tanks. Since World War II the basic features of tanks and tank tactics have remained unchanged, although there have been refinements. Tanks are vulnerable to recoilless weapons and various antitank missiles, but they remain indispensable, because of their mobility and versatile weaponry, wherever the terrain is suitable to their operation.

Tannic Acid (tannin). A yellowish amorphous powder, $C_{76}H_{52}O_{42}$, which may be extracted from powdered nutgalls. It is used in tanning, as a mordant, and in making pyrogallic acid. Decomposes at 210°C.

Tannin, Tannic Acid, or Gallotannic Acid. (Approximate empirical formula : $C_{76}H_{52}O_{46}$), colourless to pale yellow, astringent, organic substance found in a wide variety of plants. Tannin can be extracted with hot water from the bark of oak, hemlock, chestnut, and mangrove; certain sumac leaves; plant gall; coffee; tea; and walnuts. In leather making, animal skin is treated with tannin to make it resist decomposition. Tannin is also used to make Inks, as a mordant for Dyes, and in medicine as an astringent and for treating burns.

Tanning. Process by which skins and hides are made into Leather. Vegetable tanning (shown in Egyptian tomb paintings dating from 3000 B.C.) uses tannin, is usually employed for heavy leathers, and requires more than a month to complete. Mineral tanning includes alum tanning and chrome tanning, the process most common today, requiring only a few hours. In oil tanning, or chamoising, a method used by North American Indians, the pelt is treated with fats and hung to dry; the leather is usually napped on both sides. A modern tanning process employs artificial agents (syntans).

Tantalite. *See* columbite-tantalite.

Tantalum. Symbol: Ta. A heavy hard grey transition element found principally in columbite-tantalite (Fe, Mn) $(Nb,-Ta)_2O_6$. It is separated and extracted in the same way as niobium. The element is used in alloys, incandescent filaments, and as a vacuum getter. Its chemistry is very similar to that of niobium. A.N. 73; A.W. 180.95; m.pt. 2996°C; b.pt. 5425°C; r.d. 16.65; valency 2-5.

Tantalum (Ta). Metallic element, discovered in 1802 by A.G. Ekeberg. A rare, hard, malleable, blue-gray metal, it is extremely ductile and highly corrosion-resistant. Uses include electrolytic capacitors, chemical equipment, wires, abrasives, and dental and surgical instruments.

Tape Recorder. Device for recording information on strips of plastic or paper tape. In an audio tape recorder, sound is picked up by a Microphone and transformed into an electric current. The current is fed to a Transducer (in the recording head of the tape recorder), which converts the current into corresponding magnetic flux variations that magnetize the fine particles of iron, cobalt, or chromium oxides on the tape. Tape recorders are also used in conjunction with Computers to store streams of Alphanumeric symbols.

Tapestry. Hand-woven fabric of plain weave in which a design is formed by weft threads inserted into the warp with the fingers or a bobbin. The warp (linen or wool) is entirely covered by the weft (wood, silk, or metallic thread). Each area of colour is woven as a block, and the slits between blocks are later sewn up. In high-warp *(haute-lisse)* tapestry weaving, the warp is stretched on a vertical loom and the loops *(lisses),* which raise alternate warp threads to form the shed, are lifted by hand; the low-warp *(base-lise)* loom is horizontal and is equipped with foot treadles to raise the threads. A Cartoon or drawing, often by a famous artist, is copied by the weaver, who faces the underside of the fabric. Tapestries were made in ancient Egypt, Greece, China, and Peru. European tapestry weaving developed between the 10th and 17th cent. The first important French tapestries were made at Arras in the 14th cent; other important centers were Beauvais, Aubusson, Paris and Brussels. In the 1880s William Morris attempted to revive tapestry weaving using designs by Sir Edward Burne-Jones and other artists. Jean Lurcat helped revive the art in the 20th cent. Contemporary tapestries may be based on the desisgns of artists or on the weaver's original concept, and some abandon the traditional two-dimensional textile for sculptural forms.

Tapeworm. Name for parasitic flatworms in the class Cestoda, segmented worms sometimes reaching 15-20 ft. (4.6-6.1 m) in length. Tapeworms attach themselves to the intestinal wall of the host, which may be Vertebrates or Arthropods. Humans become infected with tapeworms from eating infected meat or fish. Infestation may produce no symptoms or may produce abdominal distress and weight loss. Drug treatment destroys the parasite.

Tapir. Nocturnal, herbivorous mammal (genus *Tapirus*) of the jungles of Central and South America and SE Asia. Related to the Horse and Hippopotamus, it is about the size of a donkey and piglike in appearance, with a long, flexible snout and short legs. Tapirs live in forests, browsing on twigs; they wim well and can run fast when in danger. The Central American species (T. bairdi) is threatened by the destruction of its rain-forest habitat.

Tar. Any of various thick semi-solid substances produced by decomposition of organic substance. They contain mixtures of hydrocarbons, other organic compounds, and free carbon, which gives them a black colour.

Tar and Pitch. Viscous, dark-brown to black substances, obtained by the destructive Distillation of certain organic materials, *e.g.,* Coal, Wood, and Petroleum. Although the terms *tar* and *pitch* are sometimes used interchangeably, pitch is actually a component of tar that can be isolated by heating. Tar, more or less fluid, is now used to produce Benzene and various other substances. Tar from pine wood is used to make Soap and medicinals. Coal tar derivatives are used to make Dyes, cosmetics, and synthetic flavoring extracts. Pitch tends to be more solid than tar and is used to make roofing paper, in Varnishes, as a coal-dust binder in making fuel briquettes, and as a lubricant. Asphalt is a naturally occurring pitch.

Tarragon. Perennial aromatic Old World herb *(Arrtemisia dracunculus)* of the composite family. It has long been cultivated for its leaves, used for flavoring vinegar, salads, sauces, soups, and pickles.

Tartar (argol). A red-brown crystalline deposit of potassium hydrogen tartrate obtained from wine vats.

Tartaric Acid. A white crystalline powder, $HOOC(CHOH)_2COOH$, with an acid taste. In the preparation, potassium hydrogen tartrate from wine-lees is treated with calcium hydroxide, and the precipitated calcium tartrate is decomposed by sulphuric acid. Tartaric acid is used to make tartrates and as a sequestrant and constituent of baking powders. M.pt. 170°C; r.d. 1.8.

Tartrate. Any salt or ester of tartaric acid.

Tasmanian Devil. Voracious Marsupial *(Sarcophilus harrisi)* of the Dasyure family, now found only on Tasmania. Its body is about 2 ft. (60 cm) long with a large head and weak hindquarters. Very strong for its size, it preys on animals larger than itself; it has been relentlessly hunted for its destruction of livestock and poultry.

Tautomerism. A type of isomerism in which two isomers exist in equilibrium, the two forms *(tautomers)* being easily interconvertible. A common type is keto-enol tautomerism.

Taxidermy. Process of preserving vertebrate animals in life-like form by mounting the cleaned skins on a manmade skeleton. Once employed chiefly to preserve hunting trophies and souvenirs, taxidermy is now used mainly by science museums.

Tay-Sachs Disease. Rare hereditary disease of infants, characterized by progressive mental deterioration, blindness, paralysis, epileptic seizures, and death by age four. Caused by an abnormality of fat metabolism in nerve cells, producing central Nervous System degeneration, the disease occurs almost exclusively among Jews of Eastern European descent. It can be detected in a fetus by Amniocentesis. There is no treatment.

Tea. Tree or bush, its leaves, and the beverage made from the leaves. The plant *(Thea sinensis, Camellia thea,* or *C. sinensis)* is an evergreen related to the Camellia and native to India and probably parts of China and Japan. In the wild it grows to about 30 ft. (9.1 cm) in height, but in cultivation it is pruned to 3 to 5 ft. (91 to 152 cm). Tea plants require a well-drained habitat in a warm climate with ample rainfall; the leaves are prepared by drying, rolling, and firing (heating). Black teas *(e.g.,* pekoes), unlike green teas, are fermented before firing; oolong teas are partially fermented. Tea's stimulating properties are due to caffeine, and its astringency to tannin. Grown in China since prehistoric times, tea was produced on a commercial scale there by the 8th cent. It was introduced (17th cent.) into Europe by the Dutch East India Company, and its popularity helped spur the opening of the Orient to Western commerce. In colonial America a tax on tea led to the Boston Tea Party (1773). Today tea is used by more people and in greater quantity than any beverage except water.

Teak. Tall deciduous tree *(Tectona grandis)* of the Vervain family, native to India and Malaysia but now widely cultivated in tropical areas. Teakwood is moderately hard, easily worked, and very durable. The wood contains an Essential Oil that resists the action of water and prevents the rusting of iron, and the heartwood is resistant to termites. Superior to all other woods for shipbuilding, teak is also used in furniture, flooring, and general construction.

Tear Gas (lachrymator). A substance with a vapour that causes irritation and watering of the eyes.

Technetium. Symbol: Tc. A silvery-grey radioactive transition element obtained from fission products of uranium and plutonium. Technetium was the first artificially produced element, made by bombarding molybdenum with deuterons. It is not found naturally on earth. The most stable isotope. Tc, has a half-life of 5 x 10^5 years. A.N. 43; m.pt. 2200°C; b.pt. 5030°C; r.d. 11.5; valency 2-7.

Technetium (Tc). Artificially produced radioactive element, discovered in a sample of deuteron-bombarded molybdenum in 1937 by C. Perrier and E.G. Segre. The silver-gray metal is used in radioactive tracer studies. Spectra of some stars indicate the presence of the element; the naturally occurring element has not been found on earth.

Teeth. Hard, calcified structures embedded in the bone of the jaw that perform the function of chewing food. An adult mouth contains 32 permanent teeth. A tooth consists of the crown, the portion visible in the mouth, and one or more roots embedded in a gum socket. The gums cushion the teeth, while the jawbone firmly anchors the roots. The center of each tooth is filled with soft pulpy tissue containing blood vessels and nerves. Hard, borny dentin surrounds the pulp and makes up the bulk of the tooth. The root portion has an overlayer of cementum, and the crown has a layer of enamel, the hardest substance in the body.

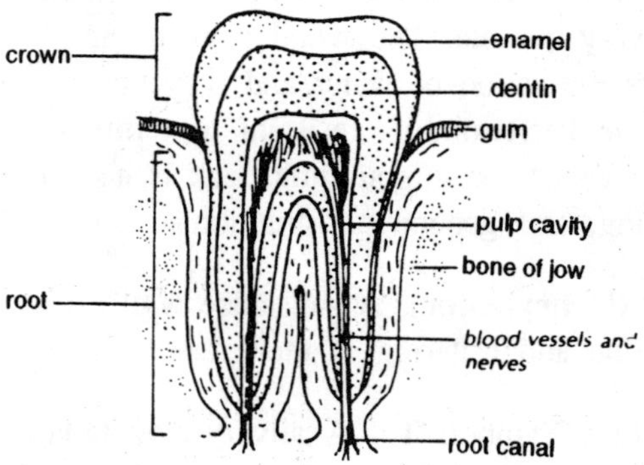

Fig. T-1. Cross section of a tooth.

Teflon. Trade name for a solid, chemically inert Polymer of tetrafluoroethylene, $F_2C{=}CF_2$. Stable up to temperatures around 572°F (300°C). Teflon is used in electrical insulation, gaskets, and in making low-adhesion surfaces, *e.g.*, in cooking utensils.

Tektite. Any of numerous small spheroidal glassy objects found in large quantities in certain parts of the world. They do not resemble terrestrial rocks and are thought to be of meteoric origin.

Telecommunications. The communication of signals, speech, pictures, or other information over long distances, as by telegraphy, radio, television, etc.

Telegraph. An apparatus for transmitting and receiving messages in the form of electrical pulses conducted along cables.

Telegraph. Electrically operated device or system for distant communication (the first ever invented) by means of visible or audible signals. The method used throughout most of the world, based in large part on the mid-19th cent. work of Samuel F.B. Morse, utilizes an Electric Circuit set up customarily by using a single overhead wire and employing the earth as the other conductor to complete the circuit. In the telegraph's simplest form, an electromagnet in the receiver is activated by alternately making and breaking the circuit. Reception by sound, with the Morse Code signals received as audible clicks, is the basis for a low-cost, reliable method of signaling. In addition to wires and Cables, telegraph messages are now sent by such means as Radio waves, Microwaves, and Communications Satellites. Telex is a telegraphy system that transmits and receives messages in printed form. Facsimile is a system for transmitting and reproducing photographs and other graphic material by wire or radio.

Telemeter. Any instrument for making distant measurements of electrical quantities, as by radio.

Telephone. Device for transmitting and receiving sound, especially speech, by means of wires in Electric Circuits. The telephones now in general use are developments of the device invented by Alexander Graham Bell and patented by him in 1876 and 1877. A modern telephone transmitter, which is essentially a carbon Microphone, contains loosely packed carbon grains. When someone speaks into the telephone, the diaphragm vibratess, causing the carbon grains to be compressed and released. This motion varies the currennt flow in the associated electric circuit. The current, when transmitted to a distant identical instrument, caues the diaphragm in it to vibrate in response to the fluctuations induced by the nearby magnetic field. Telephone lines used include ordinary open-wire lines; lead-sheathed Cables consisting of many lines; coaxial cables; and, most recently, glass fibers. Coaxial cables and fiber-optic lines are placed underground, but other cables may be either overhead or underground. Long-distance transmission of telephone messages is often accomplished by means of Radio and Microwave transmissions. In some cases microwaves are sent to an orbiting Communications Satellite, from which they are relayed back to a distant point on the earth. Sophisticated services, including automatic switching systems, automatic dialing, call forwarding, and conference calling, have been developed in recent years.

Telephoto Lens. A compound lens consisting of a converging lens system followed by a diverging lens system, designed to replace the normal lens of a camera and produce a magnified real image at the same distance from this lens.

Telescope. Any of several optical instruments used for producing magnified images of far objects. Refracting telescopes produce their magnification by combinations of lenses : reflecting telescopes use large concave mirrors.

Telescope. System of Lenses, Mirrors, or both used to gather light from a distant object and form a real optical image of it. In the refracting telescope, or refractor (invented early 17th cent.), light is bent, or refracted, as it passes through a convex objective lens so that it converges to a point — the focus — behind the lens. In a reflecting telescope, or reflector (invented 1672), light is reflected by a concave paraboloidal mirror and brought to a focus in front of the mirror. The

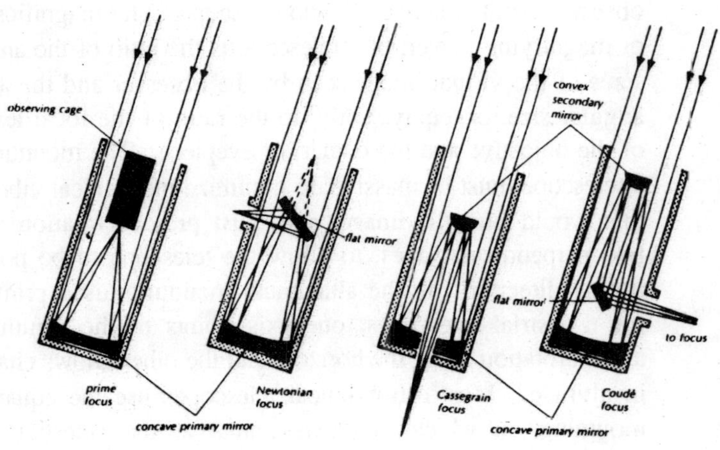

Fig. T-2. *Telescope:* Mirror arrangements for a reflecting telescope.

image is diverted to a more convenient location by one of several means. A third type of telescope, the catadioptric system, focuses light by a combination of lenses and mirrors; two examples are the Schmidt camera telescope (invented 1930), used primarily for wide-angle photography of star fields and conisting of a spherical mirror and a special correction lens used in front of the tube, and the Maksutov telescope (invented 1941), which has a spherical meniscus in place of the Schmidt's correcting plate. The size of the object's image is the product of its angular size in radians (1 radian 57°) as seen from the telescope and the focal length (the distance from the focus to the lens or mirror). The brightness of the image depends on the telescope's total light-gathering power and hence is proportional to the area of the objective lens or primary mirror, or to the square of its diameter. A telescope's resolving power, or smallest angular separation between two light points that can be unambiguously distinguished, is proportional to the ratio of the wavelength of light being observed to the diameter of the telescope. The magnification, or magnifying power, of a telescope is the ratio of the angular sizes of the virtual image seen by the observer and the actual angular size, or, equivalently, to the ratio of the focal lengths of the objective and the ocular, or eyepiece. The mounting of a telescope must be massive, to minimize mechanical vibration that would blur the image, and must provide rotation about two perpendicular axes, to allow the telescope to be pointed in any direction. In the altazimuth mounting, used primarily for terrestrial telescopes, one axis points to the zenith and allows rotation along the horizon, and the other allows changes in altitude. Most astronomical telescopes use the equatorial mounting, in which one axis points at the celestial pole. Rotation about this polar axis allows changes in right ascension or celestial longtitude; rotation about the declination axis, at right angles to it, allows changes in declination or celestial latitude. To compensate for the earth's rotation, a clock-drive mechanim is generally provided to turn the polar axis east to west at the rate of one rotation per sidereal day.

Television. Transmission and reception of still or moving images by means of electrical signals, especially by means of Electromagnetic Radiation using the techniques of Radio. One of the most widely used image pickup devices, or camera tubes, is the iconoscope (invented by Vladimir Zworykin, 1923), which consists of a thin sheet of mica upon which thousands of microscopic globules of a photosensitive silver-cesium compound have been deposited. Backed with a metallic conductor, this expanse of mica becomes a mosaic of tiny Photoelectric Cells and Capacitors. The differing light intensities of various points of a scene cause the cells of the mosaic to emit varying quantities of electrons. The cells are left with positive charges in strengths proportional to the electrons lost. An electron gun, or "scanner," passes its beam across the cells. As it does so, the charge is released, causing an electrical signal to appear on the back of the mosaic, which is connected externally to an Amplifier. The strength of the signal is proportional to the amount of charge released. In the Vidicon, another type of pickup tube, the photomissive mosaic is replaced by a photoconductive layer, resulting in increased efficiency. The scanning process, which is the essence of television accomplishment, operates as the human eye does in reading a page of printed material, *i.e.,* line by line. A complex circuit of horizontal and vertical deflection coils controls this movement and causes the electron beam to scan the back of the mosaic 30 times per econd. Two principal means of recording television programs for future use are video tape recording and kinescope. Video tape recording is similar to conventional tape recording except that because of the wide frequency range — 4.2 megahertz (MHz)— occupied by a video signal, the effective speed at which the tape passes the recording head is kept very high. Sound is recorded along with the video signal on the same tape. Kinescope is a method in which programs are recorded on motion-picture film. Appropriate changes in the signal-carrying circuitry allow kinescopes to be played back from a developed negative as well as from a positive. Systems for recording television programs on discs have been recently develope. When a television program is broadcast, the varying electrical signals are amplified and used to modulate a carrier wave; the modulated carrier is usually fed to an Antenna, where it is converted to electromagnetic waves and broadcast over a large region. The waves are sensed by antennas connected to television receivers, and the image is reconstructed essentially by reversing the pickup operation. The final image is displayed on the face of a Cathode-Ray Tube, where an electron beam scans the fluorescent face, called the "screen," line for line with the pickup scanning. The tube's inside face glows when hit by the electrons, and the visual image is reproduced. Colour

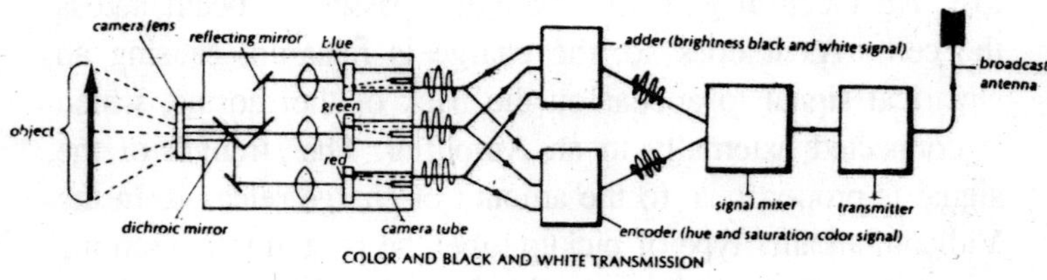

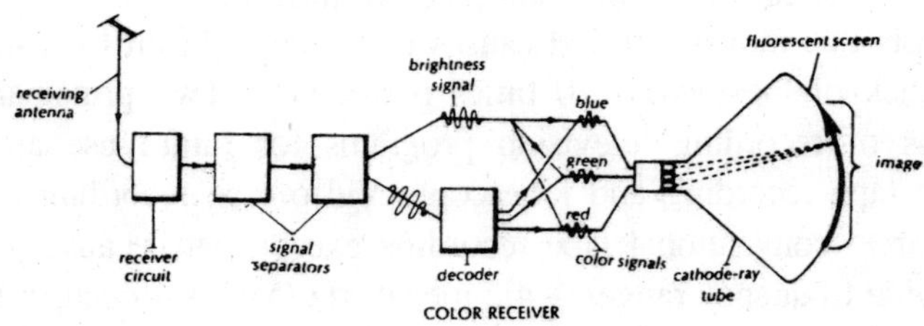

Fig. T-3. Television video transmission and reception: The camera lens focuses collected light rays into mirrors, which separate the image into its three primary colour component images. Each colour component is focused onto the face of a camera tube. The scanning beam of each tube converts the primary colour image into a colour signal. The adder combines the three colour signals to make the brightness signal. The encoder combines the signals to transmit hue and saturation information. A black and white television receiver processes only the brightness signal. A colour television receiver separates the received signal into brightness and hue and saturation components, which are recombined to produce primary colour signals for the picture tube.

television today uses as "element-sequential" system. Light from the subject is broken up into its three primary-colour components (red, blue, and green), which are simultaneously scanned by three pickups. In the receiver the signals are brought together again. Each element, or dot, on the picture tube screen is subdivided into areas of red, blue, and green phosphors. Beams from three electron guns, modulated by the three colour signals, scan the elements together in such a way that the beam from the gun using a given colour signal strikes the phosphor of the same colour.

Television. A system for electrical transmission of moving pictures, either by cable (closed-circuit television) or by radio waves (broadcast television). The visual image is converted into electrical signals by a television camera. In broadcasting, this is amplified and used to modulate the transmitted radio wave.

Television reception uses the super-heterodyne principle and the demodulated signal produces the image on the screen of a cathode-ray tube. In the receiver the spot on the screen is continually scanned across and down, forming a number of lines (625 in the British system). The received signal is used to control the intensity of the electron beam, thus forming the picture. Audio signals are transmitted along with the visual signals and the two are synchronized in the receiver.

In colour television three separate visual signals are transmitted corresponding to red, green, and blue components of the image. The cathode-ray tube used has three separate electron beams, one for each colour, and the screen is covered by groups of three coloured phosphor dots. The coloured picture is formed by an additive process involving different amounts of the three primary colours.

Telluric. 1. Relating to the planet earth.

2. *See* tellurium.

Tellurium. Symbol : Te. A silvery-white metalloid element, usually obtained from the anode sludge produced in electrolytic copper refining. The element is used in some alloys and semiconductors. Its chemistry is similar to that of sulphur: it forms *tellurous* compounds with low valency and *telluric* compounds with higher valency. Tellurium is one of the chalconide elements. A.N. 52; A.W. 127.6; m.pt. 449.5°C; b.pt. 990°C; r.d. 6.24; valency 2, 4, or 6.

Tellurium (Te). Semimetallic element, discovered in 1782 by Franz von Reichenstein. A silver-white, lustrous, brittle metalloid, it occurs in calaverite and sylvanite. It is used as an additive in steel to increase ductility and as a catalyst for petroleum cracking.

Tellurous. *See* tellurium.

Tempera. Painting method in which finely ground pigment is mixed with a base such as albumen, egg yolk, or thin glue. When used on wood panels, as it was for altarpieces and easel paintings, it was applied on a smooth, white gesso underpainting. Tempera produces clear, pure colours that resist oxidation. Known from antiquity, it was the exclusive panel medium in the Middle Ages. It was not supplanted by the more subtle oil paint until c. 1400 in N Europe and c. 1500 in Italy. A modern revival of tempera included the 19th cent. Swiss artist Arnold Bocklin and the 20th cent. Americans Shahn and Wyeth.

Temperature. Symbol : T. A measure of the hotness of a body. It two bodies have different temperatures heat will flow from the higher temperature to the lower. Heat is a measure of the total kinetic energy of the atoms or molecules in the system and temperature is a measure of their average kinetic energy. It is measured in units of degrees on a scale such as the centigrade, Fahrenheit, or Reaumur scale. The SI unit is the kelvin.

Temperature. The measure of the relative warmth or coolness of an object. The temperature of a substance measures not its head content but rather the average kinetic energy of its molecules. Temperature is measured by means of a

Thermometer or other instrument having a scale calibrated in units called degrees. A temperature scale is determeined by choosing two reference temperatures and dividing the temperature difference between these points into a certain number of degrees. The size of the degree depends on the particular temperature scale being used. The most common reference temperatures are the Melting Point of ice and the Boiling Point of water. An absolute temperature scale for which zero degree corresponds to zero average kinetic energy can be defined theoretically; the Kelvin temperature scale is an absolute scale having degrees the same size as those on the Celsius scale.

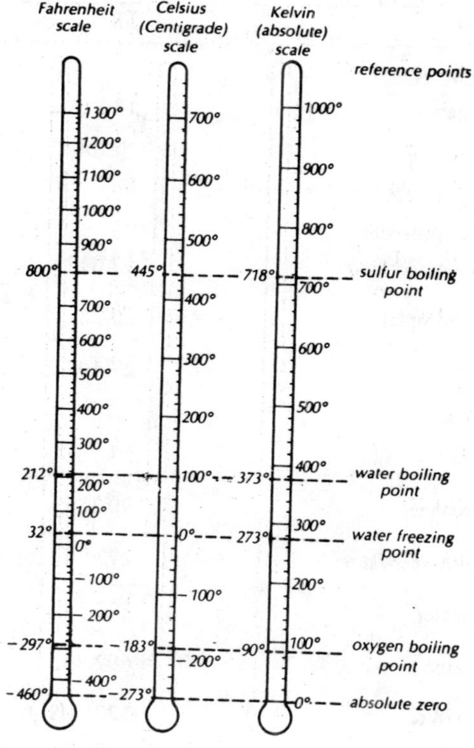

Fig. T-4. Temperature scales.

Temperature Scale. A scale for the practical measurement of temperature. Scales such as the centigrade, Fahrenheit, Reaumur, and Rankine are defined by two *fixed points* with the interval between them divided into degrees. The fixed points used on these scales are the freezing and boiling points of water. Thermodynamic temperature is measured in kelvins and practical measurements are made by reference to the *International Practical Temperature Scale,* which has 11 fixed points. Between these values temperatures are measured with a platinum resistance thermometer or, above 1337.58 K, with a pyrometer.

International Practical Temperature Scale

fixed point	*TK*	*C*
Triple point of hydrogen	13.81	—259.34
Temperature of hydrogen with vapour pressure 25.76 atmosphere	17.042	-256.108
B.pt. of hydrogen	20.28	-252.87
B.pt. of neon	27.102	-246.048
Triple point of oxygen	54.361	-218.789
B.pt. of oxygen	90.188	-182.962
Triple point of water	273.16	0.01
B.pt. of water	373.15	100
M.pt. of zinc	692.73	419.58
M.pt. of silver	1235.08	961.93
M.pt. of gold	1337.58	1064.43

Temporary Hardness. *See* hardness.

Temporary Magnetism. *See* magnet.

Tendon. Tough cord of dense connective tissue that attaches muscle to bone. If the muscle is thin and wide, the tendon may be a thin sheet. Tendons transfer muscle power over a distance, *e.g.,* forearm muscles contract and pull on tendons that pull on finger bones to produce finger movements.

Tensile Strength. The applied stress necessary to break a material under tension. It is measured in newtons per square metre, etc.

Tensimeter. An apparatus for measuring the vapour pressure of a liquid by comparing it with that of water at the same temperature.

Tensor. An abstract mathematical entity used in transformations from one coordinate system to another. Tensor analysis is a generalization of vector analysis.

Terbium. Symbol : Tb. A soft silvery-grey metallic element belonging to the lanthanide series. A.N. 65; A.W. 158.925; m.pt. 1360°C; b.pt. 3041°C; r.d. 8.23; valency 3 or 4.

Terbium (Tb). Metallic element, discovered in 1843 by C.G. Mosander. It is a silver-gray, soft, ductile, malleable Rare-Earth Metal in the Lanthanide Series. Terbium is used in lasers, semiconductor devics, and phosphors for colour-television picture tubes.

Terephthalic Acid. A white crystalline solid, $C_6H_4(COOH)_2$, made by the catalytic oxidation of xylenes. It is used to produce polyester resins and fibres such as Dacron and Terylene. Sublimes at 300° C; r.d. 1.5.

Terminal. 1. A point on an electrical circuit, battery, etc., at which a connection can be made.

2. A piece of apparatus connected to a computer for the input or output of information.

Terminal Velocity. The constant velocity achieved by a body moving through a fluid under the action of a constant force, produced when this force is balanced by the resistance of the fluid. A body falling in air attains a terminal velocity as a result of air resistance.

Terminator. The line separating the illuminated portion from the dark portion on the moon or on a planet.

Termite or **White Ant.** Soft-bodied social Insect that feeds mainly on wood. Termites live in colonies of hundreds or thousands, consisting of large numbers of nonreproductive "workers" and "soldiers" and one or a few reproductive members—"kings" and "queens". Some build huge mounds, up to 40 ft. (12.2 m) high, to house their colonies. Turning wood into cellulose by means of specialized bacteria or protozoans in the digestive tract, termites do millions of dollars of damage annually.

Tern. Sea Bird of the Old and New World, smaller than the related Gull, with long, pointed wings and forked tails. Also called sea swallows because of their graceful flight, they plunge headlong into the water to catch small fish.

Ternary. Denoting a chemical compound formed from three different elements. An example is sodium sulphate, Na_2SO_4.

Terpene. Any of a class of naturally occurring unsaturated hydrocarbons found in the essential oils of many plants. They are built up of isoprene molecular units (C_5H_8) and have formulas of the type $(C_5H_8)_n$. *Monoterpenes* contain two units, $C_{10}H_{16}$, *diterpenes* contain four, $C_{20}H_{32}$, *triterpenes* six, $C_{30}H_{48}$, etc. *Sesquiterpenes* contain three isoprene units, $C_{15}H_{24}$.

Terra-cotta [Ital. = baked earth], form of hard-baked clay widely used in the decorative arts, especially as an architectural material, either in its natural red-brown colour, painted, or with a baked glaze. Its early prevalence as a medium of artistic expression is indicated by vases, figurines, and tiles from predynastic Egypt; ancient Assyria, Persia, and China; and pre-Columbian Central America. Terra-cotta first gained architectural importance in Greece, where from c.7th cent. B.C. it was used for roof tiles and ornamental details. Its golden age was the Renaissance, when it was widely used in the architecture of N Italy and N Germany. It was later established in Italy as a sculptural material, and reached great distinction as a decorative material in the polychrome enameled reliefs of the Della Robbia family. Terra-cotta work in building and decoration spread from Italy in the 16th cent. often through migrant Italian artisans, to such countries as France and England. In the 18th cent. sculptors such as Houdon used the material for sketches. Widely used in the U.S. in modern times as an exterior covering for steel structural skeletons, terra-cotta was employed with particular skill by Louis Sullivan. Notable modern sculptures in terra-cotta include those by Maillol, Epstein, and Picasso.

Terrestrial Magnetism (geomagnetism). The magnetism of the earth. The earth has a magnetic field in which lines of force run from a point near the geographic true North Pole to one near the true South Pole, as if it contained a large bar magnet orientated at a small angle to its axis of rotation. The portion on the earth at which the field is most intense are the *magnetic North Pole* and the *magnetic South Pole*. Their points do not coincide with those of the geographic poles and also change slowly with time. At present they are lat. 76°N, long. 102°W and lat. 68°S, long. 145°E. The field at points on the earth's surface is characterized by three properties: the horizontal component, the angle of dip, and the angle of declination. Lines can be drawn on a map corresponding to equal values of these parameters: called *isodynamic, isogonal* and *isoclinal lines* respectively. The earth's magnetic field is still not fully understood but the field probably arises from the iron core of the earth and is influenced by solar phenomena and the ionosphere. It is slowly changing with time and also shows periodic annual and daily variations.

Terrestrial Planet. Any of the planets Mercury, Venus, earth, Mars, and — according to some authorities — Pluto, whose solid constitution and relatively small size cause them to differ markedly from the giant planets. The difference may result from the fact that the terrestrial planets are at a later evolutionary stage than the supposedly younger giants, though this is still uncertain.

Terrestrial Telescope. Any telescope suitable for use on the earth, as opposed to use in astronomy.

Terrier. Class of dogs originally bred to start small game and vermin from their burrows or to go to earth and kill their prey. Today they are raised chiefly as pets.

Tervalent. *See* trivalent.

Terylene. A type of polyester resin used in synthetic fibres.

Tesla. Symbol: T. An SI unit of magnetic flux density equal to a magnetic flux density of one weber per square metre. [After Nikola Tesla (1856-1943), Croatian-born U.S. electrical engineer.]

Tesla Coil. An apparatus for producing high-frequency high voltages, consisting of a transformer with its primary coil supplied through a spark gap. The high-frequency voltage will induce a discharge in a low presure of gas within a glass vacuum system and Tesla coils are used for testing such systems for leaks.

Testosterone. Principal male sex Hormone. It is necessary for the development of the external genitals in the male fetus, and at puberty its increased levels are responsible for male secondary sex characteristics *(e.g.,* facial hair).

Tetanus or **Lockjaw.** Acute infectious disease of the nervous system caused by Toxins of *Clostridium tetani* bacilus. Tetanus may follow any type of injury, including puncture wounds, animal bites, gunshot wounds, lacerations, and fractures. The toxin acts on the motor nerves and causes muscle spasms, most frequently in the jaw (lockjaw) and facial muscles. The disease is treated with an antitoxin; preventive immunization is also available.

Tetrachloroethane. A colourless non-flammable toxic liquid, $CHCl_2CHCl_2$, made by chlorination of acetylene. It is a widely used solvent. M.pt. —43°C; b.pt. 146°C; r.d. 1.6.

Tetrachloroethene (tettrachloroethylene). A colourless nonflammable liquid, $Cl_2C{:}CCl_2$, used as a solvent, particularly in dry-cleaning. M.pt. —22°C; b.pt. 121°C; r.d. 1.6.

Tetrachloromethane. *See* carbon tetrachloride.

Tetraethyl Lead (lead tetraethyl). A colourless poisonous oily liquid, $Pb(C_2H_5)_4$, with a pleasant odour. It is made by treating lead-sodium alloys with chloroethane and is a petrol additive preventing knocking in internal-combustion engines. M.pt. —136°C; b.pt. 198°C; r.d. 1.6.

Tetragonal. *See* crystal system.

Tetrahedron. A polyhedron with four faces. The faces are triangles, which are congruent and equilateral in a *regular tetrahedron.*

Tetrahydrate. A solid compound with four molecules of water of crystallization per molecule of compound, as in calcium tartrate tetrahydrate, $CaC_4H_4O_6.$ $4H_2O$.

Tetrahydrofuran. A colourless liquid heterocyclic compound, C_4H_4O, widely used as a solvent. M.pt. —65°C; b.pt. 66°C; r.d. 0.9.

Tetravalent. *See* quadrivalent.

Tetrode. A thermionic valve with four electrodes. It is a triode modified by the inclusion of a second grid, the *screen grid,* between the control grid and the anode to reduce the large grid-anode capacitance. It is held at a positive potential.

Textiles. Fabrics made from Fibers by Weaving. Knitting, Crochet, Lace-making, knotting *(e.g.,* Macrame), braiding, netting, or felting. Textiles are classified according to their component fibers, *e.g.,* Silk, Wool, Linen, Cotton, synthetics such as Rayon and nylon, and some inorganic substances, *e.g.,* Fiber Glass and Asbestos cloth. They are also classified according to their structure or weave. Yarn, fabrics, and tools for Spinning and weaving have been found among the earliest relics of human habitations, with various textiles typical of the remains of many ancient cultures, *e.g.,* linens in Egypt c. 5000 B.C., woolens in Bronze Age Scandinavia and Switzerland, cotton in India c. 3000 B.C., and silk in China c. 1000 B.C. By the 14th cent. A.D. splendid fabrics were being woven on European handlooms, and the weaving and Embroidery of Tapestries was highly developed. England and Flanders were particularly noted for their woolen textiles. The weaving of cotton textiles became important in Britain and the U.S. in the 18th and 19th cent. Since the Industrial Revolution, most textiles have been produced in factories, with highly specialized machinery, but hand techniques are still widely used. *Textile printing,* the process by which coloured designs are printed on textiles, is an ancient art. Early forms are stencil and block printing, with cylinder or roller printing developing c. 1785. More recent processes include screen printing, spray painting, and electrocoating.

Thallic. *See* thallium.

Thallium. Symbol : Tl. A soft bluish-white metallic element resembling lead, present in small amounts in some zinc ores. Thallium has two valencies, forming both thallium III *(thallous)* compounds and thallium I *(thallic)* compounds. A.N. 81; A.W. 204.37; m.pt. 303.5°C; b.pt. 1460°C; r.d. 8.27.

Thallium (Tl). Metallic element, discovered by William Crookes in 1861. A soft, malleable, lustrous, silver-gray metal, it resembles aluminum chemically and lead physically. The metal is used in certain electronic components, and its sulfate is used as an insecticide.

Thallous. *See* thallium.

Theodolite. An instrument consisting of a small telescope moving over angular horizontal and vertical scales, used for measuring angles in surveying.

Theodolite. Optical instrument used for a number of purposes in surveying, navigation, and meteorology. It consists of a telescope fitted with a spirit level and mounted on a tripod so that it is free to rotate about its vertical and horizontal axes. Theodolite measurements of the altitude and azimuth of a Weather Balloon at precise intervals are used to compute the estimated wind velocity and direction of the portion of the atmosphere through which the balloon is passing.

Theorem. A proposition that can be proved logically from a set of basic assumptions.

Theory. 1. A proposition put forward to explain observed facts but not yet established as true.

2. A body or collection of principles, methods, etc., used in explaining a fairly wide set of phenomena, as in the theory of relativity, quantum theory, etc.

Theory of Games. A methematical theory applied to situations in which there is a conflict between interests, designed to predict the optimum strategy.

Therm. A unit of energy equal to 10^5 British thermal units. It is equivalent to 1.055 056 x 10^8 joules. The therm is used for measuring heat energy, as in the calorific value of gas and other fuels.

Thermal Capacity. *See* heat capacity.

Thermal Conductivity. Symbol : λ A measure of the ability of a substance to conduct heat, euqal to the rate of flow of heat per unit area resulting from unit temperature gradient. It is given by the equation $dQ/dt = \lambda A dT/dx$, where dQ/dt is the rate of heat flow normal to an area A and dT/dx is the temperature gradient. Thermal conductivity is sometimes defined as the heat flow per unit time between two opposite faces of a unit cube of the material when these faces have unit temperature difference. It is measured in joules per second per metre per kelvin ($J\ s^{-1}\ m^{-1}\ K^{-1}$).

Thermal Diffusion. The phenomenon in which a mixture of gases subjected to a temperature gradient tends to separate so that the heavier molecules concentrate in the cooler region and the lighter molecules in the hotter region. The effect is the basis of the Clusius column for separating isotopes.

Thermal Equilibrium. Equilibrium in which there is no net flow of heat between different parts of the system. The temperature is then uniform.

Thermalize. To reduce the energies of neutrons with a moderator so that they become thermal neutrons.

Thermal Neutron. A neutron with a low kinetic energy, of the same order of magnitude as the kinetic energies of atoms and molecules. Thermal neutrons have energies of the order KT, where k is the Boltzmann constant : *i.e.* about 0.025 electronvolts at normal temperature.

Thermal Reactor. *See* fission reactor.

Thermionic Emission. Emission of electrons from a hot solid. The effect occurs when significant numbers of electrons have enough kinetic energy to overcome the solid's work function. The number of electrons increases with temperature. Thermionic emission is the process used for producing electrons in thermionic valves, X-ray tubes, and cathode-ray tubes.

Thermionic Valve. An electronic device for controlling the flow of current in a circuit by regulating the flow of electrons produced by thermionic emission. Thermionic valves are often called *electron tubes*. They consist of an evacuated or gas-filled glass or metal container into which are sealed two or more electrodes. The main electrodes are the *cathode,* which is at a negative potential and is heated so as to emit electrons, and the *anode,* which is at positive potential. The cathode may be a metal filament heated to incandescence by an electric current. Alternatively, it may be a separate electrode heated by a nearby coil (the *heater)*.

Electrons can only flow between the cathode and the anode when the anode has a positive potential with respect to the cathode. Extra electrodes *(grids)* can be introduced between the two to control this current. Types of valve are named according to the number of electrodes they have. In most applications they have been replaced by transistors although they are still used for special purposes and for handling large voltages and currents.

Thermistor. A small piece of semiconductor with an electrical resistance that decreases sharply with temperature. Thermistors can be used to measure temperature or compensate for temperature increases in electrical equipment.

Thermite. A mixture of aluminium powder with a metal oxide, usually iron III oxide, mixed in the correct proportions for reduction of the oxide by aluminium. When ignited, aluminium oxide and metal are produced: $2Al + Fe_2O_3 = Al_2O_3 + 2Fe$. The reaction is strongly exothermic and the iron is molten. Thermite is used in a technique for *in situ* welding of steel (The *Thermit process)*. It is also a constituent of some incendiary bombs. The reaction is the same as that in the Goldschmidt process for extracting metals.

Thermochemistry. The branch of chemistry concerned with the measurement and use of heats of reaction, heats of solution, etc.

Thermocouple. A temperature-measuring device formed by joining the ends of two stripes of dissimilar metals in a closed loop, with the two junctions at different temperatures. Because the voltage that arises in this circuit is proportional to the temperature difference between the junctions, the temperature at one junction can be determined if the other junction is maintained at a known temperature.

Thermocouple. A junction between two different metals used for measuring temperature by the potential differnece produced. In general, any two metals in contact develop a potential difference (the contact potential) across the junction and this increases with increasing temperature. In practice thermocouples are made by welding together the ends of two pieces of wire : suitable pairs of metals are ones that show a large increase in potential difference for small changes in temperature. They include copper/constantan, iron/constantan, and platinum/platinum-iridium. Two junctions are often used connected in series: one to measure the temperature and the other maintained at a constant low temperature. This is an application of the Seebeck effect.

Thermodynamics. The branch of science concerned with the relationship between heat, work, and other forms of energy. The theory is based on three laws :

The *first law* states that the heat absorbed by any system is equal to the work done by the system plus its change in internal energy : $\Delta q = \Delta w + dU$. It is equivalent to the idea that heat is a form of energy and an application of the law of conservation of energy. The *second law* states that heat cannot pass from a colder to a hotter body without any other external effect occurring : *i.e.* that there can be no spontaneous self-sustaining heat flow against a temperature gradient. This is only one of several more-or-less equivalent statements of the law, which is generally concerned with the availability of energy. Use of the second law and the Carnot cycle leads to the idea of entropy. The *third law* states that the entropy of a substance approaches zero as its thermodynamic temperature approaches zero. The third law is a modified form of the earlier Nernst heat theorem. Thermodynamics is used in studies of heat engines and also in studies of chemical reactions and chemical equilibrium.

Thermodynamics. Branch of science concerned with the nature of Heat and its conversion into other forms of Energy. Heat is a form of energy associated with the positions and motion of the molecules of a body. The total energy that a body contains as a result of the positions and the motions of its molecules is called its internal energy. The first law of thermodynamics states that in any process the change in a system's internal energy is equal to the heat absorbed from the environment minus the Work done on the environment. This law is a general form of the law of conservation of energy. The second law of thermodynamics states that in a system the Entropy cannot decrease for any spontaneous process. A consequence of this law is that an engine can deliver work only when heat is transferred from a hot reservoir to a cold reservoir or heat sink. The third law of thermodynamics states that all bodies at absolute zero would have the same entropy; this state is defined as having zero entropy.

Thermodynamic Temperature (absolute temperature). Symbol: T. Temperature that is a function of the internal energy possessed by a body, having a value of zero when the internal energy is zero. The unit of thermodynamic temperature is always the kelvin, which is defined in such a way that the triple point of water is 273.16 kelvins. In practice, measurements of thermodynamic temperature are made using the International Pactical Temperature Scale.

Thermoelectric Effect. Any of three effects involving direct interconversion of heat and electricity: the Seebeck effect, the Peltier effect, and the Thomson effect.

Thermoelectricity. Electricity produced by a thermoelectric effect.

Thermoluminescence. Luminescence produced when certain solid materials are heated. The effect arises when the solid contains electrons trapped at crystal defects, a condition produced by the impact of ionizing radiation. As the material is heated the electrons are freed and the energy released is emitted as photons. Thermoluminescence is used as a method of dating archaeological finds, particularly pottery. The low intensity of light produced by heating a sample is measured, thus enabling an estimate to be made of the amount of natural ionizing radiation to which the object has been subjected. This is proportional to time that has elapsed since the pottery was fired.

Thermometer. Any instrument for measuring temperature. Many different types of thermometer exist, all depending for their action on some easily measured physical property that varies with temperature. They include the mercury thermometer, alcohol thermometer, gas thermometer, and resistance thermometer. Other methods of measuring temperature involve the use of thermocouples, thermistors, and various types of pyrometer.

Thermometer. Instrument for measuring Temperature. A clinical thermometer consists of a small vacuum tube of uniform bore, with a temperature scale etched on its front. The tube is closed at one end and connected at the other with a chamber containing mercury or another liquid. When the chamber is heated, the fluid expands and rises into the tube.

Thermonuclear Bomb. *See* nuclear weapon.

Thermonuclear Reaction. *See* nuclear fusion.

Thermopile. An instrument for measuring radiant heat, consisting of a number of thermocouples connected in series with alternate junctions exposed to the radiation, the other junction being shielded. Usually the thermocouples are made up of short metal rods. The current generated by the apparatus is detected by a galvanometer.

Thermoplastic. *See* plastic.

Thermosetting. *See* plastic.

Thermos Flask. A commercial type of Dewar flask.

Thermostat. An apparatus for maintaining a constant temperature. Thermostats have a heater controlled by a temperature-sensing device such as a bimetallic strip. The term is used both for the controlling device and for the constant-temperature enclosure.

Thin film. A thin layer of solid deposited on the surface of another solid, usually by vacuum evaporation or sputtering. Thin films are used in many semiconductor devices.

Thin-layer Chromatography. *See* chromatography.

Thio-. Prefix indicating the presence of sulphur in a chemical compound.

Thiocyanate. Any salt or ester of thiocyanic acid.

Thiocyanic Acid. An unstable gaseous compound, HSCN.

Thio Ether. Any of a class of organic compounds of formula RSR', where R and R' are hydrocarbon groups. They are sometimes called organic *sulphides*.

Thionyl Chloride. A yellowish fuming liquid, $SOCl_2$, prepared by oxidation of sulphur dichloride with sulphur trioxide. It is used in organic chemistry to replace hydroxyl groups with chloine atoms, as in the preparation of acid chlorides. M.pt. —105°C; b.pt. 79°C; r.d. 1.6.

Thiosulphate. Any salt or ester of thiosulphuric acid.

Thiosulphuric Acid. An unstable dibasic acid, $H_2S_2O_3$, known only in the form of thiosulphate salts.

Thiourea (thiocarbamide). A white bitter crystalline solid, $(NH_2)_2CS$, made by heating ammonium thiocyanate. It is used as a photographic fixing agent. M.pt. 180° C; r.d. 1.4.

Thixotropy. A phenomenon shown by some non-Newtonian fluids in which the viscosity decreases as the rate of shear increases. In other words, the fluid becomes less viscous the faster it moves. Many colloidal mixtures are thixotropic, particularly gels. A useful application is in non-drip paints, which are firm when on the brush but become thinner when worked. Some lubricating oils are also thixotropic, being viscous when the lubricated parts are at rest and more mobile when movement occurs. The opposite effect to thixotropy, dilatancy, is less common.

Thomson Effect. The phenomenon in which a temperature gradient along a conductor produces a gradient of electrical potential along it. A potential difference, depending on the substance, occurs between any two points at different temperatures. [After Sir William Thomson—Lord Kelvin.]

Thomson, Sir Joseph John. 1856-1940, English physicisst. Cavendish professor of experimental physics (1884-1919) at Cambridge, he won the 1906 Nobel Prize in physics for his study of the conduction of electricity through gases. Thomson discovered (1897) the Electron and studied its charge and mass. He developed the mathematical theory of heat and electricity and worked with "positive rays" (positive ion beams), which led to a means of separating atoms and molecules according to their atomic weights. His long tenure as director of the Cavendish Laboratory at Cambridge helped make it a leading center for atomic research.

Thoria. *See* thorium dioxide.

Thorium. Symbol: Th. A soft silvery-metallic element belonging to the actinides, found in monazite ($ThPO_4$) and thorite ($ThSiO_4$). It is a radioactive element, the most stable isotope, ^{232}Th, having a half-life of 1.4 x 10^{10} years. Thorium is used in incandescent gas mantles, certain magnesium alloys, and as a source of nuclear power. A.N. 90; A.W. 232.04 m.pt. 1750°C; b.pt. 3800°C; r.d. 11.72; valency 4.

Thorium (Th). Radioactive element, discovered in 1828 by Jons Jakob Berzelius. A soft, ductile, lustrous, silver-white metal in the Actinide Series, it has 12 known isotopes. It is important for its potential conversion into the fissionable fuel uranium-233 for use in Nuclear Reactors.

Thorium Dioxide (thoria). A heavy white powder, ThO_2, obtained by reducing thorium nitrate. It is used in high temperature ceramics, gas mantles, and nuclear reactors. M.pt. 3300°C; b.pt. 4400°C; r.d. 9.7.

Thorium Series. A radioactive series beginning with thorium-232 and ending with lead-82.

Thoron. A radioisotope of radon, radon-220, produced by decay of thorium.

Thou (mil). One thousandth of an inch.

Three-body Problem. The problem of calculating the motions of three objects moving under the influence of their mutual gravitational field. In general, no exact solution is possible although the motions can be obtained by numerical methods. The problem is the simplest case of the *n*-body problem, which applies, for example, to the motion of planets around the sun.

Three Mile Island. Site of a nuclear power plant 10 mi (16 km) south of Harrisburg, Pa. On Mar. 28, 1979, failure of the cooling system of the No. 2 Nuclear Reactor led to over-heating and parttial melting of its uranium core and production of hydrogen gas, which raised fears of an explosion and dispersal of radioactivity. Thousands living near the plant left the area before the 12-day crisis ended, during which time some radioactive water and gases were released. A federal investigation, assigning blame to human, mechanical, and design errors, recommended changes in reactor licensing and personnel training, as well as in the structure and function of the Nuclear Regulatory Commission. The accident also increased public concern over the dangers of nuclear power and slowed construction of other reactors.

Thrombosis. Obstruction of an artery or vein by a blood clot, or thrombus. An arterial thrombosis is generally more serious, usually blocking the supply of oxygen and nutrients to some area of the body; a thrombus in one of the arteries leading to the heart (causing a heart attack) or to the brain (causing a Stroke) can result in death. A thrombus in the vein is known as phlebothrombosis or thrombophlebitis. Thrombosis is treated with Anticoagulants.

Thrush. Bird of the family Trudidae, found worldwide and noted for its beautiful song. Most thrushes are modestly coloured, with spotted underparts; some have bright plumage, *e.g.,* the red-breasted American Robin *(Trudus migratorius),* and the Eastern Bluebird, bright blue with a red breast. Other thrushes are the hermit thrush, Nightingale, European "blackbird," and wheatear.

Thulium (Tm). Metallic element, discovered in 1879 by P.T. Cleve. A soft, malleable, ductile, lustrous silver-white Rare-Earth Metal in the Lanthanide Series, it forms compounds with oxygen and the halogens, most of which are light green. Thulium-170 emits X-rays and is used in portable X-ray units.

Thulium. Symbol: Tm. A soft silvery-grey metallic element belonging to the lanthanide series. A.N. 69; A.W. 168.9342; m.pt. 1545°C; b.pt. 1950°C; r.d. 9.3; valency 2 or 3.

Thunder. Sound produced when a flash of Lightning passes through air, heating the adjacent air and causing it to expand rapidly. A short flash of lightning creates a relatively short crash of thunder. Rolling thunder occurs either when there is a long flash of lightning, generating thunder over a great distance, or when obstructions such as clouds, mountains, or differing layers of air cause echoes and reverberations.

Thunderstorm. Violent local atmospheric disturbance accompained by Lightning, Thunder, and heavy Rain, often by strong gusts of Wind, and sometimes by Hail. The typical thunderstorm caused by convection occurs on a hot summer afternoon when the sun's warmth has heated a large body of moist air near the ground. This air rises and is cooled by expansion. The cooling condenses the water vapor in the air, forming a cumulus Cloud. If the process continues violently, the cloud becomes immense; the summit often attains a height of 4 mi (6.5 km) above the base, and the top spreads out in the shape of an anvil as the transition to a cumulonimbus cloud occurs. The turbulent air currents within the cloud cause a continual breaking up and reuniting of the raindrops, building up strong electrical charges that result in lightning.

Thyme. Aromatic, shrubby plant (genus *Thymus*) of the Mint family. Common thyme, used as a seasoning, is the Old World *T. vulgaris,* an erect plant with grayish branches. It is cultivated mainly in Spain and France.

Thymus Gland. Mass of glandular tissue located in the neck or chest of most vertebrates. Found in the upper chest under the breastbone in humans, the thymus is essential to the development of the body's system of Immunity beginning in fetal life. The thymus processes white blood cells known as lymphocytes, which kill foreign cells and stimulate other immune cells to produce antibodies. The gland grows throughout childhood until puberty and then gradually decreases in size.

Thyratron. A type of gas-filled triode valve used in particle counters and relays. The grid is given a negative potential just large enough to prevent current flow from the cathode. If a small positive potential is superimposed on the grid potential, it reduces its magnitude and a dishcarge starts. This current flows even when the grid potential is restored because of the presence of positive ions. It is switched off by reducing the anode potential.

Thyroid Gland. Endocrine gland, situated in the neck, that regulates the body's metabolic rate. It consists of two lobes connected by a narrow isthmuss. Thyroid tissue is made up of millions of tiny, saclike follicles that store thyroid hormone in the form of thyroglobulin, a protein containing iodine. When secreted into the bloodstream, thyroglobulin is converted to thyroxine and small amounts of other similar hormones. Sufficient dietary iodine and stimulation by the Pituitary Gland are necessary for proper thyroxine production. Metabolic disorders result from oversecretion or undersecretion of the thyroid.

Tide. Alternate rise and fall of sea level in oceans and other large bodies of water. These changes are caused by the gravitational attraction of the moon and, to a lesser extent, of the sun for the earth. At any one time there are two high tides on the earth, the direct tide on the side facing the moon, and the indirect tide on the opposite side. It is believed that the indirect tide is caused by the moon actually pulling the earth away from the water on the far side. The average interval between high tides is about 12 hr. 25 min. The typical tidal range, or difference in sea level between high and low tides, in the open ocean is about 2 ft (0.6 m), but it is much greater near the coasts. The world's widest tidal range occurs in the Bay of Fundy in E Canada. Tides are also raised in the earth's solid crust and atmosphere.

Tie-dyeing. Method of textile dyeing used in Africa, India, and other parts of the world. Cotton or, sometimes, silk fabric is pleated, twisted, or folded at carefully selected points; tied or sewn in place; and dyed. The tied or sewn areas remain undyed, creating sunburst, circle, square, or other patterns. After dyeing, the thread or raffia used to tie or sew the fabric is removed, although pleats may be retained.

Tiger. Large carnivore (*Panthera tigris*) of the Cat family, found in the forests of Asia. Its yellow-orange coat features numerous prominent black stripes. Males may attain 10 ft (3 m) in length and 650 lb (290 kg) in weight. Tigers are solitary, mainly nocturnal hunters and are good swimmers but poor climbers. They have been extensively hunted for their pelts.

Timberline. Elevation above which trees cannot grow. Its location is influenced by latitude, prevailing wind directions, and exposure to sunlight. The timberline is roughly marked by the 50° F (10°C) isotherm. In general, it is highest in the tropics and descends in elevation toward the poles.

Timbre. *See* quality.

Time. The duration between two events as determined by comparison with some changing standard. Time is usually measured against a periodic process, such as the swing of a pendulum, the rotation of the earth, or the vibration of electromagnetic radiation. The SI unit of time, the second, is defined in terms of lines in the spectrum of caesium. Several astronomical measures of time also exist: solar time, which is measured with reference to the sun, sidereal time, which is measured with reference to the stars, and lunar time, measured with reference to the moon. The *calendar year* (or *civil year*) has an average length equal to that of the solar year. To obtain an exact number of days, each year is 365 days with a leap year every fourth year containing 366 days. Century years (1800, 1900, etc.) are not leap years unless they are divisible by 400.

Time and Notion Study. Analysis of the operations required to produce a manufactured article in a factory, with the aim of increasing efficiency. Each operation is studied minutely and analyzed in order to eliminate unnecessary motions and thus reduce production time and raise output. The first effort at time study was made by F.W. Taylor in the 1880s. Early in the 20th cent. Frank and Lillian Gilbreth developed a more systematic and sophisticated method of time and motion study for industry, taking into account the limits of human physical and mental capacity and the importance of a good physical environment.

Time Base. A circuit that generates a voltage that increases linearly with time at a controlled rate. Usually the.process is continually repeated: i.e. it gives a sawtooth waveform. Time bases are used in cathode-ray tubes to deflect the spot.

Time Dilation. *See* relativity.

Time Sharing. The simultaneous use of a computer by two or more separate terminals.

Tin. Symbol: Sn. A silvery-white malleable metallic element found chiefly in cassiterite (SnO_2), from which it is extracted by reduction with carbon. Below 13.2°C this metallic form *(white tin)* changes slowly into a powdery grey allotrope *(grey tin)*. It is used as a coating for steel plate (tinplate) and and in allys such as solder, Babbit metal, brass, bronze, and pewter. Tin is a reactive elment : it dissolves in hydrochloric acid to give tin II chloride, $SnCl_2$; it reacts directly with chlorine to give $SnCl_4$ and with oxygen to given SnO_2; it is dissolved by alkalis to give stannates. It forms tin IV *(stannic)* compounds, which are covalent, and the more ionic tin II *(stannous)* compounds. A.N. 50; A.W. 118.69°C; m.pt. 231.89°C; b.pt. 2270°C; r.d. 7.31 (white).

Tin (Sn). Metallic element, known and used by humans at least as early as the Bronze Age. It is a lustrous, silver-white, very soft, and malleable metal that can be rolled, pressed, or hammered into extremely thin sheets (tin foil). A tin coating, applied by dipping or electroplating, protects iron, steel, copper, and other metals from rust. Compounds of tin are used for mordants in dyeing, for weighting silk, and as reducing agents. Stannous fluoride, added to toothpastes and water supplies, prevents tooth decay. Toxic organic compounds are used as fungicides are catalysts. Cassiterite, or tinstone, is the chief ore.

Tin Chloride. Either of two chlorides of tin. *Tin II chloride* (stannous chloride) is a white crystalline solid, $SnCl_2$, obtained by the action of hydrochloric acid on the metal. It is used as a reducing agent. M.pt. 246.8°C; b.pt. 652°C; r.d. 3.95. *Tin IV chloride* (stannic chloride) is a colourless fuming liquid, $SnCl_4$, produced by direct reaction of the elements, M.pt. – 33°C; b.pt. 114°C; r.d. 2.28.

Tincture. A solution of a substance in alcohol.

Tin Hydroxide. A white substance precipitated from solutions of tin salts by alkali. The compound is amphoteric — it dissolves in exces alkali and is sometimes regarded as *stannic acid,* with formula H_4SnO_4 or H_3SnO_2. In fact it is hydrated tin IV oxide, $SnO_2.xH_2O$, which dissolves in alkali to give salts known as *stannates:* M_2SnO_3 or M_2SnO_4, where M is a monovalent metal. Hydrated tin II oxide also exists, $SnO.xH_2O$. Its salts are *stannites.*

Tin Oxide. Either of two oxides of tin. *Tin IV oxide* (stannic oxide) is a white crystalline substance, SnO_2, obtained by ehating tin in air. It is used in ceramics and as a metal polish. M.pt. 1127°C; sublimes at 1800°C; r.d. 6.6 - 9. The lower oxide, *Tin II oxide* (stannous oxide), is a black insoluble powder. SnO, obtained by precipitating the hydroxide from a solution of stannous salt and heating at 100°C. M.pt. 1080°C (600 mmHg); r.d. 6.3.

Tin Plague. The change of the white metallic form of tin into its grey allotrope at low temperatures.

Tinplate. Steel coated with a thin layer of tin to prevent corrosion.

Tinstone. *See* cassiterite

Tin Sulphide. Either of two sulphides of tin. *Tin IV sulphide* (stannic sulphide) is a yellow compound, SnS_2, precipitated from solutions of tin IV salts by hydrogen sulphide. It is used as a yellow pigment (*mosaic gold*). Decomposes at 600° C; r.d. 4.5. The lower sulphide, *tin II sulphide* (stannous sulphide), is a grey substance obtained by direct combination of the elements. Its slowly changes into tin IV sulphide and metallic tin. M.pt. 880°C; b.pt. 1230°C; r.d. 5.1.

Tint. *See* colour.

Tissue. In biology, aggregation of similar cells. In animals, the epithelial, nerve, connective, and muscle tissues are fundamental; blood and lymph are commonly classed separately as vascular tissues. Organs usually consist of several tissues. Higher plants contain meristem tissue (cells that grow, divide, and differentiate), protective tissue like cork; storage and support tissues, and vascular tissues.

Titanate. Any of a number of oxy compounds of titanium having formulas of the type $MTiO_3$ or M_2TiO_5, where M is a monovalent metal. Titanates are not salts, in the sense that they do not contain ions of the type TiO_3. Instead, they are best considered as mixed oxides. Ilmenite and perovskite are naturally occurring examples.

Titania. *See* Uranus.

Titanic. *See* Titanium.

Titanium. Symbol : Ti. A light strong lustrous white transition element, ductile when free of dissolved oxygen, found in many minerals, notably rutile (TiO_2), perovskite ($BaO.TiO_2$), and ilmenite ($FeO.TiO_2$). The ore is treated with chlorine to give the tetrachloride and this is reduced by the Kroll process.

Titanium is extensively used in strong light corrossion-resistant alloys for aircraft, missiles, and ships. The element is unreactive at ordinary temperatures but combines with most non-metals at high temperatures. It is not attacked by cold nonoxidizing acids. The most important compounds are the titanium IV *(titanic)* compounds, which are typically covalently bonded as in $TiCl_4$, and the titanates. Titanium also forms *titanous* compounds with lower valencies, as in titanium III chloride, $TiCl_3$, titanium II chloride, $TiCl_2$, and many complexes. A.N. 22; A.W. 47.90; m.pt. 1675°C; b.pt. 3620°C; r.d. 4.54; valency 2, 3, or 4.

Titanium (Ti). Metallic element, discovered in 1791 by William Gregor. It is a lustrous, silver-white, and very corrossion-resistant metal that is ductile when pure and malleable when heated. The metal and its alloys, which are light in weight and have very high tensile stregnth, are used in aircraft, spacecraft, naval ships, guided missiles, and armor plate for tanks. Titanium dioxide is used as a gemstone (titania) and paint pigment. Widely distributed in compounds (*e.g.* Rutile) in nature, titanium is present in the sun and certain other stars, in meteorites, and on the moon.

Titanium Chloride. Any of three chlorides of titanium, in which titanium has oxidation numbers 2, 3, and 4. The most important is *titanium IV chloride* (titanium tetrachloride), which is a colourless fuming liquid, $TiCl_4$, prepared by heating titanium IV oxide in a current of chlorine. It is used to make pure titanium and its salts, as a mordant, and as a polymerization catalyst. M.pt. –30°C; b.pt. 136°C; r.d. 1.8.

Titanium Dioxide. A white crystalline solid, TiO_2, occurring naturally in rutile and ilmenite. It is widely used as a pigment in paints and as a filler for paper, rubbers, and plastics.

Titanous. *See* titanium.

Titius-Bode Law. Empirical relationship between the mean distances of the planets from the sun. If each number in the series 0, 3, 6, 12, 24, ... (where a new number is twice the previous number) is increased by 4 and then divided by 10 to form the series 0.4, 0.7, 1.0, 1.6, 2.8, 5.2, 10.0, 19.6, 38.8. 77.2,....., the law holds that this series gives the mean distances of the planets from the sun, expressed in Astronomical Units. This relationship was discovered (1766) by Johann Titius and published (1772) by Johann Bode. It agreed well with the actual mean distances of the planets then known (and of Uranus and the asteroid belt, both discovered later), but not with those of the later-discovered planets Neptune and Pluto.

Titmouse. Bird of the family Paridae, which includes tits, titmice, and Chickadees. Small birds with short, pointed bills and gray and brown plumage, titmice are found chiefly in the Northern Hemisphere. They travel in flocks with other birds, *e.g.*, nuthatches and Woodpeckers, and can be taught tricks.

Titration. The determination of the Concentration of acids or bases in Solution by the gradual addition of an acidic solution of known volume and concentration to a basic solution of known volume, or vice versa, until complete neutralization (observable by the colour change in an added Indicator, such as phenolphthalein) has occurred.

Titration. An operation in which a burette is used to add controlled amounts of one solution to a measured volume of another solution. Usually an indicator is added to determine the point (called the *end point*) at which the reaction between the reagents is complete, so that neither reagent is in excess. If one solution has a known concentration (a *standard solution*) the concentration of the other can be calculated from the volumes of the two reagents. In some titrations the end point is found by continuously monitoring some physical property of the reaction mixture, such as its conductance or its colour, as solution is added from the burette. The technique is a form of volumetric analysis.

TNB. *See* trinitrobenzene.

TNT or Trinitrotoluene. $CH_3C_6H_2(NO_2)_3$, crystalline Aromatic Compound. Trinitrotoluene is a high Explosive, but, unlike Nitroglycerin, it is unaffected by ordinary shocks and must be set off by a detonator. Because it does not react with metals, it can be used in filling metal shells. It is often mixed with other explosives, *e.g.,* with ammonium nitrate to form amatol.

Toad. Certain insect-eating Amphibians of the order Anura, similar to the Frog but often more terrestrial as adults. Commonly referring to species with shorter legs, a stouter body, and thicker skin than the frog, the term *toad* is properly restricted to the so-called true toads (family Bufonidae). These are characterized by warty skins, prominent parotid glands behind the eyes, and a white, poisonous fluid exuded through the skin and from the parotid glands. Ranging from 1 to 7 in. (2.5 to 18 cm.) in size, toads inhabit cool, moist places and lay their eggs in water.

Tobacco. Plant (genus *Nicotiana*) of the Nightshade family, and the product manufactured from its leaf and used in cigars and cigarettes, snuff, and pipe and chewing tobacco. The chief commercial species *N. tabacum,* is believed native to tropical America. The tobacco plant is a coarse, large-leaved perennial, but it is usually cultivated as an annual. Tobacco requires a warm climate and rich, well-drained soil. After being picked, the leaves are cured, fermented, and aged to develop aroma. The amount of nicotine (the Alkaloid responsible for tobacco's narcotic and soothing effect) varies, depending on tobacco strain, growing conditions, and processing. The use of tobacco orginated among natives of the New World in pre-Columbian times. Introduced into Spain and Portugal in the mid-16th cent., initially as a panacea, it spread to other European countries, and by 1619 tobacco had become a leading export crop of Virginia. In recent years there has been concern over the harmful effects of nicotine, the tarry compounds, and Carbon Monoxide in tobacco smoke.

Tollen's Reagent. A solution of silver oxide in ammonia water used as a test for aldehydes, which reduce the oxide and deposit a bright mirror of silver on the wall of the test tube. Ketones do not cause this reduction.

Toluene. A colourless flammable liquid, $C_6H_5CH_3$, obtained by catalytic reforming of petroleum. It is used in aviation fuels, in the production of phenol and other chemicals, and as a solvent for paints and resins. M.pt. –94°C; b.pt. 111°C; r.d. 0.9.

Toluidene (aminotoluene). Any of three isomeric derivatives of toluene, $CH_3C_6H_4NH_2$, prepared by reduction of the appropriate nitrotoluene. The most widely used is *orthotoluidene* (2-aminotoluene), which is a pale yellow liquid used in the manufacture of dyes and saccharin. M.pt. –16°C; b.pt. 200°C; r.d 1.0.

Tolyl Group. Any of the three isomeric groups $CH_3C_6H_4$-, derived by removing a hydrogen atom from the ortho-, meta-, or para- position of the benzene ring in toluene.

Tomato. Plant (*Lycopersicon esculentum*) of the Nightshade family, and its fruit (commonly considered a vegetable because of its uses). Numerous varieties are cultivated, *e.g.*, the small cherry tomato, the yellow pear tomato, and the large, red beefsteak tomato. Popular in salads and processed into juice, catsup, and canned goods, the tomato was recognized as a valuable food onlly within the last century.

Tone. In music, a tone is distinguished from noise by its definite Pitch, caused by the regularity of the vibrations that produce it. Any tone possesses the atributes of pitch, intensity, and quality. Pitch is determined by the frequency of the vibration, measured by cycless per second. Intensity, or loudness, is determined by the amplitude, measured in decibels. Quality is determined by the overtones (subsidiary tones), the distinctive timbre of any instrument being the result of the number and relative prominence of the overtones it produces. The term *whole tone* or *whote step* refers to the interval of a major second, as in moving from one white key to the adjoining white key on the piano. *Half tone, semitone, or half step refers* to the interval of a minor second, as in moving from a white key to the adjoining black key on the piano.

Tone. An audible note with no harmonics: *i.e.* a sound of a single frequency of the type produced by a tuning fork. A pure tone has a sinusoidal waveform.

Tongue. muscular organ on the floor of the mouth in higher animals. In humans it functions primarily in chewing, swallowing, and speaking. The human tongue is covered by a mucous membrane containing small projections, or papillae, which give it a rough surface. Tiny taste organs, or buds, are found on the papillae, with many concentrated toward the back of the tongue. Sweet, sour, salty, and bitter flavors stimulate the taste cells in these buds to send impulses along associated nerves to the brain.

Tonne (metric ton). A unit of mass equal to 1000 kilograms. It is equivalent to 0.9842 (short) tons.

Tonsils. Name commonly referring to the palatine tonsils, two ovoid masses of lymphoid tissue on either side of the throat, at the back of the tongue. The pharyngeal tonsils, or adenoids, are similar masses located in the space between the back of the nose and the throat. The lingual tonsils are situated on the back of the tongue. The tonsils act as filters against disease organisms.

Topaz. Aluminum silicate mineral [$Al_2SiO_4(F,OH)_2$], used as a Gem. Commonly colourless or some shade of the yellow, the stone is transparent with a vitreous luster. Topaz crystals occur in highly acidic igneous rocks and in metamorphic rocks. Important sources include Brazil, Siberia, Burma, and Sri Lanka.

Topology. Branch of Mathematics concerned with those properties of geometric figures that are invariant under continuous transformations. A continuous transformation is a one-to-one correspondence between the points of one figure and the points of another figure such that points that are arbitrarily close on one figure are transformed into points that are also arbitrarily close on the other figure. Two figures are topologically equivalent if one can be deformed into the other by bending, stretching, twisting, or the like, but not by tearing, cutting, or folding; thus topology is sometimes popularly called "rubber-sheet geometry." A circle and a square are topologically equivalent, as are a cylinder and a sphere, but a torus (doughnut shape) is not equivalent to a sphere, because no amount of bending or twisting will change it into a sphere. Topology may be roughly divided into *point-set topology*, which considers figures as Sets of points having such properties as being open, closed, compact, connected, and so forth; *combinatorial topology,* which considers figures as combinations (complexes) of simple figures (simplexes) joined together in a regular manner; and *algebraic topology*, which makes extensive use of algebraic methods, particularly those of group theory. There is considerable overlap among these branches.

Topology. The branch of geometry concerned with general types of shape and pattern rather than with particular shapes or sizes. The basic notion is that of equivalence based on whether one figure or surface could be deformed into another without tearing it and without joining points together. Thus, a square and a circle are topologically equivalent but neither is equivalent to a figure eight.

Tornado. Dark, funnel-shaped cloud containing violently rotating air that develops below a heavy cumulonimbus cloud mass and extends toward the earth. The diameter of a tornado varies from a few feet to a mile; the rotating winds reach velocities of 200 to 300 mph (320 to 480 km/hr), and the updraft at the center may reach 200 mph (320 km/hr). In comparison with a Cyclone, a tornado covers a much smaller area but is much more violent and destructive. The atmospheric conditions required for the formation of a tornado include great thermal instability, high humidity, and the convergence of warm, moist air at low levels with cooler, drier air above. Tornadoes occurring over water are called waterspouts.

Toroidal. Having a torus shape.

Torque. Symbol T. A force or system of forces producing a turning effect, measured by its moment.

Torque or **Moment of Force.** A quantity expressing the effectiveness of a force to change the net rate of rotation of a body. It is equal to the product of the force acting on the body and the distance from its point of application to the axis around which the body is free to rotate. Units of torque include the foot-pound (or pound-foot), the dyne-centimeter, and the newton-meter.

Torr. A unit of pressure equal to the pressure that would support one millimetre of mercury. It is equivalent to 133.322 newtons per square metre. The torr is used for measuring the low pressures used in vacuum systems. [After Evangelista Torricelli (1608-47), Italian scientist.]

Torricelli, Evangelista. 1608-47, Italian physicist and mathematician. Galileo's secretary and successor as professor of philosophy and mathematics at Florence, he invented the Barometer (Torricelli tube) and improved the telescope.

Torricellian Vacuum. The vacuum produced by filling a long tube, closed at one end, with mercury and inverting the open end in a reservoir of mercury, so that the mercury column is held up by atmospheric pressure on the reservoir. The space above the mercury contains the Torricellian vacuum, in which the pressure equals the vapour pressure of mercury (1.3 x 10 mmHg).

Torsion Balance. A device in which a torque is measured by the amount by which it twists a vertical wire or fibre. The angle of twist, which is usually measured by reflecting a beam of light off a small attached mirror, is proportional to the torque and depends on the rigidity modulus of the fibre.

Torsion Pendulum. *See* pendulum.

Tortoise. Terrestrial Turtle, especially one of the family Testudinidae. Tortoises inhabit warm regions worldwide except in Australia. Most famous is the giant tortoise of the Galapagos Islands *(Testudo elephantopus)*, which can be over 4 ft (120 cm.) long and weigh over 500 lb (225 kg). Tortoises are extremely long-lived; some are known to have survived for more than 150 years.

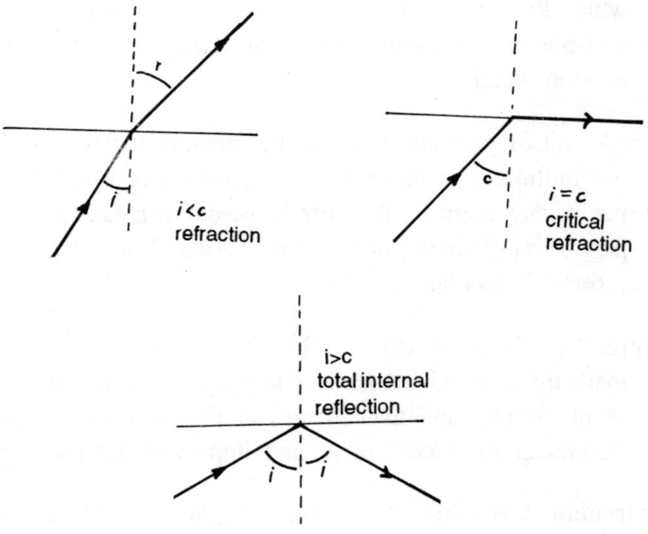

Fig. T-5. Refraction and total internal reflection

Torus. A surface or solid generated by rotating a circle about an axis that does not pass through the circle. A torus thus has the same shape as a doughnut or anchor ring.

Total Internal Reflection. Reflection of an incident ray of light back into the initial medium at an interface between this medium and one that has a lower refractive index. A ray of light travelling from glass into air is refracted provided that the angle of incidence does not exceed the value at which the angle of refraction is 90°. Angles of incidence greater than this value *(the critical angle)* give total internal reflection rather than refraction.

Totality. The point at which the sun's disc is completely obscured during a total eclipse.

Total-radiation Pyrometer. *See* pyrometer.

Totem. An object, usually an animal or plant, revered by an individual or a particular social group. A group totem represents the bond of unity and is often considered the ancestor or brother of the group's members; marriage between those of one totem is often prohibited as incest. The group's symbol and protector, the totem may be pictured on the body or masks, or carved on totem poles.

Toucan. Perching Bird of the New World tropics, related to the Woodpeckers Toucans vary in size from the jay-sized toucanets to the 24-in. (62-cm) tocos of the Amazon basin. Their enormous, often brightly coloured, canoe-shaped bills are adapted to cutting up fruits and berries.

Tourmaline. A blue or black crystalline naturally occurring borosilicate. Crystals of tourmaline show piezoelectricity and pleochroism.

Touch-me-not. Common name for any plant of the genus *Impatiens* of the Jewelweed family.

Tourmaline. Complex aluminum and boron silicate mineral $[Na,Ca)(Al,Fe,Li,Mg)_3Al_6(BO_3)_3Si_6O_{18}(OH)_4]$, used as a Gem. Colours are red, pink, blue, green, yellow, violet, and black; sometimes it is colourless. Two or more colours, arranged in zones or bands with sharp boundaries, may occur in the same stone. Tourmalines are found in pegmatite veins in granites, gneisses, schists, and crystalline limestones. Important sources include Elba, Brazil, the USSR, Sri Lanka, and parts of the U.S.

Townsend Discharge. A type of electrical discharge occurring between two electrodes in a low pressure of gas. The Townsend discharge differs from the glow discharge in having a low current (a few microamps), which is limited by placing a high external resistance in the circuit. A fairly uniform luminous plasma is produced between the electrodes and the voltage falls uniformly with distance. The voltage drop increases with increasing discharge current. [After Sir John Townsend (1868-1957), Irish physicist.]

Toxemia. Disease state caused by the presence in the blood of bacterial Toxins or other harmful substances. The term now usually applies to toxemia late in pregnancy, or pre-eclampsia, a condition once believed to be caused by toxins, but now associated with high blood pressure, protein in the urine, and edema. Toxemia can lead to eclampsia—maternal convulsions and Coma—and renal and cardiovascular damage or death if untreated.

Toxic Shock Syndrome. Acute, sometimes fatal, disease characterized by high fever, nausea, diarrhea, lethargy, blotchy rash, and sudden drop in blood pressure. Caused by a toxin-producing strain of the *Staphylococcus aureus* bacterium, the disease is mot prevalent among menstruating women using high absorbency tampons, but also affects nonmenstruating women and men.

Toxin. A poison produced by a living organism. Some toxins affect specific tissue in the host; *e.g.*, the toxin associated with Diphtheria affects mucous membranes, and that associated with Botulism destroys nerve tissue. Other toxins may produce fever, internal hemorrhage, and Shock. The presence of toxins stimulates the production of antibodies, or antitoxins, one of the body's defense mechanisms against disease.

Toy Dog. Class of very small breeds of dogs kept as pets. Some are selectively bred small forms of larger breeds; others are naturally small.

Tracer. An isotope used for investigating chemical or physical changes. An example of a physical application is in studies of diffusion. A layer of a radioactive metal isotope deposited on the surface of the normal metal will slowly diffuse in, the extent of diffusion being determined by cutting thin slices of metal and testing their radioactivity. Tracer studies are also extensively used in finding mechanisms of chemical reactions. For instance in the hydrolysis of an ester. RCOOR' + H_2O = RCOOH + R'OH, it can be shown that the oxygen from the water molecule appears in the acid rather than the alcohol. This is done by including water made from the stable ^{18}O isotope in the reaction and demonstrating that $RCO^{18}OH$ is formed. The isotopic atom used in such studies is called a *label.* Radioactive-tracer studies are usually easier than those using stable isotopes due to difficulties in analysis.

Trademark. Distinctive mark placed on merchandise to indicate its origin. Its use is the exclusive legal right of its owner.

Trailing Arbutus, Mayflower, or Ground Laurel. American wildflower *(Epigaea repens)* of the Health family. The plant blooms in early spring; its creeping stems bear clusters of fragrant pink or white flowers, sometimes hidden by the hairy evergreen leaves. It is the provincial flower of Nova Scotia and the state flower of Massachusetts.

Tranquilizer. Drug whose action on the central nervous system relieves emotional agitation. Antipsychotic drugs, or major tranquilizers, moderate symptoms of psychotic states, including agitation, delusions, and anxiety. These drugs include chlorpromazine (Thorazine), the first agent to be widely applied to mental disorders and still the standard drug. Antianxiety Drugs, or minor tranquilizers, are prescribed to relieve anxiety and tension.

Trans. Indicating the position of a group or atom in a molecule opposite to the position of a specified group or atom: *i.e.* on the opposite side of a double bond or central atom.

Transactinide Elements. Chemical elements with atomic numbers greater than 103, that of lawrencium, the last member of the Actinide Series.

Transcendental. 1. Denoting a number that cannot be the root of a polynomial equation with rational coefficients. Transcendental numbers are a type of irrational number. Examples are Π and e.

2. Denoting a mathematical function that cannot be written as a finite number of terms. Sin *x*, *e*, and log *x* are examples of transcendental functions.

Transducer. Device that accepts an input of energy in one form and produces an output of energy in sosme other form, with a known, fixed relationship between the input and output. One class of transducers consists of devices that produce an electrical output signal, *e.g.*, Microphones, Record-Player cartridges, and Photoelectric Cells. Other transsducers accept an electrical input, *e.g.*, Loudspeakers, light bulbs, and Solenoids. Transducers may be either active or passive. Active transducers require a source of energy in addition to the input signal to produce the output signal, whereas passive transducers require only an input signal.

Transducer. A device that is supplied with the energy of one system and converts it into the energy of a different system, so that the output signal is proportional to the input signal but is carried in a different form. Thus, a microphone is a transducer for converting energy carried as sound waves into electrical energy. A record-player pickup converts the vibrations of the stylus into an electrical signal. A loudspeaker converts electrical signals into sound waves.

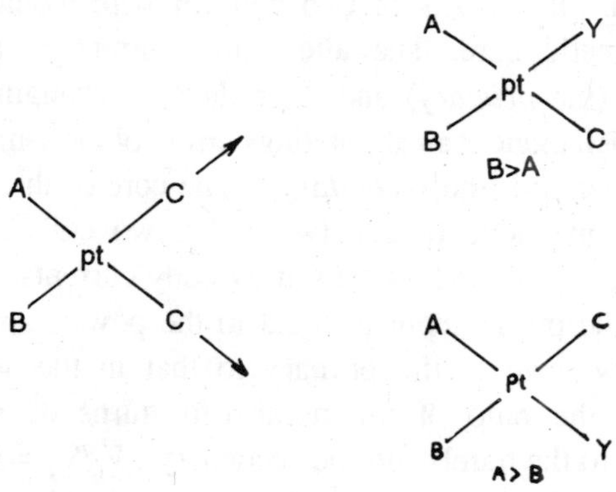

Fig. T-6. Trans substitution products.

Trans Effect. An effect observed in the substitution of square-planar inorganic complexes, in which certain ligands have the power to direct the substituent into the *trans* position. For example, if the compound $PtABC_2$ is substituted by a ligand Y to give PtABCY, the position of Y depends on the relative *trans*-directing powers of A and B. The cyanide ligand (CN^-), for example, is more efficient at causing *trans* substitution than the bromide ligand (Br). Some typical ligands, in order of their *trans*-directing power, are $CN^- > NO_2 > I^-$ $Br^- > Cl^- > NH_3 > H_2O$.

Transformation. 1. A change of a mathematical expression from one form to another by replacing the variables with new variables, which are related to the old ones by defining equations. For instance, the equation $y = x^2$ can be transformed by substituting $y_1 = y + 3$ and $x_1 = x - 2$, to give $y_1 = x_1^2 + 4x_1 + 7$. This particular transformation corresponds to a translation of axes in Cartesian coordinates.

2. A change of one nuclide into another, as by emission of an alpha or beta particle.

Transformer. A device for changing an alternating current of one voltage to one of another voltage by electromagnetic induction. It consists of two coils of wire wound around a ferromagnetic core. The alternating current is passed into one coil (the *primary*) and it produces a changing magnetic field, which induces an alternating current of the same frequency in the other coil (the *secondary*). The core of the transformer links the magnetic field between the two coils : often it is laminated to reduce loss caused by eddy currents. In a perfect system, the power input is equal to the power output and the ratio of voltage in the primary to that in the secondary is equal to the ratio of the number of turns of wire in the primary to the number in the secondary : $V_p/V_s = n_p/n_s$. If the number in the secondary is larger than in the primary the voltage is increased and the device is a *step-up transformer.* If the number in the secondary is less the device is a *step-down transformer.* If $n_p = n_s$, the voltage is unchanged. Transformers of this type are used for isolating a piece of equipment from itss power supply so that there is no direct electrical connection.

Transformer. Electrical device that transfers an alternating current or voltage from one Electric Circuit to another using electronmagnetic Induction. A simple transformer consists of two coils of wire electrically insulated from each other and aranged so that a change in the current through the primary coil will produce a change in voltage across the secondary coil. The ratio of the alternating-current (AC) output voltage to the Ac input voltage is approximately equal to the ratio of the number of turns in the secondary coil to the number of turns in the primary coil. This capability for transforming voltages is the basis for a great many applications. Transformers are classified according to their use; power transformers are used to transmit power at a constant frequency, audio transformers are designed to operate over a wide range of frequencies with a nearly constant ratio of input to output voltage, and radio-frequency transformers operate effficiently within a narrow range of high frequencies.

Transient. **1.** A sudden short-lived disturbance in a system.

2. A short-lived chemical species, such as a free radical or excited atom or molecule.

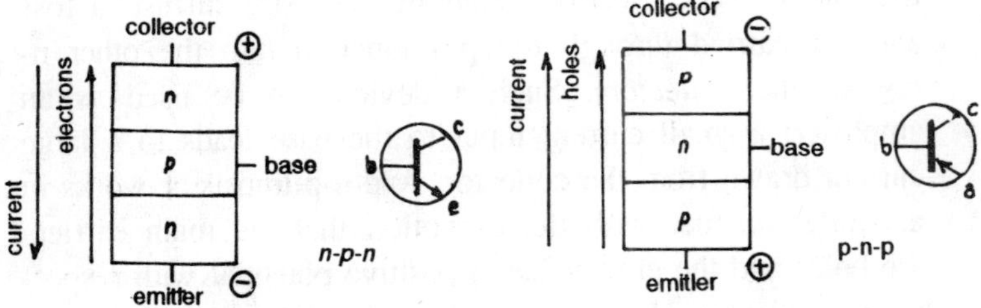

Fig. T-7. Junction transistors

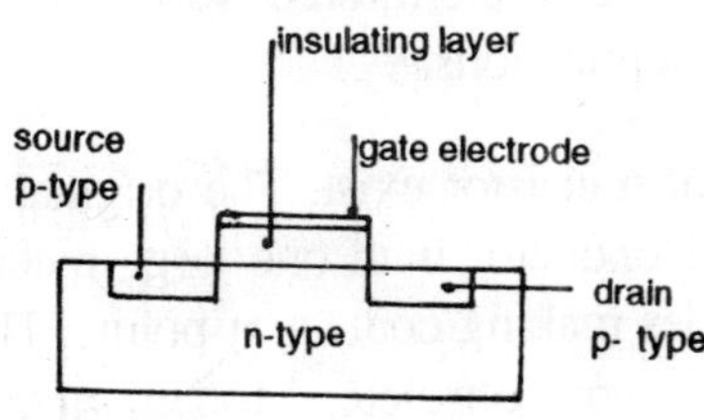

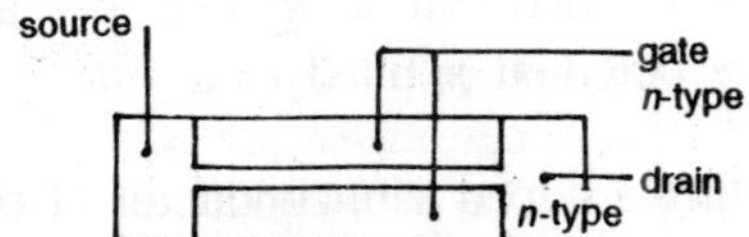

Fig. T-8. Types of field-effect transistor

Transistor. A device made of semiconducting material in which a flow of current between two electrodes can be controlled by a potential applied to a third electrode. In the most common type, known as a *junction transistor,* three pieces of semiconductor are joined together, each with a separate attached electrode. Two forms exist, designated *n-p-n* and *p-n-p* transistors according to the arrangement of n-type and p-type material.

An n-p-n transistor has a potential difference applied so that one n-region is positive and the other negative. The central p-region (the *base*) has a positive potential with respect to the negative n-region (called the *emitter*). Electrons flow from the emitter into the base and holes from the base flow to the emitter. The electrons entering the base diffuse across and are carried through the p-n junction into the other n-region (the *collector*). Such a device can be used as an amplifier; a small current input to the base leads to a large current drawn from the collector. A p-n-p transistor works in a similar manner with the exception that the main carriers are holes and the emitter has a positive potential with respect to the collector. Holes have a lower mobility than electrons and p-n-p transistors operate more slowly. Transistors of this type have replaced thermionic valves in all but the most specialized applications.

Other types of transistor exist. The original form had a single piece of semiconductor with one large metal contact and two other electrodes making contact at points. These *point-contact transistors* are now obsolete. Modern *field-effect transistors* are more complicated devices in which conduction occurs through a channel between a *source* and a *drain* and is controlled by a potential applied to a *gate.*

The gate is either a doped semiconductor of opposite type to that of the source and drain, or a metal electrode acting through an insulating layer. Field-effect transistors (*FETs*) have a very high input impedance and are used in amplifiers and integrated circuits.

Transistor. Electronic device used as a voltage and current Amplifier, consisting of Semiconductor materials that share common physical boundaries. The material most commonly used in silicon into which impurities have been introduced. In *n*-type semiconductors there is an excess of free electrons, or negative charges, whereas in p-type semiconductors there is a deficiency of electrons and therefore an excess of positive charges. Transistors are used in many applications, including Radio receivers, electronic Computers, and automatic control instrumentation (*e.g.,* in space-flight and guided missiles). Since the invention (announced in 1948) of the transistor by the American physicists John Bardeen, Walter H. Brattain, and William Shockley, many types have been designed. The *n-p-n* junction transistor consists of two *n*-type semiconductors separated by a thin layer of p-type semiconductor; the three segments are called emitter, base, and collector, respectively, and are usually sealed in glass, with a wire extending from each segment to the outside, where it is connected to an electric circuit. The transistor action is such that if the electric Potentials on the segments are properly determined, a small current between the emitter and base connections results in a large current between the emitter and collector connections, thus producing current and amplification. The *p-n-p* junction transistor, consisting of a thin layer of *n*-type semi-conductor lying between two *p*-type semiconductors, works in the same manner, except that all polarities are reversed.

Transit Instrument or **Transit.** Telescope devised to observe stars as they cross the meridian of Longitude and used for determining time. Its viewing tube swings on a rigid horizontal axis restricting its movements to the arc of the meridian. The meridian circle (a modern transit) is equipped with precisely graduated circles mounted on the horizontal axis and with stationary verniers, or reading microscopes, mounted on the fixed supports of the telescope that enable the observer to read the circles. By giving both the altitude and transit time, this instrument yields the right ascension and declination—the position on the celestial sphere—of the star.

Transition. 1. A change in energy level in an atom, ion, or molecule.

2. A change in energy level of an atomic nucleus, causing either the emission of a gamma-ray photon or the emission of a beta or alpha particle. When a particle is ejected a different nucleus is formed: *i.e.* a transformation occurs.

3. A change of a substance from one physical state to another. Transitions include changes of phase (freezing, melting, etc.), the conversion of one crystal structure into another at a particular temperature, and the onset of superconductivity in a metal at low temperature. The temperature at which a transition occurs is the *transition temperature.*

Transition Elements or **Transition Metals.** Elements of group VIII and the b groups (I through VII) of the Periodic Table, characterized by the filling of an inner *d* or *f* electron orbital as atomic number increases. Many chemical and physical properties of these elements are due to their unfilled *d* or *f* orbitals. Transition elements generally have high densities and melting points, magnetic properties, and variable valence arising from the electrons in the *d* or *f* orbitals. These metals form stable coordination complexes, or Complex Ions, many of which are highly coloured and exhibit paramagnetism.

Transition Metal. Any of a class of elements characterized by an incomplete inner shell of electrons. In the periodic table the elements of increasing atomic number have their electron shells filled in a regular way up to calcium, which has a configuration 2, 8, 8, 2. The next nine elements, from scandium to copper, have extra electrons in their penultimate shell: thus scandium has 2, 8, 9, 2, titanium has 2, 8, 10, 2, etc. The elements from titanium to copper form the first transition series: by convention scandium is not usually included in the transition metals. Two other transition series exist: from zirconium to silver and from hafnium to gold.

The transition elements are all metals, having a lustrous appearance and good electrical and thermal conductivity. They often have more than one valency and form many coordination complexes. Compounds of transition metals are often coloured and, because of unpaired electrons, many of them are paramagnetic.

Transition State. The state of highest energy occurring during the breaking and formation of bonds in a chemical reaction. In a simple case, the reaction X + YZ = XY + Z, the atom X approaches as Z leaves and the total energy increases to a maximum and then decreases again. The partially bonded system of atoms corresponding to this transition state is the *activated complex* of the reaction.

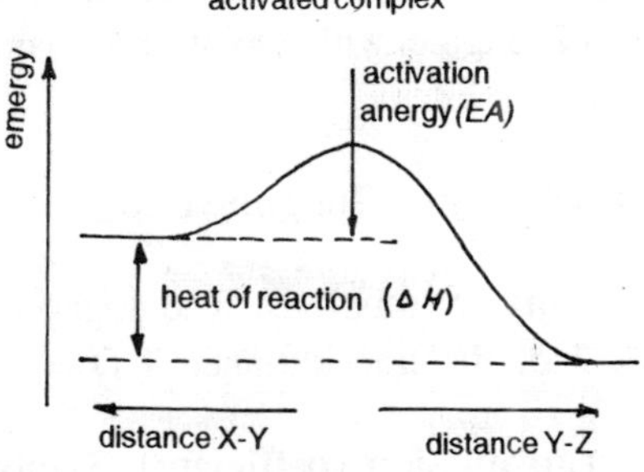

Fig. T-9. Energy profile.

Translation. Motion of a body, molecule, etc., in which every point in the body moves an equal distance and the paths of the points are parallel.

Translucent. Denoting a material that allows partial transmission of incident radiation, especially light. A substance that allows all the light to be transmitted is said to be *transparent* whereas one that absorbs all the incident light is *opaque*.

Transleithania. Austro-Hungarian Monarchy.

Transmission. In Automobiles, system for transmitting power from the engine to the wheels. The system is designed to change the high rotational speed and low torque (turning force) of the engine's crankshaft into the higher-torque rotation needed to turn the wheels over a range of speeds. A *manual transmission* consists of a system of interlocking Gear wheels arranged so that by operating a lever the driver can choose one of several ratios of speed between an input shaft turned by the engine and an output shaft that turns the wheels. For a standing start the driver selects the first, or lowest, gear, which produces high turning power at a low output shaft speed. Higher gears produce less torque at higher output shaft speeds. To allow smooth shifting from one gear to another, a clutch mechanism disengages the engine from the transmission during gear changes. The *automatic transmission* (introduced in 1939), in which gear changes are made without driver intervention, uses a fluid device called a torque converter to connect the engine with the gearbox and to control gear changes.

Transmission Coefficient. Transmittance.

Transmission Density. Symbol: D. The logarithm to base 10 of the reciprocal of the transmittance T.D = $\log_{10} 1/T$.

Transmittance (transmission coefficient). Symbol: T. A measure of the extent to which a body transmits incident light or other electromagnetic radiation, equal to the ratio of the flux transmitted to the incident flux. The transmittance depends on the path length of the radiation and on its wavelength.

Transmutation. The changing of one element into another. Transmutation occurs in radioactive decay or can be caused by bombardment of nuclei with particles.

Transparent. Translucent.

Transpiration. In terrestrial plants, loss of water by evaporation mainly through the pores (stomata) of the Leaf but also through the plant's surface cells. The pull of transpiration on the fluid in the plant is one cause of the ascent of Sap from the Roots, and thus helps provide the necessary moisture for cell functions. Desert plants are modified in ways that decrease transpiration.

Transplantation, Medical. Process by which a tissue or organ is removed and replaced by a corresponding part. Transplants usually range from those employing tissue (such as skin, bone, or cartilage) from the patient's own body (called an autograft transplant) to those involving the replacement of vital organs (such as the heart or kidney) from the body of another individual. Transplantation of complex organs calls for the surgical connection of the larger blood vessels of the donor organ to those of the recipient; connective tissue cells gradually link the graft and host tissues. Replacements for diseased or defective tissue (*e.g.,* a cornea or the heart) are generally obtained from donors who have died; large or regenerating organs or tissues such as kidney, skin, bowel, or blood can be donated by living individuals. Organs such as the heart must be transplanted as soon after the death of the donor as possible; skin, corneas, bone, and some blood fractions, however, can be stored. Transplanted tissue from another individual contains antigens that stimulate an immune response by the host's lymphocytes; therefore the main obstacle to successful transplantation is the rejection of foreign tissue by the host. To minimize rejection, an antigenic typing system (called the HLA system and similar to blood typing) is used to determine the degree of tissue compatibility between the donor and the recipient. Immunosuppressive Drugs are also used to interfere with the production of antibodies (the body's usual response to foreign substances) in the recipient. Human tissue grafting was first performed about 100 years ago by Jacques Reverdin, a Swiss surgeon. In 1902 the French surgeon Alexis Carrel developed a method of joining blood vessels that made the transplantation of organs feasible and stimulated the use of transplantation in experimental biology. The first successful transplant of a human Kidney was made by Richard

H. Lawler in Chicago in 1950; and the first human Heart transplant was peformed by the South African surgeon Christiaan Barnard in 1967. Since then, advances in biomedical engineering have also made possible the implantation of artificial body parts, such as artificial joints, and in 1982 resulted in the implantation of the first permanent artificial heart (developed principally by Robert K. Jarvik of the Univ. of Utah). Investigations continue into the development of completely artificial organs.

Transport Number. Symbol: *t*. The fraction of the total current carried by a particular type of ion when an electrolyte conducts electricity.

Transuranic Element. Any of the actinide elements with atomic number greater than 92. They do not occur naturally, being produced by nuclear reactions.

Transuranium Elements. Radioactive chemical elements with atomic numbers greater than 92 (Uranium). Only Neptunium (at. no. 93) and Plutonium (at. no. 94) occur in nature; they are produced in minute amounts in the radioactive decay of uranium. The transuranium elements of the Actinide Series were discovered as synthetic radioactive isotopes. Both American and Soviet scientists claim to have discovered independently the unstable transactinide elements 104, 105, and 106, and West German scientists reported discovering the unstable transactinide elements 107 and 109.

Transverse Wave. Wave.

Trapezium. A quadrilateral that has one pair of opposite sides parallel. The area of a trapezium is one half of the sum ($a + b$) of the lengths of the parallel sides multiplied by the perpendicular distance between them (h): *i.e.* $h(a + b)/2$.

Tree. Perennial woody plant with a single main Stem (the trunk or bole) from which branches and twigs extend to form a characteristic crown. Trees are either deciduous, *i.e.,* having broad leaves that are shed at the end of the growing season, or evergreen, having needle or scale-like leaves that are shed at intervals of 2 to 10 years. Some broad-leafed shrubs follow the conifer pattern, and the Larch sheds its needle-like leaves deciduously. Tree identification is through leaf shape and overall appearance. Trees are an important source of wood, food, and products such as resins, Rubber, Quinine, turpentine, and Cellulose.

Tree Shrew. Small, arboreal prosimian, or lower Primate, of the family Tupaiidae, found in S Asia. Tree shrews superficially resemble squirrels and are usually brown, reddish, or olive in colour, with large eyes, good vision, and flexible hands with sharp claws. They are territorial, omnivorous, and extremely active and quarrelsome.

Travelling Wave. Wave.

Triangle. A plane geometric figure with three sides. The area of a triangle is one half of the product of its base (b) and its height (h): *i.e.* $bh/2$.

Triangle of Forces. A triangle formed by representing the magnitudes of three forces acting at a point by the sides of a triangle. A triangle can only be formed for such forces when they are in equilibrium, a consequence of the rule for addition of vectors.

Triatomic. Having molecules composed of three atoms. Carbon dioxide, CO_2, and ozone, O_3, are both examples of triatomic gases.

Triazine. Any of three isomeric azines, $C_3N_3H_3$.

Triazole. Any of four isomeric heterocyclic compounds, $C_2N_3H_3$.

Tribasic. Denoting a compound with three acidic hydrogen atoms in its molecules. Orthophosphoric acid is a common tribasic acid, forming the salts M_3PO_4, M_2HPO_4, and MH_2PO_4, where M is a monovalent metal.

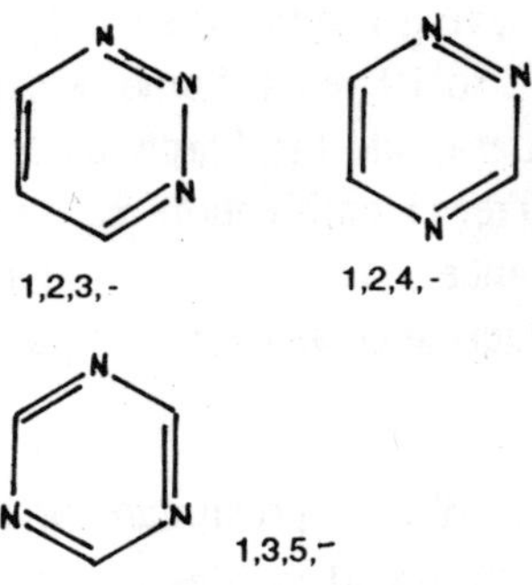

Fig. T-10. Triazines.

Triboelectricity. Static electricity produced by friction. When one material is rubbed against another it is possible for electrons to be transferred, so that one material acquires a positive charge and the other an equal negative charge. If a glass rod is rubbed with silk the glass becomes negatively charged and the silk positively charged.

Tribology. The study of friction between solid surfaces, including the origin of frictional forces and the practical problems of wear and lubrication of moving parts.

Triboluminescence. Light emission produced by friction between certain solids. Triboluminescence may be observed when some crystalline materials, such as sugar, are crushed. It results from a static electric charge generated by the friction.

Trichinosis or **Trichiniasis.** Parasitic disease caused by the roundworm *Trichinella spiralis,* following ingestion of raw or inadequately cooked meat, especially pork. The larvae mature in the intestines and are carried by the bloodstream to muscles, where they become embedded and remain. The host experiences irregular fever, profuse sweating, and muscular soreness; there symptoms usually subside soon after infestation, although vague muscle pain and fatigue may persist.

Trichloracetic Acid. A corrosive deliquescent crystalline solid carboxylic acid, CCl_3COOH, made by oxidizing chloral hydrate with nitric acid. M.pt. 57°C; b.pt. 197°C; r.d. 1.6.

Trichloroethylene. A stable toxic colourless liquid. $CHCl{:}CCl_2$, prepared by treating tetrachloroethane ($CHCl_2CHCl_2$) with alkali. It is used as a solvent and dry-cleaning agent. M.pt. – 73°C; b.pt 87°C; r.d. 1.4.

Trichloromethane. Chloroform.

Triclinic. Crystal system.

Triglyceride. A glyceride derived by substituting all three hydroxyl groups of glycerol.

Trigonal. Crystal system.

Trigonometry. The study of the properties of triangles with particular reference to the ratio of the sides of right-angled triangles. It is used in surveying, navigation, and astronomy.

Trigonometry. The study of certain mathematical relations originally defined in terms of the angles and sides of a right triangle, *i.e.*, one containing a right Angle (90°). Six basic relations, or trigonometric functions, are defined. If *A, B,* and *C* are the angles of a right triangle (C = 90°) and *a, b,* and *c* are the lengths of the respective sides opposite these angles, then six functions can be expressed for one of the acute angles, say A, as various ratios of the opposite side (*a*), the adjacent side (*b*), and the hypotenuse (*c*), as set out in the table. Although the actual lengths of the sides of a right triangle may have any values, the ratios of the lengths will be the same for all similar right triangles, large or small. It may be seen that sin B = cos A, cos B = sin A, tan B = cot A, and so forth. The values of the sine and the cosine are always between 0 and 1, the values of the secant and the cosecant are always equal to or greater than 1, and the values of the tangent and the cotangent are unbounded, increasing from 0 without limit.

The values of the trigonometric functions can be found in a set of tables or on a calculator. The notion of the trigonometric functions is extended beyond 90° (the largest angle size in a right triangle) by defining the functions with respect to Cartesian Coordinates; the functions then take on negative as well as positive values in a pattern that repeats every 360°. This repeating, or periodic, nature of the trigonometric functions leads to important applications in the study of such periodic phenomena as light and electricity. A general triangle, not necessarily containing a right angle, can also be analyzed by means of trigonometry. Spherical trigonometry, the study of triangles on the surface of a sphere, is important in surveying, navigation, and astronomy.

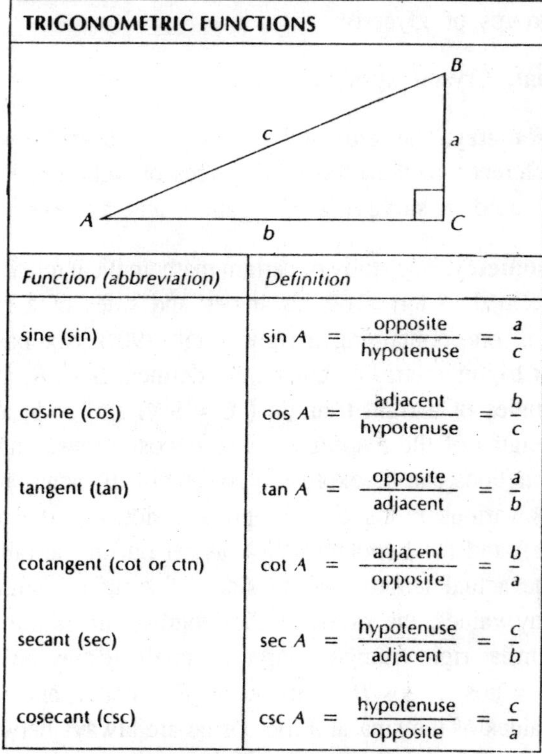

Function (abbreviation)	*Definition*
sine (sin)	$\sin A = \frac{\text{opposite}}{\text{hypotenuse}} = \frac{a}{c}$
cosine (cos)	$\cos A = \frac{\text{adjacent}}{\text{hypotenuse}} = \frac{b}{c}$
tangent (tan)	$\tan A = \frac{\text{opposite}}{\text{adjacent}} = \frac{a}{b}$
cotangent (cot or ctn)	$\cot A = \frac{\text{adjacent}}{\text{opposite}} = \frac{b}{a}$
secant (sec)	$\sec A = \frac{\text{hypotenuse}}{\text{adjacent}} = \frac{c}{b}$
cosecant (csc)	$\csc A = \frac{\text{hypotenuse}}{\text{opposite}} = \frac{c}{a}$

Fig. T-11. Trigonometric functions.

Trihydrate. A solid compound with three molecules of water of crystallization per molecule of compound, as in potassium aluminate trihydrate, $K_2Al_2\ O_4\ 3H_2O$.

Trihydric. Denoting an alcohol or phenol containing three hydroxyl groups per molecule.

Trillium or **Wake-robin.** Attractive spring wildflower (genus *Trillium*) of the Lily family, native to North America and E Asia. The single flower may be white, pink, dark red, yellow, or green, and the leaves, petals, and sepals are typically in threes.

Trimer. A compound whose molecules are formed by addition of three molecules of a simpler compound (the *monomer*). A reaction leading to formation of a trimer is a *trimerization.* Acetaldehyde, for example, trimerizes to give paraldehyde.

Trimethylamine. A colourless flammable gaseous tertiary amine, $(CH_3)_3N$, with a fishy odour, prepared by catalytic reaction between methanol and ammonia. It is strongly alkaline and is used in the manufacture of disinfectants and synthetic resins. M.pt. –117°C; b.pt. 2.8°C; r.d. 0.6 (–5°C).

Trinitrobenzene (TNB). A yellow crystalline solid, $C_6H_3(NO_2)_3$, prepared from trinitrotoluence. The compound is 1, 3, 5-trinitrotobenzene; it is used as a high explosive, being more powerful than TNT. M.pt. 122°C; r.d. 1.69.

Trinitrotoluene (TNT). A pale yellow crystalline solid, $CH_3C_6H_2(NO_2)_3$, prepared by the nitration of toluence. The compound is 2, 4, 6-trinitrotoluene: it is used as a high explosive. M.pt. 81°C; r.d. 1.6.

Triode. A type of thermionic valve containing three electrodes—an anode, a cathode, and a grid. The triode differs from the diode by the interposition of the grid, usually made of a wire spiral or mesh, between the other two electrodes. A potential applied to this electrode is used to control the flow of current between anode, and cathode—small changes in the grid potential cause large changes in the electron current. The triode valve can be used as an amplifier. If an alternating signal is applied between the grid and the cathode, a larger signal of the same frequency is produced between the anode and the cathode. The valve has the disadvantage that there is a large grid-to-anode capacitance, an effect that is reduced in the tetrode valve.

Triol. A trihydric alcohol; an alcohol containing three hydroxyl groups per molecule.

Triolein (olein). A simple glyceride of oleic acid, $C_3H_5(OOCC_{17}H_{33})_3$, found in many natural fats and oils.

Triose. A simple sugar containing three carbon atoms per molecule.

Tripalmitin (palmitin). A simple glyceride of palmitic acid, $C_3H_5(OOCC_{15}H_{31})_3$, found in many natural fats and oils.

Triphenylmethane. An aromatic hydrocarbon, $HC(C_6H_5)_3$. Its derivatives, obtained by substitution in the benzene rings, constitute an important class of dyes.

Triple Bond. Covalent bond.

Triple Point. The point at which the solid, liquid, and gas phases of a pure substance can all co-exist in equilibrium. It corresponds to definite values of pressure and temperature for each substance. The triple point of water (273.16 kelvins at 101, 325 pascals) is used in the definition of the kelvin.

Trisaccharide. A sugar with molecules containing three monosaccharide units.

Tristearin (stearin). A simple glyceride of stearic acid, $C_3H_5(OOCC_{17}H_{35})_3$, found in natural fats.

Tritiated. Indicating that a chemical compound contains tritium atoms in place of hydrogen atoms, as in tritiated methane, CT_4.

Tritium. Symbol: T, ^{3}H. The hydrogen isotope with one proton and two neutrons in its atomic nucleus. It is a radioactive substance produced by irradiation of lithium in an atomic pile. Tritium is not known to occur naturally. It is used in radioactive tracer studies. A.N. 1; A.W. 3.016; half-life 12.5 years.

Triton. A nucleus of a tritium atom, containing two neutrons and one proton.

Trivalent (tervalent). Having a valency of three.

Trochoid. A curve that is the locus of a point on the radius of a circle as that circle rolls along a straight line. If the point is at the end of the radius, *i.e.* on the circumference, then the curve is a cycloid.

Troostite. A dispersion of cementite Fe_3C in ferrite (alpha-iron) produced in steels by tempering martensite below 500°C.

Tropics. All the land and water of the earth between the Tropic or Cancer and the Tropic of Capricorn. The entire zone receives the sun's rays more directly than areas in higher latitudes, and therefore the average annual temperature is higher and seasonal changes in temperature are less. However, because of factors other than latitude (*e.g.*, distance from the ocean, prevailing winds, elevation), several different climatic types are found, including rain forest, Steppe, Savanna, and Desert.

Tropism. Response of a plant or one of its parts, involving orientation toward (positive tropism) or away from negative tropism) one or more external stimuli. For example, plant roots grow toward gravity and moisture and away from light. Other tropistic stimuli are heat, electricity, and chemical agents. The term *taxis* is applied to similar involuntary movements in animals and motile unicellular plants.

Troposphere. The lowest part of the earth's atmosphere in which the temperature decreases with increasing height. It extends to a height of about 10 kilometres and is separated from the stratosphere by a boundary region, the *tropopause.*

Tropylium Ion. The positive ion $C_7H_7^+$, found in certain salts. It has a seven-membered symmetrical aromatic ring of carbon atoms.

Trouton's Rule. The principle that the ratio of the molar heat of vaporization of any liquid to its boiling point is a constant. This value, the *Trouton constant,* is about 88 joules per mole per kelvin. The rule is equivalent to the statement that the entropy of vaporization is constant. It is not always followed, especially by liquids such as water in which hydrogen bonding occurs between the molecules.

Troy Weight. A system of weights used for gems and precious metals.

Truffle. Edible, subterranean fungus (division Fungi) found chiefly in W Europe. Truffles are small, solid, fleshy Saprophytes that usually grow close to the roots of trees in woodlands. The several species range in colour from gray or brown to nearly black and have a piquant, aromatic flavor. The truffles of the forests of Perigord, France, have long been regarded as a delicacy, and their collection is an important industry. They are hunted with dogs with hogs, which are able to scent them out underground. Thus far, truffles have not been successfully cultivated.

Trypanosome. Microscopic one-celled organism that usually lives as a parasite in the bloodstream of a vertebrate. Most undergo part of their development in the digestive tracts of insects that transmit the parasite with their bite. Some cause serious diseases in humans and animals, *e.g.,* African sleeping sickness, caused by *Trypanosoma gambiense* and transmitted by the Tsetse Fly.

Tryptophan. A sweet white crystalline essential amino acid,

$C_{11}H_{12}N_2O_2$, obtained synthetically or by hydrolysis of protein and used as a dietary supplement. M.pt. 275-90°C.

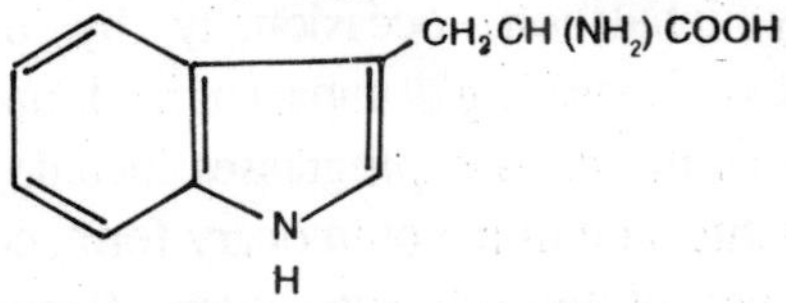

Fig. T-13. Tryptophan.

Tsetse Fly. Any of several blood-sucking African Flies. Several species transmit the Trypanosome that causes African sleeping sickness in humans.

Tsunami, Seismic Sea Wave or **Tidal Wave.** Series of catastrophic ocean waves generated by earthquakes, volcanic eruptions, or landslides beneath the sea. In the open ocean tsunamis may have wavelengths of up to several hundred miles but heights of less than 3 ft (1 m). Because this ratio is so large, tsunamis can go undetected until they approach shallow waters along a coast. Their height as they crash upon the shore mostly depends on the geometry of the submarine topography offshore, but they can be as high as 100 ft (30 m) and cause severe damage and loss of life.

Tuatara or **Tuatera.** Lizardlike Reptile (*Sphenodon punctatus),* last survivor of the order Rhynchocephalia, which flourished in the early Mesozoic era before the rise of Dinosaurs. Also called sphenodon, it lives on a few islands off New Zealand, where is is protected. The olive-coloured, yellow-speckled tuatara reaches 2 ft (60 cm) or more in length and has a spiny crest down its neck and back.

Tuber. Enlarged tip of a Rhizome (underground Stem) that stores food. Tubers, although modified, contain all the usual stem parts—Barks, Wood, pith and nodes.

Tuberculosis. Contagious disease caused by the bacterium *Mycobacterium tuberculosis,* identified by Robert Koch in 1882. Also known as TB and consumption, the disease primarily affects the lungs, although the intestines, joints, and other parts of the body may also become infected. It is spread mainly by inhalation, occasionally by ingestion through contaminated foods (*e.g.*, unpasteurized milk) and utensils. Symptoms as the disease progresses include fever, weakness, loss of appetite, and in the pulmonary form, cough and sputum. The incidence of tuberculosis—once affecting millions—has greatly decreased with improved sanitation, early detection through X-rays and skin tests, and anti-tuberculosis drugs.

Tulip. Hardy plant (genus *Tulipa*) of the Lily family, native from the Mediterranean to Japan and widely cultivated by the Dutch. Tulips, which are grown from Bulbs, have deep, cup-shaped flowers of many rich colours. Said to have been introduced into Europe from Turkey in 1554, they were objects of wild financial speculation in 17th cent. Holland.

Tumbleweed. Plant that breaks from its roots at maturity dries into a rounded tangle of branches, and rolls long distances with the wind, scattering seeds as it goes. Tumbleweeds are especially abundant in Prairie and Steppe regions. One of the most common is the Russian thistle (*Salsola pestifer*), not a Thistle but a member of the goosefoot family. Native to Asia, it is a pest on the prairies of the W U.S.

Tumor or **Neoplasm.** Tissue composed of cells that grow in an abnormal way; it may be benign or malignant. Normal tissue contains feedback controls that allow for tissue repair but do not permit expansion once a certain number of cells have developed. Tumor cells, lacking such feedback controls, proliferate and monopolize body nutrients. Benign tumors, which differ from normal tissue in structure, grow excessively and, although rarely fatal, may grow large enough to interfere with normal functioning and require surgical removal. Malignant tumors also grow excessively, but their cells lack the biological controls that normally keep cells specialized; these cells can infiltrate surrounding normal tissue and may later spread (metastasize) via Blood and the Lymphatic System to other sites. Surgery, Immunosuppressive Drugs, and radiation are primary treatments for malignant tumors.

Tuna or **Tunny.** Largest Fish of the Mackerel family. Hunted as game and food, the family includes the little tuna (10 lb/4.5 kg) of the Atlantic, the most important commercially; the bluefin tuna (*Thunnus thynnus*), the giant of bony fishes (200-500 lb/90-225 kg); and the Pacific albacore (up to 60 lb/27 kg), marketed as "whitemeat tuna."

Tungstate. Any of a number of oxy compounds of tungsten formed by dissolving tunsten trioxide in an alkali. The simplest tungstates have formula M_2WO_4 (where M is a monovalent metal) and contain WO_4^{2-} ions. Formally, they are salts of the hypothetical *tungstic acid,* H_2WO_4. If alkaline solutions of simple tungstates are acidified, *polytungstates* are produced in solution. They contain large anions of the type $(W_{12}O_{46})^{20-}$, and it is possible to obtain crystalline salts and acids (*polytungstic acids*) from such solutions.

Tungsten (wolfram). Symbol: W. A grey or white usually brittle transition element found in wolframite ($(Fe,Mn)WO_4$) and scheelite ($CaWO_4$). The ores are dissolved in sodium hydroxide and WO_3 is precipitated by acidification and reduced with hydrogen. Tungsten is used in alloys, especially high-speed steels, and in electric-lamp filaments. Its chemical properties are similar to those of molybdenum and, like molybdenum, it has a complicated chemistry, forming many compounds, particularly oxides and oxy compounds. A.N. 74; A.W. 183.85; m.pt. 3410°C; b.pt. 5927°C; r.d. 19.3; valency 2-6.

Tungsten (W) or **Wolfram.** Metallic element, first isolated in 1783 by the de Elhuyar brothers. It is a silver-white to steel-gray, very hard, ductile metal (one of the most dense) and has a higher melting point than any other metal. Tungsten is used for wires and for filaments for light bulbs and electronic tubes. Tungsten Steels are hard and strong at high temperatures. Tungsten carbide is used in place of diamond for dies and as an abrasive.

Tungsten Carbide. A hard grey powder, WC, obtained by direct combination of the elements at about 1600°. It is almost as hard as diamond and is used in making tools and in abrasives. M.pt. 2780°C; b.pt. 6000°C; r.d. 15.6.

Tungstic Acid. Tungstate.

Tunicate. Marine Chordate with a resemblance in the larval stage to Vertebrates. Familiar tunicates are the sea squirts, or ascidians, sedentary filter-feeders with cylindrical bodies, usually found attached to rocks. The tunic, or thick vest, for which they are named, is transparent or translucent and composed of Cellulose, a material extremely rare in the animal kingdom. Some tunicates have gelatinous containers called houses instead of tunics.

Tuning Fork. A metal fork with two prongs that vibrate to produce a pure tone of known pitch.

Tunnel. Underground passage, approximately horizontal, usually made without removing the overlying rock or soil. Methods of tunneling vary with the nature of the material to be cut through. In soft earth, the excavation is timbered for support as the work advances, the timbers sometimes being left as a permanent lining for the tunnel. Often two parallel excavations that will hold the side walls are constructed first, and arches connecting them are built as the material between them is extracted. Rock tunneling is accomplished with explosives and high-speed drilling machinery. Underwater tunneling through mud, quicksand, or permeable earth requires the use of a shield (devised and first used in 1825), a steel cylinder closed at its forward end, which holds rotating cutting blades. The cutting end is pushed ahead by hydraulic jacks, and excavated material is removed through openings in the shield face. River-crossing tunnels are also constructed by dredging a trench in the riverbed; then lowering prefabricated tunnel sections through the water into the trench, where divers connect them; and, finally, covering the trench and tunnel.

Tunnel Diode (esaki diode). A type of highly doped semiconductor diode that has a negative resistance over part of its operating range: *i.e.* the current flowering decreases with increasing voltage. This occurs under forward bias conditions and is caused by electrons tunnelling through the junction from the conduction band to a level of equal energy in the valence band. At higher forward-bias voltages the device behaves like a normal p-n junction. Tunnel diodes are used in ultra-high-frequency circuits and in switching circuits.

Tunnel Effect. The passage of particles through a potential barrier that they do not have sufficient energy to pass according to classical mechanics. The effect is explained by wave mechanics in terms of the probability of the particle moving through the barrier.

Turbine. Rotary engine that uses a continuous stream of fluid (gas or liquid) to turn a shaft that can drive machinery. In the *hydraulic turbines* used in hydroelectric power stations, falling water strikes a series of blades or buckets attached around a shaft, causing the shaft to rotate, this motion in turn being used to drive the rotor of an electric Generator. In a *steam turbine,* high-pressure steam forces the rotation of disks attached to a shaft. Steam turbines are used to drive most large electric generators and ship propellers. The term *gas turbine* is usually applied to a unit whose essential components are a Compressor, a combustion chamber, and a turbine. The turbine drives the compressor, which feeds high-pressure air into the combustion chamber; there it is mixed with a fuel and burned, providing high-pressure gases to drive the turbine. In a turboprop engine, the turbine is used to turn a propeller as well as the compressor. In a turbojet engine, the gases first drive the turbine and then are expelled from the engine to provide propulsion. Gas turbines are used mainly as aircraft engines.

Turbulent Flow. Flow of a fluid in which the motion is irregular and the velocity at any point can vary with time in magnitude and direction.

Turnip. Garden vegetable of the same genus (*Brassica*) of the Mustard family as the Cabbage, native to Europe. Grown for its edible green leaves (greens) and its nutritious rounded primary root, it is also used as stock feed. The two principal kinds are the white (*B. rapa*) and the yellow (*B. napobrassica*), also known as rutabaga or Swedish turnip.

Turpentine. An oily liquid extracted from pine resin. It consists chiefly of pinene, $C_{10}H_{16}$, and is used as a solvent.

Turpentine. Yellow to brown semifluid oleoresin exuded from the sapwood of pines, firs, and other confiers. It consists of an Essential Oil (oil of turpentine) and a type of resin called rosin. Commercial turpentine is oil of turpentine with the rosin removed. When pure, it is a colourless, transparent, oily liquid with a penetrating odour and characteristic taste. Turpentine is used chiefly as a solvent and drying agent in paints and Varnishes.

Turquoise. Hydrous aluminum and copper sulfate mineral [$Al_2(OH)_3PO_4$. H_2O + Cu]. Usually found in microscopic crystals, it is opaque with a waxy luster, varying in colour from greenish gray to (Gem-quality) sky blue. Because of their porosity, the gem varieties absorb dirt and grease, changing the coluor to an unattractive green; exposure to heat or sunlight can also harm the colour. The finest specimens are from Iran; other sources are the Sinai peninsula and the SW U.S.

Turtle. Reptile (order Chelonia) with an armorlike shell and strong, beaked, toothless jaws. Turtles are found throughout most of the temperate and tropical world. The land-living species are commonly called Tortoises; the name *terrapin* is generally applied to large freshwater or brackish water species. Turtles range in length from a few inches to over 7 ft (11 m); many specimens have lived over 50 years in captivity. The 200 to 300 species of turtle are classified in 12 families. The largest family of the Northern Hemisphere is the family of common freshwater turtles (Emydidae). Turtles are the oldest living group of reptiles, dating back to the time of the earliest Dinosaurs.

Tutsi or **Watutsi.** Cattle-raising people of central Africa; they are also known as Watusi or Batusi. An aristocratic people, they are a minority in both Rwanda and Burundi, countries that are former Tutsi kingdoms. The Tutsi, who probably originated in Ethiopia, long held the peasant Hutu in feudal subjugation. In the 1970s, despite much iintegration of Tutsi and Hutu culture, many members of both tribes died in bloody fighting in Burundi. The Tutsi are spectacularly tall, often over 7 ft (2.1 m) in height.

Tweeter. A loudspeaker for use at high audiofrequencies. *Compare* woofer.

Twilight. Period between sunset and total darkness or between total darkness and sunrise. Civil, nautical, and astronomical twilight occurs when the sun's center is between 0° and, respectively, 6°, 12°, and 18° below the horizon.

Twin. A crystal that has grown in two branches, both crystal lattices sharing a common plane so that one branch of the crystal is the mirror image of the other. The process of growing in this way is called *twinning.*

Tyndall Effect. The scattering of light by small particles in its path. The Tyndall effect causes a light beam to be visible in a dusty atmosphere, colloidal solution, etc. [After John Tyndall (1820-93), English physicist.]

Tyndall, John. 1820-93, English physicist, science lecturer, and writer; b. Ireland. He was professor (1853-87) and superintendent (1867-87) at the Royal Institution, London. He investigated light, sound, and radiant heat and studied Alpine glaciers. The scattering of light by Colloids, known as the Tyndall effect, is named for him.

Type. For Priniting, was invented in China. Related devices like seals and stamps for making impressions on clay were used in Babylon and elsewhere. Movable types made from metal molds were used in Korea a half-century before the European invention of movable type attributed to Johann Gutenberg. There is no evidence, however, that the European invention was not independent. The first dated European printing from movable type is a papal indulgence printed at Mainz, Germany (1454). Johann Fust, using Gutenberg's press, printed the first dated European book, a psalter (1457). The Mazarin Bible, completed on the press no later than 1455, however, is thought to be the first book printed in Europe. The type used in these beginnings was of the kind known as black letter or Gothic (*e.g.,* modern Old English or German), derived from popular handwriting styles. Roman type was used by several printers before Nicolas Jenson improved it so as to establish it as standard. Italic was first used by Aldus Manutius, who also introduced small capitals. Type characters are usually made by pouring metal into previously cut matrices, by photomechanic techniques, and, less frequently, by using plastic and other synthetic materials. Types may be handset or cast by machine. Famous type designers in addition those named above include John Baskerville, Giambattista Bodoni, William Caslon, Francois Ambroise Didot, Claude Garamond, Frederic William Goudy, Robert Granjon, William Morris, Bruce Rogers, and Geofroy Tory.

Type Metal. Any alloy of lead (60%), antimony (30%), and tin (10%). It expands when it solidifies, a property that makes it suitable for casting type and other intricate work.

Typewriter. Instrument for producing printed letters by manual operation. The first practical commercial typewriter was invented in the U.S. in 1867 by Christopher Sholes and his colleagues, and was manufactured by the gunsmith Philo Remington in 1874. This early model had only capital letters; a shift-key model appeared in 1878. The electric typewriter came into use c.1935. In the 1960s some machine designs replaced type levers with a type-surfaced metal ball that moves rapidly across a stationary paper holder. New computer-controlled typewriters can store the data to be typed and reproduce it automatically.

Typhoid Fever. Acute generalized infection caused by *Salmonella typhosa.* The main sources of infection are contaminated water or milk and food handlers who are carriers. Symptoms include high fever, rose-coloured spots on the abdomen and chest, and diarrhea or constipation. Complications, especially in untreated patients, may be numerous. The disease is treated with the Antibiotic chloramphenicol, typhoid Vaccination is a valuable oreventive measure.

Typhus. Any of a group of infectious diseases caused by rickettsias, microorganisms classified between bacteria and viruses. Symptoms include fever and the early onset of rash and headache. Typhus is treated with Antibiotics and can be prevented by Vaccination.

UHF. Ultra-high frequency.

Ultracentrifuge. A centrifuge designed to work at very high angular speeds, so that the centrifugal force produced is large enough to cause sedimetation of colloids. The rate at which the sedimentation occurs depends on the size of the particle: the apparatus can be used to investigate sizes of colloidal particles and molecular weights of proteins and other macromolecules.

Ulcer. Inflamed, open sore, usually slow to heal, on the skin or mucous membranes. It may develop as a result of injury, prolonged bed rest, vascular disease, or unknown reasons; therapy is directed at the underlying cause. *Peptic ulcer* occurs in the mucous membrane of the intestinal tract in areas accessible to acid secreted by the stomach. The exact cause is unknown, but increased acid secretion, tissue vulnerability to acid secretion, and emotional disturbance are believed to play a part. Antacids and drugs blocking gastric acid secretions are effective treatments.

Ultra-high Frequency (UHF). A frequency in the range 0.3 gigahertz to 3 gigahertz.

Ultramicroscope. A microscope in which colloidal particles can be observed by using dark-field illumination.

Ultrasonic. Above audible frequencies: *i.e.* above 20 kHz.

Ultrasonics. The study of the properties and applications of ultrasound.

Ultrasonics. The study and application of Sound waves with frequencies greater than 20,000 cycles per second, *i.e.,* beyond the range of human hearing. Ultrasounds are commonly produced by piezoelectric transducers. They are used for nondestructive testing, and for the cleaning of fine machine parts and surgical instruments. In medicine, Ultrasound devices are used to examine internal organs without surgery. Ultrasonic whistles are audible to dogs and are used to summon them.

Ultrasound. Propagated vibrations with a frequency above that of audible sound: *i.e.* above 20 kHz. Ultrasound has many applications including flaw detection of metals, medical diagnosis, depth finding, cleaning, dispersion of one liquid in another, and coagulation of some dispersions.

Ultrasound. In medicine, a technique that uses sound waves to study hard-to-reach body areas. In scanning with ultrasound, high-frequency sound waves are transmitted to the area of interest and the returning echoes recorded. First developed in World War II to locate submerged objects, the technique is now widely used in virtually every branch of Medicine, *e.g.,* in obstetrics to study the fetus, in cardiology to detect heart damage, in Ophthalmology to detect retinal problems. It is noninvasive, involves no radiation, and avoids the possible hazards—such as bleeding, infection, or reactions to chemicals—of other diagnostic methods.

Ultraviolet Radiation. Electromagnetic radiation with wavelengths between 13 and 397 nanometres. Ultraviolet radiation lies between light and X-rays in the electromagnetic spectrum. It is produced by emission from excited atoms and ions and can be generated by gas-discharge tubes. Large amounts are produced by the sun, although the atmosphere absorbs radiation with wavelengths below 200 nm.

Ultraviolet radiation causes ionization of matter and also induces certain chemical reactions. It can split some molecules into free radicals and initiate polymerization processes. In the human body it converts ergosterol into vitamin D and in plants it induces photosynthesis.

The ultraviolet region is arbitrarily divided into the *near ultraviolet* (397-200 nm) and the *far ultraviolet* (200-13 nm). Radiation with wavelengths shorter than about 200 nm lies in the vacuum ultraviolet.

Ultraviolet Radiation. Invisible Electromagnetic Radiation with frequencies (about 10^{15} to 10^{18} Hz) between that of visible violet light and X rays; it ranges in wavelength from about 400 to 4 nanometers. Ultraviolet (UV) radiation can be detected by the Fluorescence it induces in certain substances and by its blackening of photographic film. Most of the UV component of sunlight is absorbed by the Ozone layer of the atmosphere. UV radiation can also be produced artificially in arc lamps. Vitamin D in humans is produced by the action of UV radiation on ergosterol, a substance present in the human skin.

Ultraviolet Spectrum. An emission or absorption spectrum in the ultraviolet region. Ultraviolet emission occurs from gas discharges and is caused by electron transitions in atoms, molecules, and ions. It gives information on energy levels and molecular structure. Ultraviolet absorption is used in Rydberg spectroscopy. Most ultraviolet radiation is absorbed by glass so the radiation is dispersed by quartz prisms or by reflection gratings.

Umbra. 1. A region of complete shadow.

2. The central darker portion of a sunspot.

Umbriel. Uranus.

Uncertainty Principle. Heisenberg uncertainty principle.

Unconscious. In psychology, that aspect of mental life not subject to recall at will. Sigmund Freud regarded the unconscious as a vast portion of the mind, including the instinctual drives and the repressed residue of unacceptable experiences and desires. C.G. Jung added the concept of an inherited unconscious to the Freudian view. Although behaviourists reject the idea of the unconscious, most modern psychologists accept the concept to mean latent, or unretrieved, memories and ideas.

Uniaxial Crystal. A crystal with one optic axis.

Unicorn. Fabulous equine beast with a horn in the middle of its forehead. It was once considered native to India and was reportedly seen throughout the world. Pure whiite, it was been used as a symbol of virginity and, in Iconography, is associated with the Virgin Mary and with Christ. Hunting the unicorn was a Tapestry subject in the late Middle Ages and Renaissance.

Unidentified Flying Object (UFO) or **Flying Saucer.** An object or light phenomenon reportedly seen in the sky whose appearance, trajectory, and general dynamic and luminescent behavior do not suggest a logical, conventional explanation. Although many alleged sightings have been interpreted as reflections of the sun's rays from airplanes, as weather balloons, or as various meteorological phenomena, some sightings remain unexplained by investigators in terms of known phenomena.

Unified-field Theory. A theory that seeks to explain gravitational and electromagnetic interactions and the strong and weak nuclear interactions in terms of a single set of equations. The approach is an extension of the general theory of relativity in which the gravitational field arises as a consquence of the curvature of space-time. A unified-field theory would involve discovering a geometry of space-time from which the electromagnetic and nuclear interactions also arise naturally.

Uniformitarianism. The doctrine that past geological changes in the earth were brought about by the same causes as those now taking place. As first advanced in 1785 by James Hutton, it stressed the slowness and gradualness of rates of change. It was initially overshadowed by the doctrine of catastrophism and was opposed because it seemed contrary to religious beliefs. In the 19th cent. it gained support through the efforts of Sir Charles Lyell.

Unimolecular. Relating to one molecule. A *unimolecular layer* is an adsorbed layer one molecule (or atom) thick. A *unimolecular reaction* is one in which single molecules of reactant undergo a change: *i.e.* the transition state does not involve more than one molecule. Such reactions must be distinguished from first-order reactions.

Unit. A magnitude of a physical quantity used for measurements of other magnitudes of the physical quantity by comparison. In defining a unit, a standard is adopted which, by agreement, is taken to have a specified value. For example, a particular piece of platinum has been adopted as the SI unit of mass—the kilogram. This is an example of a *primary standard* of a unit: *i.e.* the standard that defines the unit. Such a standard does not have to be an actual object: the metre for example is defined in terms of the wavelength of light from a particular spectral transition. Standards of this type, based on atomic constants or on the properties of substances, are preferable as they cannot change with time and can be reproduced in different places. Any measurement is effectively a comparison of the object or system measured with the standard. Masses, for example, can be compared by weighing with a balance. Measurements are usually not made against the primary standard but against *secondary standards* of predetermined value.

Units defined by reference to primary standards are *base units*, and other units—*derived units*—can be defined by reference to these base units. For example, the SI units of length and mass—the metre and the kilogram—are base units, whereas the unit of density—the kilogram per cubic metre—is a derived unit. In making physical measurements, systems of units are employed, determined by a choice of certain base units. Imperial, CGS, and MKS units are examples. These have now been replaced for most scientific work by SI units.

Unitary Symmetry. A relationship between sets of elementary particles whereby they show a special symmetry between their isospin quantum numbers and their hypercharge. These quantum numbers are related by the operations of a certain group (called SU3). Using this idea, particles can be classified into multiplets that can be thought of as different states of the same particle. All particles in the same multiplet have the samc parity and spin. The relationship between particles in a multiplet is shown on a graph. Unitary-symmetry theory was used to predict the existence of the omega-minus particle. The extension of unitary-symmetry theory to a larger symmetry group (SU4) leads to the concept of charm.

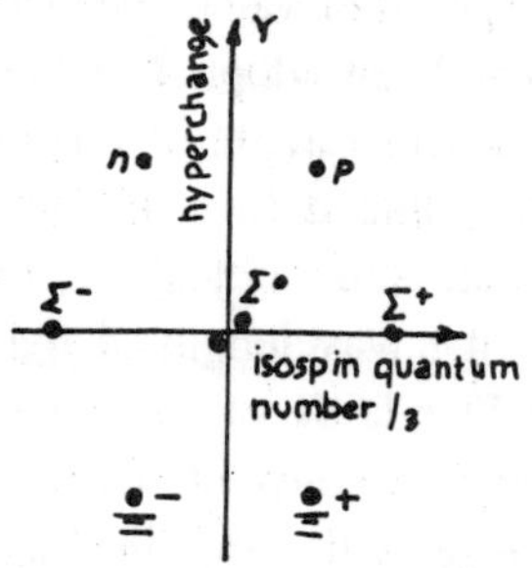

Fig. U-1. Octet of baryons.

Unit Cell. The smallest repeated unit in a crystal lattice. In sodium chloride, for example, the unit cell is a cube with eight ions at its corners.

Unit Pole. A magnetic pole of unit strength. The unit pole in electromagnetic units is defined as the pole that repels an identical magnetic pole with a force of one dyne when placed one centimetre away in vacuum.

Unit Vector. A vector with unit magnitude and specified direction.

Univalent. Monovalent.

Universal Gas Constant. Gas constant.

Universe. The whole of space and all the matter (galaxies, etc.) contained in it. The age of the universe is estimated as about $5 \times 10^9 - 50 \times 10^9$ years and it is thought to contain 10^{41} kilograms of matter. Its origin is still under discussion.

Unnilhexium (Unh). Artificial radioactive Transuranium Element. Claims for its production have been made by a Soviet group at Dubna, who in 1974 bombarded lead with chromium ions to obtain the Unh^{259} and Unh^{260} isotopes, and groups in California at the Lawrence Berkeley and Lawrence Livermore Laboratories, who, also in 1974, bombarded californium with oxygen ions to obtain the Unh^{263} isotope.

Unnilpentium (Unp). Artificial radioactive Transuranium Element. Claims for its production have been made by a Soviet groupt at Dubna, who in 1967 bombarded americium with neon to obtain apparently the Unp^{360} and unp^{261} isotopes, and by an American group at the Univ. of California, Berkeley, who in 1970 bombarded californium with nitrogen nuclei to obtain the Unp^{260} isotope. The names *nielsbohrium* and *hahnium* were originally proposed, respectively, by the Dubna and Berkeley groups.

Unnilquadium (Unq). Artificial radiactive Transuranium Element. Claims for its production have been made by a Soviet group in Dubna, who in 1964, bombarded plutonium with neon ions to obtain apparently the Unq^{260} isotope, and by an American group at the Univ. of California, Berkeley, who in 1969 bombarded californium with carbon nuclei to obtain the Unq^{257} and unq^{259}, and possibly the Unq^{258}, isotopes. The names *kurchatovium* and *rutherfordium* were originally proposed, respectively, by the Dubna and Berkeley groups.

Unsaturated. Denoting a chemical compound that is not saturated: *i.e.* one that contains double or triple bonds. Unsaturated compounds, such as olefines and acetylenes, tend to form compounds by addition rather than by substitution unless they are aromatic compounds.

Unstable. **1.** Denoting a chemical compound that is not stable.

2. Equilibrium.

Upper Atmosphere. The atmosphere from a height of about 30 kilometres upwards: *i.e.* that part investigated by rockets and artificial satellites rather than by balloons.

Urals or **Ural Mountains.** Mountain system extending north-south c. 1,500 mi (2,400 km) across the W USSR, part of the traditional border between Europe and Asia. Narodnaya (6,212 ft/1,893 m), in the north, is the highest point. Except in the polar and northern sections, the mountains are densely forested, wth lumbering an important industry. They are also enormously rich in minerals, including coal, iron ore, aluminum, copper, manganese, potash, and oil. Huge industrial centers, part of the Urals industrial area (c. 290,000 sq mi/751,500 sq km), are located at Sverdlovsk, Magnitogorsk, Chelyabinsk, Nizhni Tagil, and Perm. During World War II entire industries were transferred to the Urals from the W USSR.

Uranium (U). Radioactive metallic element, discovered in oxide form in Pitchblende by M.H. Klaproth in 1789. A silver-white, hard, dense, malleable, ductile, highly reactive metal in the Actinide Series, it occurs naturally as a mixture of three Isotopes. Because of a constant decay rate, the age of urnium samples can be estimated. The rare uranium-235 isotope is the only naturally occurring fission fuel for Nuclear Energy. Breeder reactors convert the abundant but nonfissionable uranium-238 into fissionable plutonium-239. Uranium-235 and plutonium-239 are also practicable fissionable nuclei for Atomic Bombs.

Uranium. Symbol: U. A heavy ductile malleable silvery metallic element belonging to the actinides, found in several ores, notably pitchblende (UO_2). The element is a mixture of three isotopes: ^{234}U, ^{235}U (0.7%), and ^{238}U (99%). The most stable isotope, ^{238}U, has a half-life of 4.5 x 10^9 years and is used in breeder reactors as a source of the plutonium isotope, ^{239}Pu. Uranium enriched with ^{235}U (half-life 7.1 x 10^8 years) is used as a fuel in fission reactors. A.N. 92; A.W. 238.03; m.pt. 1150°C; r.d. 18.68.

Uranium Dioxide. Uranium oxide.

Uranium Hexafluoride. A volatile white crystalline solid, UF_6, obtained by successive hydrofluorination and fluorination of uranium dioxide. It is used in gas-diffusion processes for separating the isotopes of uranium. M.pt. 64.5°C; r.d. 5.1.

Uranium Oxide. Any of several oxides of uranium. The most important is *uranium IV oxide* (uranium dioxide), a black crystalline solid, UO_2, made by oxidation of uranium ore to *uranium VI oxide* (UO_3) followed by reduction with hydrogen. It is used to pack nuclear fuel rods. M.pt. 3000°C; r.d. 10.9.

Uranium Series. A radioactive series beginning with uranium-238 and ending with lead-82.

Uranus. The seventh planet in order of succession outwards from the sun; the third of the giant planets with a dense atmosphere of hydrogen and methane. Orbiting the sun once in 84.01 years at a mean distance of 2870 million kilometres, the planet has a diameter of 47 100 km, a mass 14.5 times that of the earth, and an average relative density of 1.7. Its period of axial rotation is about 10 hours 45 minutes and its temperature is probably as low as –200°C. One of its most interesting features is that, unlike any other planet, the inclination of its equator to the plane of its orbit is 98°, giving earthbound observers a view of the planet as if it were lying on its side: its lines of latitude cross the disc vertically. Uranus has five satellites: *Oberon, Titania, Ariel, Umbriel,* and *Miranda.*

Uranus. In astronomy, 7th Planet from the sun, at a mean distance of 1.7832 billion mi (2.8696 billion km). It has a diameter of 32,490 mi (52,290 km) and is a gaseous planet with an atmosphere composed mostly of hydrogen, helium, methane, and ammonia. Detcted on Mar. 13, 1781, by Sir William Herschel, it was the first planet discovered in modern times with the aid of a telescope. The extreme (98°) inclination of Uranus's equatorial plane with respect to its orbital plane gives the planet a retrograde rotation, *i.e.,* a rotation opposite to the direction of revolution. Uranus has five known natural satellites: *Titania,* the largest, with a diameter of 650 mi (1,040 km), and *Oberon* were discovered by Herschel in 1787; *Ariel* and *Umbriel,* by William Lassell in 1851; and *Miranda,* by Gerard Kuiper in 1948. On Mar. 10, 1977, astronomers detected a system of narrow rings (now estimated to be nine in number) of small, dark particles orbiting around the planet. The *Voyager* 2 space Probe is scheduled to fly byUranus in 1986.

Uranyl. Containing the ion UO_2^{2+} or the divalent group UO_2: The uranyl ion is a greenish-yellow colour and is present in such salts as uranyl nitrate, $UO_2(NO_3)_2$.

Urea (carbamide). A white powder, $CO(NH_2)_2$, manufactured by heating ammonia with carbon dioxide at high pressure and allowing the ammonium carbamate ($NH_4CO_2NH_2$) formed to decompose. It is present in urine. Urea is used as fertilizer, in animal feeds, and in urea-formaldehyde resins. M.pt. 133°C; r.d. 1.3.

Urea. Organic compound, formula $CO(NH_2)_2$, that is the principal end product of nitrogen metabolism in most mammals. Most of the nitrogen in protein eventually appears in urea, which is concentrated by the Kidneys and excreted in urine.

Urea-formaldehyde Resin. A synthetic aminoresin made by reacting urea with formaldehyde. The process yields a water-soluble syrup which is used as an adhesive. The syrup can be dried to a powder which, when mixed with filler, is used in moulding a strong hard white thermosetting plastic.

Urethane. A white crystalline solid, $CO(NH_2)(OC_2H_5)$, made by heating ethanol with urea nitrate. M.pt. 49°C; b.pt. 180°C; r.d. 0.9.

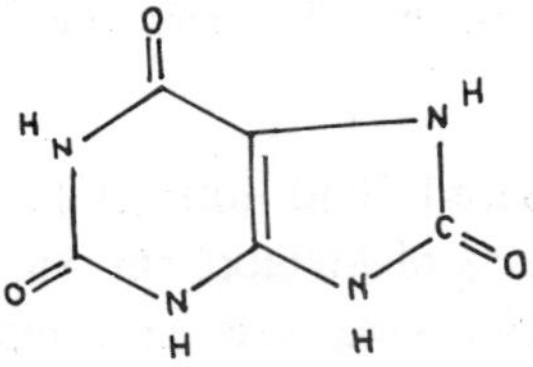

Fig. U-2. Uric acid.

Uric Acid. A colourless odourless tasteless powder, $C_5H_4N_4O_3$, occurring in small quantities in urine. Decomposes on heating; r.d. 1.87.

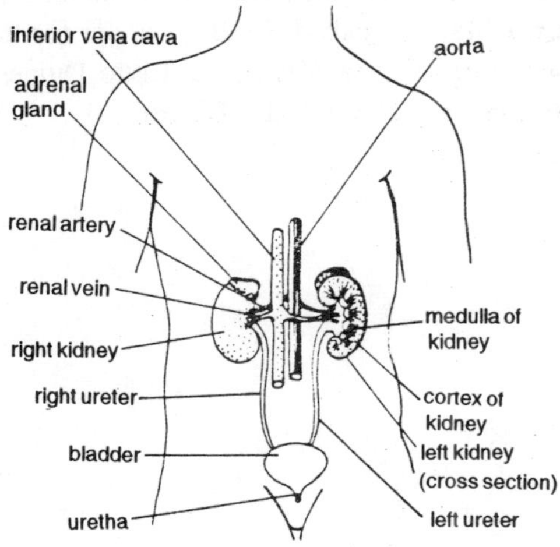

Fig. U-3. Urinary system.

Urinary System. Group of organs of the body concerned with excretion of urine. In humans, the urinary system includes the kidneys, ureters, bladder, and urethra. Blood containing waste products enters the Kidneys through the renal artery. The kidneys purify the blood by filtering out waste products, which, together with water, form Urine. From the kidneys, urine passes through thick-walled tubes called ureters into the bladder, a muscular sac that temporarily stores the urine and contracts to expel it. Urine released from the bladder flows into the urethra, the canal carrying it to the outside of the body.

Urine. Clear, amber-coloured fluid formed by the Kidneys and carrying waste products of Metabolism out of the body. Urine is 95% water, in which urea, uric acid, mineral salts, toxins, and other waste products are dissolved. It may also contain ordinary substances used by the body but excreted by the kidneys when excessive amounts are present in the bloodstream. Analysis of urine is important in detecting diseases.

Ursa Major and **Ursa Minor.** (Lat., = the great bear; the little bear], two conspicuous northern Constellations. In the U.S. part of Ursa Major is called the Big Dipper (or the Drinking Gourd) and part of Ursa Minor, the Little Dipper. Polaris is at the extreme end of the Little Dipper.

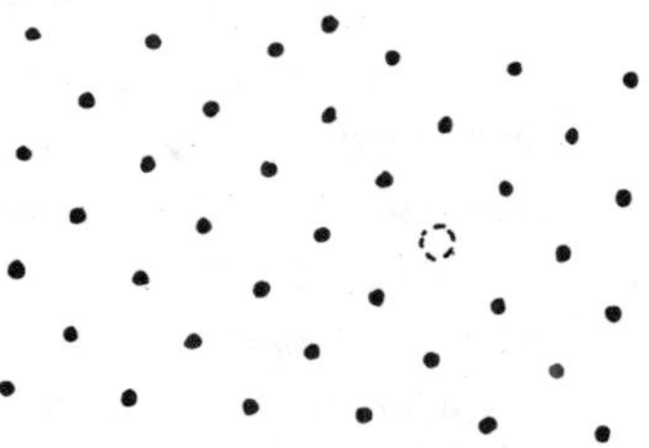

Fig. V-1. Vacancy.

Vacancy (Schottky defect). A type of defect in crystals in which position that is normally occupied by an atom, ion, or molecule is unoccupied.

Vaccination. Inoculation of a substance (vaccine) into the body for the purpose of producing active Immunity against a disease. The vaccine is usually a weakened culture of the agent causing the disease. Vaccination was used in ancient times in China, India, and Persia, and was introduced to the West in the late 18th cent. by E. Jenner. Vaccinations are used today to prevent such diseases as Smallpox, Diphtheria, Poliomyelitis, Rabies, and Typhoid.

Vacuum. Theoretically, space devoid of matter. A perfect vacuum has never been obtained; the best man-made vacuums contain less than 10^5 gas molecules per cc, compared to about 30 x 10^{18} molecules per cc for air at sea level. In intergalactic space, the regions with the highest vacuum are estimated to contain less than one molecule per cc. Several kinds of *vacuum* *pumps* have been devised, *e.g.*, the ion pump, which ionizes gas molecules and draws them to a charged collector, and the cryogenic pump, which condenses gas molecules on an extremely cold surface of a container.

Vacuum. A region containing gas at a pressure below atmospheric pressure. A vacuum system is a sealed apparatus evacuated by a pump. The ultimate pressure achieved depends on the pumping speed and on the presence of adsorbed gas, volatile materials, leaks, and other source of gas in the system. Grades of vacuum are classified into *soft* (or *low*) vacuum (down to about 10^{-4} mmHg) and *hard* (or *high*) vacuum (below 10^{-4} mmHg). An *ultra-high vacuum* has a pressure below about 10^{-9} mmHg. Vacuum technology is important in the manufacture of light bulbs, thermionic valves, cathode-ray tubes, and similar devices and is also used in industrial processes such as freeze drying and vacuum evaporation. Vacuum systems are also necessary in many branches of scientific research

Vacuum Distillation. Distillation of liquids under reduced pressure, thus lowering the boiling point of the liquid and allowing a lower temperature to be used. The technique is valuable for distilling substances that would decompose or explode at higher temperatures.

Vacuum Evaporation. A process for depositing thin films of solid on a surface by evaporating the material at high temperature in a vacuum. Under these conditions atoms leave the hot material and travel directly to the cool surface without colliding with other molecules in the gas. They condense on the cool surface, gradually building up a layer of solid.

Vacuum Flask. *See* Dewar flask.

Vacuum ultraviolet. The part of the ultraviolet region of the spectrum in which the radiation is absorbed by air. Experiments using this radiation have to be performed in evacuated apparatus. The region includes ultraviolet radiation with wavelengths less than about 200 nm.

Valence or **Oxidation State.** Combining capacity of an Atom expressed as the number of single bonds the atom can form or the number of electrons an Element gives up or accepts when reacting to form a compound. The valence of an atom is determined by the number of electrons in the outermost, of valence, electron shell. An atom exists in its most stable configuration when its outermost shell is completely filled; in combining with other atoms, it thus tends to gain or lose valence electrons in order to attain a stable configuration. The valence of many elements is determined from their ability to combine with hydrogen or to replace it in compounds.

Valence Band. The energy band of a solid that is occupied by the valence electrons of the atoms forming the solid.

Valence Bond. A single linkage between two atoms in a molecule, considered to be a shared pair of electrons between the atoms. The valence-bond approach to the theory of chemical bonding is based on the idea that a molecule is held together by definite linkages between pairs of atoms. In the alternative approach the molecule is thought of as a collection of atomic nuclei with the electrons moving in the resultant field of these nuclei.

Valence Electron (valency electron). An electron in the outer shell of an atom that participates in the chemical bonding when the atom forms compounds.

Valency (valence). The combining power of an element, atom, ion, or radical, equal to the number of hydrogen atoms that the atom, ion, etc., could combine with or displace in forming a compound. For instance, carbon has a valency of four, as displayed in methane, CH_4. Many elements exhibit more than one valency; phosphorus has a valency of three in the trichloride, PCl_3, and five in the pentachloride, PCl_5. The valency of an ion is equal to its charge: the ammonium ion, NH_4^+, has a valency of one. The valency of a radical is the number of hydrogen atoms with which it is capable of combining: the methyl radical, CH_3, has a valency of one. In general, the valency of a chemical entity is the number of single covalent or electrovalent bonds that it can make.

Valeric Acid. *See* pentanoic acid.

Valine. A white crystalline essential amino acid. $(CH_3)_2CHCH(NH_2)COOH$, occurring in proteins. It is used as a dietary supplement.

Valve. An electron tube for controlling a flow of current by motion of electrons between electrodes. Valves are either evacuated or gas-filled.

Valve Voltmeter. A type of voltmeter in which the low input voltage, which may be either direct or alternating, is amplified by a valve circuit. The output current is proportional to the input voltage and is registered by a meter calibrated in millivolts. The instrument draws very little current because of its high input impedance.

Vampire. In folklore, animated corpse that sucks the blood of humans, usually enslaving its victims and making them vampires. Vampires could be warded off with charms and killed by a stake driven through the heart. Literature's most famous vampire is Count Dracula, in Bram Stoker's *Dracula* (1897).

Vanadate. Any of a large number of oxy compounds of vanadium in its + 5 oxidation state. Simple compounds with formulas of the type M_3VO_3 *(orthovanadates)*, MVO_3 *(metavanadates)*, and $M_4V_2O_7$ *(pyrovanadates)* can be prepared. They are formally salts of the hypothetical acids *orthovanadic acid*, H_3VO_4, *metavanadic acid*, HVO_3, and *pyrovanadic acid*, $H_4V_2O_7$. However, the behaviour of vanadates is quite complicated. When vanadium V oxide, V_2O_5, is dissolved in alkali, the solution contains simple orthovanadate ions VO_4^{3-} and more complicated polymeric anions. The $[V_3O_9]^{3-}$ ion, for example, has a ring containing three vanadium atoms and three oxygen atoms. A large number of different polymeric vanadates *(polyvanadates)* can be precipitated from such solutions. Polyvanadates are also formed by dissolving V_2O_5 in acid solutions: they contain ions of the type $V_{10}O_{25}^{6-}$. Any vanadate can be regarded as derived from a corresponding *vanadic* or *polyvanadic acid.*

Vanadic Acid. *See* vanadate.

Vanadinite. A red, orange, brown, or yellow mineral, $Pb_5Cl(VO_4)_3$, used as an ore of lead and vanadium.

Vanadium. Symbol: V. A silvery-white ductile transition element found in many minerals, especially patronite (VS_4, carnotite $[K_2(ClO_2)_2 (VO_4)_2{\cdot}3H_2O]$, and vanadinite $[Pb_5Cl(VO_4)_3]$. The pure metal can be obtained by the van Arkel-de Boer process or by reduction of the oxide with calcium. Usually it is produced alloyed with iron for use in steels.

Chemically the element resists corrosion, although it combines with oxygen and other nonmetals at high temperature. It does not dissolve in alkalis or nonoxidizing acids. Vanadium forms compounds with several valencies, the higher oxidation states being more important. Vanadium IV compounds tend to be covalently bonded: in particular they include the vanadyl compounds and complexes. Vanadium V compounds are also covalent, as in the molecular halides and oxyhalides and the vanadates. Numbers of vanadium II and III complexes are also known, including the hydrated cations $[V(H_2O)_6]^{3+}$ and $[V(H_2O)_6]^{2+}$. A.N. 23; A.W. 50.94; m.pt. 1890°C; b.pt. 3380°C; r.d. 6.1; valency 2-5.

Vanadium (V). Metallic element, discovered in 1801 by A.M. del Rio. It is a soft, silver-gray metal that adds strength, toughness, and heat-resistance to Steel alloys. Vanadium compounds, especially the pentoxide, are used in ceramics, glass, and dyes and are important as catalysts in the chemical industry.

Vanadium Oxide. Any of three oxides, *vanadium III oxide*, V_2O_4, *vanadium IV oxide*, V_2O_4 or *vanadium V oxide* (vanadium pentoxide), V_2O_3. Vanadium V oxide is a yellow to brown crystalline solid, extracted from vanadium minerals. It is a starting material for other vanadium compounds and a widely used catalyst, particularly for the oxidation of sulphur dioxide. M.pt. 690°C; r.d. 3.4.

Vanadyl. Containing the ion VO^{2+} *(vanadyl ion)* or the divalent group = VO *(vandyl group).* The vanadyl ion is a bright blue colour.

Van Allen belts (radiation belts). Two belts of charged particles surrounding the earth in the region of the equator at heights of 2000-5000 and 13 000-19 000 kilometres. They are mainly protons and electrons, originating in the solar wind and in cosmic rays, trapped by the earth's magnetic field. [After James Alfred Van Allen (b. 1914), U.S. physicist.]

Van Allen Radiation Belts. Two belts of Radiation outside the earth's atmosphere, extending from c. 400 to c.40,000 mi (c. 650 to c. 650,000 km). The region of the belts is called the magnetosphere. The high-energy protons and electrons, of which the belts are composed, circulate along the earth's magnetic lines of force. These particles are probably emitted by the sun in its periodic solar flares and, after traveling across space, are captured by the earth's magnetic field. They are probably responsible for the Aurora seen about both polar regions, where the belts dip into the upper regions of the atmosphere. The belts were discovered by detectors (developed by American physicist James Van Allen and colleagues) aboard *Explorer 1,* the first U.S. artificial satellite.

Van Arkel-de Boer Process. A process for obtaining pure specimens of certain elements by forming a volatile iodide and then decomposing the vapour on an incandescent tungsten filament.

Van de Graaff generator. A type of electrostatic generator in which charge is accumulated on a metal sphere from an endless belt of fabric. The belt is mounted vertically with its top end inside a hollow metal sphere and charge is applied to the belt from points at the bottom, which are connected to an external source of high potential. This charge is collected by other points attached to the sphere. The device is capable of developing potentials of millions of volts. It is often used as a high voltage source for a linear accelerator *(Van de Graaff accelerator)*. [After R.J. van de Graaff (1901-67), U.S. physicist.]

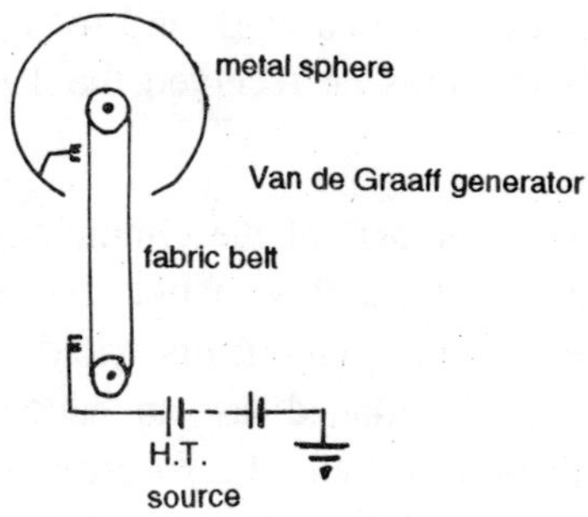

Fig. V-2. Van de Graaff generator.

Van der Waals Equation. A modified form of the ideal gas law: $(P + a/v^2)(v-b) = RT$, where a and b are constants for a particular gas. The equation was designed to apply to real gases. The term a/v^2 corrects for a reduction in pressure caused by attraction between molecules and the b term is a correction for the size of the molecules, being equal to about four times their actual volume. [After J.D. van der Waals (1837-1923), Dutch physicist.]

Van der Waals Forces. The attractive forces existing between molecules. Van der Waals forces are responsible for the cohesion of molecules in liquids and nonionic crystals. They act over short distances, being proportional to the seventh power of the distance, and are responsible for the correction to the pressure in van der Waals' equation. They are caused by attraction between dipoles and, in nonpolar molecules, by dispersion forces.

Van der Waals, Johannes Diderik. (Van der vals), 1837-1923, Dutch physicist. His theory of corresponding states (1880) contained an equation of state (now named for him) for homogeneous substances in terms of pressure, volume, and temperature (see Gas Laws); unlike the ideal gas law, his equation contains constant factors (different for each real substance) to account of the fact that molecules are of finite size and experience weak forces of mutual attraction (van der Waals forces). For this work and for the discovery of the law of binary mixtures he received the 1910 Nobel Prize in physics.

Vanilla. Vine (genus *Vanilla)* of the Orchid family, native to hot damp climates of Central America but now cultivated in other tropical regions. The fruits yield vanilla, a popular flavoring usually marketed as an alcoholic extract. The commercial plant is usually *V. fragrans;* the source of the flavor is an aromatic essence, vanillin. Vanilla flavoring, now usually artificially synthesized, is also obtained from the tonka bean, which contains coumarin, a substance with a vanillalike aroma.

Van't Hoff Factor. Symbol: *i*. The ratio of the total number of entities (ions, molecules, etc.) of solute in a solution to the number that would be present if the solute consisted of undissociated molecules. The factor is used in the equation for osmotic pressure when the solute is an electrolyte: $\Pi V = iRT$. It reflects the fact that osmotic pressure depends on the number of entities present in solution. [After J.H. van't Hoff (1852-1911), Dutch physical chemist.]

Van't Hoff's Isochore. A relation between the equilibrium constant*(K)* of a reaction and the temperature *(T):*

$$\frac{d \log K}{d\,T} = \frac{\Delta H}{RT}$$

where H is the enthalpy of the reaction.

Van't Hoff, Jacobus Hendricus. (Vant hof), 1852-1911, Dutch physical chemist. His studies in molecular structure laid the foundation of stereochemistry. For his work in chemical dynamics and osmotic electrical conductivity he won (1901) the first Nobel Prize in chemistry.

Van't Hoff's Law. The principle that the osmotic pressure of solution is equal to the pressure that the solute world exert if it were an ideal gas at the same temperature as the solution and occupying the same volume. Thus the osmotic pressure (Π) is given by the equation $\Pi V = RT$.

Vapour. A substance in the gaseous state at a temperature below its critical temperature, so that it could be liquefied by pressure alone.

Vapour Density. The density of a vapour under specified conditions. The vapour density is often taken as a relative density—the ratio of the weight of a given volume of a gas or vapour to the weight of the same volume of hydrogen measured under the same conditions. The vapour density of a substance is equal to one half of its molecular weight and measurement of vapour density is a method of determining molecular weights. The principle techniques are Victor Meyer's method and Dumas' method.

Vapour Pressure. The pressure of vapour produced by a solid or liquid. In particular, the term is used for the *saturated vapour pressure,* which is the pressure of vapour in equilibrium with the solid or liquid. Thus, if a solid is placed in an enclosed space the pressure of its vapour increases until a steady state is reached in which the number of molecules leaving the solid is equal to the number returning to the solid from the gas phase. The pressure at which this occurs is the saturated vapour pressure. It depends on the temperature.

Variable. A quantity that can take a range of values in a mathematical expression. A *dependent variable* is one whose values are determined by other *independent variables.* For example, in $y = 7x^3 + 2x$, y is the dependent variable and x the independent variable.

Variable Star. Star that varies, either periodically or irregularly, in the intensity of its light. Variable stars are grouped into three broad classes: the pulsating variables, the eruptive variables, and the eclipsing variables. In *pulsating variables*, which account for more than half of the known variable stars, slight instabilities cause the star alternatively to expand and to contract, resulting in changes in absolute luminosity and temperature. The pulsating variables can be subdivided into short-term, long-term, semi-regular, and irregular variables. *Short-term variables* have periods ranging from less than one day to more than 50 days among them are the important but rare Cepheid Variables. Commoner short-term variables are the RR Lyrae stars; about 2,500 of this type are known in our galaxy. They have periods of less than one day, and all have roughly the same intrinsic brightness, which makes them another useful distance indicator. The *long-term variables* are the most numerous of the pulsating stars. They are red giant and supergiant stars with periods ranging from 80 to 300 days. *Semiregular variables* are stars whose periodic variations are occasionally interrupted by sudden bursts of light. *Irregular variables* show no periodicity in their variations in brightness. The *eruptive variables* are highly unstable stars that suddenly and unpredictably increase in brightness. *T. Tauri stars* are the least violent of these explosive stars. *Novas* are small, very hot stars that suddenly increase thousands of times in luminosity; their decline in luminosity is much slower, taking months or years. Supernovas, upon exploding, increase millions of times in brightness and are totally destroyed. *Eclipsing variables* are not true (intrinsic) variables but rather are Binary Star systems.

Variac. A device for producing a variable alternating voltage from a fixed input voltage. It is essentially a toroidal winding on a core, tapped by a moving brush.

Variance. The square of the mean deviation of a statistic.

Variation. *See* declination.

Varicose Vein. Superficial vessel that is abnormally twisted, lengthened, or dilated, seen most often on the legs and thighs and usually attributed to inefficient valves within the vein. Conditions such as Pregnancy or obesity, or increased pressure from prolonged standing, reduce the support of tissues surrounding the veins, causing them to dilate. Treatment includes use of support hosiery and, if necessary, surgical removal.

Variometer. A variable inductor, usually consisting of two coils connected in series and arranged so that one can be rotated inside the other, thus changing the self-inductance of the pair of coils.

Varistor. A resistor, usually of semiconductor material, that does not obey Ohm's law: *i.e.,* the current is not directly proportional to the voltage.

Varnish. Solution of gum or of natural or synthetic Resins that dries to a thin, hard, usually glossy film; it may be transparent, translucent, or tinted. Oil varnishes are made from hard gum or resin dissolved in oil. Spirit varnishes, *e.g.,* Shellac, are usually made of soft gums or resins dissolved in a volatile solvent. Enamel is varnish with added Pigments. Lacquer may be either a synthetic or natural varnish. As a decorative or protective coating, varnish has been used for oil paintings, for string instruments, and, by the ancient Egyptians, for mummy cases.

Varve. In geology, pair of thin sedimentary layers formed annually by seasonal climatic changes. Usually found in glacial lake deposits, varves consist of a coarse-gained, light-coloured summer layer and a finer-gained, dark-coloured winter layer. Varves, and the pollen they contain, are useful for interpreting recent climatic history.

Vasectomy. Minor surgical procedure for sterilization of the male, involving excision of the vas deferens, the thin duct that carries sperm cells away from the testicles. It is performed as a permanent method of Birth Control.

Vaseline. A colourless or pale yellow semisolid containing a mixture of petroleum hydrocarbons, used as an inert protective coating, particularly in pharmacy.

Vat Dyes. Any of a class of dyes that can be reduced to a soluble leuco form and then oxidizéd to the coloured dye after application to the material. Indigo is an example of a vat dye.

Vault. Curved ceiling over a room. It is generally composed of separate units of a material such as bricks, tiles or blocks, so shaped that when assembled their weight can be concentrated on the proper supports. (Vaults may also be formed of homogeneous material, *e.g.,* concrete.) These separate units exert not only the downward pressure of their weight, but also side thrusts that must be met with resistance in the form of thickened walls or of Buttresses, as in Romanesque and Gothic architecture. Vaults may also be designed so that their thrusts oppose or counteract. The Egyptians used brick vaulting for drains. The Chaldeans and Assyrians appear to have made use of high domes and barrel vaults. The Greeks used no vaults. Etruscan technique was absorbed by the Romans, who, using concrete, developed a mature system. Roman vaults were perfectly rigid, devoid of external thrust, and required no buttresses. Thus they could be placed over vast spaces. Medieval systems developed from Roman vaulting. The tunnel, or barrel vault, spans two walls like a continuous arch. The cross, or groined vault, at the intersection of a two barrel vaults, forms four arched opening. Ribs to strengthen the groins and sides of vault appeared in the 11th cent., and became gradually the organic supporting skeleton of Gothic architecture. The pointed Arch was used a vaulting oblong compartments. In England these developments culminated in the 15th cent. in the perpendicular style. Renaissance architects returned to the basic Roman forms. Modern vaulting is largely accomplished with reinforced concrete.

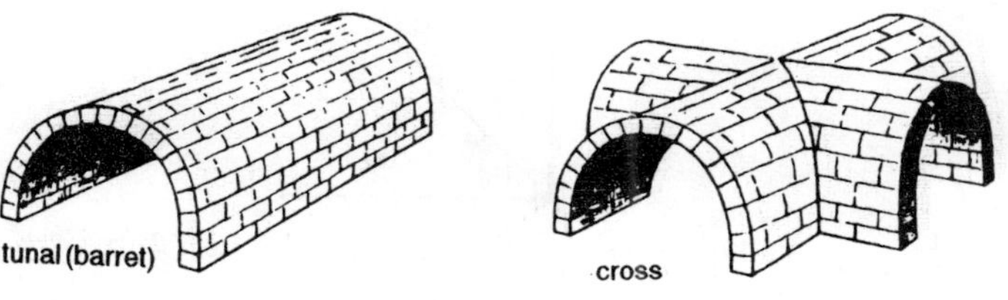

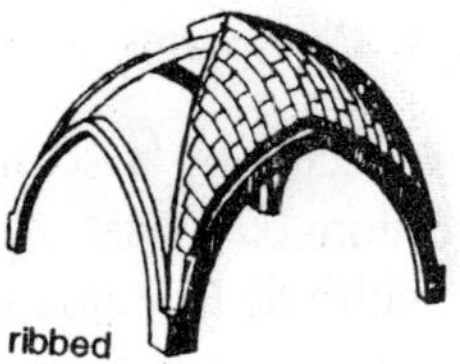

Fig. V-3. Vaults.

Vavilov, Nikolai Ivanovich. 1887-1943, Russian botanist and geneticist. Trying to trace the origins of various crops by locating areas with the greatest number and diversity of their species, the reported (1936) Ethiopia and Afghanistan as the birthplaces of agriculture and hence of civilization. He divided cultivated plants into those domesticated from wild forms, *e.g.*, oats and rye, and those known only in cultivation, *e.g.* corn. He reportedly died a Soviet concentration camp after losing favour to Trofim Lysenko, whose theories he opposed.

Vector. A quantity that is specified both by its magnitude and its direction. Velocity, acceleration, and force are examples of vector quantities. Mass and electric charge, on the other hand, are scalars, having only magnitude. Usually symbols for vectors are printed in italic boldface type or written with an arrow or a bar above them. A vector can be represented by a line with the same direction as the quantity and a length equal to its magnitude.

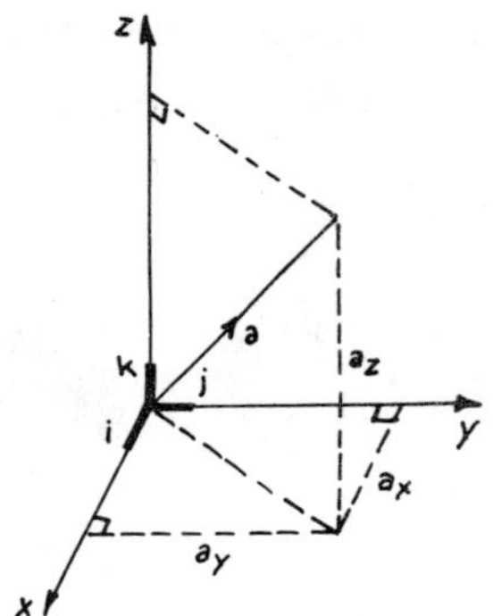

Fig. V-4. Components of vector.

Multiplication of a vector by a scalar gives another vector with the same direction but with a magnitude equal to its original magnitude multiplied by that of the scalar. It is often convenient to express a vector as a *unit* vector, specifying the direction, multiplied by a number that is its magnitude. In three dimensions, a vector a can be represented as the sum of three vectors directed along the axes of a Cartesian coordinate system. Thus $a = i\,a_x + j\,a_y + k\,a_z$ where *i, j,* and *k* are unit vectors. The modulus of *a* is $(a_x^2 + a_y^2 + a_z^2)^{1/2}$. Vectors can be added together and multiplied together in two ways.

Veda. [Skt., = knowledge], Oldest scriptures of Hinduism and the most ancient religious texts in an Indo-European language, still accepted to some extent by all Hindus as the authoritative statement of the essential truths of Hinduism. It is the literature of the Aryans who invaded NW India c.1500 B.C. Influenced by indigenous religious ideas, it was compiled from c.1000 to c.500 B.C. The Veda consists of four types of literature. The *Samhita*, the basic compilation of prayers and hymns, ascribed to inspired seer-poets, includes a threefold canonical body of hymns—the *Rig-Veda, Sama-Veda,* and *Yajur-Veda*—

and the *Atharva-Veda,* a collection of magic spells. The *Brahmanas* are prose explanations of the sacrifice. The *Aranyakas* contain instructions for meditation as the mental performance of sacrifice. The Upanishads are works of mysticism and speculation. Vedic writings expressed the idea of a single underlying reality embodied in Brahaman, the absolute Self. The Vedic sacrifice, with its invocation of any of 33 Vedic gods through Mantras or hymns, became increasingly elaborate with the passage of time, and the sacrifice came to be regarded as the fundamental agency of creation.

Vedanta. [Skt., = the end of the Veda], One of the six classical system of Indian Philosophy. The term refers to the teaching of the Upanishads (the last section of the Veda) and also to the knowledge of its ultimate meaning. By extension the term refers to schools based on the Brahma Sutras of Badarayana (early centuries A.D.), which summarize Upanishadic doctrine. Most important is the nondualist (*advaita*) Vendanta of Shankara (A.D. 788-820), who taught that spiritual liberation is achieved not by ritual action but by eradication of ignorance—*e.g.,* the belief that the illusory multiplicity of the world is real—and by attainment of the knowledge of Brahman, the absolute Self. The qualified nondualist Ramanuja (1017-1137) advocated devotion as the means of salvation, and held that the world and souls are real but depend on God. The dualist Vadanta of Madhva (1197-1276) asserted the permanently separate reality of the world, souls, and God. Vedanta is still a living tradition, with profound influence on the intellectual and religious life of India. Prominent modern Vedantists include Swami Vivekananda, Aurobindo Ghose, and Sarvepalli Radhakrishnan.

Vegetable. Popular name for many food plants and for their edible parts. No clear distinction exists between a vegetable and a Fruit. Vegetables are valuable sources of vitamins, minerals, starches, and proteins.

Vegetarianism. Theory and practice of eating only fruits, vegetables, cereal grains, nuts, and seeds, and excluding meat, fish, or fowl, and often eggs and diary products. The basis of the practice may be religious, ethical, economic, or nutritional, and its followers differ in strictness of observance. Practiced since ancient times by religious groups, vegetarianism as a separate movement developed in the mid-19th cent., particularly in England and the U.S.

Veld or **Veldt.** Grassy, undulating Plateaus of South Africa and Zimbabwe, ranging in elevation from c.500 to c.6,000 ft (150 to 1,830 m). Abundant crops of potatoes and maize are grown, and large cattle herds are grazed; the area also has industrial and mining centers.

Velocity. Symbol: v. A measure of rate of motion, equal to the distance moved per unit time. Velocity is a vector quantity: its value at a point is determined both by its magnitude (dx/dt) and its direction. The velocity of a particle travelling in a curved path has the direction of the tangent to the path at the point considered.

Venereal Disease. Any of several infectious diseases almost always transmitted through sexual contact. These diseases include Gonorrhea, Syphilis, nongonococcal urethritis, and genital Herpes Simplex. Changes in sexual behaviour over the past two decades have contributed to an increased incidence in venereal disease worldwide. Public health programs and use of Antibiotics have reduced the severity of the diseases, but epidemics, particularly of gonorrhea, still occur.

Venetian White. A white pigment consisting of a mixture of lead carbonate and barium sulphate.

Venus. The second planet in the solar system in order of succession outwards from the sun. Its mean distance from the sun is about 108.2 million kilometres. Of all the planets, Venus approaches closest to the earth: at inferior conjunction it comes as near as 42 million km. Moving with a mean velocity of some 35 km per second, it takes 225 earth days to orbit the sun. In size, Venus is comparable with our own planet. Its mass and gravity are about four-fifths those of the earth, its diameter (excluding atmosphere) is about 12,100 km, and its mean relative density is 5.1. Venus supports an extremely dense atmosphere consisting mainly of carbon dioxide (about 95%) and minute amounts of nitrogen, water vapour, carbon monoxide, oxygen, hydrochloric and hydrofluoric acids, and ammonia. The atmosphere has an outer cloud covering of 100% and is impenetrable to normal optical instruments. The period of axial rotation is 243 earth days. Venus's motion being retrograde. The outer clouds, perhaps made of ice, have a temperature of –43°C, but this temperature rises quickly to + 330°C just below the clouds. The temperature increases enormously on descending through the atmosphere and it is believed to reach 480°C at the surface. It is believed that the terrain of Venus is a rocky desert that probably glows dimly by virtue of its own red heat. Atmospheric pressure is believed to be about 95 atmospheres. The eccentricity of Venus's orbit is 0.0067 and its inclination is about 3° to the plane of the ecliptic. The brightest object in the sky after the moon, Venus is observable as a morning or evening star.

Venus. In astronomy, 2d Planet from the sun, at a mean distance of 67.2 million mi (108.2 million km). It has a diameter of 7,519 mi (12,100 km); terrains that can be subdivided into lowland plains (20%), rolling uplands (70%), and highlands (10%); and an atmosphere whose chief gas is carbon dioxide. The entire surface of Venus is opaquely covered by a dense cloud layer of concentrated sulfuric acid. The atmospheric pressure at the surface of the planet is 90 times that of earth. Because its greatest Elongation is 47°, Venus can never be seen much longer than 3 hr after sunset or 3 hr before sunrise. It has no known satellites. Space Probes that have encountered Venus include *Mariners 2, 5,* and *10* (1962, 1967, and 1974); *Pioneer-Venus 1* and *2* (1978); and numerous Soviet Venera spacecraft.

Venus. In Roman religion, goddess of vegetation; identified from the 3rd cent. B.C. with the Greek Aphrodite. In imperial times she was worshiped as Venus Genetrix, mother of Aeneas; Venus Felix, bringer of good fortune; Venus Victrix, bringer of victory; and Venus Verticordia, protector of feminine chastity. Among the famous sculptures of the goddess are the *Venus of Milo* (Louvre) and the *Venus of Medici* (Uffizi, Florence).

Venus's-flytrap. Insectivorous or carnivorous bog plant (*Dionaea muscipula*) native to the Carolina savannas, now widely cultivated as a novelty. The leaves, borne in a low rosette, resemble bear traps. They are hinged at the midrib, each half bearing sensitive bristles; when a bristle is touched by an insect, the halves snap shut and the marginal teeth interlock to imprison the insect until it is digested.

Verdigris. Any of various greenish basic salts of copper. True verdigris is basic copper acetate, a green or blue solid of variable composition used as a paint pigment. The term is often applied to green patinas formed on metallic copper. Copper cooking vessels can form a coating of basic copper carbonate, $CuCO_3.Cu(OH)_2$. The verdigris formed on copper roofs and domes is usually basic copper sulphate, $CuSO_4.3Cu(OH)_2H_2O$, while basic copper chloride, $CuCl_2.Cu(OH)_2$, may form in regions near the sea.

Vermicide. A substance used for killing worms.

Vermiculite. A hydrated silicate of magnesium, aluminium, and iron belonging to the micas. When heated quickly it can expand by up to twenty times its original volume. This exfoliated form is used in insulation, packing, and light-weight concrete, and as a filler for paints and plastics, a soil conditioner, and a compost for plants.

Vermilion. *See* mercury sulphide.

Vernal Equinox. *See* equinox.

Vernier. A short scale moving alongside the main scale of a measuring instrument, used to obtain a more accurate value of the measurement. The vernier scale is constructed so that it has ten divisions that correspond in distance to nine divisions on the main scale. The pointer on the instrument is the zero on the vernier scale. When this falls between two graduations of the main scale, the next decimal place in the reading is obtained by noting the number of the graduation on the vernier that coincides with a graduation on the main scale. [After Pierre Vernier (1580-1637), French mathematician and soldier.]

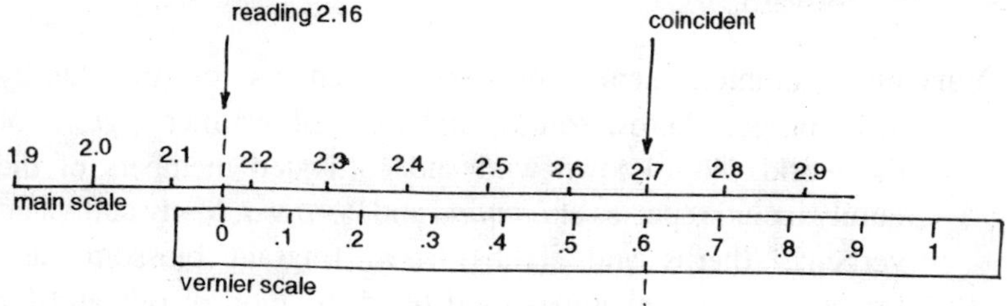

Fig. V-5. Vernier scale.

Versed Sine. One minus the cosine of an angle: vers $\theta = 1 - \cos\theta$.

Vertebrate. Any animal having a backbone or spinal column. All vertebrates belong to the subphylum Vertebrata of the phylum Chordata. The five classes of vertebrates are Fish, Amphibians, Reptiles, Birds, and Mammals. Vertebrates are comparatively large, have a high degree of specialization of their parts, and are bilaterally symmetrical. All have an interior skeleton, a brain enclosed in a cranium, a closed circulatory system, and a heart divided into two, three, or four chambers; most have two pairs of appendages modified as fins, limbs, or wings in the different classes. Animals without backbones are called Invertebrates.

Vertex. A point of intersection of two sides of a plane figure.

Vertical Circle. The great circle on the celestial sphere that passes from the observer's zenith through a given celestial body or a specified celestial point.

Vertical Takeoff and Landing Aircraft (VTOL). Craft capable of rising and descending vertically from and to the ground, thus requiring no runway. Some examples are the Airship and the Balloon, which are very inefficient in their forward motion, and the Helicopter, which has very limited performance. Convertiplanes are VTOL craft that can fly horizontally with same effectiveness as a conventional airplane. Their vertical lift during takeoff is provided by a rigid rotor that spins about a vertical axis, by tilted propellers, or, as in some jet VTOL craft, by vanes that can direct the engine thrust upward or forward.

Vervain. Common name for some members of the family Verbenaceae, herbs, shrubs, and trees of warmer regions of the world. Well-known wild and cultivated members of the family include species of *Lantana* and *Verbena*. Many cultivated verbenas (herbs and shrubs) have fragrant blossoms and leaves that are sometimes used for distillation of oils and for tea, as are those of the similar lemon verbena (*Lippia citriodora*). Wild American species are usually called vervains. Economically, the most important member of the family is Teak.

Very High Frequency (VHF). A frequency in the rage 30 megahertz to 300 megahertz.

Very Low Frequency (VLF). A frequency in the range 3 kilohertz to 30 kilohertz.

Vesalius, Andreas. 1514-64, Flemish anatomist. At the University of Padua he produced his chief work, the notably illustrated *De humani corporis fabrica* (1543), based on studies made by dissecting human cadavers. His discoveries in anatomy overthrew many hitherto uncontested doctrines of Galen and caused criticism to be directed against himself. He was physician to Emperor Charles V and his son Philip II.

Vesicant. Producing blisters.

Veterinary Medicine. Diagnosis and treatment of animal diseases. The first veterinary school opened in France in 1761; veterinary schools arose in the U.S. about the time of the Civil War. Veterinary research has made important contributions to medicine in general; for example, forms of Vaccination devised for animals were found to be effective for humans. Since World War II the development of live and modified live virus vaccines, Antibiotics, and sulfonamides has led to the prevention and control of once incurable animal diseases.

VHF. *See* very high frequency.

Vibrating-reed Electrometer. A type of electrometer in which a low steady potential difference is converted to an alternating potential by connecting it to a capacitor, one of the plates of which is mechanically vibrated. The alternating potential can then be amplified.

Vibration. Rapid oscillatory motion, as of a tuning fork, stretched string, etc.

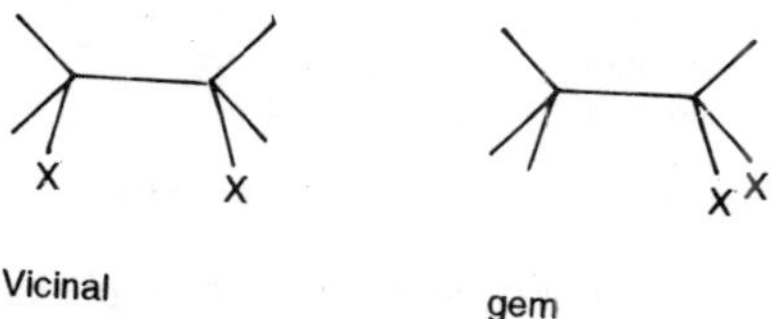

Fig. V-6. Vicinal and gem positions in molecules.

Vicinal. Denoting positions on adjacent atoms in a molecule. In the names of compounds the abbreviation *vic-* is used. *Compare* gem.

Victor Meyer's Method. A technique for measuring the vapour density of volatile liquids and solids. A weighed amount of the sample is put in a small stoppered tube, which is dropped through a long tube into a bulb surrounded by a constant temperature bath at a temperature above the boiling point of the sample. The tube is then sealed. The sample is vaporized and the vapour forces out the stopper of the sample bottle and displaces air from the tube, out of the side arm and into a graduated collection tube. Thus, the volume V of vapour produced by a mass m of sample at the temperature of the bulb is obtained. The density of vapour at this temperature is given by *mp/V (p-p),* where p is the pressure at which the volume is measured and p' is the vapour pressure of water at the temperature at which the air is collected, After Victor Meyer (1848-97), German chemist.]

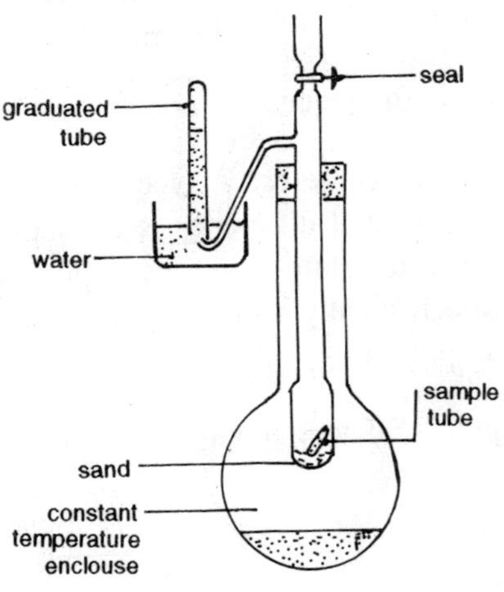

Fig. V-7. Victor Meyêr apparatus.

Videodisc. A disc, similar to the record in a phonograph system, that is capable of producing both pictures and sound. A videodisc cannot record television programs off the air for later playback, unlike a video cassette recorder (VCR). Videodiscs, however, generally produce pictures that are clearer in detail and truer in colour than those produced by VCRs, and they also offer better sound quality. Two quite different videodisc systems have been developed: one uses a mechanical stylus that senses varying patterns of electrical capacitance imprinted in grooves on the disc surface, whereas the other uses a laser to read a pattern of microscopic pits cut in a spiral pattern on the inside surface of the disc.

Video Frequency. A frequency in the range used for transmitting television signals, lying between 10 hertz and 2 megahertz.

Vidicon. A type of television camera tube in which the image is produced on an insulating screen coated with a photoconducting layer. A positive potential is applied and each part of the mosaic acts as a small capacitor storing charge. The amount of charge stored depends on the resistance of the screen at that point, which in turn depends on the amount of light falling on the photoconductor. The back of the screen is scanned with an electron beam, thus releasing the charge and causing a current to flow. For each element of the screen the discharged current depends on the amount of incident light.

Videotex. A computer-based system that allows a user to retrieve and display Alphanumeric and pictorial information at home on a Personal Computer. By linking the personal computer to the videotex data base, it is possible to provide access to information on a scale that is virtually unlimited. There are two forms of videotex systems. One-way *teletext* systems permit the selection and display of such general information as airline schedules, traffic conditions, and traditional newspaper content. *Viewdata* sytems are more specific and provide for two-way, or interactive, communication. Thus, specific questions may be researched by indexing into the appropriate data base: *e.g.,* bank balances can be verified and bills paid, and travel and hotel reservations can be made. Current viewdaata systems include Prestel (Britain), Teletel (France), and the more advanced but still experimental Telidon (Canada). Because videotex makes the home computer part of a network caontaing a larger, more powerful Computer, additional computing power and data-storage facilities can be made available to the user.

Vinegar. A dilute impure solution of acetic acid made by fermentation of ethanol, particularly in cider, wine, or malted barley liquor. It is used as a preservative and food flavouring.

Vinegar. Sour liquid consisting mainly of acetic acid and water, produced by the action of bacteria on dilute solutions of Ethanol derived from previous yeast Fermentation. The alcoholic liquor used, *e.g.*, cider or wine, determines the characteristic colour and flavour. Acetic fermentation may be impeded by an excessive growth of mother of vinegar, a slimy mass of bacteria, or of the parasitic vinegar eel, a minute, threadlike worm. Vinegar is used as a salad dressing, as a preservative, as a mild disinfectant, and, in cooking, as a fiber softener.

Vinyl Acetate. A colourless liquid $CH_3COOCH:CH_2$, manufactured by catalytic reaction between acetic acid and ethylene or acetylene. It is the starting material for polyvinyl acetate resins: the pure liquid must be stabilized as it polymerizes on exposure to light. M.pt. –100°C; b.pt. 73°C; r.d. 0.9.

Vinyl Chloride. *See* chloroethene.

Vinyl Cyanide. *See* acrylonitrile.

Vinyl Ether. A colourless volatile flammable liquid, CH_2: $CHOCH:CH_2$, prepared by treating dichloroethyl ether with alkali and used as an anaesthetic. B.pt. 28°C; r.d. 0.8.

Vinyl Group. The monovalent group CH_2CH-. derived from ethylene.

Vinylidene Group. The divalent group $CH_2:C$ =, derived from ethylene.

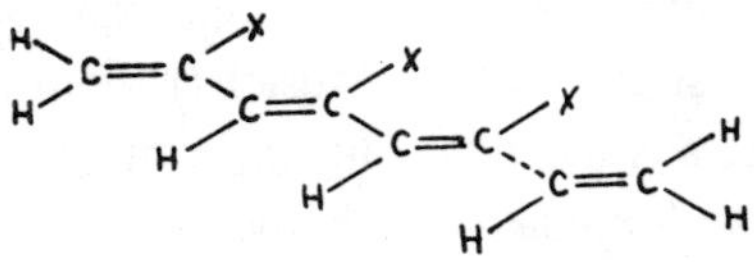

Fig. V-8. Vinyl polymer.

Vinyl Resin. Any of a class of synthetic resins made by polymerizing a vinyl compound. The polymerizations are usually carried out in the pure liquid phase or in emulsions. The most important vinyl polymers are polyvinyl chloride, polyvinyl acetate, and various copolymers of vinyl chloride with vinyl acetate. The vinyl resins belong to the class of ethenoid resins. The term *vinyl* is used for any vinyl resin or material produced from such a resin.

Violet. Common name for some members of the family Violaceae, chiefly perennial herbs (and sometimes shrubs) found on all continents. Violets (genus *Viola* and similar related species) are popular florists', garden, and wild flowers. Violets have fragrant, deep purple to yellow or white blossoms. Various species, especially the sweet, or English, violet (*V. odorata*), have been used in perfumes, dyes, and medicines, and have been candied. The common pansy was derived from the Old World *V. tricolor*. Its irregularly shaped, variously coloured flowers have five velvety petals. A violet is the floral emblem of three states (New Jersey, Rhode Island, and Wisconsin). Some unrelated plants are also called violets, *e.g.*, the African Violet of the Gesneria family.

Viper. Poisonous Snake of the family Viperidae, found in Eurasia and Africa. Characterized by erectile, Hypodermic fangs, vipers range in size from under 1 ft (30 cm) to nearly 6 ft (2 m) and often have zigzag or diamond markings. Best known are the European Asp (*Vipera aspis*), native to S Europe, and the common European viper, or adder (*V. berus*), found throughout Europe and N Asia. The Pit Vipers of the Americans belong to different family.

Virchow, Rudolf. (Fir'kho), 1821-1902, German pathologist. He taught at the Univ. of Wurzburg and then directed the Pathological Inst. in Berlin. A founder of cellular pathology, he contributed to nearly every branch of medicine and to anthropology, and introduced sanitary reforms in Berlin. Elected a member of the Prussian lower house and later of the Reichstag, he was a leader of the liberal Progressive party opposed to Bismarck.

Virginia Creeper. Woody vine (*Parthenocissus quinquefolia*) of the Grape family, popular as a wall covering. It has. blueblack berries and clings disk-tipped tendrils, with some branches hanging free. The five-fingered leaves are sometimes confused with the three-fingered Poison Ivy.

Virgin Neutron. A neutron that has not experienced a collision.

Virial Equation. An equation of state representing real gases: $pV = A + Bp + Cp^2 + Dp^3 + \ldots$, where A, B, C, D, etc., are empirical constants (the *virial coefficients* of the gas).

Virtual Image. *See* image.

Virtual particle. A particle thought of as existing for a very brief period of time in an interaction between two other particles. Thus, an electromagnetic interaction between two charged particles can be described by emission and absorption of *virtual photons.* The strong interaction between nucleons in the nucleus is described in terms of the exchange of *virtual pions.* Another particle, the *W-particle* or *intermediate vector boson,* has been postulated as the virtual particle responsible for the weak interaction between particles.

Virtual Work. The work that would be done if a body or system subjected to a set of forces were given an infinitesimal displacement. The system is in equilibrium only if the virtual work is zero.

Virus. Organism composed mainly of Nucleic Acid within a protein coat, ranging in size from 100 to 2,000 Angstroms; they can be seen only with an electron microscope. During the stage of their life cycle when they are free and infectious, viruses do not carry out the usual functions of living cells, such as Respiration and growth; however, when they enter a living plant, animal, or bacterial cell, they make use of the host cell's chemical energy and protein and nucleic acid synthesizing ability to replicate themselves. Some, like the bacteriophages, have protein tails used for infecting the host. Viral nucleic acids are single or double standard and may be DNA or RNA. After viral components are made by the infected host cell, virus particles are released; the host cell is often dissolved. Some viruses do not kill cells but transform them into a cancerous state; some cause illness and then seem to disappear, while remaining latent and later causing another, sometimes much more severe, form of disease. Viruses, known to cause cancer in animals, are suspected of causing cancer in humans; they also cause measles, mumps, yellow fever, poliomyelitis, influenza, and the common cold. Viral infections are not presently treatable with drugs, but Vaccination is useful in effecting immunization against many.

Viscometer. Any instrument of apparatus for measuring the viscosity of a fluid. The main types depend for their action on the flow of fluid through a narrow tube; the torque transmitted through the fluid from a rotating cylinder to a concentric stationary cylinder; the fall of a sphere through the fluid; and the damping of a vibrating body by the surrounding fluid.

Viscose. The viscous brown solution of cellulose xanthate in sodium hydroxide that is used in making rayon.

Viscose Rayon. *See* rayon.

Viscosity. The property of liquids and gases of resisting flow. It is caused by forces between the molecules of the fluid. In streamline flow the fluid can be thought of as containing parallel layers, which move at different rates. The resistance caused by the viscous force is a tangential force opposing the motion of the layer.

Newton's law of viscosity states that this force per unit area is proportional to the velocity gradient between layers: *i.e.* $F = \eta A dv/dx$. The shear stress is proportional to the rate of shear strain. Fluids that show this behaviour are called *Newtonian fluids (compare* non-Newtonian fluid). The constant η is the *coefficient of viscosity* of the fluid, usually called the *dynamic viscosity* or simply the *viscosity*. It is measured in newton-seconds per square metre or in poise. The *kinematic viscosity* is the dynamic viscosity divided by the density of the fluid. It is given the symbol ν ($= \eta/\rho$) and is measured in square metres per second or in strokes.

Viscous. Having a high viscosity.

Visible Spectrum. 1. The spectrum of wavelengths in visible light.

2. An emission or absorption spectrum in the visible region. Light emission is caused by electron transitions in atoms and molecules or by vibration of atoms in incandescent solids. It gives information on the energy levels and on molecular structure.

Visual Binary. *See* binary.

Visual-display Unit. A device on the output of a computer that can produce a display on a cathode-ray tube in the form of characters or diagrams. It may be fitted with a light pen.

Vitamin. Any of a group of chemical compounds that are essential constituents of the diet of animals. In general, vitamins are not synthesized in the body, although vitamin D is produced in humans by the action of ultraviolet light on ergosterol. Chemically, the compounds are unrelated and they have a variety of functions in metabolism—many of them act as coenzymes. Their absence from the diet produces certain deficiency diseases: vitamin C prevents scurvy,l vitamin D prevents rickets, etc.

Vitamin C. *See* ascorbic acid.

Vitreous. Glass-like; having the appearance or structure of a glass.

Vitreous Humour. *See* eye.

Vitriol. 1. Any of several sulphate salts, such as copper sulphate (*blue vitriol*), iron II sulphate (*green vitriol*), or zinc sulphate (*white vitriol*).

2. *See* sulphuric acid.

Vivisection. Use of living animals for scientific research. Rats, mice, guinea pigs, and rabbits are those used most frequently. The practice of vivisection dates back to ancient times and has contributed to progress in understanding Anatomy, Physiology, and Genetics. Regulatory agencies include the National Research Council in the U.S. and the Laboratory Animal Bureau in Britain.

VLF. *See* very low frequency.

Vodka. Traditional spirituous drink of Russia, the Baltic states, and Poland; now consumed internationally. Containing over 90% alcohol (usually diluted before marketing), vodka has little or no odour or taste. It is distilled from rye and barley malt or from the cheaper maize or potatoes.

Voice. In music. Singing voices are classified by ranges as soprano and contralto (or alto), the high and low female voices, with mezzo-soprano as an intermediate classification; and as tenor and bass, the high and low male voices, with baritone as intermediate classification. The sound of the castrato (a male singer with an artificially high voice, the result of castration in boyhood), for which many 17th and 18th cent. soprano and alto roles were intended is approached today by the countertenor. Choral music generally requires a range of about an octave and a half for each voice; a solo singer must have at least two octaves. Over the centuries great changes have taken place in the art of singing within Western musical culture, and modern singers can only approximate the timbre of earlier eras. Gregorian chant may have been sung with a nasal timbre resembling Oriental technique. Bel canto, the virtuosic art of vocal technique that flourished from the 17th to 19th cent., has been revived in the 20th cent.

Volatile. Capable of vaporizing easily. Volatile substances, such as ether and camphor, have high vapour pressures.

Volcano. Aperture in the crust of a planet or natural satellite through which gases, Lava, and solid fragments are discharged. The term is also applied to the conical Mountain (cone) built up around the vent by ejected matter. Volcanoes are described as active, dormant, or extinct. On earth about 500 are known to be active. Belts of volcanoes are found along the crest of the mid-ocean ridge system and at converging crustal plate boundaries. There are also isolated volcanoes not associated with crustal movements (*e.g.*, the Hawaiian Islands); Eruptions range from the quiet type (Hawaiian) to the violently explosive (*e.g.*, Pelee and Krakatoa). The *Mariner 9* space probe revealed (1971) that the planet Mars has a number of volcanoes, including the largest in the solar system, Olympus Mons. The *Voyager 1* space probe photographed (1979) at least eight active volcanoes on Io, a satellite of Jupiter.

Vole. Mouselike Rodent (called field or meadow mouse in North America) related to the Lemming. Found in Eurasia, N Africa, and North America, most voles are from $3^1/_2$ to 7 in. (9 to 18 cm) long and have gray or brown coats and short tails. They eat mainly grasses but also feed on seeds and insects; they nest in dense growth or shallow burrows.

Volt (V). Unit of voltage or, more technically, of electric Potential and Electromotive Force. It is defined as the difference of electric potential existing across the ends of a conductor having a resistance of 1 Ohm when the conductor is carrying a current of 1 Ampere.

Volt. Symbol: V. The SI of electric potential, potential difference, or electromotive force, equal to the potential difference between two points on a conductor carrying a steady current of one ampere when the power dissipated between the points is one watt. [After Count Alessandro Volta (1745-1827), Italian physicist.]

Volta, Alessandro, Conte. 1745-1827, Italian physicist. He invented the so-called Volta's pile (or voltaic pile), the electrophorus (a device to generate static electric charges), and the voltaic

Cell. The Volt, a unit of electrical measurement, is named for him.

Voltage. Symbol: *V*. Potential difference or electromotive force as measured in volts.

Voltage Divider. *See* potential divider.

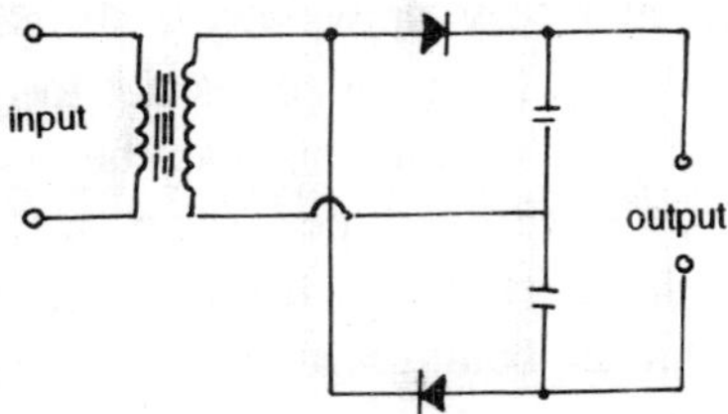

Fig. V-9. Voltage-doubler circuit.

Voltage doubler. An electrical circuit for producing an output voltage twice as large as the input voltage. The usual form has two rectifiers.

Voltaic Cell. *See* cell.

Voltameter (coulombmeter). An electrolytic cell used for measuring a quantity of current electricity by electrodeposition of a metal, usually copper or silver. The cell contains a salt of the metal and, on passing the current, metal is deposited on the cathode. The total quantity of electricity passed can be calculated from the increase in weight of the cathode and the electrochemical equivalent of the metal.

Voltmeter. Instrument used to measure, in Volts, differences in electric Potential, or voltage, between two points in a circuit. Most voltmeters are based on the d'Arsonval Galvanometer and are of the analog type, *i.e.,* they give voltage readings that can vary over a continuous range, as indicated by a scale and pointer. Digital voltmeters, which provide voltage readings represented by groups of digits, are, however, becoming increasingly common. A voltmeter is often combined with an Ammeter and an Ohmmeter in a multipurpose instrument.

Volta's Pile. An early battery invented by Volta, consisting of a pile of cells, each having a zinc and copper disc separated by a cloth disc soaked in sulphuric acid.

Voltmeter. Any of several instruments used to measure voltage. The term is usually used for meters that give a direct reading on a scale calibrated in volts, millivolts, etc. All voltmeters are designed so that they draw very little current from the circuit. When the required accuracy is not too great, moving-coil and moving-iron instruments can be used to measure potential difference. Both types have a high resistance incorporated so that they do not draw very much current, the moving-coil voltmeter being used for direct voltages and the moving-iron voltmeter being suitable for both direct and alternating voltages. Various types of electrometer are also used as voltmeters. The electrostatic electrometer is calibrated in the kilovolt range. Other types work by electronic amplification. The valve voltmeter has a thermionic-valve circuit with a measuring instrument in its output: most modern electrometers have solid-state amplifiers and often a digital output. Potential differences can also be measured by potentiometer and by a cathode-ray oscilloscope.

Volume. Symbol : V. A measure of the amount of space occupied by an object or solid figure. It is measured in cubic metres, cubic feet, etc.

Volumetric Analysis. Any of various forms of quantitative analysis, all of which depend on measurements of volumes. Volumetric analysis of gases depends on volume changes measured in a eudiometer. A mixture of gases can be analysed by selectively absorbing different components and measuring the volume decrease: oxygen, for example, can be absorbed by pyrogallol. Another method of gas volumetric analysis is Dumas' method of finding the percentage of nitrogen in organic compounds. Volumetric analysis of solutions and liquids is performed by finding the amount of unknown solution that reacts chemically with known amounts of standard solution.

Von Neumann, John. (Noi'man), 1903-57, American mathematician; b. Hungary; came to U.S., 1930. He was associated with the Inst. for Advanced Study after 1933 and was appointed (1954) a member of the U.S. Atomic Energy Commission. A founder of the theory of games, he also made fundamental contributions to quantum theory and to the development of the atomic bomb. Von Neumann was a leader in the design and development of high-speed electronic computers.

Vulcanite (ebonite). A hard black insulating material made by vulcanizing rubber with sulphur.

Vulcanization. The process of improving the qualities of natural or synthetic rubber by heating it with sulphur. Temperatures of about 150° C are used and the reactions are promoted by accelerators. Carbon black is also usually added.

Vulgar Fraction. *See* fraction.

Vulture. Large Bird of prey of temperate and tropical regions, also called Buzzard. Old World vultures, of the family Accipitridae, are allied to Hawks and Eagles; the American vultures and Condors belong to the family Cathartidae. American vultures have no syrinx and are thus voiceless, emitting only weak hisses. They feed mainly on carrion. Most have dark plumage and small, naked heads. Normally solitary, they gather in crowds to feed.

Waksman, Selman Abraham. 1888-1973, American microbiologist; b. Russia; came to U.S., 1910. As a microbiologist (1918-54) at the New Jersey Agricultural Experiment station, he studied with his colleagues, the role of microorganisms in the decomposition of organic matter, the origin and nature of humus, and the production of substances detrimental to some bacteria. For his discovery of streptomycin and of its value in treating tuberculosis he won the 1952 Noble Prize in physiology or medicine.

Walking Stick or **Stick Insect**. Extremely long-bodied, slow-moving, herbivorous Insect, principally Oriental and tropical in distribution. Waling sticks have green, gray, or brown bodies that closely resemble twigs or grass stems, comouflaging them against predators. Most are wingless and have long antennae. Some species, the longest insects in the world, achieve a length of over 1 ft (33 cm).

Wallace, Alfred Russel. 1823-1913, English naturalist. From his study of comparative biology in Brazil and the East Indies, he evolved a concept of Evolution similar to that of Charles Drawin. His special contribution to the evidence for evolution was in biogeography; he systematized the science and wrote *The Geographical Distribution of Animals* (2 vol., 1876) and a supplement, *Island Life* (1881).

Wall Effect. Any effect caused by the inside wall of a container in an experimental apparatus or other device. An example of a wall effect is adsorption and recombination of free radicals on the inside wall of the reaction vessel in flash-photolysis experiments.

Walnut. Common name for some members of the family Juglandaceae, chiefly deciduous trees of the north temperate zone, typified by large, aromatic compound leaves. Several species are commercially important for their lumber and edible nuts (usually encased in leathery or woody hulls). The family includes the Hickory and pecan, and the walnuts (genus *juglans*). The dark-coloured wood of the black walnut (J. nigra) of E North America, and the English, or Persian, walnut (J. regia), native to W Asia, is unusually hard and durable; it is valued for furniture, paneling, musical instruments, and other uses. The nut of the English walnut is usually the walnut sold commercially. The butternut, or white walnut (J. cinerea), also of E North America, is valued for its timber; its sweet, oily nut is not comemrcially important.

Walrus. Marine Mammal (*Odobenus rosmarus*) found in arctic seas. Largest of the fin-footed mammals, it has long tusks, light-brown wrinkled hide, and bristly cheek pads; an adult male may weight up to 3,000 lb (1,400 kg). Walruses live on beaches near shallow water in herds of about a hundred animals and eat mainly shellfish. Hunted for blubber, hides, and ivory, they are now endangered.

Wampum. Beads or disks made by North American Indians from mollusk shells, used for money, ornaments, or ceremonial exchange during treaties. Wampum is Algonquian for "white string of beds," but there is also a more valuable purple variety. It was especially prized in the Eastern Woodlands and Plains areas, reaching inland tribes through trade. Wampum belts often had pictographic designs worked into them.

Warfarin. A colourless odourless tasteless crystalline solid, $C_{19}H_{16}O_2$. It is widely used a rodenticide. M.pt. 161°C.

Washing Soda. *See* sodium carbonate.

Wasp. Winged Insect of the same order (Hymenoptera) as the Ant and Bee. Wasps have biting mouthparts and a thin stalk attaching the thorax to the abdomen ("wasp waist"); females have a sting, which they can use repeatedly, for paralyzing prey. Most species are solitary, but paper wasps, hornets, and yellow jackets are social. Social wasps usually are divided into three castes : egg-laying queens, workers (sexually undeveloped females), and drones, or males. They build papery nests consisting of either a single comb (paper wasps) or a large group of combs inside a papery sheath (hornets and yellow jackets). The nest may be underground. In all solitary wasps the female seals a single egg in a nest with paralyzed prey for food; potter, or mason, wasps and mud-dauber wasps are solitary wasps that build nests of mud.

Wassermann, August Von. 1866-1925, German physician and bacteriologist. He directed experimental therapy and serum research (1906-13) at the Koch Inst. and experimental therapy (from 1913) at the Kaiser Wilhelm Inst. He developed inoculations against cholera, typhoid, and tetanus, and devised (1906) the *Wassermann test* for diagnosing syphilis. Positive reaction when blood or spinal fluid is tested indicates the presence of antibodies formed as a result of syphilis infection.

Waste Disposal. Generally, the disposal of waste products resulting from human activity. Once a routine concern, the disposal of waste has become a pressing problem in the 20th cent. because of the growth of population and industry and the toxicity of many new industrial byproducts. Traditionally, waste was often dumped into nearby strems. Man-made sewers, which date back at least to the 6th cent. B.C. in Rome, were widely introduced in U.S. cities in the mid-19th cent. In the absence of sewerage, waste has often been stored in underground cesspools, which leach liquids into the soil but retain solids, or in simple sewage tanks, *e.g.*, septic tanks, in which organic matter disintegrates. Raw sewage is now commonly treated before being discharged as effluent, usually by reducing solid components to a semi-liquid sludge. Although sludge may be processed, *e.g.*, as fertilizer, it has often been buried or dumped at sea, a practice denounced by many environmentalists. Most *solid waste* in the U.S., such as municipal refuse, is deposited in open dumps. More sophisticated methods of disposal include the sanitary landfill, where waste is spread thin and seperated by layers of tamped earth, and special incinerators that burn combustible waste while generating steam (for heating) and/or gases (to run turbines). The recycling of noncombustible products such as glass and metals, *e.g.* aluminum cans, is growing and offers long-range hope for disposal. *Toxic wastes* include many chemicals, some generated in substantial quantities as common indsutrial byproducts, *e.g.*, heavy metals (notably mercury, lead, and cadmium), certain hydrocarbons, and some poisonous organic solvents. Such substances are usually sealed in metal drums and deposited underground or in the ocean, but the containers have often corroded and leaked their contents, polluting the land and water supply. One of the greatest modern hazards is *radioactive waste*, produced in increasing amounts as byproducts of research and nuclear-power generation and particularly dangerous because many kinds remain lethal for thousands of years. These materials are now stored deep underground, in caves or old salt mines, a solution that has encountered many technical difficulties and public objections to specific sites. A process to solidify nuclear waste matter and reduce its potential danger as a contaminant was being tested in the early 1980s.

Water. Odourless, tasteless, transparent liquid that is colourless in small amounts but exhibits a bluish tinge in large quantities. It is the most abudant liquid on earth. In solid form (ice) and liquid form it covers about 70% of the earth's surface. Chemically, water is a compound of hydrogen and oxygen whose formula is H_2O. The two H—O bonds form an angle of about 105°—an arrangement that results in a polar molecule, because there is a net negative charge toward the oxygen end (the apex) of the V-shaped molecule and a net positive charge at the hydrogen ends. Consequently, each oxygen atom is able to attract two nearby hydrogen atoms of two other water molecules. These hydrogen bondings keep water liquid at ordinary temperatures. Because water is a polar compound, it is a good solvent. Because of the hydrogen bondings between molecules, the latent heats of fusion and of evaporation, and the Heat Capacity of water, are all unusually high. For these reasons water serves both as a heat-transfer medium (*e.g.*, ice for cooling and steam for heating) and as a temperature regulator (the water in lakes and oceans helps regulate the climate). Water is chemically active, reacting with certain metals and metal oxides to form bases, and with certain oxides and nonmetals to form acids. Although compeletely pure water is a poor conductor of electricity, it is a much better conductor than most pure liquids because of its self-ionization, *i.e.* the ability of two water molecules to react to form a hydroxyl ion (OH^-) and a hydronium ion (H_3O^+).

Water. A colourless odourless tasteless liquid, H_2O. Water is the commonest hydrogen compound, covering almost three-quarters of the earth's surface and found in all living matter and in many minerals. It is used as a standard for several units and physical quantities, in particular for the fixed points of the centigrade scale, the relative densities of liquids and solids, and the definitions of the calorie and the litre. Water has several important properties. The solid form (ice) is less dense than the liquid at 0°C: the densities are 916.8 kg/m^3 and 999.8 kg/m^3 respectively. This is why ice floats on water and why frozen water pipes often burst when they begin to thaw. Water also has a maximum density (1000 kg/m^3) at 4°C, unlike most liquids, which have their maximum density at their melting point.

The molecule of water are polar and the liquid is held together by bydrogen bonds. The polar nature of water makes it an excellent solvent for many compoounds, particularly for salts, bases, and other ionic solids.

Water Bug. Air-breating aquatic Bug, with a sharp sucking beak. Waterbugs are usually found on or below the surface of quiet streams and ponds. Giant water bugs are the largest bugs and among the largest insects; one South American from attains a length of over 4 in. (10 cm). Other water bugs include the water boatmen, back-swimmers, water scorpions, and water striders.

Watercolour Painting. In its wider sense, all pigments mixed with water rather than oil, including Fresco and Tempera as well as aquarelle (the process now commonly called watercolour). Gouache and distemper are also watercolours, but prepared with gluey bases. The earliest existing paintings, found in Egypt, are watercolours. Gouache was used by Byzantine and Romanesque artists, and the illuminated manuscripts of the Middle Ages were produced with watercolour. The medium was used during and after the Renaissance by Durer, Rembrandt, Van Dyck, and others to tint and shade drawings and woodcuts. In the 18th cent. the modern aquarelle grew into a complete painting technique. Quick and easy to apply, the aquarelle's colours are at once brilliant, transparent, and delicate. The medium was particularly popular in England, where its greatest masters were Constable and J.M.W. Turner. Many 19th cent. painters used watercolour extensively, largely for landscapes. In France they included Daumier, Delacroix, and Gericault, and later, Cezanne and Dufy. Among notable American watercolourists are J.S. Sargent, Winslow Homer, Prendergast, Marin, and Sheeler.

Watercress. Hardy perennial European herb (*Nasturtium officinale*) of the Mustard family, widely naturalized in North America, found in or around water. It is cultivated commercially for its small, pungent leaflets, used as a peppery salad green or garnish. The ornamental plant whose common name is Nasturtium is unrelated.

Water Equivalent. Symbol: *w*. The mass of water that would have the same heat capacity as a given object.

Waterfall. Sudden drop in a stream formed where it passes over a layer of harder rock to an area of softer, more easily eroded rock. As a stream grow older, the waterfall, by undercutting and erosion, normally moves upstream and loses height until it becomes a series of rapids and finally disappears. Because of the waterpower that waterfalls can provide, many cities are located near them. Notable examples are Niagara Falls and Victoria Falls.

Water Gas. Colourless, poisonous gas that burns with an intensely hot, bluish flame. It is mainly a mixture of carbon monoxide and hydrogen and is almost entirely combustible. The gas is made by treating white-hot, hard coal or coke with a blast of steam, forming carbon monoxide and hydrogen. It is important in the preparation of hydrogen, as a fule in the making of steel, and in other industrial processes, *e.g.*, the Fischer-Tropsch Process.

Water Gas. A mixture of hydrogen and carbon monoxide made by passing steam over hot coke: $C + H_2O = H_2 + CO$. It is used as a fuel and as a raw material for the manufacture of some organic chemicals. The reaction is endothermic and water gas is sometimes made at the same time as producer gas, either by using a mixture of steam and air or alternate blasts of steam and air to cool and heat the coke.

Water Lily. Common name for some members of the family Nymphaeaceae, freshwater perennials found throughout most of the world. Often having large shield-shaped leaves and showy blossoms of various colours, the faamily includes water lilies, lotuses, and pond lilies (general *Nymphaea, Nelumbo,* and *Nuphar*, respectively). Well-known species include the blue or white Egyptian lotus of the genus *Nymphaea*, the national emblem of Egypt, and the Indian lotus of the genus *nelumbo*, sacred to several Oriental religions. Many members of the family have edible seeds or Tubers.

Watermelon. Plant (*Citrullis vulgaris*) of the Gourd family, native to Africa. The fleshy, juicy fruit, which can be red to pink, yellow, or white, is eaten fresh; the rind is pickled; and the seeds are also eaten. One white-fleshed variety, the citron melon, is used like citron in preserving.

Water Moccasin or **Cottonmouth.** Venomous Snake (*Ancistrodon piscivorus*) of the swamps and bayous of the S U.S.A. Pit Viper, it has a heat-sensitive organ for detecting warm-blooded prey. The average length is 3 to 4 ft (90 to 120 cm). If started it raises its head and shows the white interior of its mouth—hence the name *cottonmouth.*

Water of Crystalization. Water combined in the form of molecules in definite proportions in the crystals of many substances. The crystalline solid is a *hydrate* and can be further described as a *monohydrate, dihydrate,* etc., according to the number of water molecules present. The molecules are usually bound by coordination or by hydrogen bonding. Copper sulphate, for example, forms a pentahydrate, $CuSO_4.5H_2O$, in which four water molecules are coordinated to the copper II ion to give $[Cu(H_2O_4]^{2+}$ and the fifth is held by hydrogen bonds to sulphate ions. The four molecules are lost at 100°C but the fifth is not removed until the solid is heated to 250°C. Strongly bound water of crystalization of this type is sometimes called *water of constitution.*

Watershed. Elevation or divide separating the Catchment Area, or drainage basin, of one river system from another. The Rocky Mts. and the Andes form a watershed between westward-flowing and eastward-flowing streams. The term is also often used as synonymous with drainage basin.

Water Wheel. Ancient device for utilizing the power of flowing or falling water. The oldest known wheel, probably first used in the Middle East, was essentially a grindstone mounted a top a vertical shaft whose vaned or paddled lower end dipped into a swift stream. In the 1st cent. B.C. a horizontal shaft came into use; the wheel attached to it had radial vanes around its edge. Today the water wheel has been largely replaced by the Turbine.

Watt. Symbol: W. The SI unit of power, equal to one joule per second. In an electric circuit the power dissipated in watts is equal to the current in amperes multiplied by the potential difference in volts. [After James Watt (1736-1819), Scottish engineer.]

Wattage. Symbol: *W*. Power as measured in watts.

Watt, James. 1736-1819, Scottish inventor. Repairing a Newcomen Steam Engine, he devised improvements that resulted in a new type of engine (patented 1769) with a separate condensing chamber, an air pump to bring steam into the chamber, and insulated engine parts. He also perfected a rotary engine. Watt coined the term *horsepower*.

Wattless. *See* reactive.

Wattmeter. An instrument for measuring electrical power directly in watts.

Wave. A periodic change in some property or physical quantity through a medium or space, occurring in such a way that there is a regular periodic variation of the property with direction and usually with time. The varying property can be a physical displacement, such as that of a stretched string or the surfce of water. Alternatively, it can be a quantity, such as a temperature, pressure, or strength of electromagnetic field.

Sound, for instance, is propagated by change is pressure (or density) of a gaseous medium. At any instant of time the pressure will vary along a particular direction. A graph of pressure against distance would have the form of a sine curve, with pressure increasing and decreasing regularly. If the behaviour of the medium is considered over a period of time, the pressure at each point along this direction will also rise and fall periodically with time. The overall effect is of the periodic displacement with distance travelling through the medium with constant velocity. In this way energy is carried through the medium.

This is an example of a *travelling wave,* in which the whole periodic displacement moves. Under certain conditions a *standing or stationary wave* can be obtained. In this case, the property does not change with time at any particular point in the medium.

Sound moves as a *longitudinal wave, i.e.* one in which the local displacement is in the direction of motion of the wave. A vibrating stretched string and electromagnetic radiation are both examples of a *transverse wave,* in which the displacement is perpendicular to the direction of motion.

Wave. In physics, the transfer of Energy by some form of regular vibration, or oscillatory motion, either of some material medium or by the variation of intensity of the field vectors of an electromagnetic field. In longitudinal, or compressional, waves the vibration is in the same direction as the transfer of energy; in transverse waves the vibration is at right angles to the transfer of energy. The amplitude of a wave is its maximum displacement. Thye distance between successive crests or successive troughs is the wavelength λ of a wave. One full wavelength of a wave represents one complete cycle, that is, one complete vibration in each direction. All waves are referenced to an imaginary synchronous motion in a circle; thus one complete cycle is divided into 360 degress. The phase is that part of the cyle, expressed in degrees, that is completed at a certain time. The various phase relationships between combining waves determine the type of interference that takes place. The frequency ν of a wave is equal to the number of crests (or troughs) that pass a given fixed point per unit of time. The period T of a wave is the time lapse between the passage of successive crests (or troughs). The speed v of a wave is determined by its wavelength and its frequency according to the equation v $+\lambda\nu$. Because the frequency is inversely related to the period *T*, this equation also takes the form v + λ/*T*.

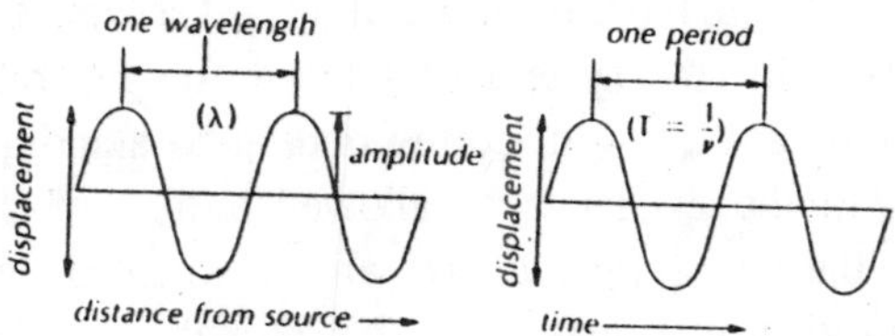

Fig. W-1. Wave diagrams showing the displacement of a wave plotted against distance from source and against time. A wave travels one wavelength during one period.

Waveform. The graph obtained by plotting the value of a periodic quantity against time.

Wave Front. A surface that is the locus of points having the same phase in a wave motion. Usually, the normal to the wave front at any point has the direction of propagation of the wave at that point.

Wave Function. Symbol: ψ. The amplitude of the wave associated with a particle as obtained from teh Schrodinger equation. The wave function is a mathematical expression: a function of the position of the particle. Its physical significance is that the square of the value of ψ at any point, ψ^2, is proportional to the probability of finding the particle at that point.

Waveguide. A hollow conducting tube of rectangular cross section used to guide microwaves propagated along it.

Wavelength. Symbol: λ. The distance between adjacent points with the same phase in a wave: *i.e.* the distance between adjacent peaks. The wavelength of a wave is related to its velocity, *c*, and frequency, *v*, by $\lambda = c/v$.

Wave Mechanics. A form of quantum mechanics in which particles (electrons, protons, etc.) are regarded as de Broglie waves, so that any system of particles (such as the electrons orbiting the nuclei of atoms) can be described by a wave equation—the Schrodinger equation. The basic problems in wave mechanics are those of formulating and solving the Schrodinger equation for the system of interest. In general, only certain solutions are allowed, corresponding to standing waves. This leads to the existence of allowed values of the energy of each particle (*eigenvalues*) and associated wave functions (*eigenfunctions*) that give the probabilities of finding the particle at different points in space.

Wavemeter. An instrument for measuring the wavelength of radio waves.

Wave Number. Symbol: σ. The number of single waves in unit distance, equal to the reciprocal of the wavelength.

Wave Theory. *See* light.

Wax. Any of various naturally occurring solid or semisolid water-insoluble substances that are soft and can be melted or softened at fairly low temperatures. Substances secreted by plants or animals, such as beeswax, are mixtures of esters of high-molecular weight fatty acids. Mineral waxes, such as paraffin wax, are usually mixtures of high-molecular weight aliphatic hydrocarbons, especially alkanes.

Wax. Substance secreted by glands on the abdomen of the bee and known commonly as beeswax; also various substances resembling beeswax. Chemically, waxes are complex mixtures of esters, fatty acids, free alcohols, and higher hydrocarbons. They are usually harder and less greasy than fats, but like fats they are less dense than water and insoluble in it. Waxes can be obtained from plants (*e.g.*, carnauba wax from palm leaves) or animals (*e.g.*, lanolin from wool fibers and spermaceti from sperm whales). Paraffin and ozocerite are mineral waxes composed of hydrocarbons. Japan wax and Bayberry wax are composed chiefly of fats.

Waxwing. Any of three species (genus *Bombycilla*) of perching songbirds of the Northern Hemisphere. Waxwings have crests (raised only in alarm) and brownish-gray plumage with flecks of red pigment resembling sealing waax on the wings. The species are the cedar waxwing; the Bohemian, or greater, waxwing; and the Japanese waxwing, found only in NE Asia.

Weak Interaction. A type of interaction between elementary particles, occurring at short range and having a magnitude about 10^{10} times weaker than the electromgnetic force. Weak interactions occur in certain decay processes, in which a particle is thought of as breaking into virtual particles that interact by weak interaction and then separate as true particles. In such decays, strong interactions cannot be responsible since strangeness would not be conserved. It has been postulated that weak interaction are caused by a heavy virtual particle—the intermediate vector boson.

Weasel. Small, lithe, carnivorous Mammal of the family Mustelidae, which also includes the Mink, Ferret, Wolverine, Skunk, Badger, and Otter. All members of the family have scent glands, used chiefly for territorial marking but sometimes for defense. Found in Eurasia, N Africa, and the Americas, weasels are characterized by long bodies and necks, short legs, small ears, and dense, lustrous fur (usually brown with white underparts). They hunt small animals at night, often killing more than they can eat.

Weather. State of the atmosphere at a given time and place with regard to temperature, air pressure, humidity, wind, cloudiness, and precipitation. The term *weather* is restricted to conditions over short periods of time; conditions over long periods are referred to as Climate. The study of weather is called Meteorology.

Weather Balloon. Balloon, usually helium-inflated, used in the measurement and evaluation of atmospheric conditions.

Atmospheric-pressure and other meteorological information may be sent by radio from the balloon; monitoring of its movement provides information abount winds at its flight level.

Weathering. Term for the processes by which Rock at the surface of the earth is disintegrated and decomposed by the action of atmospheric agents, water, and living things. Some of the processes are mechanical, *e.g.,* the impact of running water. Others are chemical, *e.g.*, oxidation, hydration, or carbonization. Weathering aids in the formation of soil and prepares materials for Erosion.

Weather Satellite. Artificial Satellite used to gather data on a global basis for improvement of weather forcasting. Information is provided about cloud cover, storm location, temperature, and heat balance in the earth's atmosphere. The first experimental weather satellite was *Tiros 1*, lauched by the U.S. in 1960. The U.S. National Operational Meterological System consists of both polar-orbiting and geostationary satellites. The U.S. Dept. of Defense operates two sunsynchronous polar-orbiting satellites that provide global meteorological data for military purposes. The USSR also operates polar-orbiting weather satellites.

Weaving. The art of forming a fabric by interlacing threads at right angles. One of the most ancient arts, once practiced chiefly by women, weaving sprang up independently in different parts of the world and is often mentioned in early literature. Archaeological remains have included textiles woven from Cotton, Silk, and Wool. Silk weaving reached Byzantium from China in the 6th cent. A.D., and cotton and silk weaving were brought to Spain by the Moors in the 8th and 9th cent. Flanders was known for its wool weaving by the 10th cent., and Flemish weavers brought to England by William the Conquerer gave impetus to the art there. In medieval France Tapestry weaving was highly developed. Weaving was a household industry in the American colonies. The Spinning and weaving inventions of the 18th cent. made possible the transition of today's huge, mechanized textile industry, but fine woven textiles are still made by hand. On the Loom, the warp threads are held under tension and are raised of lowered alternately to create a shed through which the weft (or woof) thread is inserted. Woven fabrics are classified by their structure. The basic weaves are plain, or tabby, in which the weft passes over alternate warp threads; twill, with a diagonal design made by interlacing two to four warp threads; and satin, with threads floating on the surface to reflect light, giving the fabric luster. Pile weaves, such as those used in Carpets, have a surface of cut or uncut loops.

Weber. Symbol: Wb. The SI unit of magnetic flux, equal to the flux linking a circuit of one turn that produces an e.m.f. of one volt when reduced uniformly to zero in one second. It is equal to 10^8 maxwells. [After Wilhelm Edward Weber (1804-91), German physicist.]

Weed. Any wild plant, especially those competing with crop plants for soil nutrients, water, and space. Control methods include the use of various Herbicides and soil cultivation. Plants cultivated in one area may become weeds when introduced into another area.

Week. Period of time shorter than the month, commonly seven days. The seven-day week is said to have originated in ancient times in W Asia, probably in Mesopotamia. It is thought to have been a planetary week predicated on the concept of the influence of the planets: the sun, the moon, and five bodies recognized today as Planets—Mars, Mercury, Jupiter, Venus, and Saturn. The Hebrew week is based chiefly on the Sabbath, which comes every seventh day. In the Roman Empire the planetary week was at first preeminent, and the use of planetary names, based on the names of pagan deities, continued even after Constantine I made (c.321) the Christian week, beginning on Sunday, official. The Roman names for the days pervaded Europe; in most languages the forms are translations from

Latine or attempts to assign corresponding names of divinities. The Latin names, their translations, the English equivalents, and their derivations follow : *dies solis* [sun's day], Sunday; *dies lunae* [moon's day], Monday; *dies Martis* [Mars' day], Tuesday [Tiw's day]; *dies Mercurii* [Mercury's day] Wednesday [Woden's day]; *dies Jovis* [Jove's or Jupiter's day], Thursday [Thor's day]; *dies Veneris* [Venus' day], Friday [Frigg's day]; and *dies Saturni* [Saturn's day], Saturday.

Weight. 1. Symbol: W. The force exerted on a body by gravitational attraction by the earth. The weight of a body always acts towards the earth's centre and depends on its distance from the earth's surface and on the local value of g—the acceleration of free fall. Weights can be expressed in the usual units of force such as newtons: the weight of a mass m is then given by mg. Sometimes units such as the gram-weight and pound weight are used. The term *weight* is often used loosely for mass.

2. A number expressing the reliability assigned to a measurement. If a series of experimental values, $x_1, x_2, x_3,...,$ has been obtained for a quantity, all the values will not necessarily be of equal worth. A *weighted mean* can be used, $(w_1x_1 + w_2x_2 + ...)/(w_1 + w_2 + ...)$. In which w_1, w_2, etc. are the weights given to each value. The weight of a value is a measure of the confidence that can be placed in it and is either obtained from the probable error or assigned intuitively.

Weight. Measure, expressed in pounds or grams, of the force of gravity on a body. Because the weights of different bodies at the same location are proportional to their masses, weight is often used as a measure of Mass. Unlike the mass, the weight of a body depends on its location in the gravitational field of the earth or of some other astronomical body.

Weightlessness. The absence of any observable effects of Gravitation. Apparent weightlessness is encountered by astronauts in a spacecraft orbiting the earth. The gravitational force acting on the astronauts and other objects in the spacecraft is exactly cancelled by the centrifugal force arising from the common accelerated motion of the spacecraft, astronauts, and spacecraft objects.

Weight Thermometer. A small glass (or silica) bulb with a narrow tube attached, used for determining the coefficients of expansion of liquids. The bulb is filled with the sample and weighed so that the mass of liquid can be determined. It is then heated to a known temperature and the mass of liquid expelled by expansion is found by weighing. This enables the coefficient of apparent expansion of the liquid to be calculated.

Welding. Process for joining separate pieces of metal in a continuous metallic bond. In cold-pressure welding, high pressure is applied at room temperature. Forge welding (or forging) is done by means of hammering, with the addition of heat. In most processes, the points to be joined are melted, additional molten metal is added as a filler, and the bond is allowed to cool. In the Thomson process, melting is caused by resistance to an applied electric current. Another process is that of the atomic hydrogen flame, in which hydrogen molecules passing through an electric arc are broken into atoms by absorbing energy. Outside the arc the molecules reunite, yielding heat to weld the material in the process.

Werner, Abraham Gottlob. 1750-1815, German geologist. He was the first to classify minerals systematically. Although his theory of neptunism (that the earth's lands precipitated out of an original world ocean) is now rejected, geology is indebted to him for applying chronology to rock formations and for his precise definitions.

Westinghouse, George. 1846-1914, American inventor and manufacturer; b. Central Bridge, N.Y. His railroad inventions included the air Brake (1868) and automatic signal devices. He formed companies in 1869 and 1882 to manufacture them. He also pioneered in introducing into the U.S. the high-tension alternating-current system for transmission of electricity. Over 400 patents were credited to Westinghouse.

Weston Cell (cadmium cell). A type of primary cell used as a standard of e.m.f. It has a mercury anode covered with a paste of mercurous sulphate and a cadmium-amalgam cathode. The electrolge is a saturated colution of cadmium sulphate and crystals of cadmium sulphate are present in the cell. Its e.m.f. is 1.0186 V at 20°C. [After Edward Weston (1850-1936), American electrical engineer.]

Wet-and Dry-Bulb Hygrometer. A type of hygrometer consisting of two thermometers placed side by side, one having its bulb covered with moist cloth. The wet thermometer indicates a lower temperature because it is cooled by evaporation of the water vapour, a process that depends on the amount of moisture present in the air. Thus the difference in the readings of the two thermometers gives an indication of the relative humidity, which can be found from special tables.

Wetting Agent. A substance that causes a liquid to spread when in contact with a solid. Wetting agents act by lowering surface tension.

Whale. Aquatic Mammal of the order Cetacea, found in oceans worldwide. Adapted to an entirely aquatic life, they have a fishlike shape, nearly hairless skin, an insulating layer of blubber, and flipperlike forelimbs; the tail is flattened horizontally and used for propulsion. Surfacing to breathe, whales expel air through a dorsal blowhole. They have good vision and excellent hearing, and many species use echolocation for underwater navigation. Most whales travel in schools, often migrating thousands of miles. There are two major groups of whales, the toothed, including the sperm whale, the beluga (or white) whale, and the narwhal; and the toothless, or baleen, including the right, humpback, and blue whales. Species vary greatly in size and include the largest animal and has ever lived, the blue whale, upto 100 ft (30 m) long and 150 tons. Commercial whaling has reduced all large species to endangered status. The intelligence and communication systems of the while are subjects of scientific investigation.

Whaling. The hunting of whales is thought to have been first pursued from land by thhe Basques as early as the 10th cent. Whaling on a large scale was first organized at the beganning of the 17th cent., largely by the Dutch. By the mid-17th cent. it had developed greatly in America, centering first in Nantucket and then New Bedford, the greatest whaling port in the world until the decline (c.1850) of the industry. With the capture (1712) of a sperm whale by a Nantucket fisherman the superior qualities of Sperm Oil were discovered, and American whalers went farther and farther in search of the whale—into the S Atlantic and then the Pacific, sometimes on voyages of three to four years. Herman Melville's *Moby Dick* is a classic account of a whaling voyage. The advent of the Civil War, a drop in the demand for sperm oil, and a steadily decreasing number of whales caused the decline of whaling, but the development of a harpoon with an explosive head and of the factory ship that can completely process a whale inaugurated a new whaling era in the 20th cent. Japan and the USSR are the principal nations now involved in the industry. International efforts to protect the whale have made progress, but several species have been seriously depleted.

Wheat. Cereal plant (genus *Triticum*) of the Grass family, a major food and an important commodity of the world Grain market. The plant is an annual (probably derived from a perennial) that grows best where the weather is cool during its early growth. Modern wheat varieties are usually classified as winter wheats (fall-planted and unusually winterhardy) and spring wheats (spring-planted). Hard-kerneled varieties (hard wheats) yield flour with a high gluten content, used to make breads; soft-kerneled ones (soft wheats) are starchier, and their flour is used to make cakes and biscuits. Durum wheat is a very hard-kerneled wheat used in Pasta products. Wheat is also used in the manufacture of Beer and Whiskey, and the grain, the Bran (the residue from milling), and the rest of the plant are valuable livestock feed. Wheat was one of the first grains to be cultivated: bread wheat was grown in the Nile R. valley by 5000 B.C. Major modern wheat-producing areas include the USSR, the U.S., Canada, China, W Europe, India, Pakistan, and Australia.

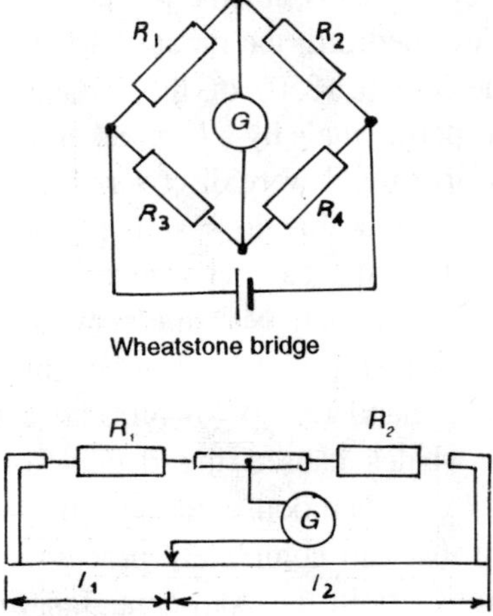

Fig. W-2. Forms of Wheatstone bridge.

Wheatstone Bridge. A network of resistances used for the measurement of resistance. The resistances are adjusted until the galvanometer has a zero reading. The bridge is then balanced and $R_1/R_2 = R_3/R_4$. If three of the resistances are known the value of the fourth fan be measured. A common form of the bridge has a length of resistance wire with a sliding contact attached to the galvanometer. R_1 is a standard resistance and R_2 is the unknown one. At balance, $R_2/R_1 = l_2/l_1$. [Named after Sir Charles Wheat-stone (1802-75), English physicist.]

Wheatstone, Sir Charles. 1802-75, English physicist and inventor. He was coinventor of an electric telegraph and also invented an automatic transmitter, an electric recording device, and an automatic telegraph. Wheatstone is credited with inventing the concertina, improving the stereoscope and dynamo, and popularizing a method for the measurement of electric resistance using a network now known as the Wheatstone bridge.

Wheel and Axle. A simple machine consisting of a wheel and an axle joined together. The effort is applied to a rope attached to the rim of the wheel and the load is carried by a rope attached to the axle. The mechanical advantage is equal to the radius of the wheel divided by that of the axle.

Whelk. Large marine Snail having a thick-lipped spiral sheel with many protuberances, found in temperate waters. Whelks are scavengers and carnivores, feeding on crabs, lobsters, and other shellfish. The largest species, the knobbed whelk, grows up to 16 in. (40.6 cm).

Whippet. Small, slender Hound; shoulder height, 18-22 in. (45.755.8 cm); weight, c.20 lb (9 kg). Its smooth coat is usuallly white, tan, or gray. It was developed in England in the mid-18th cent. and used for coursing hares in an enclosed areas. Today it is a race dog and a pet.

Whirlpool. Revolving current in an ocean, river, or lake. It may be caused by the configuration of the shore, irregularities in the bottom of the water body, the meeting of opposing currents or tides, or wind action onn the water. There are no true whirlpools really dangerous to shipping.

Whiskey. Spirituous liquor distilled from a fermented mash of grains, usually rye, barley, oats, wheat, or corn, and matured in wood casks, usually for three or more years. Inferior grades are made from potatoes, beets, or other roots. Scotch whiskey, usually a blend, takes its dry, somewhat smoky flavour from the barley malt, cured with peat, used in its preparation. The somewhat similar Irish whiskey, for which no peat is used, has a full, sweet taste. American whiskeys, classified as rye or a bourbon (a corn liquor), are higher in flavour and deeper in colour than Scotch or Irish whiskeys. Canadian whiskey, characteristically light, is produced from cereal grain only. First distilled in monasteries in 11th cent. England, whiskey has been manufactured commercially since the 16th cent.

Whistler. A whistling noise of descending pitch interfering with radio reception. It is caused by radio waves produced by lightning flashes and reflected off the ionosphere.

White Dwarf. Any of a class of dwarf stars with low magnitudes that have diameters on the scale of that of the earth but masses commensurate with that of the sun. Matter in them is closely packed and totally ionized, consisting of electrons and atomic nuclei. Thus the densities of such stars are enormously high. It is generally believed that white dwarfs represent the closing stage in the life cycles of certain starts.

White Dwarf. In astronomy, star that is abnormally faint for its white-hot temperature. It has the mass of the sun, the radius of the earth, and a central density about one million times that of water. White dwarfs have exhausted their nuclear fuel and represent one of the final stages of Stellar Evolution. About 500 white dwarfs have been discovered; the first of these was the companion of the bright star Sirius.

Whitehead, Alfred North. 1861-1947, English mathematician and philosopher. He taught mathematics at the University of London (1911-24) and philosophy at Harvard University (after 1924). *Principia Mathematica* (3 vol., 1910-13), which he wrote with Bertrand Russell, is a landmark in the study of logic. Whitehead's inquiries into the structure of science provided the background for his metaphysical work. His "philosophy of organism" (for which he devoloped a special vocabulary) viewed the universe as consisting of processes of becoming, and God as interdependent with the world and developing from it. His many works include *Science and the Modern World* (1925) and *Process and Reality* (1929).

White Lead. *See* lead carbonate.

White Light. Light that contains all wavelengths in the visible range. Strictly speaking, the wavelengths should all have equal intensities but the term is used to describe light containing different intensities while still appearing white to the eye.

White Noise. Noise that has a continuous wide range of frequencies of uniform energy.

White Spirit. A mixture of hydrocarbons boiling in the range 150-200° C, obtained by distillation of petroleum. It is used as a solvent, particularly for paint, and as fuel for cigarette lighters.

White Vitriol. *See* zink sulphate.

Whitney, Eli. 1765-1825, American inventor of the Cotton Gin; b. Westboro, Mass. He completed (1793) a model gin that rapidly separated the fiber of short-staple cotton from the seed. Whitney was unable to meet the demand for the machine, and his 1794 patent received legal protection only from 1807 to 1812. Thus the invention, which created great wealth for others who copied his model, gained him little. In 1798 he began producing the first firearms with standardized, interchangeable parts.

Whooping Cough or **Pertussis.** Highly communicable, infectious disease, predominantly of childhood. The early stage is manifested by symptoms of an upper respiratory infection; after about two weeks, a series of paroxysmal coughs are followed by a characteristic high-pitched "whoop" as a breath is taken. A serious disease, whooping cough may give rise to such complications as Pneumonia, convulsions, and brain damage; infants should be immunized against the disease as early as possible.

Wiedemann Franz Law. The principle that the ratio of thermal conductivity to electrical conductivity has the same value for all metals at a given temperature. The ratio is directly proportional to the thermodynamic temperature for pure metals at normal temperatures.

Wiel Displacement Law. The principl that, for the radiation emitted from a black body, the product of the wavelength at which maxium emission occurs and the thermodynamic temperature is a constant : *i.e.* λ max T = C. The maximum of a curve of wavelength against temperature is displaced towards shorter wavelengths as the temperature is increased. [After Wilhelm Wien (1864-1928), German physicist.]

Wiener, Norbert. 1894-1964, American mathematician and educator; b. Columbia, Mo. Known for his theory of Cybernetics and his contributions to the development of computers and calculators, Wiener also did research in probability theory and the foundations of mathematics.

Wig. Arrangement of artificial or human hair worn to conceal baldness or as part of a costume, either theatrical, ceremonial, or fashionable. In ancient Egypt elaborate wigs were worn as protection from the sun. Roman women often wore blonde wigs. In 17th cent. Europe, wigs of long, curled hair were popular. Since human hair was expensive and hard to obtain, horse's or goat's hair was often used. From c.1690 to 1794 scented pomade and white powder were used on wigs. Elaborate men's wigs were supplanted by small wigs with curls over the ears and a queqe tied in back, still worn in English courts. By 1788 men began to wear their own hair tied in back. After 1800, wigs again became part of women's fashions, although many men today wear short wigs (toupees) to conceal baldness.

Wigner Effect. A change in the crystal sructure of a solid as a result of irradiation. It is observed in graphite bombarded with neutrons : the material changes size due to displacement of carbon atoms from their lattice positions. [After Eugene Paul Winger (b. 1902), U.S. physicist born in Hungary.]

Wigner Nuclide. Either of a pair of nuclides that are isobars for which each nuclide has an odd mass number and a neutron number differing by one from its proton number. Tritium (^{3}H) and helium-3 (^{3}He) are a pair of Winger nuclides.

Wildlife Refuge. Haven or sanctuary for animals in an area of land or land and water set aside and maintained for their preservation and protection. In the U.S. game laws were passed in various states as early as the late 17th cent. Modern wildlife-conservation policy began with a conference called by Pres. Theodore Roosevelt in 1908 to inventory the nation's natural resources. Today the U.S. Wildlife Refuge System comprises area to protect big game (*e.g.*, bison, bighorn sheep, and elk); small game; waterfowl; and colonial nongame birds such as pelicans, terns, and gulls. Most numerous are waterfowl refuges, which supply areas for breeding, wintering, resting, and feeding along major flyways during Migration. Other countries throughout the world also maintain parks, refuges, and game preserves. The International Union for the Conservation of Nature, with more than 30 member nations, monitors wildlife conditions worldwide.

Wild Rice. Tall aquatic plant (*Zizania aquatica*) of the Grass family of a genus separate from common Rice (*Oryza*). Wild rice is a hardy annual with broad blades, reedy stems, and large terminal panicles. It grows best in shallow water along the margins of ponds and lakes in the N U.S. and S Canada; certain varieties also grow in the South. The seed is harvested by primitive methods and has never been cultivated with success. It was an important food of certain Indian tribes, especially in the Great Lakes region.

Wilkins, Maurice Hugh Frederick. 1916-, Irish biophysicist; b. New Zealand. He successfully extracted fibers from a gel of DNA, which, when analyzed by X-ray diffraction, showed a helical molecular structure. For this work he shared the 1962 Nobel Prize in physiology or medicine with James Watson and Francis Crick, who, on the basis of Wilkins' results and other scientific information, built a model of the DNA molecule.

Will-σ-the-wisp (ignis fatuus). A pale flickering moving flame that is sometimes seen at night over marshy ground. It is caused by combustion of marsh gas (methane) produced by decomposing vegetation.

Willow. Common name for some members of the family Salicaceae, deciduous trees or shrubs of worldwide distribution. The family comprises the willows and poplars and is typified by male and female flowers borne in catkins on separate plants. The narrower-leaved willows (genus *Salix*) flourish in cold, wet ground. The weeping willow (*S. babylonica*), native to China, and the North American pussy willow (*S. discolor*), with silky catkins are cultivated as ornamentals. The poplars (genus *Populus*), including the cottonwoods and aspens, usually have heart-shaped or ovate leaves. The white, or silver, poplar (*P. alba*), a popular ornamental native to Eurasia, is now naturalized inn North America. The cottonwoods, also called poplars, typically have seeds covered with fibrous coats, which when released at maturity clump together in cottony balls. The quaking aspen (*P. termuloides*), of the mountains of W U.S., has leaves that quiver in the slightest breeze. The yellow poplar, or tulip trees, is an unrelated plant of the Magnolia family.

Wilson, Charles Thomson Rees. 1869-1959, Scottish physicist. He was Jacksonian professor of nnatural philosophy at Cambridge. Noted for his studies of atmospheric electricity, he devised a method for protecting barrage balloons from lightning during World War II. For his invention of the Wilson cloud chamber for studying the activity of ionized particles, he shared with Arthur Compton the 1927 Nobel Prize in physics.

Wilson Cloud Chamber. *See* Cloud chamber. [After C.T.R. Wilson (1869-1959), Scottish physicist.]

Wilson, Edmund Beecher. 1856-1939, American zoologist, b. Geneva, III. He taught at Columbia Univ. (1891-1928), where he initiated genetic research. His main work was on the function of the cell in heredity and on thhe role of the chromosomes (including the significance of the sex chromosome). His major writing was *The Cell in Development and Inheritance* (1896).

Wilson, Edward O(sborne). 1929-, American biologist and leading proponent of Sociobiology; b. Birmingham, Ala. Educated at the Univ. of Alabama and at Harvard, he joined the Harvard faculty inn 1956 and later became professor of zoology. His exhaustive study of ants and other social insects (on which he is the world's chief authority) led to publication of *Sociobiology* (1975), a controversial work on the genetic factors in human behaviour.

Wimshurst Machine. A type of elecctrostatic generator having two circular discs of glass or other insulating material with radial strips of tinfoil stuck to their sides. The discs rotate in oppositive directions and the charge is produced by friction with tinfoil brushes and collected by metal points. [After J.Wimshurst (1836-1903), English scientist.]

Wind. Flow of air parallel to theh earth's surface. The direction of wind is indicated by a weather vane. A wind is named according to the direction from which it is blowing, *e.g.*, a wind blowing from the north is a north wind. Wind velocity is measured by means of a cup anemometer, an instrumennt with three or four small hollow metal hemispheres set so that they catch the wind and revolve about a vertical rod; an electrical device records the revolutions of the cups and thus the wind velocity. Winds are caused by the unequal heating of the earth's surface by the sun. Warmer air expands, becomes lighter, and rises. Cooler air rushes in from surrounding areas to fill the empty space. This process continues, creating a steady flow of air called a convection current. This, along with the rotation of the earth and other secondary factors, causes the basic planetary wind systems that circle the earth, bringing constant changes in the weather.

Windaus, Adolf. (Vin'dous) 1876-1959, German chemist. A professor of chemistry and director of the chemistry laboratories at the Univ. of Gottingen, he won the 1928 Nobel Prize in chemistry for his work on sterols, especially in relation to vitamins. He discovered and synnthesized vitamin D_3, the component of vitamin D most important in preventing rickets.

Windmill. Mechanical device that harnesses wind power in order to pump water, grind grain, power a sawmill, or drive an electrical generator. Windmills were probably not known in Europe before the 12th cent., but thereafter they became the chief source of power until the advent of the Steam Engine during the Industrial Revolution. The operational apparatus of the typical Dutch windmill is a four-to-six-armed structure that carries sails made of light wood or canvas. Revolving in the wind, the sail mechanism turns a shaft that operates the pump, millstone, or saw. Modern windmills are made of various lightweight materials. New types, such as the wind turbine used to generate electricity, are designed to turn even in light winds. Their output—depending on their size and on wind speeds—can vary from 1 kw to the 2,500 kw produced by the newest experimental model.

Window. 1. A range of wavelengths over which a given material tansmits electromagenetic radiation. The atmosphere, for example, has three windows : an optical window, letting through light and ultraviolet radiation in the range 760-200 nm; an infrared window in the range 8-11 μm; and a radio window in the range 8 mm-20m.

2. A sheet of material allowing the passage or electromagnetic of particle radiation. Thin sheets of metal are used as windows in X-ray tubes and Geiger counters.

Wind Tunnel. Device for studying the interaction between a solid body and an airstream. A wind tunnel simulates the conditions of an aircraft in flight by causing a high-speed stream of air to flow past a model of the aircraft (or part of an aircraft) being tested. The model is mounted on wires so that lift and drag forces on it can be determined by measuring the tension in the wires. Pressures on the model surface are measured through small flush openings on the surface. Wind tunnels are also used to study the effect of wind on other objects, such as automobiles, buildings, and bridges.

Wine. Alcoholic beverage made by Fermentation of the juice of the grape. Wines are distinguished by colour, flavour, bouquet (aroma), and alcoholic content. They may be red (when the whole crusted grape is used), white (using the juice only), or rose (when skins are removed after fermentation has begun). Wines are also classified as dry (when grape sugar ferments completely) or sweet (when some sugar remains). There are three main types of wine: natural (still), fortified, and sparkling. The alcoholic content of natural wine comes from fermentation. Fortified wine (*e.g.,* Sherry, port, Madeira) has brandy or other spirits added to it. Sparkling wine (Champagne is the best known) is fermented a second time after bottling. Wine is differentiated by the variety of grape, climate, location and soil of the vineyard, and treatment of the grapes before and during wine making. Fermentation starts when wine yeasts on the skins of ripe grapes come in contact with the grape juice (called must). Run off into casks, the new wine then undergoes a series of chemical processes, including oxidation, precipitation of proteins, and fermentation of chemical compounds, that create characteristic bouquet. After periodic clarification and aging in casks, the wine is ready to be bottled. The world's leading wine producer is France, with outstanding products from Bordeaux and Burgundy, the Loire and Rhone valleys, and Alsace. Other major producers are Italy, Spain, Germany, and the U.S. In the U. S., California is the leading wine-producing region, making European-type wines from grapes of the Old World species. The term *wine* is also applied to beverages made from other plants, *e.g.*, dandelion and elderberry.

Wisteria. Woody, twining vine (genus *Wistaria*) of the Pulse family, cultivated and highly esteemed for the beautiful pendant clusters of lilac, white, or pink flowers. Two species are native to the U.S., but the showier Asian species are commonly cultivated.

Witchcraft. Exercise of supernatural powers through occult arts such as Magic, sorcery, and satanism. Its origins lie in the belief in separate powers of good and evil in ancient pagan cults and in religions including Gnosticism and Zoroastrianism. The christian church persecuted witches from the 14th to the 18th cent. : under the Spanish Inquisition, up to 100 alleged witches were burned in a day; in 1692, 20 persons were executed as witches in Salem, Mass. The practice of witchcraft persists.

Witch hazel. Common name for some members of the family Hamamelidaceae, trees and shrubs found mostly in Asia. The family includes the winter hazels (genus *Corylopsis*), the witch hazels (genus *Hamamelis*), the sweet gums (genus *Liquidambar*), and the witch elders (genus *Fothergilla*). The American witch hazel (*H, virginiana*), whose leaves resemble those of the Hazel, is a fall-blooming shrub or small tree found E of the Rockies. The name *witch hazel* is also applied to an astringent liniment obtained from the leaves and bark of the plant. Sweet gums are characterized by their star-shaped leaves, their brilliant fall colouring, and their round fruits with hornlike projections. Their hard wood is used for cabinetmaking.

Witherite. A white of greyish mineral consisting of barium carbonate, Ba CO_3, used as a source of barium compounds.

Witwatersrand or the **Rand.** Rich gold-mining area and chief industrial region of South Africa, in the Transvaal, between the Vaal and Olifants rivers. It includes Soweto ιownship and the cities of Johannesburg, Benoni, Boksburg, Springs, and Germiston. The Rand traditionally produces about one third of the world's total gold output annually. Other major industries include steel-milling, machine-building, diamond-cutting, and the manufacture of textile and furniture.

Wohler, Friedrich. (Vo'ler), 1800-1882. German chemist. Professor at the Univ. of Gottingen, he devised a method for isolating Aluminum that he also used to isolate Beryllium and Yttrium. Wohler's synthesis of urea opened a new era in organic chemistry and contributed to the theory of isomerism. He also contributed to the chemistry of metabolism.

Whler's Synthesis. The production of urea by evaporating a solution of ammonium cyanate (NH_4) NCO, first performed in 1828 by the German chemist Friedrich Wohler (1800-82). The reaction was first laboratory synthesis of an organic compound from inorganic materials.

Wolf. Carnivorous Mammal (genus *Canis*), related to the Jackal and Dog. Three wolf species are generally recognized: the gray wolf (*C. lupus*), the red wolf (*C. niger*), and the prairie wolf, or Coyote. The gray wolf, also called timber wolf in North America, resembles a German Shepherd, with a shaggy coat, erect ears, and a bushy tail; the male is usually about 3 ft (90 cm) high at the shoulder and weights about 100 lb (45 kg). Ruthless hunting has nearly exterminated the gray wolf.

Wolfram. *See* tungsten.

Wolframite. A naturally occurring tungstate of iron and manganese, (Fe,- Mn) WO_4, used as an ore of tungsten.

Wollaston Prism. A prism for obtaining plane-polarized light. It is made from two triangular prisms of quartz or calcite, each cut so that their optic axes are at right angles to one another. The prism is designed to be used for ultraviolet radiation. [Named after W.H. Wollaston (1766-1828), English physicist.]

Wollaston, William Hyde. 1766-1828, English scientist. His achievements include the discovery (1802) of the dark lines (Fraunhofer lines) in the solar spectrum; the invention of the reflecting goniometer and the camera lucida; the discovery of the elements Palladium and Rhodium; and the establishment of the equivalence of galvanic and frictional electricity. He endowed the Wollaston science medal, awarded annually by the Geological Society, London.

Wollaston Wire. Very thin wire (diameters as small as 1 μm) made by encasing platinum wire in a layer of silver, drawing the wire out, and then dissolving the silver with acid.

Wolverine or **Glutton.** Heavy, short-legged, bearlike Mammal (*Gulo gulo*) related to the Weasel, c.3-3$^1/_2$ ft (91-106 cm) long. It inhabits the mountains of North America and Eurasia near the timberline, feeding on many different animals. Its long, dark-brown, frost-proof fur is prized by Eskimos as trim for hoods and cuffs.

Wombat. Shy, nocturnal Marsupial of Australia and Tasmania, related to the Koala. Thick-set, with a large head, short legs, a short tail, and a shuffling gait, it is about 3 ft (91.5 cm) long. Wombats live in burrows in forests and grasslands and eat grass, roots, and bark.

Wood. In botany, tissue (xylem) forming the buld of the Stem of a woody plant. Xylem conducts the Sap upward from the Root system to the Leaf, stores food, and provides support; it is composed of various cell types specialized for each function. Xylem is formed in the growing season by the Cambium. In temperate regions, cells formed in the spring or wet season are larger and thinner-walled than those formed in the summer of dry season; this results in conspicuous Annual Rings. Conifer wood (softwood) has a uniform, nonporous appearance; deciduous trees have a vessel-permeated xylem with a complex appearance, often called hardwood. Freshly cut wood contains much moisture and is dried (seasoned) in the sun or in kilns before use.

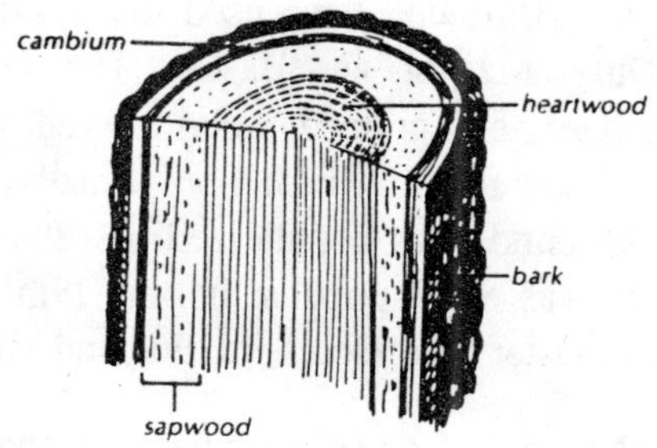

Fig. W-3. Wood: Cross section of a woody stem.

Wood Alcohol. Methanol.

Woodchuck or Groundhog. Species of marmot. This large Rodent, found throughout most of Canada and the NE U.S., has a heavyset body (about 2 ft/60 cm long, excluding the tail) and is covered with coarse, brownish hair. According to superstition the groundhog leaves its burrow on Feb. 2, Groundhog Day, and returns underground for six weeks if it sees its shadow, indicating six more weeks of winter.

Woodcut and Wood Engraving. Prints made from designs cut in relief (where material is cut away to leave the design that is printed) on wood, in contrast to Intaglio methods (where the lines that are incised are printed) such as Engraving and Etching. Woodcutting, the oldest form of printmaking, is done by cutting soft wood with a knife along the grain. Woodcuts appeared in Europe at the beginning of the 15th cent. At that time the same artist designed and carved the block. Later the cutting was often performed by specialists. Used for the block books that preceded printing, woodcuts were often employed as book illustrations after invention of the printing press. During the Renaissance the most eminent woodcut designerswere Durer and Hans Holbein, the Younger. With the increasing popularity of engraving on metal, interest in woodcuts declined. It was not revived until the 1890s, with the prints of such artists as Gauguin and Munch. Many artists of the 20th cent. have used the woodcut; they include Derain, Dufy, Maillol, and Baskin. The *wood engraving* is made on hard, end-grained wood carved with a graver or burin. The medium, in which lines usually print white on a black background, was popularized in 18th cent. England by Bewick. It was very popular in the 19th cent., in France (where its master was Dore), in England, and in the U.S.

Woodpecker. Name for some members of the family Picidae, climbing Birds found in much of the world. Woodpeckers typically have sharp, chisellike bills for pecking holes in trees, where they find insects. Among the North American species are the sociable downy woodpecker (*Picus pubescens*); the hairy woodpecker (*P. villosus*); and the redcrested or pileated woodspecker (*Hylotomus pileatus*). The flickers (genus *Melanerpes*), the only brown-backed woodpeckers, sometimes capture insects in the ground. The sapsuckers, which drink sap through small holes, can damage trees.

Wood's Metal. A low-melting alloy of bismuth (50%), lead (25%), tin (12.5%), and cadmium. M.pt. 71°C.

Woodward, Robert Burns. 1917-79, American chemist. A professor at Harvard Univ., he was one of the first to determine the strucutre of such organic compounds as penicillin, strychnine, terramycin, and aureomycin. For his synthesis of organic compounds (*e.g.*, quinine, patulin, cholesterol, cortisone, strychnine, lysergic acid, lanosterol, reserpine, chlorophyll, and tetracycline) he won the 1965 Nobel Prize in chemistry.

Woofer. A loudspeaker for use at low audio frequencies. *Compare* tweeter.

Wool. Fiber from the fleece of the domestic Sheep. Wool is warm, absorbent, elastic, strong, and crease-resistant. It was probably the first Fiber to be made into cloth, since herding flocks was the first step in the change from primitive culture and wool is found only on domesticated sheep. Egyptian, Babylonian, and Peruvian archaeological remains have yielded fragments of woolen fabrics. England, with a heritage of sheep raising the Weaving from its early history and from Roman influence, became Europe's great wool-producing country, and wool was the staple of British industry until the 18th cent., when Cotton began to supplant it. In the American colonies sheep raising began at Jamestown, and Spinning and weaving were important domestic industries. The first factory in America using water power to weave wool was opened at Hartford, Conn., in 1788. The U.S. now produces a substantial amount of the world's wool; other major producers include Australia, the USSR, and Great Britain. In wool making, sheep are sheared with mechanical clippers. The wool is then sorted according to fineness, crimp, length of fiber, and felting qualities, and dirt and lanolin are removed. Wool may be bleached or dyed as fleece, yarn, or cloth. Woolen cloth is woven from short-staple fibers spun into soft yarn. Worsted yarn produces cloth with a hard, smooth texture, *e.g.*, gabardine, serge.

Word. A number of bits processed as a single unit in a computer. Typical word lengths are 12, 32, 48, and 64 bits.

Work. Symbol: w. Energy transferred to or from a body or system, thus causing a change in its energy. Work always involved an applied force moving a certain distance and is measured by the product of the force and the distance it moves along its line of action. If the force is perpendicular to the direction of movement, no work is done. Work, like other forms of energy, is measured in joules.

Work. In physics, transfer of Energy by a force acting against a resistance or a body and resulting in displacement. Work W has a magnitude equal to the scaler product of the force F and the distance d of the resulting movement; thus $W = Fd \cos \theta$, where θ is the angle between the directions of the force and the movement. The foot-pound (English Units of Measurement), the erg (cgs system), and the joule (mks system) are the units of work or energy expended, respectively, by a 1-lb force acting through a distance of 1 ft, by a 1-dyne force through 1 cm, and by a 1-newton force through 1 m. One foot-pound equals 1.356 joules; 1 erg equals 10^{-7} joules.

Work Functioin. Symbol : ϕ The minimum energy necessary to remove an electron from a metal at absolute zero.

Work Hardening. *See* strain hardening.

Working Dog. Class of dogs raised either to herd cattle and sheep, as draft animals, as wartime message carriers, in police and rescue work, as guardians, or as Guide Dogs.

Worm. Term for various unrelated invertebrates with soft, often long and slender bodies. The most primitive are the Flatworms, including the Planarians, Flukes, and Tapeworms. Ribbon Worms are colorful marine carnivores, and several loosely related groups of worms (*e.g.*, horsehair worms, Nematodes, and Rotifers) are classed as Roundworms. The segmented or Annelid works include the Earthworms and Leeches. Although it has no taxonomic validity, the term *worm* is also used for the Shipworm (a type of Clam), some insect larvaae (*e.g.*, the armyworm and inchworm), and the acorn worm.

W-particle. *See* virtual particle.

Wren. Small plump, perching songbird of the primarily New World family Trogiodytidae. Its plumage is usually brown or reddish above and white, gray, or buff below. Among the best singers are the canyon, Carolina, and winter wrens. Wrens are valuable insect destroyers.

Write. To enter information into the store of a computer.

Wrought Iron. A form of iron containing a very small amount of carbon, produced by melting pig iron with iron oxide to remove the carbon and hammering the hot solid metal to force out liquid slag. Wrought iron contains inclusions of slag with a fibrous structure produced by the hammering. It is less liable to rus than carbon steels and can easily be forged and welded. It has now largely been replaced by various types of steel.

Wrought Iron. Commercially purified Iron. In the Aston process, pig iron is refined in a Bessemer converter and then poured into molten iron-silicate slag. The resulting semisolid mass is passed between rollers that squeeze out most of the slag. Wrought iron is tough, malleable, ductile, and corrosion-resistant, and melts only at high temperature. It is used to make ornamental ironwork, rivets, bolts, pipes, chains, and anchors.

Wundt, Wilhelm Max. (Voont), 1832-1920, German physiologist and phychologist. In 1878 he founded at Leipzig the first laboratory for experimental psychology. His experimental method remains permanent contribution to psychology.

Wurtzite. A naturally occurring form of zinc sulphide, ZnS.

Wurtz Reaction. A chemical reaction for t56he synthesis of hydrocarbons by the action of sodium an alkyl halides: 2RCl + 2Na = 2NaCl + RR. The similar reaction with aryl halides or with a mixture of an alkyl and an aryl halide to give an aromatic hydrocarbon is called the *Wurtz-Fittig reaction.* [After Adolphe Wurtz (1817-84), French chemist.]

Xanthate. Any salt or ester of a xanthic acid. Cellulose xanthate is used in making viscose rayon.

Xanthic Acid. Any of a class of organic acids with the general formula CS(OR)(SH), where R is usually an alkyl group. The term is often used to denote *ethylxanthic acid,* $CS(OC_2H_5)(SH)$.

Xenon. Symbol: Xe. A colourless odourless tasteless nonflammable gaseous element present in the atmosphere, from which it is extracted as a by-product in the liquefaction of air. Xenon is used in light bulbs, fluorescent lamps, and in lasers. It is one of the inert gases. A.N. 54; A.W. 131.3; m.pt. -111.8°C; b.pt. -108°C; r.d. 4.53 (air = 1).

Xenon (Xe). Gaseous element, discovered spectroscopically in 1898 by William Ramsay and M.W. Travers. It is a rare, colourless, odorless, tasteless Inert Gas used in certain photographic-flash lamps, in high-intensity arc lamps for motion-picture projection, in high-pressure arc lamps to produce ultraviolet light, and in numerous radiation-detection instruments.

Xenon Fluoride. Any of a group of crystalline solids obtained by reacting xenon and fluorine. The compounds XeF_2, XeF_4, and XeF_6 are readily hydrolysed powerful fluorinating agents.

Xerography. A process for copying documents by an electrostatic method. Any electrostatically charged plate is used, usually coated with selenium. Ultraviolet light is passed through the document onto the plate, which it discharges. In this way an electrostatic image is produced depending on the amount of radiation falling on different parts of the plate. A dark powder with opposite charge to that of the plate is then applied. It sticks to the charged areas from which it can be transferred to paper. The powder, which consists of a mixture of graphite and a thermosetting resin, is fixed on the paper by heat.

Xi Particle. Symbol X An elementary particle with either zero charge and a mass 2572 times that of the electron or a negative charge and a mass 2585 times that of the electron. The two xi particles are classified as hyperons.

X-radiation. Electromagnetic radiation composed of X-rays.

X-ray. Invisible, highly penetrating Electromagnetic Radiation of much shorter wavelength (higher frequency) than visible light. The wavelength range for X rays is from about 10^{-8} m to about 10^{-11} m. Discovered in 1895 by Wilhelm Roentgen, X rays are produced by the impact of high-energy electrons (of several thousands of eV) on a metal anode in a highly evacuated glass bulb. Among other applications, X rays are used in medicine and in the study of crystal structure.

X-ray Astronomy. The study of sources of X-rays in space. The radiation is absorbed by the earth's atmosphere and investigations in X-ray astronomy are carried out by means of counters mounted in space probes. The results show numbers of discrete X-ray sources together with a continuous X-ray background. The discrete sources, many of which cannot be detected by optical or radio astronomy, may be caused by high-temperature emission from plasmas or by synchrotron, radiation.

X-ray Background. A fairly isotropic continuous background of X-rays observed in space. Its origin is still uncertain: one possible mechanism is the interaction of fast electrons in cosmic rays with low-energy photons in the microwave background, producing X-ray photons by an inverse Compton effect. Contributions may also come from numbers of discrete X-ray sources.

X-ray Crystallography. The determination of the structure of crystals and molecules by use of X-ray diffraction. Several experimental techniques exist applicable to powders, fibres, single crystals, etc. The sample is placed in a beam of monochromatic X-rays and the diffraction pattern recorded on a photographic plate. A typical pattern from a single crystal consists of a symmetrical array of spots, corresponding to diffraction from different crystal planes. The crystal structure can be calculated from the positions of the spots. It is also possible, by measuring the intensities of diffracted X-rays, to compute a map of electron density within molecules in the crystal.

X-ray Diffraction. Diffraction of X-rays by crystals. The wavelengths of X-rays are the same order of magnitude as the interatomic distances in crystals and the periodic crystal lattice can act like a grating and produce a diffraction pattern from the X-rays.

X-rays (Rontgen rays). Electromagnetic radiation with wavelengths in the range $10^{-7} - 10^{-10}$ metre. X-rays lie between ultraviolet radiation and gamma rays in the electromagnetic spectrum. They are produced by bombarding a solid target with a beam of high-energy electrons. In this process an incident electron knocks one of the inner electrons from an atom, forming a vacancy in an inner shell. An electron from an outer orbit fills this vacancy, the excess energy being emitted as an X-ray photon.

Similar emission can occur when atoms are hit by other X-ray photons or by gamma rays or when inner electrons are lost by nuclear capture. The X-ray spectrum produced by this process (*X-ray fluorescence*) contains lines with wavelengths characteristic of the target substance. In X-ray tubes, these are superimposed on the continuous bremsstrahlung created by deceleration of the incident electrons.

X-rays affect photographic plates and cause some ionization in matter. Their most useful property is their ability to penetrate matter and they are extensively used for medical diagnosis and treatment and for the detection of flaws in metal parts.

X-ray Spectrum. A spectrum of X-rays emitted by bombardment of a sample with electrons or photons. Each element emits X-rays with characteristic wavelengths (*characteristic X-rays*) and these can be used as a method of analysis. The spectrum is taken by using a crystal spectrometer.

X-ray Star. A star that emits X-rays.

X-ray Tube. A tube for producing X-rays, typically consisting of an evacuated container in which an electron gun is used to direct high-energy electrons onto a metal anode (the *target* or *anti-cathode*). The X-rays leave the tube through a thin metal window.

X Unit (Siegbahn unit). Symbol: XU. A unit of length equal to $1.002\ 02 \times 10^{-13}$ metre. It is used for expressing wavelengths of X-rays and gamma rays.

Xylene (dimethyl benzene). Any of three colourless toxic flammable isomers, $C_6H_4(CH_3)_2$, obtained by fractional distillation of petroleum. The mixture of isomers is used as aviation fuel and as a solvent. Individual isomers are used to make many organic compounds. B.pt. 135-145°C (mixture); r.d. 0.9 (mixture).

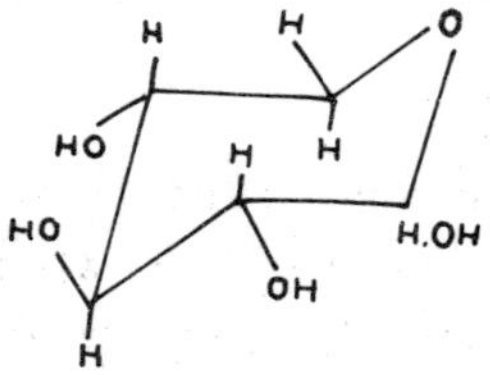

H H O
HO H H
HO H.OH
OH
H

Fig. X-1. D-xylose.

Xylose (wood sugar). A white crystalline monosaccharide, $C_5H_{10}O_5$, extracted by hydrolysis of wood or straw. The naturally occurring substance is dextrorotatory. M.pt. 144°C; r.d. 1.53.

Xylyl Group. The univalent group $CH_3C_6H_4CH_2$-.

X-Y Recorder. A type of pen recorder having two inputs, one displacing the pen in the *x* direction and the other in the *y* direction, so that a graph can be drawn of one varying electrical signal against another.

Yak. Bovine Mammal (*Bos grunniens*) of Tibet and adjacent regions. Oxlike in build, with humped shoulders, a long, thick coat, and large, upcurved horns, it has been hunted to near extinction. The wild yak may reach 65 in. (165 cm) at the shoulder. Today yaks live in isolated herds at elevations above 14,000 ft (4,300 m). Long domesticated, they are a source of milk, meat, and leather, and are used as draft animals.

Yam. Common name for some members of the family Dioscoreaceae, tropical and subtropical climbing herbs and shrubs with starchy Rhizomes often cultivated for food. The thick rhizomes, especially of the genus *Dioscorea,* often weigh up to 30 lbs (13.6 kg) and are an important food source for humans and livestock. The Sweet Potato, often erroneously called yam, belongs to the Morning Glory family.

Yard. An imperial unit of length equal to 0.9144 metres. It was formerly defined as the distance, at 62°F, between lines scribed on two gold plugs in a standard bronze bar.

Yaws. Tropical infection of the skin, caused by a spirochete, *Treponema pertenue,* transmitted by contact with infected persons or their clothing, and by insects. An ulcerating lesion appears at the site of contact, with similar lesions later appearing all over the body; if untreated, lesions penetrate the bones of the feet. The disease can be treated with Penicillin.

Year. The time taken for the earth to move around the sun. It is measured in various ways.

Year. Time required for the earth to complete one orbit about the sun. The *tropical*, or *solar, year* measured relative to the sun is the time (365 days, 5 hr, 48 min, 46 sec of mean Solar Time) between successive vernal Equinoxes. The *sidereal year* is the time (365 days, 6 hr, 9 min, 9.5 sec of mean solar time) required for the earth to complete an orbit of the sun relative to the stars; it is longer than the tropical year because of the Precession of the Equinoxes. The *anomalistic year* is the time (365 days, 6 hr, 13 min, 53.0 sec of mean solar time) required for the earth to go from its perihelion point once around the sun and back to this point; its greater length is due to the slow rotation of the earth's orbit as a whole.

Yeast. Name for certain microscopic, unicellular Fungi, and for commercial products consisting of masses of yeast cells. Yeasts consist of oval or round cells that reproduce mainly by budding (a small outgrowth on the cell's surface increases in size until a wall forms to separate the new individual, or bud) but also by means of Spores. They are used in alcoholic Fermentation and to leaven bread. Brewer's yeast is high in B-complex vitamins and is used as a dietary supplement. Certain other fungi are also sometimes called yeasts.

Yellow Fever. Acute infectious disease caused by a virus transmitted by the bite of the female *Aedes aegypti* mosquito. It causes fever, chills, prostration, Jaundice, and, in severe cases, internal hemorrhage, Coma, and death. Endemic to many tropical and subtropical areas, the disease was once prevalent in the Caribbean. In 1900 a team headed by Walter Reed proved the mosquito-transmission theory of the Cuban physician C.J. Finlay, and U.S. surgeon general W.C. Gorgas led mosquito-eradication and sanitation procedures to control the disease before construction of the Panama Canal could begin. The disease now usually occurs only in sporadic outbreaks, and immunization is an effective preventive measure.

Yew. Evergreen tree or shrub (genus *Taxus*) of the family Taxaceae, somewhat similar to Hemlock but bearing red berrylike fruits instead of true cones. Of somber appearance, with dark green leaves, yews have been associated with death and funeral rites since antiquity. The wood of the English yew (*T. baccata*) was used for the longbows of English archers, and the wood of several species is still so used.

Yield Point. The stress beyond which a material undergoes a relatively large plastic deformation for small increases in stress. Beyond this point, the material no longer obeys Hooke's law.

Ylide. A type of compound in which the molecule has both a positive and negative change, as in $R_2C\text{-}PR_3$, where R and R^1 are organic groups.

Yoga [Skt., = union]. General term for spiritual disciplines, followed for centuries by devotees of both Hinduism and Buddhism, to attain higher consciousness and liberation from ignorance, suffering, and rebirth. It is also the name of one of the six orthodox systems of Indian Philosophy. *Raja yoga* [royal yoga] was expounded by Patanjali (2d cent. B.C.), who divided the practice into eight stages, the highest of which is Samadhi, or identification of the individual consciousness with the Godhead. Hindu tradition in general recognizes three main types of yoga: *jnana yoga,* the path of wisdom and discrimination; *bhakti yoga,* the path of love and devotion to a personal God; and *karma yoga,* the path of selfless action. *Hatha yoga,* widely practiced in the West, emphasizes physical control and postures. *Kundalini yoga,* associated with Tantra, is based on the physiology of the "subtle body". It attempts to open centers of psychic energy called *chakras,* said to be located along the spinal column, and to activate the *kundalini,* a force located at the base of the spine. Yoga is usually practiced under the guidance of a guru, or spiritual teacher. Contemporary systems of yoga stress attaining spiritual realization without withdrawing from the world, as the older traditions taught.

Young, John Watts. 1930-, American astronaut, b. San Francisco. When he commanded the first orbital test flight (apr. 12-14, 1981) of the space shuttle *Columbia,* he set a record for the greatest number of orbital spaceflights (five) by an individual. Young also served as pilot of *Gemini 3* (Mar. 23, 1965), command pilot of *Gemini 10* (July 18-21, 1966), command-module pilot of the *Apollo 10* lunar-orbit mission (May 18-26, 1969), and commander of the *Apollo 16* lunar-landing mission (Apr. 16-27, 1972), on which he became the ninth person to walk on the moon.

Young's Modulus. Symbol: E. A modulus of elasticity applied to the stretching or compression of a material, equal to the ratio of the applied stress to the strain produced. [After Thomas Young (1773-1829), English physicist.]

Young's Slits. An experiment to demonstrate interference of light by placing a screen with two fine slits cut in it in front of a monochromatic light source. The slits act as coherent sources of light of equal intensity and produce interference fringes.

Young, Thomas. 1773-1829, English physicist, physician, and Egyptologist. He was professor of natural philosophy (1801-3) at the Royal Institution, where he presented the modern physical concept of energy, and was elected (1811) a staff member of St. George's Hospital, London. He stated (1807) a theory of color vision known as the Young-Helmholtz theory and described the vision defect called Astigmatism. Reviving the wave theory of Light, Young applied it to refraction and dispersion phenomena. He also established a coefficient of elasticity (Young's modulus) and helped to decipher the Rosetta Stone.

Ytterbium. Symbol: Yb. A soft silvery metallic element belonging to the lanthanide series. A.N. 70; A.W. 173.04; m.pt. 824°C; b.pt. 1193°C; r.d. 7; valency 3.

Ytterbium (Yb). Metallic element, discovered by J.C.G. de Marignac in 1878. A soft, malleable, ductile, lustrous, silver-white Rare-Earth Metal in the Lanthanide Series, it forms many compounds. It is widely distributed in a number of minerals, but has no commercial uses.

Yttrium. Symbol: Yt. A silvery ductile metallic element. Yttrium falls into group IIIb of the periodic table but is usually classified with the lanthanide elements. A.N. 39; A.W. 88.905; m.pt. 1523°C; b.pt. 3337°C; r.d. 4.46; valency 3.

Yttrium (Yt). Metallic element, first isolated by Friedrich Wohler in 1828. Yttrium is an iron-gray Rare-Earth Metal. Its oxide is used in making red phosphors for color-television picture tubes.

Yucca. Stiff-leaved, stemless or treelike Succulent (genus *Yucca*) of the Lily family, native mainly to Mexico and the SW U.S. Yuccas produce a large stalk of white or purplish blossoms. They are pollinated by the yucca moth, and in its absence they rarely produce Fruit. Species include the Adam's-needle (*Y. filamentosa* and others), the Joshua tree (*Y. Brevifolia*), and the Spanish dagger (*Y. gloriosa*). Roots of some species are used for soap.

Yukawa Hideki. 1907-81, Japanese physicist. In 1935 he predicted the existence of the meson, an Elementary Particle that is heavier than an electron but lighter than a proton and that carries the nstrong nuclear force between particles in the atomic nucleus. After particles corresponding to his prediction were discovered in 1947, Yukawa was awarded the 1949 Nobel prize for physics.

Yukawa Potential. A potential used to explain the short-range strong interactions occurring between nucleons in the atomic nucleus. It is assumed that the potential energy is proportional to $[\exp(-\mu r)]/r$, where r is the distance apart and μ is a constant. [After Hideki Yukawa (b. 1907), Japanese physicist.]

Z

Zebra. Herbivorous, hoofed African Mammal (genus *Equus*), distinguished from other Horses by its striking pattern of alternating white and dark brown (or black) stripes. Standing about 4 ft (120 cm) high, the zebra has a heavy head, a stout body, and a short, thick mane. It inhabits open plains or brush country in herds of up to 1,000, often mixing with other grazing animals, *e.g.*, Antelope, and can run at speeds of up to 40 mi (60 km) per hr.

Zebu. Domestic Mammals of the Cattle family, *Bos indicus,* found in E Asia, India, and Africa. It has a large, fatty hump and is fawn, gray, black, or bay. The zebu has great endurance and is used in India for riding and draft purposes. In the U.S. zebus are called Brahman cattle.

Zeeman Effect. The splitting of single lines in a spectrum into two or more components with slightly different wavelengths when the sample is subjected to a magnetic field. The *normal Zeeman effect* involves splitting into three components and is caused by changes of the energy levels of the orbiting electrons due to interaction of the magnetic moment of the orbit with the applied field. The *anomalous Zeeman effect* is more complex and involves the spin of electrons. In the Zeeman effect, the magnitude of the splitting is proportional to the strength of the field. It can be used in determining the magnetic field associated with stars or with sunspots. [After Pieter Zeeman (1865-1943), Dutch physicist.]

Zener Diode. A semiconductor diode that gives a sudden increase in the reverse current when the reverse voltage increases above a certain value. Zener diodes can be used for regulating the voltage in a circuit. [After Clarence Melvin Zener (b. 1905), U.S. physicist.]

Zenith. The point on the celestial sphere directly above an observer.

Zeno of Elea. C. 490-c.430 B.C., Greek philosopher of the Eleatic school founded by Parmenides. Zeno's only known work, extant in fragmented form, uses a series of paradoxes to show the error of commonsense notions of time and space, thereby demonstrating Parmenides' doctrine that motion and multiplicity are logically impossible. Contemporary thinkers have shown renewed interest in the problems Zeno raised.

Zeolite. Any of a class of natural or synthetic silicates of sodium, potassium, and calcium, having a three-dimensional crystal structure containing water molecules held in cage-like cavities in the lattice. When heated they lose water and the framework of the crystal can then trap other molecules of suitable size. Zeolites are used for separating mixtures by selective adsorption: for this reason they are known as *molecular sieves*.

Zero Point Energy. The energy of vibration of atoms at absolute zero of temperature, equal to $hv/2$, where h is the Planck constant and v the frequency of oscillation.

Zeta. An experimental fusion reactor at Harwell. It has a torroidal reaction chamber in which the plasma is contained by a magnetic field produced by a solenoidal winding. The name is an acronym for *Zero Energy Thermonuclear Apparatus*.

Ziegler-Natta Catalyst. Any of a class of related catalysts that promote the stereospecific polymerization of ethylene and other olefines. Typically, they consist of mixtures of transition-metal halides, such as titanium II chloride, with nontransitionmetal alkyls, especially aluminium alkyls. [After Karl Ziegler (b. 1897), German chemist, and Giulio Natta (b.1903), Italian chemist.]

Zinc. Symbol: Zn. A bluish-white lustrous brittle metallic element found in zincite (ZnO), zinc blende (ZnS), smithsonite ($ZnCO_3$), and hemimorphite [$Zn_4Si_2O_7(OH)_2.H_2O$]. The metal is obtained by roasting the ore and reducing the oxide by heating with carbon. It is used in many alloys, such as brass, bronze, nickel silver, German silver, and solder, and in galvanizing iron.

Zinc is a fairly reactive electropositive element, reacting with oxygen, sulphur, the halogens, and other nonmetals. It is amphoteric, dissolving in nonoxidizing acids and also in alkalis to form zincates. Most of its compounds contain Zn^{2+} ions although some are covalently bonded, as in the zinc alkyls. A.N. 30; A.W. 65.37; m.pt. 419.6°C; b.pt. 907°C; r.d. 7.13; valency 2.

Zinc (Zn). Metallic element. A lustrous, bluish-white, fairly reactive metal, zinc is ductile and malleable when heated. It is commercially important in Galvanizing iron and in the preparation of certain alloys, *e.g.*, Brass and sometimes Bronze. It is used for the negative plates in certain electric batteries, and for roofing and gutters. Zinc compounds are numerous and widely used. Zinc is essential for growth of plants and animals.

Zincate. A salt containing the ion ZnO_2^{2-}, produced by dissolving zinc in an alkali. Zincates are formally salts of amphoteric zinc hydroxide, $Zn(OH)_2$, and are probably more correctly regarded as hydrated zinc hydroxide anions.

Zinc Blende. Naturally occurring zinc sulphide, ZnS.

Zinc Chloride (butter of zinc). A poisonous deliquescent white crystalline solid, $ZnCl_2$, prepared by adding hydrochloric acid to zinc or zinc oxide. It is used for galvanizing iron, as a catalyst, as a dehydrating agent, and as a soldering flux. M.pt. 290°C; b.pt. 732°C; r.d. 2.9.

Zinc Hydroxide. A white amphoteric solid, $Zn(OH)_2$, precipitated by addition of alkali to zinc salts.

Zincite. A red or orange naturally occurring form of zinc oxide, ZnO, used as an ore of zinc.

Zinc Oxide (Chinese white, philosopher's wool). An odourless white powder, ZnO, which turns yellow when hot. It is made by vapour-phase oxidation of pure zinc or by roasting zinc oxide ore with coal followed by oxidation in air. Zinc oxide is a filler and pigment in plastics and paints, and is a widely used mild antiseptic in ointments. M.pt. 1975°C; r.d. 5.5.

Zinc Sulphate (white vitriol). A colourless efflorescent crystalline solid, $ZnSO_4.7H_2O$, manufactured by roasting zinc blende with subsequent leaching. It is used as a styptic, a mordant, and as a preservative for wood. Loses water above 200°C; decomposes above 500°C; r.d. 1.9.

Zinc Sulphide. A yellow-white solid, ZnS, occurring naturally in wurtzite, sphalerite, and zinc blende. It is used as a pigment for paints and glass and as a phosphor in cathode-ray tubes. Sublimes at 1180°C; r.d. 3.98 (wurtzite).

Zircon. A brown, red, or colourless naturally occurring silicate of zirconium. $ZrSiO_4$, used as a gemstone and a source of zirconium and zirconium oxide.

Zirconium. Symbol: Zr. A greyish-white lustrous metallic element found principally in zircon ($ZrSiO_4$) and baddeleyite (ZrO_2), from which it can be extracted by the Kroll process. Zirconium is used in nuclear reactors as a neutron absorber, as a vacuum getter, and in alloys. Its chemistry is similar to that of titanium, its main compounds being covalently bonded.

A.N. 40; A.W. 91.22; m.pt. 1852°C; b.pt. 4377°C; r.d. 6.51; valency 2, 3, or 4.

Zirconium (Zr). Metallic element, discovered in oxide form by M.H. Klaproth in 1789. It is a very strong, malleable, ductile, lustrous, silver-gray metal that is extremely resistant to heat and corrosion. Zirconium is used in flashbulbs and to clad uranium fuel for nuclear reactors; its compounds are used as refractory material in furnaces, crucibles, and ceramic glazes.

Zodiac. A band on the celestial sphere containing the paths of the planets, the sun, and the moon. It is about 18° wide (*i.e.* 9° each side of the ecliptic) and is divided into 12 equal parts named after constellations.

Zodiac. Zone of the sky that includes about 8° on either side of the Ecliptic. The apparent paths of the sun, the moon, and the major planets all fall within this zone. The zodiac is divided into 12 equal parts of 30° each, each part being named for a Constellation and represented by a sign. Because of the Precession of the Equinoxes, the equinox and solstice points have each moved westward about 30° in the last 2,000 years; thus the zodiacal constellations, which were named in ancient times, no longer correspond to the segments of the zodiac represented by their signs. The zodiac is of importance in Astrology.

Zodiacal Light. Faint light seen in the west just after sunset or in the east just before sunrise. It is caused by reflection of sunlight from small meteoric particles around the sun.

Zodiacal Light. Faint band of light sometimes seen just after sunset or before sunrise, extending up from the horizon at the sun's setting or rising point. It is caused by reflection of sunlight from tiny dust particles concentrated in the plane of the Ecliptic.

Zone Refining. a method of purifying solid materials, used especially for semiconductors. A long bar of the sample is heated at a point near one end so that a small region of the bar is melted. The molten phase is held in position by surface tension and impurities in the neighbouring regions tend to concentrate in the melt. By moving the heater, this molten zone is slowly moved along the bar, so that the impurities are transferred to one end. Impurity levels as low as 1 part in 10^{10} can be achieved by using several passes.

Zoology. In biology, the study of animals. Early efforts to classify animals were based on physical resemblance, habitat, or economic use. The invention of the microscope and the use of experimental techniques expanded zoology as a field and established many of its branches, *e.g.,* Cytology, Histology, embryology, Physiology, and Genetics. Modern zoology studies cell structure and function, as well as psychological, anthropological, and ecological aspects of animals.

Zoon Lens. An adjustable compound lens used in cinematic or television cameras to change the magnification continuously without losing the sharpness of the image.

Zwitterion (ampholyte ion). An ion that has both a positive and negative charge. Examples are found in the amino acids: glycine, (CH_2NH_2COOH), for example, can form the zwitterion $CH_2NH_3^+COO^-$ by ionization of the carboxyl group and protonation of the amino group.

Zworykin, Vladimir Kosma. 1889-1982, American physicist and inventor; b. Russia. He came to the U.S. in 1919 and joined (1929) the Radio Corp. of America, becoming vice president in 1947. He and his co-workers developed the iconoscope, a scanning tube for the Television camera, and the kinescope, a cathode-ray picture tube in the television receiving apparatus.